INTERNATIONAL CENTRE FOR MECHANICAL SCIENCES

COURSES AND LECTURES - No. 299

MATHEMATICAL PROGRAMMING METHODS IN STRUCTURAL PLASTICITY

EDITED BY

D. LLOYD SMITH
IMPERIAL COLLEGE, LONDON

SPRINGER-VERLAG WIEN GMBH

Le spese di stampa di questo volume sono in parte coperte da
contributi del Consiglio Nazionale delle Ricerche.

This volume contains 132 illustrations.

Originally published by Springer Verlag Wien-New York in 1990

In order to make this volume available as economically and as rapidly as possible the authors' typescripts have been reproduced in their original forms. This method unfortunately has its typographical limitations but it is hoped that they in no way distract the reader.

ISBN 978-3-211-82191-6 **ISBN 978-3-7091-2618-9 (eBook)**
DOI 10.1007/978-3-7091-2618-9

PREFACE

Civil engineering structures tend to be fabricated from materials that respond elastically at normal levels of loading. Most such materials, however, would exhibit a marked and ductile inelasticity if the structure were overloaded by accident or by some improbable but naturally occurring phenomenon. Indeed, the very presence of such ductility is an important safety provision for large scale constructions where human life is at risk. In the mathematical theory of plasticity there is found a material model which is both fairly simple and generally representative of the essential characteristics of observed behaviour. It must be considered unrealistic, therefore, not to include the effects of plasticity in the comprehensive evaluation of safety in a structural design.

Mathematical Programming (MP) is concerned with seeking to optimise a function whose variables are also required to satisfy additional conditions or constraints. Such constraints may be represented by systems of equations, or of inequalities, connecting the variables, and they would include the possibility that some of the variables might be constrained in sign. It is clear that the search for a solution to a system of inequalities, or of equations, without the optimising of a function is a special case of the above problem.

During the second world war, the need to address complex problems of planning and resource allocation for economic as well as military purposes coincided with the increasing availability of large-scale computing facilities. This provided the climate for the rapid developlment of MP, and, since the mathematical description of plasticity necessarily involves inequality conditions, it was not long before the link between these two subjects was established.

In the ensuing twenty-five years there has been a thorough investigation of the role of MP in all the major problem types in structural plasticity. The appeal of MP in this context has been twofold: computational and theoretical. MP can provide a ready fund of general algorithms. More importantly, it invests applications with a refined mathematical formalism, rich in fundamental theorems, which often gives additional insight into known results and occasionally leads to new ones.

The conspectus offered by the lectures at the C.I.S.M. Advanced School in Udine, and contained in this book, is hardly a complete one. It can be said, however, that these lectures address a selection of the most interesting and potentially most useful applications of MP in structural plasticity: the ultimate strength and elastoplastic deformability of sections, framed and continuum structures; the ability of a structure to shake down or to adapt itself to respond elastically to a complex programme of loading; the assessment of practical upper bounds on specific measures of deformation for a structure responding either quasi-statically or dynamically to such a complex loading programme; the evolutive dynamic behaviour of structures with rigid-plastic and elastic-plastic constitutive laws; the static response and instability of an elastoplastic structure in a regime of large displacements; the use of stochastic and fuzzy programming methods for representing the role of uncertainty in the assessment of ultimate strength.

In the best traditions of applied mechanics, the following lectures successfully implement applications from a foundation of fundamental mechanics. Their success, if I may be permitted to say, is achieved through a systematic and consistent mathematical modelling of both structure and material. In this regard, particular attention has been given to the mesh and nodal network modelling of structural systems, to the virtues of representing general non linear material behaviour through asymptotic expansions, and to the achievement of a consistent two-field approximation of stresses and strain rates in the finite-element modelling of continuum plasticity.

Beyond the interest that the subject generates as a fertile field for the nurturing of research projects is the conviction that its educational value thoroughly befits a university discipline. This genuine pedagogic concern is amply reflected in the lectures of my colleagues, the distinguished lecturers at the C.I.S.M. Advanced School, through their many years of experience of giving graduate classes in structural plasticity. While the lectures begin with a brief survey of the most important results of MP, of particular interest is the unifying pedagogic role that complementarity theory brings to a broad class of problems in structural mechanics.

In the long term, the enthusiasm with which a subject becomes popularly embraced is determined, it seems to me, by the scope it affords for computational and theoretical development. Practical espousement is largely dependent upon the availability of efficient commercial software. Some notable achievements for plastic limit analysis and design are described in one of the following lectures, but the consensus must be that not nearly enough has yet been done on software and algorithm development, particularly for large-scale structural systems. On the theoretical front, however, the subject is vibrantly healthy. The ability of MP and complementarity theory to encompass a class of problems much wider than those of structural plasticity, has already been mentioned. Furthermore, the exploration of links between finite-dimensional complementarity problems and the variational inequalities of continua seems to be providing an ideal mathematical basis for mechanical problems in which the essential characteristics (material or structural) are non-smooth.

As I write these prefatory remarks, I am constantly reminded of one whose duty it should have been. The C.I.S.M. Advanced School in MP Methods in Structural Plasticity was conceived by the late John Munro and was to be coordinated by him. Sadly, Professor Munro died on 27 February 1985. The subject of his Chair - Civil Engineering Systems - was the vehicle through which he sought to broaden perspectives in civil engineering education. His finest original researches resulted from a special ability to distil engineering wisdom by reducing a problem to its simplest terms: if that required the questioning of established axioms, he was prepared to do it. The passing of this refined and sensitive man has removed a singular spirit and, for many, a most respected friend.

A central figure in MP Methods in Structural Plasicity is Giulio Maier. While unable to take an active part in the Advanced School, he remained a constant source of encouragement and advice to this writer: to Professor Maier I am much in debt. In my closing address at the Advanced School I conveyed the sincere thanks to the lecturers to the secreterial staff of C.I.S.M., especially Miss Elsa Venir, and to the participants who, through informal and interactive discussions, made a valuable contribution to the success and enjoyment of the meeting. Time has not changed those sentiments.

David Lloyd Smith

CONTENTS

CHAPTER 1

MATHEMATICAL PROGRAMMING

D. Lloyd Smith
Imperial College, London, U.K.

ABSTRACT

A brief review is made of those basic aspects of the theory of mathematical programming which are of most relevance in the mathematical description of the theory of structures with plastic constitutive laws. Firstly, the classical problem of optimisation is used to introduce the method of Lagrange multipliers. The same method is then applied to a mathematical program with inequality constraints, and the necessary optimality criteria of Karush, Kuhn and Tucker are obtained. With the imposition of convexity, the concept of duality in mathematical programming is described, and the solutions of the consequent dual mathematical programs are related to the saddlepoint property of the associated Lagrange function.

INTRODUCTION

Mathematical programming provides an appealing formalism for encoding the behaviour of discrete structures formed from materials in which plasticity is a constitutive component. It can be associated with any basis for the discretisation of a structure, especially that of the finite element; it offers a refined mathematical foundation, rich in theorems which present interesting and useful structural interpretations; it has a small collection of well-constructed algorithms through which numerical solutions may be obtained. The integration of such features sets practical problem-solving in structural plasticity within a strong scientific discipline: one could hardly ask for more.

Optimal design provides an obvious field of application for mathematical programming. In the absence of viable alternative analytical methods, it has proved really quite successful; and yet one might feel a sense of surprise that optimisation has not so far had a stronger impact upon engineering design. In engineering plastic analysis, a considerable body of literature [1,2] attests to the vigour with which mathematical programming is being used to explore an ever-widening range of problem classes, to unify their theoretical foundation and to obtain general results of relevance in practical design.

For those problems of plastic analysis most often addressed in relation to practical design, other, quite different, methods have achieved popular currency. The nonlinear boundary value problem of incremental elastoplastic analysis, for instance, is usually solved by an *ad hoc* numerical scheme in which a *linear elastic* solver is employed iteratively within each increment of the control parameter. There is no clear evidence that such procedures are more efficient than that of a properly formulated mathematical programming approach. What is clear, however, is that a considerable body of research effort has been expended on the development of reliable and professionally written computer software for the analysis of linear elastic systems. To extend the range of application of the procedures to elastoplastic systems is an entirely sensible capitalisation on the previous investment of effort; and the availability of good software ensures its own reward. The lesson is transparent: for whatever appealing features it may have, mathematical programming will only receive widespread professional utilisation through the development of high quality applications software.

Structural mechanics is a consistently logical and intellectually challenging scientific discipline. It uses appropriate mathematical apparatus to describe, as closely as possible, all the relevant physical laws that must be obeyed by a structural system which is subjected to a specific disturbance in its environment, and to deduce general results or theorems pertaining to its response. It demands frequent development to accord with innovation in structural form, construction materials and in mathematics. In the longer term, the progressive use of particular mathematical apparatus in structural mechanics may be determined by its scope for encompassing a broader class of relevant problems. We may observe that the mathematical descriptions of structural plasticity, of the contact between elastic bodies and of the movement of bodies against frictional forces have certain features in common. The setting of such disparate physical problems within a generalising mathematical context often conveys greater physical insight and certainly widens the circle of interaction between those who will provide the impetus for advancement. The developing and generalising field of non-smooth mechanics[3] displays an elegant, if rather sophisticated, mathematical apparatus within which mathematical programming may be seen as a natural vehicle for the discrete representation of some problems. On the other hand, it is in the very nature of engineering that its exponents will always attempt to devise solutions which are at once simple and physically motivated.

A thoroughly entertaining and well-informed perspective is given by Maier[4] on the status of mathematical programming in structural mechanics: it is both critical of past effort and perceptively forward-looking.

What now follows is a brief review of mathematical programming in which emphasis is laid upon those general results of most relevance to structural mechanics. Nonlinear programming has a considerable variety of useful algorithms[5]. One therefore tends to select an algorithm on the basis of its efficacy for solving a specific nonlinear problem or class of problems. No description of such algorithms is attempted here, although, later in this volume, Teixeira de Freitas[6] describes a gradient method which has proved particularly suited to the solution of nonlinear problems in elastoplastic structures.

THE MATHEMATICAL PROGRAM

Mathematical programming is part of the general theory of optimisation. It involves the minimisation (or maximisation) of a function of a finite number of real variables when those variables are themselves constrained to satisfy a finite number of supplementary conditions.

Let $(x_1, x_2, ..., x_q, ..., x_n)$ be a set of n variables which take real values. A *mathematical program* is constituted in the following statements: find values $(\bar{x}_1, \bar{x}_2, ..., \bar{x}_n)$ of the variables $(x_1, x_2, ..., x_n)$ which minimise the *objective function*

$$z(x_1, x_2, \ldots\ldots, x_n) \tag{1.1}$$

while satisfying the m *inequality constraints*

$$g_i(x_1, x_2, \ldots\ldots, x_n) \leq 0, \qquad i = 1, 2, \ldots, m \tag{1.2}$$

the p *equality constraints*

$$h_k(x_1, x_2, \ldots\ldots, x_n) = 0 \qquad k = 1, 2, \ldots, p \tag{1.3}$$

and the q *sign constraints* or *non-negativity restrictions*

$$x_r \geq 0 \qquad r = 1, 2, \ldots, q \tag{1.4}$$

which may be imposed on some of the variables; the remaining variables are said to be *unrestricted*. The functions z( ), g_i( ) and h_k( ) are real functions of the n variables, but they are not necessarily differentiable, nor are they necessarily continuous.

The mathematical program is a *linear program* if $z(\ )$, $g_i(\ )$ and $h_k(\ )$ are all continuous linear functions of the variables; otherwise it is a *nonlinear program*. If z is a continuous quadratic function while $g_i(\ )$ and $h_k(\ )$ are continuous linear functions of the variables, the mathematical program is a *quadratic program*. Linear and quadratic programs, which have a special importance in structural plasticity, will be considered in more detail subsequently.

Let $\mathbf{x}$ represent the column vector of variables x_j where $j = 1, 2, ..., n$. Such a vector $\mathbf{x}$ may be considered as the position vector of a point in an n-dimensional euclidean space. A vector $\mathbf{x}$ which satisfies all the constraints (1.2) to (1.4) is said to be a *feasible solution* of the mathematical program. The set X of all the feasible solutions is called the *feasible region*.

Figure 1.1 suggests an example of two variables x_1, x_2 required to satisfy three inequality constraints $g_i(x_1, x_2) \le 0$ and no equality constraints. The directions of increasing values of the constraint functions $g_i(x_1, x_2)$ are indicated by arrows, and the boundary of the feasible region is hatched. Contours of constant values of the objective function z are also shown, and the optimal (minimising) solution is at the point $(\bar{x}_1, \bar{x}_2)$. At the optimal solution, two of the constraints are satisfied with equality, $g_1(\bar{x}_1, \bar{x}_2) = 0$, $g_2(\bar{x}_1, \bar{x}_2) = 0$: they are said to be *active* or *binding* constraints. The third constraint is satisfied with strict inequality, $g_3(\bar{x}_1, \bar{x}_2) < 0$: it may be described as *inactive* or *passsive*.

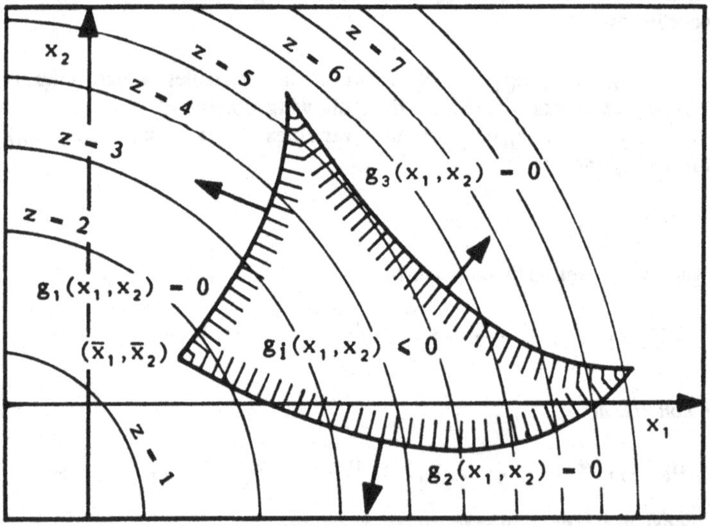

Figure 1.1

In essence, all constraints of the mathematical program can be represented in the form of the inequalities (1.2), as may quickly be demonstrated. An equality constraint $h_k(\mathbf{x}) = 0$ may be replaced by two inequality constraints $h_k(\mathbf{x}) \le 0$, $-h_k(\mathbf{x}) \le 0$, and a sign constraint $x_r \ge 0$ has the alternative form $-x_r \le 0$.

It should also be noted that the maximisation of $z(\mathbf{x})$ can be achieved by the device of minimising $-z(\mathbf{x})$ and observing that

$$\text{Maximum } [z(\mathbf{x})] = -\text{Minimum } [-z(\mathbf{x})]$$

OPTIMALITY CRITERIA FOR UNCONSTRAINED MINIMISATION

The classical problem of optimisation is:
find values $(\bar{x}_1, \bar{x}_2, \ldots, \bar{x}_n)$ of the variables $(x_1, x_2, \ldots, x_n)$ which minimise the continuous and differentiable function $z(x_1, x_2, \ldots, x_n)$; that is

$$\text{Minimise } z = z(\mathbf{x}). \tag{1.5}$$

It is seen that the variables are unrestricted and that there are no supplementary conditions to be satisfied: it is a problem of *unconstrained optimisation*.

Define the *gradient* of $z(\mathbf{x})$ as the column vector of first partial derivatives

$$\nabla z(\mathbf{x}) = [\partial z(\bar{\mathbf{x}})/\partial x_j], \qquad j = 1, 2, \ldots, n. \tag{1.6}$$

and form the system of n (possibly nonlinear) equations at $\bar{\mathbf{x}}$

$$\nabla z(\bar{\mathbf{x}}) = 0. \tag{1.7}$$

If $z(\mathbf{x})$ has a local minimum at $\bar{\mathbf{x}}$, then $\bar{\mathbf{x}}$ must satisfy conditions (1.7): they are *necessary conditions* for a minimum. The converse is not true, so the conditions are not sufficient. A solution $\bar{\mathbf{x}}$ of (1.7) is said to be a *stationary point* of $z(\mathbf{x})$: it locates a local minimum or a local maximum of the function, or a point which is neither of these. One would need to investigate all the solutions of (1.7) in order to determine the overall or global minimum.

Suppose that $z(\mathbf{x})$ is twice differentiable and define the *Hessian matrix* of second partial derivatives by the square matrix

$$\mathbf{H}(\mathbf{x}) = [\partial^2 z(\mathbf{x})/\partial x_i \partial x_j], \qquad i, j = 1, 2, \ldots, n \tag{1.8}$$

Let superscript T denote transposition. If $\bar{\mathbf{x}}$ satisfies (1.7) and

$$Q = \mathbf{s}^T \mathbf{H}(\bar{\mathbf{x}})\, \mathbf{s} > 0 \tag{1.9}$$

for any nonzero column vector $\mathbf{s}$, then $z(\mathbf{x})$ has a local minimum at $\bar{\mathbf{x}}$: relations (1.7) and (1.9) therefore give *sufficient conditions* for a local minimum. For a local maximum, the direction of the inequality relation in (1.9) should be reversed.

The expression Q given in (1.9) is quadratic in $(s_1, s_2, \ldots, s_n)$: it is said to be a *quadratic form*. If $Q > 0$ $(Q < 0)$ for any values $(s_1, s_2, \ldots, s_n)$, excepting the case when all the values are zero, the quadratic form and its square symmetric matrix **H** are *positive definite* (*negative definite*). If $Q \geq 0$ $(Q \leq 0)$, there being at least one set of values $(s_1, s_2, \ldots, s_n)$, not all zero, for which $Q = 0$, then Q and **H** are *positive semidefinite* (*negative semidefinite*). If Q is positive for some $\mathbf{s}$ and negative for others, then Q and **H** are *indefinite*.

Hancock's[7] account of the historical development of classical optimisation gives proofs of the above necessary and sufficient conditions, and it includes an informative discussion of the resolution of the case in which **H** is semidefinite. An elementary and lucid description of the above results is given by Courant[8].

Such theoretical considerations have a bearing upon the related problem of constrained optimisation. In practice, unconstrained optimisation is usually performed by direct numerical search. Avriel[5], Walsh[9] and Fletcher[10] present useful surveys of appropriate methods.

OPTIMALITY CRITERIA FOR EQUALITY CONSTRAINED MINIMISATION

The second problem of classical optimisation is:
find values $\bar{\mathbf{x}}$ of the variables $\mathbf{x}$ which minimise the function $z(\mathbf{x})$ subject to p supplementary conditions $h_k(\mathbf{x}) = 0$. That is,

$$\text{Minimise } z = z(x), \quad h_k(x) = 0, \qquad k = 1, 2, \ldots, p < n \tag{1.10}$$

In problem (1.10), the functions $z(\ )$ and $h_k(\ )$ will be assumed continuous and differentiable, and the variables $\mathbf{x}$ unrestricted: this is *equality constrained optimisation*.

The constraints $h_k(\mathbf{x}) = 0$ are p equations in n variables x_j. An obvious strategy is to use the equations to eliminate p of the variables from $z(\mathbf{x})$. What would remain is the unconstrained minimisation of a function of $(n - p)$ independent variables. Such an elimination is only possible if the *Jacobian matrix*

$$\nabla h_k(x) = [\partial h_k(x)/\partial x_j], \quad k = 1, 2, \ldots, p \quad j = 1, 2, \ldots, n \tag{1.11}$$

of n rows and p columns is of full rank p. Even then, the elimination may well prove extremely difficult to accomplish.

An elegant alternative for transforming the original problem into one of unconstrained minimisation, without the need for eliminating some of the variables, is the method of Lagrange multipliers. Let $\bar{\mathbf{x}}$ be a local minimum of problem (1.10). In any move to a neighbouring point $\mathbf{x} = \bar{\mathbf{x}} + \mathbf{dx}$ which also satisfies the constraints, the change in z, as represented by the total differential dz, is

$$dz = \sum_j \frac{\partial z(\bar{x})}{\partial x_j} dx_j = 0 \tag{1.12}$$

The increments dx_j of the feasible move are not independent since the neighbouring point $\mathbf{x}$ must continue to satisfy the constraints. Therefore the changes in value of the constraint functions must be

$$dh_k = \sum_j \frac{\partial h_k(\bar{x})}{\partial x_j} dx_j = 0 \tag{1.13}$$

The interdependencies among the dx_j are completely specified by equations (1.13).

Let us introduce a set of (as yet) undetermined parameters $(\mu_1, \mu_2, \ldots, \mu_p)$ which shall be known as *Lagrange multipliers*. Each of the p equations (1.13) may be multiplied, in turn, by the corresponding multiplier and added to (1.12), giving

$$\sum_j \left[\frac{\partial z(\bar{x})}{\partial x_j} + \sum_k \mu_k \frac{\partial h_k(\bar{x})}{\partial x_j} \right] dx_j = 0. \tag{1.14}$$

We remind ourselves that p of the increments dx_j are dependent and $(n - p)$ are independent. There are exactly p Lagrange multipliers; they may now be set at such values $\mu_k = \bar{\mu}_k$ as will cause the expressions in brackets, corresponding to the dependent dx_j, to vanish. Each of the remaining dx_j is independent and therefore arbitrary; for each, the corresponding expression in brackets must necessarily vanish in order that (1.14) be satisfied. The resulting n equations, together with the p constraints

$$\frac{\partial z(\bar{x})}{\partial x_j} + \sum_k \bar{\mu}_k \frac{\partial h_k(\bar{x})}{\partial x_j} = 0, \quad j = 1, 2, \ldots, n. \tag{1.15}$$

$$h_k(\bar{x}) = 0 \qquad k = 1, 2, \ldots, p. \tag{1.16}$$

give $(n + p)$ equations in $(n + p)$ variables $\bar{x}$ and $\bar{\mu}$.

Let us define the following function as the *Lagrangian*

$$L(x, \mu) = z(x) + \sum_k \mu_k h_k(x) \tag{1.17}$$

A stationary point $(\bar{x}, \bar{\mu})$ of this unconstrained function must satisfy

$$\nabla_x L(\bar{x}, \bar{\mu}) = \left[\frac{\partial L(\bar{x}, \bar{\mu})}{\partial x_j} \right] = 0 \tag{1.18}$$

$$\nabla_\mu L(\bar{x}, \bar{\mu}) = \left[\frac{\partial L(\bar{x}, \bar{\mu})}{\partial \mu_k} \right] = 0 \tag{1.19}$$

and these equations coincide with (1.15) and (1.16) respectively.

If the function $z(x)$ of the equality constrained optimisation problem (1.10) has a local minimum at $\bar{x}$, then $\bar{x}$ and some $\bar{\mu}$ must satisfy (1.18) and (1.19): they are *necessary conditions* for an equality constrained local minimum, and they coincide with the conditions for a stationary point of the unconstrained Lagrangian. Again, Courant[8] describes the development with simplicity and precision.

Implicit in the above statement is the existence of the Lagrange multipliers – that values $\bar{\mu}$ can be calculated from (1.18) and (1.19). Their existence is assured if the Jacobian matrix of (1.11) is of full rank p. If it is not, the multipliers may be calculated, but not necessarily uniquely, by means of Caratheodory's device of introducing a supernumerary multiplier μ_0 into the Lagrangian.

$$L(x, \mu) = \mu_0 z(x) + \sum_k \mu_k h_k(x) \tag{1.20}$$

The value $\bar{\mu}_0 = 1$ is normally chosen for μ_0. However, when the rank of the Jacobian is less than p, finite values $\bar{\mu}_k$ of the Lagrange multipliers will only exist if we take $\bar{\mu}_0 = 0$. Hadley[11] gives a clear and detailed account of such matters.

Let $H_{\mathbf{x}}(\bar{\mathbf{x}}, \bar{\mu}) = [\partial^2 L(\bar{\mathbf{x}}, \bar{\mu})/\partial x_i \partial x_j]$ be a Hessian matrix of the Lagrangian. If $\bar{\mathbf{x}}$ and $\bar{\mu}$ satisfy (1.18) and (1.19), and if

$$s^T H_x(\bar{x}, \bar{\mu})\, s > 0 \qquad (1.21)$$

for every nonzero s such that

$$s^T \nabla h_k(x) = 0 \qquad (1.22)$$

then $z(\mathbf{x})$ has a local minimum at $\bar{\mathbf{x}}$ when subject to the equality constraints $h_k(\mathbf{x}) = 0$: these are the *sufficient conditions* for an equality constrained local minimum which are given by Avriel[5], Hancock[7] and Walsh[9]. For a maximum, reverse the inequality in (1.21). Beveridge and Schechter[12], on the other hand, caution us in the use of the Lagrangian for the inference of sufficient conditions.

KARUSH-KUHN-TUCKER OPTIMALITY CRITERIA

We shall now describe the optimality criteria for mathematical programming. Consider the mathematical programming problem with m inequality constraints:

$$\text{Minimise } z = z(x); \quad g_i(x) \leq 0, \quad i = 1, 2, \ldots, m. \qquad (1.23)$$

Let $z(\ )$ and $g_i(\ )$ be continuous and differentiable functions and let the variables $\mathbf{x}$ be unrestricted. The *ith* inequality constraint may be converted to an equation by the addition of a *slack variable* $x_{n+i} = s_i^2$ where, although x_{n+i} must be non-negative, s_i remains unrestricted; thus

$$g_i(x) + s_i^2 = 0 \qquad (1.24)$$

Now the modified problem is equality constrained, and its Lagrangian is

$$M(x, \lambda, s) = z(x) + \sum_i \lambda_i [g_i(x) + s_i^2], \qquad (1.25)$$

where $(\lambda_1, \lambda_2, \ldots, \lambda_m)$ is a set of Lagrange multipliers. If $z(\mathbf{x})$ has a local minimum at $\bar{\mathbf{x}}$ subject to the equality constraints (1.24), then $\bar{\mathbf{x}}$ and some $\bar{\lambda}$ satisfy the stationarity conditions

$$\nabla_x M = \left[\frac{\partial M}{\partial x_j} \right] = \nabla_x z(\bar{x}) + \sum_i \bar{\lambda}_i \nabla_x g_i(\bar{x}) = 0 \qquad (1.26)$$

$$\nabla_\lambda M = \left[\frac{\partial M}{\partial \lambda_i} \right] = [\, g_i(\bar{x}) + \bar{s}_i^2 \,] = 0, \quad i = 1, 2, \ldots, m \qquad (1.27)$$

$$\nabla_s M = \left[\frac{\partial M}{\partial s_i} \right] = [\, 2\, \bar{\lambda}_i\, \bar{s}_i \,] = 0, \quad i = 1, 2, \ldots, m \qquad (1.28)$$

Consider the m relations (1.27) and the m relations (1.28). Suppose that $\bar{s}_i \neq 0$. Then the strict inequality $g_i(\bar{x}) < 0$ applies, and the *ith* constraint is inactive or passive; it could be discarded from the problem formulation without affecting the result. Also, since $\bar{s}_i \neq 0$ we have that $\bar{\lambda}_i = 0$. Now suppose that $\bar{s}_i = 0$. Clearly, the *ith* constraint is now active or binding, $g_i(\bar{x}) = 0$, and $\bar{\lambda}_i$ can be zero or non-zero.

It is therefore seen that the only essential constraints to the problem are those which are satisfied as strict equalities. In spite of this, the minimising solution differs from that of the equality constrained problem in one important detail. The Lagrange multipliers $\bar{\mu}_k$ associated with the equality constraints $h_k(\bar{x}) = 0$ are unrestricted: those multipliers $\bar{\lambda}_i$ which associate with the inequality constraints $g_i(\bar{x}) \leq 0$ are non-negative, as is shown by Beveridge and Schechter[12] and by Rao[13] among many others.

As a simple example, consider first the equality constrained problem:

$$\begin{aligned} \text{Minimise } z &= x_1^2 + (x_2 - 2)^2 + 4, \\ h_1 &= x_2 - x_1 = 0, \\ h_2 &= x_1 - 2 = 0. \end{aligned} \tag{1.29}$$

The solution must be located at the intersection of the two constraints, the only feasible point, where $\bar{x}_1 = 2, \bar{x}_2 = 2$ and $\bar{z} = 8$. At this point, the gradients are

$$-\nabla_x z(2,2) = [-4 \quad 0]^T, \qquad \nabla_x h_1(2,2) = [-1 \quad 1]^T, \qquad \nabla_x h_2(2,2) = [1 \quad 0]^T$$

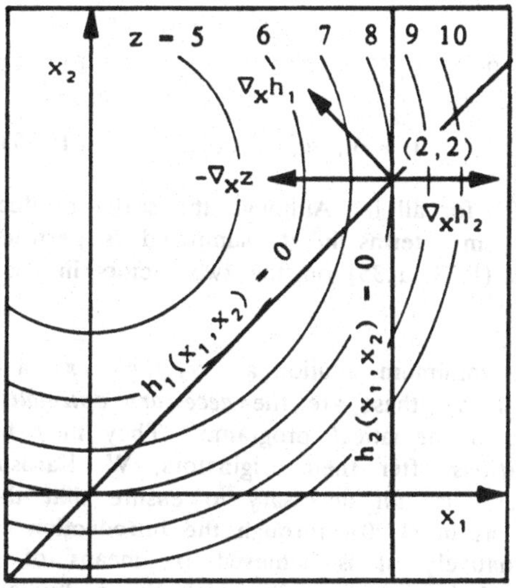

Figure 1.2

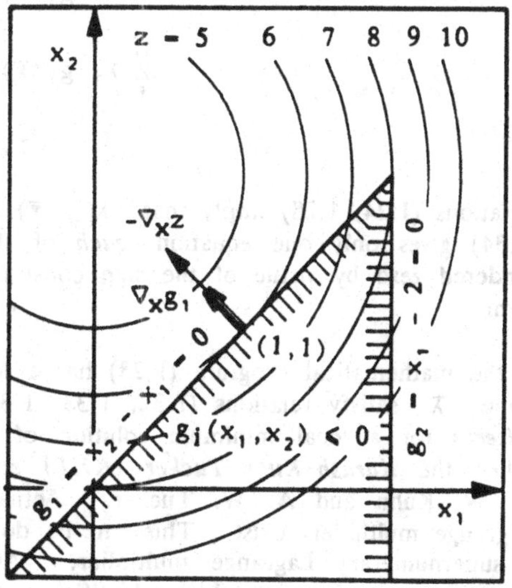

Figure 1.3

where $-\nabla_x z$ points in the direction of greatest rate of *decrease* in z, while $\nabla_x h_1$ and $\nabla_x h_2$ indicate the directions of greatest rate of increase in the constraint functions. We may obviously write the two conditions (1.15) as

$$-\nabla_x z(2,2) = (0)\ \nabla_x h_1(2,2) + (-4)\ \nabla_x h_2(2,2) \tag{1.30}$$

from which it is seen that $\bar{\mu}_1 = 0$, $\bar{\mu}_2 = -4$. This outcome is shown in Figure 1.2.

Now consider the same problem with the constraint relations replaced by inequalities. The feasible region is now an open triangular set of points, and the previous solution, although still feasible, does not give a minimum value of z. This is achieved at the point $\bar{x}_1 = 1$, $\bar{x}_2 = 1$ where $\bar{z} = 6$. At this point, the gradients are

$$-\nabla_x z(1,1) = [-2\ \ -2]^T, \qquad \nabla_x g_1(1,1) = [-1\ \ -1]^T, \qquad \nabla_x g_2(1,1) = [1\ \ \ 0]^T$$

Constraint $g_2 = x_1 - 2 \leq 0$ is not active, however, and so $\bar{\lambda}_2 = 0$ is required to satisfy (1.28). We will find that the two conditions (1.26) must be

$$-\nabla_x z(1,1) = (2)\ \nabla_x g_1(1,1) + (0)\ \nabla_x g_2(1,1) \tag{1.31}$$

from which it is seen that $\bar{\lambda}_1 = 2$, $\bar{\lambda}_2 = 0$. This situation is shown in Figure 1.3.

Returning now to the general problem (1.23), we should observe that the additional conditions $\lambda_i \geq 0$ allow (1.26, 1.27, 1.28) to be written:

$$\nabla_x z(\bar{x}) + \sum_i \bar{\lambda}_i\ \nabla_x g_i(\bar{x}) = 0 \tag{1.32}$$

$$g_i(\bar{x}) \leq 0, \qquad i = 1, 2, \ldots, m \tag{1.33}$$

$$\sum_i \bar{\lambda}_i\ g_i(\bar{x}) = 0 \tag{1.34}$$

$$\bar{\lambda}_i \geq 0, \qquad i = 1, 2, \ldots, m \tag{1.35}$$

Relations (1.24, 1.28) imply that $\bar{\lambda}_i\ g_i(\bar{x}) = 0$ for all i. Although the scalar product (1.34) gives only one equation, *each* of the m terms in its summand is perforce rendered zero by virtue of the sign constraints (1.33, 1.35) on the two factors in each term.

If the mathematical program (1.23) has a local minimum solution at $\bar{x}$, then $\bar{x}$ and some $\bar{\lambda}$ satisfy relations (1.32, 1.33, 1.34, 1.35): these are the *necessary optimality criteria* for a local minimum solution of the mathematical program. They may be called the *Karush-Kuhn-Tucker (KKT) conditions* after their originators, W. Karush, H. W. Kuhn and A. W. Tucker. Actually, it is still necessary to ensure that the Lagrange multipliers exist. This can be done, as in (1.20), through the introduction of a supernumerary Lagrange multiplier. Alternatively, it is achieved by means of a *constraint qualification* which identifies and excludes situations in which the multipliers would become infinitely large. Mangasarian[14] presents an extensive theoretical discourse on these matters.

A local minimum for problem (1.23) may occur at a point in the interior of the feasible region X – where none of the constraints is active, all the Lagrange multipliers vanish and $\nabla_x z(\bar{x}) = \mathbf{0}$ as in unconstrained optimisation – or at a point on the boundary of X. In this latter event, we may give the KKT conditions a *geometrical characterisation*: at a local minimum point $\bar{x}$ on the boundary of X, the negative gradient vector $-\nabla_x z(\bar{x})$ of the objective function z is contained within the cone formed by the gradient vectors $\nabla_x g(\bar{x})$ of the *ACTIVE* constraints.

Sufficient conditions, based upon second partial derivatives and similar to those given above for equality constrained optimisation, are proved by McCormick[15].

A more expansive representation of a mathematical program is one which specifically displays equality and inequality constraints, together with non-negativity restrictions on some of the variables. Consider the problem

$$\text{Minimise } z(x,y); \quad g_i(x,y) \leq 0, \quad i = 1, 2, \ldots, m$$
$$h_k(x,y) = 0, \quad k = 1, 2, \ldots, p$$
$$x_r \geq 0. \quad r = 1, 2, \ldots, q \tag{1.36}$$

in which z, g_i and h_k are continuous and differentiable functions of the non-negative variables x_r and the unrestricted variables y_j. Firstly, write the inequalities $x_r \geq 0$ in standard form $-x_r \leq 0$; then add slack variables s_i^2 and t_r^2 to the respective sets of inequalities to convert them to equalities. Then the Lagrangian for the modified problem becomes

$$M(x,y,\lambda,\mu,\nu,s,t) = z(x,y) + \sum_i \lambda_i \, [g_i(x,y) + s_i^2]$$
$$+ \sum_k \mu_k \, h_k(x,y) + \sum_r \nu_r \, [t_r^2 - x_r] \tag{1.37}$$

where λ_i, μ_k and ν_r are Lagrange multipliers, and the usual conditions for a stationary point of M follow

$$\nabla_x M = \nabla_x z(\bar{x},\bar{y}) + \nabla_x g(\bar{x},\bar{y}) \, \bar{\lambda} + \nabla_x h(\bar{x},\bar{y}) \, \bar{\mu} - \bar{\nu} = 0 \tag{1.38}$$

$$\nabla_y M = \nabla_y z(\bar{x},\bar{y}) + \nabla_y g(\bar{x},\bar{y}) \, \bar{\lambda} + \nabla_y h(\bar{x},\bar{y}) \, \bar{\mu} = 0 \tag{1.39}$$

$$\nabla_\lambda M = [g_i(\bar{x},\bar{y}) + \bar{s}_i^2] = 0 \tag{1.40}$$

$$\nabla_\mu M = [h_k(\bar{x},\bar{y})] = 0 \tag{1.41}$$

$$\nabla_\nu M = [\bar{t}_r^2 - \bar{x}_r] = 0 \tag{1.42}$$

$$\nabla_s M = [2 \, \bar{s}_i \, \bar{\lambda}_i] = 0 \tag{1.43}$$

$$\nabla_t M = [2 \, \bar{t}_r \, \bar{\nu}_r] = 0 \tag{1.44}$$

In (1.38, 1.39), $\nabla_x g(\bar{x},\bar{y}) = [\partial g_i(\bar{x},\bar{y})/\partial x_r]$ *et cetera* denote the appropriate Jacobian matrices evaluated at the point $(\bar{x},\bar{y})$.

We have already seen that the Lagrange multipliers associated with the original inequality constraints must be non-negative for a minimisation problem. Accordingly, we must set $\bar{\lambda} \geq 0$ and $\bar{\nu} \geq 0$, and these additional relations allow (1.38 to 1.44) to be written

$$\nabla_x z(\bar{x},\bar{y}) + \nabla_x g(\bar{x},\bar{y})\,\bar{\lambda} + \nabla_x h(\bar{x},\bar{y})\,\bar{\mu} - \bar{\nu} = 0 \qquad (1.45)$$

$$\nabla_y z(\bar{x},\bar{y}) + \nabla_y g(\bar{x},\bar{y})\,\bar{\lambda} + \nabla_y h(\bar{x},\bar{y})\,\bar{\mu} = 0 \qquad (1.46)$$

$$h(\bar{x},\bar{y}) = 0 \qquad (1.47)$$

$$g(\bar{x},\bar{y}) \leq 0 \qquad \bar{\lambda}^T g(\bar{x},\bar{y}) = 0 \qquad \bar{\lambda} \geq 0 \qquad (1.48,49,50)$$

$$\bar{x} \geq 0 \qquad \bar{\nu}^T \bar{x} = 0 \qquad \bar{\nu} \geq 0 \qquad (1.51,52,53)$$

These are the required KKT conditions. The multipliers $\bar{\nu}$ can easily be eliminated from the display, but their appearance more clearly informs of the effects af non-negative variables on the necessary conditions for a local minimum.

The additional conditions $\bar{\lambda} \geq 0$, $\bar{\nu} \geq 0$ imposed on the multipliers associated with the inequality constraints are necessary for the achievement of a local minimum. For a local maximum, it is necessary that $\bar{\lambda} \leq 0$, $\bar{\nu} \leq 0$. More commonly, non-negative multipliers $\bar{\lambda}$, $\bar{\nu}$ are retained in the necessary conditions for a local maximum of function z in (1.36) by writing the Lagrangian (1.37) with *negative* signs before each of the summations.

Triads of relations such as (1.48,49,50) or (1.51,52,53) express the concept of *complementarity*: for each i, if $g_i(\bar{x},\bar{y}) < 0$ then $\bar{\lambda}_i = 0$, or if $\bar{\lambda}_i > 0$ then $g_i(\bar{x},\bar{y}) = 0$: for each r, if $\bar{x}_r > 0$ then $\bar{\nu}_r = 0$, or if $\bar{\nu}_r > 0$ then $\bar{x}_r = 0$. In mechanics, several quite different physical phenomena are most conveniently described within discrete element models by such relations: steel cables under axial load; beams and plates on continuous, flexible foundations; certain laws of friction between contacting surfaces; certain constitutive laws for plastic flow in solids. Whether a cable is taught or slack; whether some element of a beam is in contact with its foundation or has lifted off; whether contacting surfaces are in relative sliding motion or are not; whether a solid material element is activated into plastic flow or is not; these are dichotomies that complementarity naturally embraces and effectively resolves.

We may feel an air of dissatisfaction with the theoretical developments so far described. The optimality conditions relate to local minima only; if the global minimum is required, it must be found by comparison of all the local minima. Furthermore, while a solution $\bar{x}$ of the mathematical program (1.36) must satisfy the KKT conditions, the converse is not generally true: only if $\bar{x}$ satisfies the KKT conditions together with some additional sufficient conditions dependent on second partial derivatives, will it locate a local minimum of the mathematical program. The KKT conditions and the mathematical program are not, therefore, *equivalent* mathematical formalisms. A resolution of the concerns to which we have just alluded may be had through the imposition of convexity on the objective and constraint functions.

CONVEX SETS AND FUNCTIONS

The feasible region of a mathematical program comprises a set X of points $\mathbf{x}$ for which the coordinates x_j $(j = 1,2,\dots,n)$ of each point satisfy the constraints. To be more precise, we should say that $\mathbf{x}$ is a member or an element of the set X $(\mathbf{x} \in X)$ and that the set X is contained within the n-dimensional real Euclidean space R^n $(X \subset R^n)$. This simply means that each of the n coordinates x_j is a real number. The set X is said to be a *convex set* if, for every $\mathbf{x}^1 \in X$ and $\mathbf{x}^2 \in X$, all points on the straight line segment connecting $\mathbf{x}^1$ and $\mathbf{x}^2$ also belong to X. That is,

$$\{w_1 x^1 + w_2 x^2\} \in X; \quad w_1 + w_2 = 1, \; w_1 \geq 0, \; w_2 \geq 0. \tag{1.54}$$

In the plane R^2, Figure 1.4 shows convex sets while Figure 1.5 shows sets which are not convex. Note that a feasible region formed from linear equality constraints is convex, while one formed from nonlinear equality constraints is generally not convex.

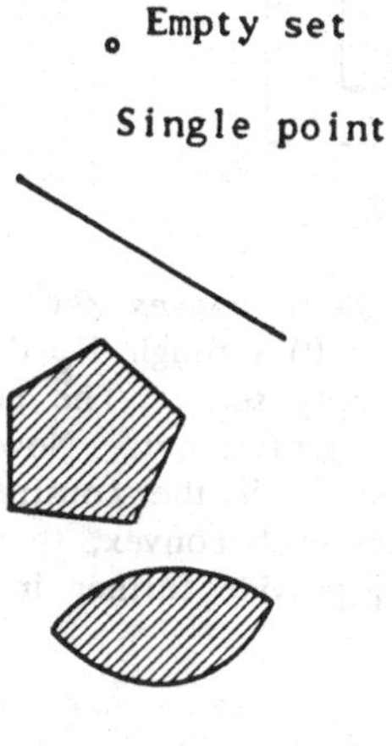

Figure 1.4

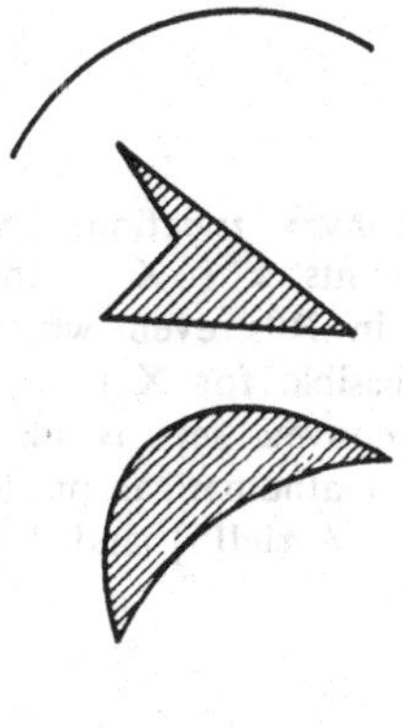

Figure 1.5

A real function $f(\mathbf{x})$ takes a value from the set R of real numbers (possibly including the extensions $\pm\infty$) for each $\mathbf{x}$ in some set $D \subset R^n$. The function $f(\mathbf{x})$ is said to be a *convex function* if the chord or straight line segment connecting any two points on the surface of $f(\mathbf{x})$ in R^{n+1} lies on or above the surface; that is, for every $\mathbf{x}^1 \in D$ and $\mathbf{x}^2 \in D$, and for convex combination parameters $w_1 \geq 0$, $w_2 \geq 0$, $w_1 + w_2 = 1$,

$$w_1 \, f(x^1) + w_2 \, f(x^2) \geq f[w_1 \, x^1 + w_2 \, x^2]. \tag{1.55}$$

If (1.55) is satisfied as a strict inequality for every $\mathbf{x}^1$ and $\mathbf{x}^2$ $(\mathbf{x}^1 \neq \mathbf{x}^2)$, then $f(\mathbf{x})$ is said to be a *strictly convex function*. Clearly, a convex function need not be differentiable nor even continuous.

If $f(\mathbf{x})$ is a convex function, then $-f(\mathbf{x})$ is said to be a *concave function*.

The concept of a convex function can be associated with that of a convex set in the following manner. The *epigraph* E_g of the function $g(\mathbf{x})$ is the set of points which lie on and above the surface of $g(\mathbf{x})$; that is, E_g is the set of points $(\mathbf{x},\mu)$ such that

$$E_g = \{(x,\mu): x \in D,\ \mu \in R,\ \mu \geq g(x)\} \subset R^{n+1} \qquad (1.56)$$

For $g(\mathbf{x})$ to be a convex function, it is necessary and sufficient that E_g be a convex set. A convex function and its epigraph are indicated in Figure 1.6.

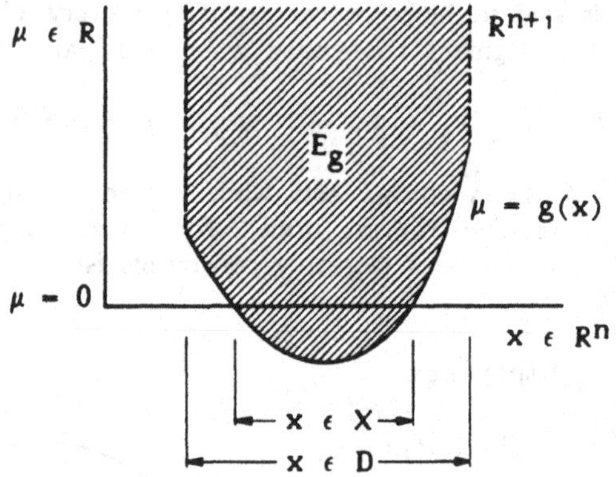

Figure 1.6

Let $g(\mathbf{x})$ be a convex function, then E_g must be a convex set. It follows that the level subset of points $\mathbf{x} \in X$ in the section $(\mu = 0)$ through E_g defined by $g(\mathbf{x}) \leq 0$ is a convex set in R^n, even when X is the empty set. The converse, however, is not true: it is possible for X to be convex when $g(\mathbf{x})$ is not. Now, the intersection of any number of convex sets is also a convex set. If the constraint functions $g_i(\mathbf{x})$ appearing in the mathematical program (1.23) are each convex, then the feasible region is a convex set. Avriel[5] and Mangasarian[14] provide further information on convex sets and functions.

OPTIMALITY CRITERIA FOR CONVEX MATHEMATICAL PROGRAMMING

We now restrict the problem of mathematical programming to the minimisation of a convex function defined on a convex set or feasible region. The mathematical program

$$\begin{aligned} \text{Minimise } z(x,y); \quad & g_i(x,y) \leq 0, && i = 1, 2, \ldots, m \\ & h_k(x,y) = 0, && k = 1, 2, \ldots, p \\ & x_r \geq 0. && r = 1, 2, \ldots, q \end{aligned} \qquad (1.57)$$

satisfies this restriction when z and g_i are convex functions over the feasible region, and when h_k are *linear* functions. It is called a *convex program*.

The following important results may be applied in convex programming.

Global minimum: If z has a local minimum at $(\bar{x},\bar{y})$ in the feasible region, then it is a global minimum; that is, $\bar{z} = z(\bar{x},\bar{y})$ is the smallest value of z to be found in the entire feasible region. There may, of course, be other points $(\bar{x},\bar{y})$ where z takes the same value $\bar{z}$. Such points form a level subset of E_z at $\mu = \bar{z}$ which is necessarily convex, and the intersection of this subset with the feasible region forms a convex set of alternative optimal solutions.

Unique solution: Let z(**x**,y) be strictly convex over the feasible region. If z has a local minimum at $(\bar{x},\bar{y})$ in the feasible region, then $(\bar{x},\bar{y})$ is the unique solution to the convex program (1.57).

Sufficiency of KKT conditions: If there are vectors $(\bar{x},\bar{y},\bar{\lambda},\bar{\mu},\bar{\nu})$ which satisfy the KKT conditions (1.45 to 1.53), then $(\bar{x},\bar{y})$ is a global minimum solution of the convex program (1.57).

Assuming always that the Lagrange multipliers exist, this last result means that the KKT conditions are both necessary and sufficient for a global optimal solution of the convex program. We may now say that the convex program (1.57) and the KKT problem, which demands the solving of the mathematical structure (1.45 to 1.53), are ***equivalent mathematical formalisms***: all solutions $(\bar{x},\bar{y})$ are common to both. The optimal design of engineering structures presents itself generically as a problem of mathematical programming, but most problems in structural mechanics do not. If the vectorial relations of equilibrium, compatibility and material constitution of some problem in structural mechanics can be marshalled into the form displayed by the KKT problem, and if the appropriate requirements of convexity are satisfied, then the mechanics problem is equivalently described by a convex program. We may, of course, solve the structural problem by solving the convex program; but we may also infer a theoretical result from the minimising nature of the program – that the generically vectorial problem of structural mechanics may be represented equivalently by a variational or minimum principle.

Finally, we shall note that, if z is a non-constant *concave* function and the constraint set X remains convex, then any minimising solution $(\bar{x},\bar{y})$, if it exists, must lie on the boundary of the feasible region X.

LAGRANGIAN SADDLEPOINT OPTIMALITY CRITERIA

The Lagrangian associated with the mathematical program (1.23) will be written

$$L(x,\lambda) = z(x) + \sum_i \lambda_i \, g_i(x), \tag{1.58}$$

from which the slack variables s_i^2 are now excluded. This function is defined for all values of the variables x_j $(j = 1,2,\ldots,n)$ for which the functions z and g_i are defined, and for all values $\lambda_i \geq 0$ $(i = 1,2,\ldots,m)$. Further optimality criteria for mathematical programming are associated with the following problem.

Saddlepoint problem: Find a point $(\bar{x},\bar{\lambda})$ with $\bar{\lambda} > 0$, if it exists, such that

$$L(\bar{x},\lambda) < L(\bar{x},\bar{\lambda}) < L(x,\bar{\lambda}) \tag{1.59}$$

for all x and all $\lambda > 0$.
The saddlepoint problem gives rise to quite different, but interesting and useful, optimality criteria which will now be stated.

Sufficient condition: If $(\bar{x},\bar{\lambda})$ is a saddlepoint, in the sense of (1.59), of the Lagrangian (1.58), then $\bar{x}$ locates a local minimum solution of the mathematical program (1.23).

This result does not require any convexity assumptions, so it can be applied to any nonlinear program, even one with nonlinear equality constraints. More generally, it does not require that any of the functions be differentiable. The converse, however, is not generally true: a local minimum solution of the mathematical program need not be associated with a saddlepoint of its Lagrangian. To remedy this, we must impose convexity on the functions z and g_i.

Necessary condition: Let z and g_i be convex functions, and let g_i satisfy a constraint qualification to ensure the existence of the Lagrange multipliers λ_i. If $\bar{x}$ solves the mathematical program (1.23), then $\bar{x}$ and some $\bar{\lambda}$ locate a saddlepoint (1.59) of the Lagrangian, and $\bar{\lambda}_i\ g_i(\bar{x}) = 0$ for each i $(i = 1,2,\ldots,m)$.

Again, differentiablity of the functions z and g_i is not essential for this result. The imposition of convexity provides conditions which are both necessary and sufficient for a solution of the mathematical program (1.23). We may say that the mathematical program and the saddlepoint problem are equivalent problems in the sense that an $\bar{x}$ which solves one of the problems also solves the other, and *vice versa*. With the requirements of convexity and differentiablity satisfied by z and g_i, it is shown by Mangasarian[14] that the KKT problem (1.32 to 1.35) and the saddlepoint problem (1.59) are also equivalent.

For the mathematical program (1.36), the Lagrangian may now by written

$$L(x,y,\lambda,\mu) = z(x,y) + \sum_i \lambda_i\ g_i(x,y) + \sum_k \mu_k\ h_k(x,y) \tag{1.60}$$

for all y and μ, and for all $x > 0$ and $\lambda > 0$. Its saddlepoints $(\bar{x},\bar{y},\bar{\lambda},\bar{\mu})$, if they exist, are such that

$$L(\bar{x},\bar{y},\lambda,\mu) < L(\bar{x},\bar{y},\bar{\lambda},\bar{\mu}) < L(x,y,\bar{\lambda},\bar{\mu}) \tag{1.61}$$

for all y and μ, and for all $x > 0$ and $\lambda > 0$. The necessary optimality conditions stated above are only valid if z and g_i are convex and if h_k are linear functions.

The Lagrangian of a convex program is a convex-concave function; if (1.36) is a convex program, then its Lagrangian (1.61) is convex in x,y and concave in λ,μ. The search for a saddlepoint of such a function may be based on a minimax strategy since

$$z(\bar{x},\bar{y}) = \min_{x>0,y}\ \max_{\lambda>0,\mu}\ L(x,y,\lambda,\mu) \tag{1.62}$$

We may care to note that this type of extremum occurs in structural mechanics as a discrete representation of a mixed variational principle in the mechanics of a solid continuum. Such a principle is one in which the static and kinematic fields receive simultaneous variations and the correct fields are identified by the stationarity of an associated scalar functional. Washizu[16] describes the Hellinger-Reissner principle of elastostatics among other principles of this type.

DUALITY IN MATHEMATICAL PROGRAMMING

A mathematical program may require the minimisation of a scalar function $z(\mathbf{x})$ of variable vector $\mathbf{x}$ over a feasible region or constraint set X. It often happens in problems with a physical motivation that, associated with this first or, as we may call it, *primal* mathematical program, there occurs a second, requiring the maximisation of a scalar function $w(\lambda)$ of variable vector λ over a constraint set Λ. There may be no obvious connection between the two programs; their objective functions are different and their variable vectors are unrelated. On the other hand, suppose that

$$z(x) \geq w(\lambda), \qquad \text{for all } x \in X,\ \lambda \in \Lambda. \tag{1.63}$$

This is the most basic result of duality, often called *weak duality*, and the maximisation program which satisfies it may be called a *dual program*.

The optimal values of z and w must also respect the same inequality, Min z $\geq$ Max w. If they differ, there is said to be a *duality gap*: if they are equal, then

$$\underset{x \in X}{\text{Min}}\ z(x) = z(\bar{x}) = M = w(\bar{\lambda}) = \underset{\lambda \in \Lambda}{\text{Max}}\ w(\lambda) \tag{1.64}$$

which is usually referred to as *strong duality*. Dual programs for which (1.64) is satisfied share a unique optimal objective function value M, and they may be used to calculate upper and lower bounds on M. They may also have considerable theoretical interest. In structural mechanics, for instance, dual programs have the potential for giving discrete representation to the established complementary variational principles of continuum mechanics; and they afford a means for the divining of new principles.

Linear programming has provided the model for this type of duality. The minimising and maximising nature of the programs was quickly linked by von Neumann to his own saddlepoint criterion for an optimising strategy in the theory of games. Primal-dual linear programs have the following features:

* They exhibit strong duality.
* The structure of the dual is completely inferable from that of the primal.
* The primal variables do not appear in the dual program, and *vice versa*.
* The primal variables are the dual Lagrange multipliers, and *vice versa*.
* The operation of inferring the structure of the dual program from that of the primal is *reflexive*: the dual of the dual program is the primal.

For the general nonlinear program, considerable effort has been expended in the search for a dual program that also exhibits these additionally desirable features. It must be said that most success has attended those investigations where the primal program has been assumed convex.

Where should one begin to look for a maximising program that will, at least, satisfy (1.63)? We have noted that the Lagrangian of a convex program is a convex-concave function and that the search (1.62) for a saddlepoint has both a minimising and a maximising character. It is hardly surprising, therefore, that attention quickly fell upon the Lagrangian in the quest to find a dual program.

Let the primal be the convex program

$$\text{Minimise } z = z(x); \quad g_i(x) \leq 0, \quad i = 1,2,\ldots,m. \tag{1.65}$$

Then, in virtue of convexity, (1.65) is equivalent to a saddlepoint problem founded on its Lagrangian $L(\mathbf{x},\lambda)$ which may be represented in the minimax form

$$\min_{x} \max_{\lambda\geq 0} L(x,\lambda) = z(\bar{x}) = \max_{\lambda\geq 0} \min_{x} L(x,\lambda). \tag{1.66}$$

Suppose that a function $w(\lambda)$ is defined by

$$w(\lambda) = \min_{x} L(x,\lambda). \tag{1.67}$$

Then the second relation in (1.66) establishes strong duality for the program

$$\text{Maximise } w = w(\lambda); \quad \lambda_i \geq 0, \quad i = 1,2,\ldots,m. \tag{1.68}$$

in which $w(\lambda)$, if it can be constructed, is concave. Lasdon[17] describes the construction with examples. Unfortunately, the objective function $w(\lambda)$ of the dual program must itself be constructed by optimisation, as (1.67) demands. Since $L(\mathbf{x},\lambda)$ is convex in $\mathbf{x}$ and is unconstrained, its minimum may be found by setting its gradient vector to the zero vector. Thus, we may rewrite (1.68)

$$\text{Maximise } L = L(x,\lambda); \quad \nabla_x L(x,\lambda) = 0, \quad \lambda \geq 0, \tag{1.69}$$

where the maximisation must now extend over both $\mathbf{x}$ and λ. This is the dual program of Wolfe. Its objective function contains both primal and dual variables and is no longer concave. The theorems which concern the existence of optimal solutions to each of the two programs, and which are modelled on those for linear programming, are considered in detail by Mangasarian[14].

The same approach may be employed in the case of the convex program (1.57). The dual program may be written down immediately

$$\text{Maximise } w = w(\lambda,\mu,\nu); \quad \lambda \geq 0, \ \mu \text{ unrestricted}, \ \nu \geq 0, \tag{1.70}$$

where the objective function w requires minimisation of the Lagrangian (1.60) over $\mathbf{x}\geq\mathbf{0}$ and $\mathbf{y}$:

$$w(\lambda,\mu,\nu) = \min_{x\geq 0,y} L(x,y,\lambda,\mu) = \min_{x,y} \{ L(x,y,\lambda,\mu) - \nu^T x \} \tag{1.71}$$

The Lagrange multiplier vector $\nu\geq\mathbf{0}$ allows $\mathbf{x}$ to be treated as unrestricted in the second relation of (1.71), and the minimum may then be found by differentiation. It requires that

$$\nabla_x L(x,y,\lambda,\mu) - \nu = 0, \qquad \nabla_y L(x,y,\lambda,\mu) = 0,$$

which then allows the dual program to be rewritten in Wolfe's form:

$$\text{Maximise } L = L(x,y,\lambda,\mu) - x^T \nabla_x L(x,y,\lambda,\mu) \;;$$

$$\nabla_x L(x,y,\lambda,\mu) \geq 0, \quad \nabla_y L(x,y,\lambda,\mu) = 0, \quad x \geq 0, \quad \lambda \geq 0. \qquad (1.72)$$

The dual programs of linear programming, as well as those of quadratic programming, can be obtained from this form.

An alternative approach to duality has been taken by others, notably Rockafellar[18], which is based upon the theory of convex functions. It is more general and more elegant than the minimax approach, and it has the potential to invest this duality with all the features of linear programming duality which may be thought so desirable. There is an inherent geometrical duality in the description of a function $u = f(\mathbf{x})$, where u is a real number, $u \epsilon R$, and $\mathbf{x}$ is an n component vector, $\mathbf{x} \epsilon R^n$. It may be described as a locus of points $[\mathbf{x},u]$ in R^{n+1}; or it may be conceived as an envelope of tangent planes – more properly hyperplanes – $[\xi,\eta]$, where ξ is the gradient vector of each plane and η is the intercept which that plane makes on the u coordinate axis. This geometric duality is subsumed in the Legendre transformation, which, as Courant and Hilbert[19] show, provides a symmetric change of variables in terms of which complementary energy functionals may be defined for continuum problems of solid mechanics and thermomechanics in the calculus of variations. The same duality pervades the concept of a conjugate convex function which forms the basis for the alternative approach to duality in convex mathematical programming.

Consider once more the convex program

$$\text{Minimise } z = z(x) \;; \quad g_i(x) \leq 0, \qquad i = 1,2,\ldots,m. \qquad (1.73)$$

The sensitivity of the program and the stability of its solutions can be tested by giving variation to its structure through the imposition of a scheme of perturbations. Let v_i be a set of variable perturbation parameters. Then, among other possibilities, one form of the perturbed convex program is

$$\text{Minimise } z = z(x) \;; \quad g_i(x) \leq v_i, \qquad i = 1,2,\ldots,m. \qquad (1.74)$$

An unconstrained minimisation problem may be had by enveloping the constraints with the objective function in the convex function

$$F(x,v) = \begin{cases} f(x) & \text{if } \; g_i(x) \leq v_i, \\ +\infty & \text{otherwise} \end{cases} \qquad (1.75)$$

The conjugate function is defined in terms of vectors ξ and λ by the following rule:

$$F^*(\xi,\lambda) = \sup_{x,v} \left[\; \xi^T x + \lambda^T v - F(x,v) \;\right] \qquad (1.76)$$

where the supremum – the least upper bound – allows for cases in which a strict maximum does not exist. The conjugate function is also convex. For any feasible $\mathbf{x}$ and $\mathbf{v}$, (1.76) therefore gives

$$F^*(\xi,\lambda) \geq \xi^T x + \lambda^T v - F(x,v). \tag{1.77}$$

Although **v** and **ξ** are important in the consideration of stability, we are quite at liberty to assign them any values we may choose. Let us choose **v=0**, **ξ=0**. Then

$$F(x,0) \geq -F^*(0,\lambda), \tag{1.78}$$

and, because **x** and **λ** are independent, we have the familiar weak duality which suggests the primal-dual programs

$$\min_x \{ F(x,0) \}, \qquad \max_\lambda \{ -F^*(0,\lambda) \} \qquad (1.79,\ 1.80)$$

The primal program (1.79), which has **v=0**, should coincide with the convex program (1.73). Again, the construction of the dual centres on calculating the maximum or supremum in (1.76). For a given primal program, however, different schemes of perturbations tend to give different dual programs, an interesting feature. It may be shown that the operation of forming the dual is reflexive, and that, when convex programs with a certain extreme behaviour are excluded, strong duality is exhibited. An excellent review of these concepts, with easy-to-follow examples, is given by Avriel[5], while Whittle[20] offers an attractive exploration of Lagrangian methods in the same geometrical setting.

REFERENCES

1. Maier, G. and Munro, J., Mathematical programming applications to engineering plastic analysis, Applied Mechanics Reviews, **35** (1982) 1631–1643.

2. Maier, G. and Lloyd Smith, D., Update to: Mathematical programming applications in engineering plastic analysis, in: Applied Mechanics Update 1986 (Eds. C. R. Steele and G. S. Springer), The American Society of Mechanical Engineers 1986, 377–383.

3. Panagiotopoulos, P. D., Inequality Problems in Mechanics and Applications, Birkhauser, Basel 1985.

4. Maier, G., Mathematical programming applications to structural mechanics: some introductory thoughts, Eng. Struct., **6** (1984) 2–6.

5. Avriel, M., Nonlinear Programming: Analysis and Methods, Prentice-Hall 1976.

6. Teixeira de Freitas, J. A., A gradient method for elasto-plastic analysis of structures, in this volume.

7. Hancock, H., Theory of Maxima and Minima, Dover Publications 1960.

8. Courant, R., Differential and Integral Calculus, vol. 2, Blackie, Glasgow 1936.

9. Walsh, G. R., Methods of Optimisation, Wiley 1979.

10. Fletcher, R., (Ed.), Optimisation, Academic Press 1969.

11. Hadley, G., Nonlinear and Dynamic Programming, Addison Wesley 1964.

12. Beveridge, G. S. G. and Schechter, R. S., Optimisation: Theory and Practice, McGraw-Hill 1970.

13. Rao, S. S., Optimisation: Theory and Applications, Wiley Eastern 1978.

14. Mangasarian, O. L., Nonlinear Programming, McGraw-Hill 1969.

15. McCormick, G. P., Second order conditions for constrained minima, SIAM J. Appl. Math., **15** (1967) 641–652.

16. Washizu, K., Variational Methods in Elasticity and Plasticity, 2nd. ed., Pergamon 1975.

17. Lasdon, L. S., Optimisation Theory for Large Systems, Macmillan 1970.

18. Rockafellar, R. T., Convex Analysis, Princeton 1970.

19. Courant, R. and Hilbert, D., Methods of Mathematical Physics, vol. 1, Interscience 1953.

20. Whittle, P., Optimisation Under Constraints, Wiley 1971.

CHAPTER 2

LINEAR PROGRAMMING

D. Lloyd Smith
Imperial College, London, U.K.

ABSTRACT

Linear programming is the branch of optimisation which is probably most often used in connection with problems of structural engineering – in the generic modelling of some types of problem or in the approximation of other more complicated ones. A review is made of the Simplex algorithm for linear programming in its two-phase form, and the program solution is related to the Karush-Kuhn-Tucker optimality conditions. The duality of linear programming is described in terms of the associated Lagrangian, and the optimal solution of the dual linear program is obtained from the Simplex solution of the primal problem through the Simplex multipliers.

INTRODUCTION

In the field of structural mechanics, the mathematical models tending to find most application are those invested with considerable linearity. For such models, there are usually standardised and easily implementable solution processes, and the theoretical basis is well-endowed with useful theorems. This is certainly true of linear elastic structural mechanics. It is also true of the optimal design of perfectly plastic structures; here a linear objective function represents the weight or the cost of the structure and is to be minimised. The natural or generic mathematical description of such a problem is as a linear program. Plastic limit analysis, or the determination of the collapse load of a structure, is a problem for which the vectorial laws of structural mechanics have an initially different mathematical description; but it is a description that, under certain conditions, implies equivalence with dual linear programs. Linear programming is the only effective mathematical apparatus for representing and solving these problems of structural plasticity.

THE LINEAR PROGRAM

A linear program (LP) is a mathematical program which has a linear objective function and linear constraint functions. The *standard form* of the LP is:

$$\text{Minimise } z = \mathbf{c}^T \mathbf{x} \; ; \qquad \mathbf{A}\,\mathbf{x} = \mathbf{b}, \qquad \mathbf{x} \geq \mathbf{0}, \tag{2.1}$$

wherein (a)each of the variables x_j $(j = 1,2,\ldots,n)$ is non-negative, (b)each of the constants b_i $(i = 1,2,\ldots,m)$ is non-negative, and (c)each of the m constraints is an equality. In (2.1), $\mathbf{A}$ is a rectangular matrix formed from constant elements a_{ij}, $\mathbf{c}^T$ is a row vector having constant elements c_j and $m < n$.

The standard form is used to present an LP to the Simplex algorithm for calculation. Of course, linear programs need not naturally appear in this standard form, but they may easily be converted. If the *i*th constraint is $\mathbf{a}_i^T \mathbf{x} \leq b_i$, where $\mathbf{a}_i^T$ is a row vector of constants and $b_i \geq 0$, then we may write $\mathbf{a}_i^T \mathbf{x} + x_s = b_i$ by adding a slack variable $x_s \geq 0$. In similar manner, the constraint $\mathbf{a}_i^T \mathbf{x} \geq b_i$, in which $b_i \geq 0$, may be written $\mathbf{a}_i^T \mathbf{x} - x_s = b_i$ by the subtraction of a surplus variable $x_s \geq 0$. If variable x_j is unrestricted, we may write $x_j = x_j^+ - x_j^-$, where $x_j^+ \geq 0$ and $x_j^- \geq 0$; we therefore introduce two new non-negative variables for each unrestricted one.

The feasible region of an LP is bounded by "flat surfaces" or hyperplanes. It is therefore convex. Optimal solutions of an LP must necessarily lie on the boundary of the feasible region. Indeed, if an optimal solution occurs at an interior point of one of these hyperplanes, then optimal solutions occur at *all* points on that hyperplane. It follows that, to find an optimal solution for an LP, one need only check solutions which correspond to "corner points" or vertices of the feasible region.

The equality constraints of the standard form (2.1) are m simultaneous linear algebraic equations in n variables $(m < n)$. If we first ignore the non-negative nature of the variables, we may find a solution of these equations by a simple strategy. Choose m of the variables $\mathbf{x}$ to form a set of *basic variables* $\mathbf{x}_b$; the remaining $(n - m)$ variables will then be known as the *non-basic variables* $\mathbf{x}_n$. If the non-basic variables are set to zero, it may be possible to solve the system of equations for the basic variables. Such a solution is called a *basic solution*. Now remember that the variables should be non-negative: if none of the values in a basic solution is negative, we shall call it a *basic feasible solution*. The "corner points" or vertices of the feasible region correspond with the set of basic feasible solutions.

THE SIMPLEX ALGORITHM

For the solving of LP problems, the Simplex algorithm has almost universal use. It may be described in two phases; the first phase is required to find a first basic feasible solution, while the second constructs a sequence of basic feasible solutions for which the corresponding objective function values form a non-increasing sequence. This process converges in a finite number of iterations on a minimising solution of the LP, and it may be interpreted in geometrical terms as a sequence of moves from one vertex of the feasible region to an adjacent one. Since the sequence of objective function values is monotonic (non-increasing), only a small subset of the vertices is explored.

The full tableau form of the Simplex algorithm obtains a basic solution for each iteration by simple elimination, and the tableau displays it in the *canonical form*

$$\begin{bmatrix} \mathbf{I} & \bar{\mathbf{A}} \end{bmatrix} \begin{bmatrix} \mathbf{x}_b \\ \mathbf{x}_n \end{bmatrix} = \bar{\mathbf{b}} \qquad (2.2)$$

The basic solution is $[\ \mathbf{x}_b = \bar{\mathbf{b}},\ \mathbf{x}_n = \mathbf{0}\ ]$, and $\bar{\mathbf{b}} \geq \mathbf{0}$ is necessary for the solution also to be feasible.

To be specific, we shall consider an LP having two equality constraints, presented by Dantzig[1] in his excellent book on linear programming. It is already in standard form.

Minimise z; where (2.3)

$$\left[\begin{array}{ccccc} 5 & -4 & 13 & -2 & 1 \\ 1 & -1 & 5 & -1 & 1 \\ \hline 1 & 6 & -7 & 1 & 5 \end{array}\right] \begin{bmatrix} x_1 \\ x_2 \\ x_3 \\ x_4 \\ x_5 \end{bmatrix} = \left[\begin{array}{c} 20 \\ 8 \\ \hline z \end{array}\right]$$

$$x_1 \geq 0, \quad x_2 \geq 0, \quad x_3 \geq 0, \quad x_4 \geq 0, \quad x_5 \geq 0.$$

Phase I

In order to construct a first basic feasible solution for LP(2.3), we add an *artificial variable* to each constraint, giving m artificial variables in all: thus x_6 is added to the first constraint and x_7 to the second, and a canonical form is thereby created with x_6 and x_7 as basic variables. With all other variables set to zero, x_6 = 20 and x_7 = 8; but a feasible solution to (2.3) requires that both artificial variables be zero. The

Simplex algorithm is used to drive these variables to zero, if it is possible. To accomplish this task, we construct a measure of infeasiblity. The *infeasibility function* f is the sum of all the artificial variables; in the present case

$$f = x_6 + x_7. \tag{2.4}$$

Initially, f = 20 + 8 = 28. The artificial variables are also non-negative; they can be driven to zero by minimisation of the function f. If the minimum value of f is zero, we shall have found a first basic feasible solution: if not, then no feasible solution exists and the feasible region is empty. This latter outcome is often the result of an incorrectly posed problem or of errors in the assembly of data.

To begin, then, let us insert the equation (2.4) into LP (2.3) as its fourth row, and eliminate x_6 and x_7 from it by subtracting each of the constraint rows. Since they are variable, let us also move f and z to the left-hand side to accompany the variables **x**. These operations result in the starting tableau – Tableau 0 of the solution display set out in Table 1. For each tableau, the set of current basic variables is indicated by the marks ⊙ which appear under the columns associated with these variables.

The Phase I objective is to minimise f, and the steps of the algorithm are:

1. Find the minimum coefficient $\bar{d}_c$ among the coefficients $\bar{d}_j$ of the variables x_j (j = 1,2,...,m+n), including the artificial variables, in the f row – row 4 – of the tableau. If $\bar{d}_c \geq 0$, the current value of f is minimal: go to step 6.

2. If $\bar{d}_c < 0$, then x_c is selected as a new basic variable. Column j = c is the *pivot column*, and the mark □ appears under this column in each tableau.

3. Find the minimum value $\theta_r = \min \theta_i = \min[\ \bar{b}_i/\bar{a}_{ic}\ ] = \bar{b}_r/\bar{a}_{rc}$ of the ratio θ_i between the constant $\bar{b}_i$ and the corresponding element $\bar{a}_{ic}$ of the pivot column among all the constraint rows i (i = 1,2,...,m). Only pivot column elements for which $\bar{a}_{ic} > 0$ should be considered in calculating θ_i. In the case of equal minima, r is usually chosen as the index of the first constraint row which exhibits the minimum value of θ_i. Constraint row i = r is the *pivot row*, and the mark □ appears to the right of this row in each tableau. The basic variable in that row will become non-basic.

4. Transform the tableau by performing a *pivot operation*, centred on the pivot $\bar{a}_{rc}$. This is just Gauss-Jordan elimination: it associates a canonical form of the tableau with the new set of basic variables.

5. Return to step 1 and repeat the sequence of steps.

6. If the minimum value of f is f = 0, then proceed to Phase II: the current basic solution is feasible. Otherwise STOP – the problem is infeasible.

TABLEAU 0 (PHASE I)

x_1	x_2	x_3	x_4	x_5	x_6	x_7	$-z$	$-f$		
5	-4	13	-2	1	1				20	□ $\theta_1 = \frac{20}{13}$
1	-1	5	-1	1		1			8	$\theta_2 = \frac{8}{5}$
1	6	-7	1	5			1		0	
-6	5	-18	3	-2				1	-28	
		□			⊙	⊙				

TABLEAU 1 (PHASE I)

x_1	x_2	x_3	x_4	x_5	x_6	x_7	$-z$	$-f$		
$\frac{5}{13}$	$\frac{-4}{13}$	1	$\frac{-2}{13}$	$\frac{1}{13}$	$\frac{1}{13}$				$\frac{20}{13}$	$\theta_1 = 20$
$\frac{-12}{13}$	$\frac{7}{13}$		$\frac{-3}{13}$	$\frac{8}{13}$	$\frac{-5}{13}$	1			$\frac{4}{13}$	□ $\theta_2 = \frac{1}{2}$
$\frac{48}{13}$	$\frac{50}{13}$		$\frac{-1}{13}$	$\frac{72}{13}$	$\frac{7}{13}$		1		$\frac{140}{13}$	
$\frac{12}{13}$	$\frac{-7}{13}$		$\frac{3}{13}$	$\frac{-8}{13}$	$\frac{18}{13}$			1	$\frac{-4}{13}$	
		⊙		□		⊙				

TABLEAU 2 (PHASE I)

x_1	x_2	x_3	x_4	x_5	x_6	x_7	$-z$	$-f$		
$\frac{1}{2}$	$\frac{-3}{8}$	1	$\frac{-1}{8}$		$\frac{1}{8}$	$\frac{-1}{8}$			$\frac{3}{2}$	
$\frac{-12}{8}$	$\frac{7}{8}$		$\frac{-3}{8}$	1	$\frac{-5}{8}$	$\frac{13}{8}$			$\frac{1}{2}$	□ $\theta_2 = \frac{4}{7}$
12	-1		2		4	-9	1		8	
					1	1		1	0	
	□	⊙		⊙						

Table 1
(continued over)

TABLEAU 3 (PHASE II) OPTIMAL TABLEAU

x_1	x_2	x_3	x_4	x_5	x_6	x_7	$-z$	$-f$	
$\frac{-1}{7}$		1	$\frac{-2}{7}$	$\frac{3}{7}$	$\frac{-1}{7}$	$\frac{4}{7}$			$\frac{12}{7}$
$\frac{-12}{7}$	1		$\frac{-3}{7}$	$\frac{8}{7}$	$\frac{-5}{7}$	$\frac{13}{7}$			$\frac{4}{7}$
$\frac{72}{7}$			$\frac{11}{7}$	$\frac{8}{7}$	$\frac{23}{7}$	$\frac{-50}{7}$	1		$\frac{60}{7}$
	⊙	⊙							

Table 1

Two iterations are required to minimise f, and Tableaux 1 and 2 of Table 1 record the calculations. Phase I terminates with a first basic feasible solution

$$x_3 = \frac{3}{2}\,, \qquad x_5 = \frac{1}{2}\,, \qquad z = -8\,, \qquad f = 0. \tag{2.5}$$

Phase II

In Phase II we revert to the original objective – that of minimising z. The f row – row 4 – is no longer needed and may be deleted. Also, the artificial variables are no longer needed; but they will be retained in the Phase II tableaux because they provide some information that will later be of great use. The first basic feasible solution is used to initiate an "improving" sequence of basic feasible solutions. Here, the word "improving" is used in the sense that the corresponding sequence of values of z is monotonically decreasing – that is, between any two successive tableaux, z does not increase.

The steps of the Simplex algorithm in Phase II are almost the same as those given for Phase I.

1. Find the minimum coefficient $\bar{c}_c$ among the coefficients $\bar{c}_j$ of the variables x_j ($j = 1,2,\ldots,n$), now excluding the artificial variables, in the z row – row 3 – of the tableau. If $\bar{c}_c \geq 0$, then STOP – the current value of z is minimal.

2. If $\bar{c}_c < 0$, then x_c is selected as a new basic variable. Column $j = c$ is the pivot column.

3. Find the minimum value $\theta_r = \min \theta_i = \min[\, \bar{b}_i/\bar{a}_{ic} \,] = \bar{b}_r/\bar{a}_{rc}$ of the ratio θ_i between the constant $\bar{b}_i$ and the corresponding element $\bar{a}_{ic}$ of the pivot column among all the constraint rows i ($i = 1,2,\ldots,m$). Only pivot column elements for which $\bar{a}_{ic} > 0$ should be considered in calculating θ_i. In the case of equal minima, r is usually chosen as the index of the first constraint row which exhibits the minimum value of θ_i. Constraint row r is the pivot row, and the basic variable in that row will become non-basic.

If $\bar{a}_{ic} \leq 0$ for all i (i = 1,2,...,m), then STOP – the minimum value of z is *unbounded*. That is, min $z \to -\infty$.

4. Transform the tableau by performing a pivot operation, centred on $\bar{a}_{rc}$.

5. Return to step 1 and repeat the sequence of steps.

Phase II terminates in Tableau 3 of Table 1 with an optimal solution

$$x_3 = \frac{12}{7}, \quad x_2 = \frac{4}{7}, \quad z = \frac{-60}{7}. \tag{2.6}$$

The columns of Tableau 3 which correspond with the artificial variables are shown shaded as a reminder that they are no longer eligible as pivot columns in Phase II.

This form of the full tableau presentation of the Simplex calculations is that suggested by Dantzig[1]. It is clear that a compaction of each tableau may be had by omitting the columns associated with the current set of basic variables. A slightly different way of presenting the calculations, employing compact tableaux, is described by Gass[2] and by Hadley[3]. For large-scale LP calculations, a more refined and systematic matrix presentation of the algorithm is used. It is called the *Revised Simplex Method*; it is described by the authors mentioned above and, particularly, by Orchard-Hays[4]. While the Simplex algorithm is of almost universal use in the solution of LP problems, a new and quite different type of algorithm has been reported by Karmarkar[5]. It seems to show some promise, but has yet to find its way into common engineering use.

KARUSH-KUHN-TUCKER OPTIMALITY CONDITIONS

An LP is a convex mathematical program. It is therefore a special case of the convex program (1.57), set out in Chapter 1, for which the KKT conditions are those given in (1.45 to 1.53). Consider the LP

$$\text{Minimise } z = \mathbf{c}^T \mathbf{x}\ ; \quad \mathbf{A}\,\mathbf{x} \geq \mathbf{b}\ , \quad \mathbf{x} \geq \mathbf{0}. \tag{2.7}$$

We may write the main constraints in the usual form

$$\mathbf{g}(\mathbf{x}) = \mathbf{b} - \mathbf{A}\,\mathbf{x} \leq \mathbf{0}, \tag{2.8}$$

and, for convenience, let us introduce a set of slack variables $\mathbf{u}$ into (2.8)

$$\mathbf{g}(\mathbf{x}) + \mathbf{u} = \mathbf{b} - \mathbf{A}\,\mathbf{x} + \mathbf{u} = \mathbf{0}, \qquad \mathbf{u} \geq \mathbf{0}. \tag{2.10}$$

LP(2.7) may be obtained from the convex program (1.57) by removal of unrestricted variables $\mathbf{y}$ and equality constraints $\mathbf{h}(\ ) = \mathbf{0}$. Differentiation of $z(\mathbf{x})$ and $\mathbf{g}(\mathbf{x})$ gives

$$\nabla_x z(\mathbf{x}) = \mathbf{c}, \qquad \nabla_x \mathbf{g}(\mathbf{x}) = -\mathbf{A}^T, \tag{2.11, 2.12}$$

from which the KKT problem (1.45 to 1.53) may be specialised for the LP. We obtain

$$\begin{bmatrix} 0 & -A^T \\ A & 0 \end{bmatrix} \begin{bmatrix} \bar{x} \\ \bar{\lambda} \end{bmatrix} - \begin{bmatrix} \bar{\nu} \\ \bar{u} \end{bmatrix} = \begin{bmatrix} -c \\ b \end{bmatrix} \qquad (2.13)$$

$$\bar{x} \geq 0 \qquad \bar{\nu} \geq 0$$
$$\bar{\nu}^T \bar{x} + \bar{\lambda}^T \bar{u} = 0$$
$$\bar{\lambda} \geq 0 \qquad \bar{u} \geq 0$$

where $\bar{\lambda}$ and $\bar{\nu}$ are particular values of the Lagrange multipliers. It should be noted that, in virtue of the non-negativity of its factors, the orthogonality condition

$$\bar{\nu}^T \bar{x} + \bar{\lambda}^T \bar{u} = 0 \qquad (2.14)$$

implies that each of its component products $\bar{\nu}_r\bar{x}_r$ and $\bar{\lambda}_i\bar{u}_i$ also vanish for all r and i. The KKT problem (2.13) has a special structure: it comprises simultaneous linear algebraic equations supported by complementarity conditions. This mathematical form has become known as a *linear complementarity problem (LCP)*; a mathematical discussion of it is given by Murty[6] and by Lemke[7,8]. The KKT optimality conditions of an LP give rise to a particular type of LCP – one in which the variables $\bar{x}$ and the Lagrange multipliers $\bar{\lambda}$ are uncoupled, and the matrix is skew-symmetric. It could be made symmetric if the second column vector were to contain $[\ \bar{\nu}^T \ \ -\bar{u}^T\ ]^T$, the Lagrange multipliers of the non-negativity restrictions and the surplus (negative slack) variables associated with the main constraints.

Because any LP is a convex program, LP(2.7) is equivalent to the KKT problem (2.13): values $\bar{x}$ which solve (2.13) also solve the LP, and *vice versa*. This may be of considerable use for some problems in structural mechanics. Suppose that such a problem does not present itself naturally (or generically) as an LP, but that its vectorial relations of structural mechanics – equilibrium, compatibility and material constitution – can be arrayed in the form (2.13). Then we may infer that the problem can be equivalently represented and solved as an LP. Furthermore, the objective of minimising z(x) invites interpretation as a (discrete) variational principle of structural mechanics.

The KKT problem equivalent to any minimising LP may be obtained directly from (1.45 to 1.53). Alternatively, one may make inferences from (2.13). For the LP in standard form, the constraints are equalities

$$\text{Minimise } z = c^T x \ ; \quad A x = b \ , \quad x \geq 0. \qquad (2.15)$$

The slack variables u therefore vanish, and the Lagrange multipliers λ associated with them are now unrestricted. As a consequence, one of the sets of complementarity conditions disappears, and the KKT problem can be written in the symmetric form displayed in (2.16). This LP and corresponding KKT problem are particularly useful in structural mechanics since they can be employed in representing the problem of plastic limit analysis.

$$\begin{bmatrix} 0 & A^T \\ A & 0 \end{bmatrix} \begin{bmatrix} \bar{x} \\ \bar{\lambda} \end{bmatrix} + \begin{bmatrix} \bar{\nu} \\ 0 \end{bmatrix} = \begin{bmatrix} c \\ b \end{bmatrix} \qquad (2.16)$$

$$\bar{x} \geq 0 \qquad \bar{\nu}^T \bar{x} = 0 \qquad \bar{\nu} \geq 0$$

Suppose the LP is one in which the variables $\mathbf{x}$ are unrestricted

$$\text{Minimise } z = c^T x \ ; \quad A\, x \geq b \ . \qquad (2.17)$$

With unrestricted variables $\mathbf{x}$, there are no non-negativity constraints and the associated Lagrange multipliers ν must vanish. Now the other set of complementarity conditions disappear, and the form of the KKT problem may be inferred from (2.13) once more.

DUALITY IN LINEAR PROGRAMMING

The duality evidenced by a convex program has been explored through the nature of the saddlepoint property of its Lagrangian. If we define a function $w(\lambda)$ by

$$w(\lambda) = \min_x L(x,\lambda) \qquad (2.18)$$

in terms of the Lagrangian $L(\mathbf{x},\lambda)$ of the minimising convex program (2.19), then

$$\text{Minimise } z = z(x) \ ; \quad g(x) \leq 0 \qquad (2.19)$$

$$\text{Maximise } w = w(\lambda) \ ; \quad \lambda \geq 0. \qquad (2.20)$$

are dual mathematical programs. Consider, as an example, the LP

$$\text{Minimise } z = c^T x \ ; \quad A\, x \geq b \ , \quad x \geq 0. \qquad (2.21)$$

The Lagrangian for this program may be written

$$L(x,\lambda,\nu) = c^T x + \lambda^T [\ b - A\, x\ ] + \nu^T [\ -x\ ], \qquad \lambda \geq 0,\ \nu \geq 0 \qquad (2.22)$$

It is convex in the variables $\mathbf{x}$, which, through the inclusion of the last term, may be considered unrestricted. Thus, its minimum value over all values of $\mathbf{x}$ may be found by differentiation to be located such that

$$\nabla_x L(x,\lambda,\nu) = c - A^T \lambda - \nu = 0. \qquad (2.23)$$

We shall now re-order the terms in the Lagrangian (2.22)

$$L(x,\lambda,\nu) = b^T \lambda - x^T [\ A^T \lambda + \nu - c\ ], \qquad \lambda \geq 0, \quad \nu \geq 0. \qquad (2.24)$$

Since the minimum value of the Lagrangian occurs when the relations of (2.23) are satisfied, we see from (2.24) that

$$w(\lambda) = \min_{x} L(x,\lambda,\nu) = b^T \lambda \qquad (2.25)$$

The dual LP follows directly from (2.20). However, the result (2.25) necessitates the imposition of (2.23) as supplementary constraints; they may be written as inequalities by removing the negative slack variables $\nu \geq 0$. Therefore, the primal-dual LPs are

$$\text{Minimise } z = c^T x \; ; \quad A\,x \geq b \; , \quad x \geq 0. \qquad (2.21)$$

$$\text{Maximise } w = b^T \lambda \; ; \quad A^T \lambda \leq c \; , \quad \lambda \geq 0. \qquad (2.26)$$

The expression (2.24) may be described as the Lagrangian of the maximising LP(2.26). From it we see that the Lagrange multipliers associated with the main inequality constraints of (2.26) are $\mathbf{x} \geq \mathbf{0}$: the variables of the primal LP are the Lagrange multipliers of the dual LP, and *vice versa*.

The effect of variations in the form of the primal LP on that of the dual problem are:

Equality constraints. If the primal main constraints are equalities – that is, the primal slack variables **u** vanish – then the corresponding Lagrange multipliers (dual variables) λ are unrestricted.

$$\text{Minimise } z = c^T x \; ; \; A\,x = b \; , \; x \geq 0. \qquad (2.27)$$

$$\text{Maximise } w = b^T \lambda \; ; \; A^T \lambda \leq c. \qquad (2.28)$$

Unrestricted variables. If the primal variables $\mathbf{x}$ are unrestricted, then the corresponding Lagrange multipliers (dual slack variables) ν vanish – that is, the dual main constraints are equalities.

$$\text{Minimise } z = c^T x \; ; \; A\,x \geq b. \qquad (2.29)$$

$$\text{Maximise } w = b^T \lambda \; ; \; A^T \lambda = c, \; \lambda \geq 0. \qquad (2.30)$$

A useful observation concerning the optimal solutions to primal and dual LPs may be made from equation (2.14). If primal variable $\bar{x}_r$ is non-zero, then the rth dual constraint is active ($\bar{\nu}_r = 0$); contrarily, if the rth dual constraint is passive ($\bar{\nu}_r$ non-zero), then $\bar{x}_r = 0$ in the primal. The result $\bar{\lambda}_i \bar{u}_i = 0$ allows similar statements to be made about the primal main constraints and the dual variables in the optimal solutions of both LPs. These observations are referred to as *complementary slackness*.

The fundamental results of LP duality are formalised in a famous theorem given by Gale, Kuhn and Tucker[9] which we shall simply state.

Weak duality. If $\mathbf{x}$ is a feasible solution of the primal LP and λ is a feasible solution to the dual LP, then $c^T \mathbf{x} \geq b^T \lambda$.

Strong duality. If $\bar{\mathbf{x}}$ is an optimal solution of the primal LP, then there is a $\bar{\lambda}$ which is optimal for the dual LP, and *vice versa*. Additionally, $c^T \bar{\mathbf{x}} = b^T \bar{\lambda}$.

Unbounded dual. If the primal LP has no feasible solution while the dual LP has, then $\max w \to +\infty$.
Unbounded primal. If the dual LP has no feasible solution while the primal LP has, then $\min z \to -\infty$.
Existence of Optima. If the primal LP and the dual LP both have feasible solutions, then there is an $\bar{x}$ and a $\bar{\lambda}$ which are optimal for the primal and dual LPs respectively.

LP duality gives rise to an equivalence between the primal-dual LPs and the corresponding KKT problem: vectors $\bar{x}$, $\bar{\lambda}$ with corresponding slack vectors $\bar{u}$, $\bar{\nu}$ that solve (2.21) and (2.26) also solve KKT problem (2.13), and *vice versa*. Such an equivalence is extremely useful. If a problem in structural mechanics has a vectorial description that can be identified with one of the KKT problems discussed in this section, then it can also be represented by either of the associated primal-dual LPs. They may provide alternative physical interpretations of the problem, which often adds insight. There may be computational advantages in duality. The computational effort required by the Simplex algorithm increases substantially with the number of main constraints, and increases much less with the number of variables. One can choose to solve whichever of the two LPs has the fewer constraints, and, as we shall soon see, obtain the solution of the other as a by-product.

THE SIMPLEX MULTIPLIERS

The variables appearing in the primal and dual LPs are different and may have quite different physical meanings. Whichever LP is solved by the Simplex algorithm, it may be desirable to have optimal values of the variables in the other LP without our having to solve it. We shall indicate how this may be done by employing the example (2.3); its solution sequence, as generated by the Simplex algorithm, is set out in Table 1.

The primal LP(2.3) has the structure of LP(2.27). The dual LP(2.28) has unrestricted variables, and therefore gives:

Maximise w; where

$$\begin{bmatrix} 5 & 1 \\ -4 & 1 \\ 13 & 5 \\ -2 & -4 \\ 1 & 1 \\ \hline 20 & 8 \end{bmatrix} \begin{bmatrix} \lambda_1 \\ \lambda_2 \end{bmatrix} \begin{matrix} \\ \\ \leq \\ \\ \\ = \end{matrix} \begin{bmatrix} 1 \\ 6 \\ -7 \\ 1 \\ 5 \\ \hline w \end{bmatrix} \qquad (2.31)$$

The coefficients $\bar{c}_j$, which appear in the z row of the Simplex tableaux for the solution of the primal LP, provide the information which identifies whether the current tableau contains an optimal solution. The row of coefficients $\bar{c}_j$ in the current tableau has been obtained from the corresponding row of coefficients c_j in Tableau 0 through the subtraction of a linear combination of the constraint rows of Tableau 0, in which the coefficients are a_{ij}. Utilising the data given in the z row and the constraint rows of Tableau 0, we may infer that the values of coefficients in the z row of any succeeding tableau may be expressed by:

$$\begin{bmatrix} \bar{c}_1 \\ \bar{c}_2 \\ \bar{c}_3 \\ \bar{c}_4 \\ \bar{c}_5 \\ \hdashline \bar{c}_6 \\ \bar{c}_7 \\ \hdashline -z \end{bmatrix} = \begin{bmatrix} 1 \\ 6 \\ -7 \\ 1 \\ 5 \\ \hdashline 0 \\ 0 \\ \hdashline 0 \end{bmatrix} - \begin{bmatrix} 5 & 1 \\ -4 & -1 \\ 13 & 5 \\ -2 & -1 \\ 1 & 1 \\ \hdashline 1 & 0 \\ 0 & 1 \\ \hdashline 20 & 8 \end{bmatrix} \begin{bmatrix} \pi_1 \\ \pi_2 \end{bmatrix} \tag{2.32}$$

where the parameters π_1 and π_2 are called *Simplex multipliers* and their values change from one tableau to the next. The Simplex algorithm terminates with an optimal solution when

$$\bar{c}_1 \geqslant 0, \quad \bar{c}_2 \geqslant 0, \quad \bar{c}_3 \geqslant 0, \quad \bar{c}_4 \geqslant 0, \quad \bar{c}_5 \geqslant 0. \tag{2.33}$$

These inequalities make the current values $\bar{\pi}_1$ and $\bar{\pi}_2$ of the Simplex multipliers form a feasible solution to the dual LP, as inspection of (2.31) will soon show. However, the primal solution is optimal; it follows from the last equation in (2.32), and from strong duality, that

$$\min z = 20\,\bar{\pi}_1 + 8\,\bar{\pi}_2 = \max w. \tag{2.34}$$

Therefore, the current values of the Simplex multipliers give an optimal solution for the dual LP. Furthermore, we see from rows 6 and 7 of (2.32) that

$$\bar{c}_6 = -\bar{\pi}_1 = -\bar{\lambda}_1, \qquad \bar{c}_7 = -\bar{\pi}_2 = -\bar{\lambda}_2. \tag{2.35}$$

An optimal value of the ith dual variable is given by the negative of the coefficient $\bar{c}_{n+i}$ corresponding to the artificial variable x_{n+i} in the optimal tableau for the primal LP. From Tableau 3 of Table 1, we find the optimal dual solution in the shaded region of the z row.

$$\bar{\lambda}_1 = \frac{-23}{7}, \qquad \bar{\lambda}_2 = \frac{50}{7}. \tag{2.36}$$

This ability to locate an optimal solution to the dual LP from the solving of the primal LP makes the algorithm particularly useful. In plastic limit analysis, for example, the Simplex multipliers enable optimal solutions to be found for the kinematic variables $\bar{x}$ (a collapse mechanism) and the static variables $\bar{\lambda}$ (a set of safe stress resultants consistent with the collapse mechanism) through the solution of a single LP.

ALTERNATIVE OPTIMAL SOLUTIONS

Clearly, the objective function $z(\mathbf{x})$ of an LP cannot be strictly convex. While the global optimum value of $z = z(\bar{\mathbf{x}})$ is found by the Simplex algorithm, which terminates with some optimal solution $\bar{\mathbf{x}}$, there may well be a convex set of solutions $\bar{\mathbf{x}}$ which will supply the same minimum value of z. The existence of these alternative optimal solutions may be inferred from the z row of the optimal tableau. In this row, the coefficients must satisfy $\bar{c}_j \geq 0$ for all variables x_j, excepting artificial variables. If x^b_j is a basic variable, then $\bar{x}^b_j \geq 0$ and $\bar{c}^b_j = 0$. Let x^n_j be a non-basic variable in the optimal tableau, then $\bar{x}^n_j = 0$ and $\bar{c}^n_j \geq 0$. An alternative basic solution is constructed by making one of the x^n_j a basic variable. If $\bar{c}^n_j > 0$, then the making basic of x^n_j would cause z to increase , and the alternative basic solution could no longer be optimal. If $\bar{c}^n_j = 0$, then z would neither increase nor decrease as x^n_j becomes basic, and the alternative basic solution must remain optimal.

REFERENCES

1. Dantzig, G. B., Linear Programming and Extensions, Princeton 1963.

2. Gass, S. I., Linear Programming, Methods and Applications (3rd Ed.) McGraw-Hill 1969.

3. Hadley, G., Linear Programming, Addison-Wesley 1962.

4. Orchard-Hays, W., Advanced Linear Programming Computing Techniques, McGraw-Hill 1968.

5. Karmarkar, N., A new polynomial-time algorithm for linear programming, Combinatorica, **4** (1984) 373-396.

6. Murty, K. G., Complementarity problems, in: Mathematical Programming for Operational Researchers and Computer Scientists (Ed. A. G. Holzman), Dekker 1981, 173-196.

7. Lemke, C. E., Recent results on complementarity problems, in: Nonlinear Programming (Eds. J. B. Rosen, O. L. Mangasarian and K. Ritter), Academic Press 1970, 349-384.

8. Lemke, C. E., A survey of complementarity theory, in: Variational Inequalities and Complementarity Problems (Eds, R. W. Cottle, F. Giannessi and J-L. Lions), Wiley 1980, 213-239.

9. Gale, D., Kuhn, H. W. and Tucker, A. W., Linear programming and the theory of games, in: Activity Analysis of Production and Allocation (Ed. T. C. Koopmans), Wiley 1951.

CHAPTER 3

QUADRATIC PROGRAMS AND COMPLEMENTARITY

D. Lloyd Smith
Imperial College, London, U.K.

ABSTRACT

The optimality criteria associated with a convex minimising quadratic program are shown to form a special type of linear complementarity problem, one that has a symmetric matrix. The dual quadratic programs of Dorn are derived, as are the symmetric dual programs of Cottle which are especially useful in elasto-plastic structural analysis. Cottle's duality theorem is stated, and attention is given to the joint solution and to the question of uniqueness of solution. Wolfe's algorithm for solving quadratic programs is discussed. The Wolfe-Markowitz algorithm for solving a parametric quadratic program is introduced, and its use in solving the parametric linear complementarity problem associated with elasto-plastic structural analysis is described.

INTRODUCTION

A quadratic program (QP) requires the minimisation (or maximisation) of a quadratic objective function subject to equality or inequality constraints in which the constraint functions are *linear*. QPs abound in discrete structural models for static and dynamic problems in elasto-plasticity. Again, the regularising feature is that of linearity. It happens that the KKT necessary and sufficient optimality conditions for a QP are almost entirely linear; in fact, they form a linear complementarity problem, just a little different in structure from the one that arises in linear programming. The solution of this LCP gives an indirect means of solving a QP.

Typically, a QP may be presented in the manner

$$\text{Minimise } z = \frac{1}{2} x^T D x + c^T x \ ; \quad A x \geq b \ , \quad x \geq 0 \ . \tag{3.1}$$

There are m linear main constraints $\mathbf{A}\,\mathbf{x} \geq \mathbf{b}$ in n non-negative variables $\mathbf{x}$. The objective function z is seen to contain a quadratic form with a square coefficient matrix $\mathbf{D}$ and a linear form with coefficient (row) vector $\mathbf{c}^T$. It could also contain a constant term, but that would have no effect on the optimal solution $\bar{\mathbf{x}}$, and would simply shift the optimal value of z by the same constant. It is sufficiently general that $\mathbf{D}$ be assumed symmetric, since $2\mathbf{D} = (\mathbf{D} + \mathbf{D}^T) + (\mathbf{D} - \mathbf{D}^T)$ is the sum of a symmetric and a skew-symmetric matrix and $\mathbf{x}^T(\mathbf{D} - \mathbf{D}^T)\mathbf{x} = 0$ for any $\mathbf{x}$.

Firstly, we must discuss the convexity of (3.1). Since linear functions are both convex and concave, the convexity of the QP turns on that of the quadratic form in its objective function. If $\mathbf{D}$ is positive semidefinite, that is

$$Q = x^T D x \geq 0 \quad \text{for all } x, \tag{3.2}$$

then Q is a convex function. If $\mathbf{D}$ is positive definite, that is

$$Q = x^T D x > 0 \quad \text{for all } x \neq 0, \tag{3.3}$$

then Q is a strictly convex function. We shall assume throughout that any minimising QP is convex.

The constraints of a QP are linear, and the feasible region X is therefore bounded by "flat surfaces" or hyperplanes, as in linear programming. However, because of the nonlinear nature of the objective function, optimal solutions $\bar{\mathbf{x}}$ of a QP may lie in the interior of X or on its boundary, but not necessarily at a vertex.

KARUSH-KUHN-TUCKER OPTIMALITY CRITERIA

Since QP(3.1) is convex, the KKT conditions are both necessary and sufficient for an optimal solution $\bar{\mathbf{x}}$. We may establish these optimality criteria from the modified Lagrangian of (3.1) in the manner set out in Chapter 1. The KKT conditions (1.45 to 1.53) for the general convex program may easily be specialised for QP(3.1) which has no equality constraints $\mathbf{h}(\ ) = \mathbf{0}$ and no unrestricted variables y. Firstly, the main constraints must be written in the standard form $\mathbf{g}(\mathbf{x}) \leq \mathbf{0}$, and, for our later convenience, we shall add slack variables $\mathbf{u} \geq \mathbf{0}$:

$$g(x) + u = b - A\,x + u = 0, \qquad u \geq 0. \tag{3.4}$$

The necessary differentiation simply gives:

$$\nabla_x z(x) = D\,x + c\ , \qquad \nabla_x g(x) = -A^T\ , \tag{3.5}$$

in which we remember that D is symmetric. The KKT conditions (1.45 to 1.53) now become:

$$\begin{bmatrix} D & -A^T \\ A & 0 \end{bmatrix} \begin{bmatrix} \bar{x} \\ \bar{\lambda} \end{bmatrix} - \begin{bmatrix} \bar{\nu} \\ \bar{u} \end{bmatrix} = \begin{bmatrix} -c \\ b \end{bmatrix} \tag{3.6}$$

$$\bar{x} \geq 0 \qquad \bar{\nu} \geq 0$$
$$\bar{\nu}^T\,\bar{x} + \bar{\lambda}^T\,\bar{u} = 0$$
$$\bar{\lambda} \geq 0 \qquad \bar{u} \geq 0$$

where $\bar{\lambda}$ and $\bar{\nu}$ are particular sets of values of the Lagrange multipliers λ and ν, associated with the main constraints and non-negativity restrictions respectively. As was observed in linear programming, these necessary and sufficient optimality criteria have the characteristic form of a linear complementarity problem (LCP). In its standard form, an LCP may be written

$$M\,z - w = q$$
$$z \geq 0 \qquad z^T\,w = 0 \qquad w \geq 0 \tag{3.7}$$

where **M** is a NxN matrix, **z** and **w** are complementary variable N-vectors and **q** is a vector of constants. For the QP(3.1), the KKT conditions constitute a special LCP in which **M** exhibits a *bisymmetric* form, as shown in (3.6).

It will have been observed that the matrix of (3.6) contains a square, null submatrix. We may replace that null submatrix by a symmetric positive semidefinite matrix **E** in the following manner. Consider a QP which contains non-negative variables **x** and also a set of unrestricted variables **y**:

$$\text{Minimise } z = \frac{1}{2}\,x^T D\,x + \frac{1}{2}\,y^T E\,y + c^T x\ ;$$
$$A\,x + E\,y \geq b\ , \qquad x \geq 0. \tag{3.8}$$

For this QP, equations (1.45) and (1.46) become

$$[\ D\,\bar{x} + c\ ] + [\ -A^T\ ]\ \bar{\lambda} - \bar{\nu} = 0, \tag{3.9}$$

$$[\ E\,\bar{y}\ ] + [\ -E\ ]\ \bar{\lambda} \quad = 0 \tag{3.10}$$

Carefully notice that, if $\mathbf{E}$ is positive definite, then $\bar{y} = \bar{\lambda}$ is the only solution of (3.10): however, if $\mathbf{E}$ is only positive semidefinite, then, although $\bar{y} = \bar{\lambda}$ is still a solution, there may now be other $\bar{\lambda} \neq \bar{y}$ which satisfy (3.10). We may replace $\mathbf{E}\bar{y}$ by $\mathbf{E}\bar{\lambda}$ for the main constraints (1.48), and then the KKT conditions can be written

$$\begin{bmatrix} -\mathbf{D} & \mathbf{A}^T \\ \mathbf{A} & \mathbf{E} \end{bmatrix} \begin{bmatrix} \bar{x} \\ \bar{\lambda} \end{bmatrix} + \begin{bmatrix} \bar{\nu} \\ -\bar{u} \end{bmatrix} = \begin{bmatrix} \mathbf{c} \\ \mathbf{b} \end{bmatrix} \tag{3.11}$$

$$\bar{x} \geq 0 \qquad \bar{\nu} \geq 0$$
$$\bar{\nu}^T \bar{x} + \bar{\lambda}^T \bar{u} = 0$$
$$\bar{\lambda} \geq 0 \qquad \bar{u} \geq 0$$

The QP(3.8) and corresponding KKT conditions were introduced by Cottle[1]. We have chosen to display LCP(3.11) in a form which is symmetric but non-standard. It happens that the fundamental mechanical laws – equilibrium, compatibility and material constitution – of elastoplastic structures can be marshalled into this form in several different formulations with either $\mathbf{D} = \mathbf{0}$ or $\mathbf{E} = \mathbf{0}$ [3].

DUALITY IN QUADRATIC PROGRAMMING

Let us consider first the QP(3.1)

$$\text{Minimise } z = \frac{1}{2}\, x^T D\, x + c^T x \;;\quad A\, x \geq b\,,\quad x \geq 0\,. \tag{3.1}$$

The Lagrangian for this program, including the constraints $\mathbf{x} \geq \mathbf{0}$, can be written

$$L(x,\lambda,\nu) = \frac{1}{2}\, x^T D\, x + c^T x + \lambda^T[\, b - A\, x\,] + \nu^T[\, -x\,]\,. \tag{3.12}$$

in which $\lambda \geq \mathbf{0}$ and $\nu \geq \mathbf{0}$ are Lagrange multipliers, and $\mathbf{x}$ may be considered unrestricted. Our aim is to construct a function

$$w(\lambda,\nu) = \min_{x} L(x,\lambda,\nu)\,, \tag{3.13}$$

and, since $\mathbf{x}$ is unrestricted and $L(\mathbf{x},\lambda,\nu)$ is convex in $\mathbf{x}$, we may locate the minimum conditions by differentiation, in the usual manner

$$\nabla_x L(x,\lambda,\nu) = D\, x + c - A^T\lambda - \nu = 0\,. \tag{3.14}$$

In these conditions, we remember that $\mathbf{D}$ is positive semidefinite; it means that while

$$D\, x = D\, \mu \tag{3.15}$$

is satisfied by $\mu = x$, it may also be solved by some $\mu \neq x$. In virtue of (3.15), and in order to impose the conditions (3.14), we re-arrange the Lagrangian

$$L(\mu,\lambda,\nu) = -\frac{1}{2}\mu^T D\,\mu + b^T\lambda - x^T[-D\,\mu - c + A^T\lambda + \nu\,]. \qquad (3.16)$$

From this, by imposing the minimising conditions (3.14), we see that

$$w(\mu,\lambda) = -\frac{1}{2}\mu^T D\,\mu + b^T\lambda\,, \qquad \text{if} \quad -D\,\mu - c + A^T\lambda \leq 0 \qquad (3.17)$$

The primal and dual QPs are therefore

$$\text{Minimise } z = \frac{1}{2}x^T D\,x + c^T x\;; \quad A\,x \geq b\,, \quad x \geq 0\,. \qquad (3.1)$$

$$\text{Maximise } w = -\frac{1}{2}\mu^T D\,\mu + b^T\lambda\;; \quad -D\,\mu + A^T\lambda \leq c, \quad \lambda \geq 0. \qquad (3.18)$$

This form of QP duality was first obtained by Dorn[2]. The dual variables μ are not required to be non-negative in the establishing of duality; if **D** is positive definite, however, then the optimal solutions will be such that $\bar{\mu} = \bar{x}$, and x is non-negative.

Consider now the QP(3.8)

$$\text{Minimise } z = \frac{1}{2}x^T D\,x + \frac{1}{2}y^T E\,y + c^T x\;; \qquad (3.8)$$
$$A\,x + E\,y \geq b\,, \qquad x \geq 0.$$

The Lagrangian for this program is

$$L(x,y,\lambda,\nu) = \frac{1}{2}x^T D\,x + \frac{1}{2}y^T E\,y + c^T x$$
$$+ \lambda^T[\,b - A\,x - E\,y\,] + \nu^T[\,-x\,]\,. \qquad (3.19)$$

in which $\lambda \geq 0$ and $\nu \geq 0$ are the Lagrange multipliers associated respectively with the main constraints and with the non-negativity restrictions, and in which x is now to be considered as unrestricted. This function is convex in x and y and is minimised over those variables when

$$\nabla_x L(x,y,\lambda,\nu) = D\,x + c - A^T\lambda - \nu = 0\,, \qquad (3.20)$$

$$\nabla_y L(x,y,\lambda,\nu) = E\,y - E\,\lambda = 0. \qquad (3.21)$$

We also recall that **D** is positive semidefinite, and we may write

$$D\,x - D\,\mu = 0 \qquad (3.22)$$

The Lagrangian may be rearranged

$$L(x,y,\lambda,\mu,\nu) = -\frac{1}{2}\,\mu^T D\,\mu - \frac{1}{2}\,\lambda^T E\,\lambda + b^T\lambda$$
$$- x^T[\,-D\,\mu - c + A^T\lambda + \nu\,] - y^T[\,E\,\lambda - E\,y\,]. \qquad (3.23)$$

Provided that the minimising conditions (3.20) and (3.21) are satisfied, the dual objective function is

$$w(\mu,\lambda) = -\frac{1}{2}\,\mu^T D\,\mu - \frac{1}{2}\,\lambda^T E\,\lambda + b^T\lambda\,, \qquad (3.24)$$

and the primal and dual QPs are therefore

Primal (3.8)	Dual (3.25)
Min: $z = \frac{1}{2}x^T Dx + \frac{1}{2}y^T Ey + c^T x$	Max: $w = -\frac{1}{2}\mu^T D\mu - \frac{1}{2}\lambda^T E\lambda + b^T\lambda$
$Ax + Ey \geq b$	$-D\mu + A^T\lambda \leq c$
$x \geq 0$	$\lambda \geq 0$
(3.8)	(3.25)

These are the symmetric dual QPs proposed by Cottle[1]. When $\mathbf{E} = \mathbf{0}$ the dual QPs of Dorn are obtained, and when both $\mathbf{E} = \mathbf{0}$ and $\mathbf{D} = \mathbf{0}$ the symmetric dual linear programs are recovered.

Because (3.8) is a convex program, there is an equivalence between the primal-dual QPs (3.8 and 3.25), the KKT problem (3.11) and the saddlepoint problem associated with the Lagrangian (3.19). It may be noticed that, in the symmetric display of the KKT problem (3.11), the slack variable column contains the slack variables ν associated with the dual ($\leq$) constraint relations and the surplus variables $-\mathbf{u}$ associated with the primal ($\geq$) constraint relations.

The re-arrangement (3.23) may be described as the Lagrangian of the maximising QP (3.25). The two forms (3.19) and (3.23) show that the primal and dual QPs have the same Lagrangian, and that the variables of the primal QP are the Lagrange multipliers of the dual QP, and *vice versa*.

The effect of variations in the form of the primal QP(3.8) on that of the dual problem (3.25) and on that of the KKT problem (3.11) are:
Equality constraints. If the *i*th primal main constraint is an equality – that is, the primal slack variable u_i vanishes – then the corresponding Lagrange multiplier (dual variable) λ_i is unrestricted.
Unrestricted variables. If the *r*th primal variable x_r is unrestricted, then the corresponding Lagrange multiplier (dual slack variable) ν_r vanishes – that is, the *r*th dual main constraint is an equality.

The formal statement of QP duality for (3.8) and (3.25) given by Cottle[1] is similar to the model for linear programming:

Weak duality. If $(\mathbf{x},\mathbf{y})$ and (μ,λ) are feasible solutions of the primal and dual QPs respectively, then $z(\mathbf{x},\mathbf{y}) \geq w(\mu,\lambda)$.
Duality. If $(\bar{\mathbf{x}},\bar{\mathbf{y}})$ is an optimal solution of the primal QP, then there is a $\bar{\lambda}$ such that $(\bar{\mathbf{x}},\bar{\lambda})$ is an optimal solution of the dual QP, and min z = max w. Also $\mathbf{E}\bar{\lambda} = \mathbf{E}\bar{\mathbf{y}}$.
Converse. If $(\bar{\mu},\bar{\lambda})$ is an optimal solution of the dual QP, then there is a $\bar{\mathbf{x}}$ such that $(\bar{\mathbf{x}},\bar{\lambda})$ is an optimal solution of the primal QP, and min z = max w. Also $\mathbf{D}\bar{\mathbf{x}} = \mathbf{D}\bar{\mu}$.
Unbounded dual. If the primal QP has no feasible solution while the dual QP has, then max $w \to +\infty$.
Unbounded primal. If the dual QP has no feasible solution while the primal QP has, then min $z \to -\infty$.
Existence of optima. If both primal and dual QPs have feasible solutions, than both have optimal solutions.
Joint solution. If either QP has an optimal solution, then there exists a $(\bar{\mathbf{x}},\bar{\lambda})$ which is an optimal solution for both the primal and the dual QPs.

In applications, $\mathbf{x}$ and μ refer to the same physical entities – perhaps kinematic variables. Different physical quantities – perhaps static variables – are represented by $\mathbf{y}$ or equivalently by λ. It is because the primal and dual QPs can lead to different *values* being obtained for the same variables that we employ different notations for the primal and dual variables. Cottle's theorem shows that, if either QP is solvable, then, among the optimal solutions, there is always a solution $\bar{\mathbf{x}} = \bar{\mu}$, $\bar{\mathbf{y}} = \bar{\lambda}$ which solves both QPs jointly. If $\mathbf{D}$ is positive definite, then $\bar{\mathbf{x}} = \bar{\mu}$ uniquely: if $\mathbf{E}$ is positive definite, then $\bar{\mathbf{y}} = \bar{\lambda}$ uniquely.

The symmetric LCP (3.11), together with Cottle's symmetric primal and dual QPs, provide a most useful way of displaying the various formulations of structural elasto-plasticity; see, among others, Lloyd Smith[3] and Teixeira de Freitas on elasto-plasticity in skeletal structures in Chapter 9 of this volume. In this structural context, it is convenient to display LCP(3.11) in a non-standard form in which the symmetric matrix heralds the existence of dual QPs in exactly the same way that the *self-adjoint* nature of a problem in continuum mechanics points to the existence of *complementary variational principles*.

WOLFE'S ALGORITHM

There are several popular and useful algorithms for quadratic programming, among them the Simplex method of Beale[4] and the gradient method of Fletcher[5]. One of the simplest algorithms for solving a QP is that due to Wolfe[6]. It is, however, an indirect solution: the problem it actually addresses is not the QP, but rather its associated KKT problem. This is, of course, a linear complementarity problem. More generally, therefore, Wolfe's algorithm solves LCPs.

Wolfe's algorithm was devised for an LCP similar to (3.6) with $\bar{\mathbf{u}} = \mathbf{0}$ and $\bar{\lambda}$ unrestricted. It was given in two forms - a short form and a long form. It is the short form that we shall here call Wolfe's algorithm : it is only guaranteed to converge when $\mathbf{D}$ is positive definite.

The essential idea is to establish a basic solution of the almost linear system (3.6), using Phase I of the simplex algorithm, but enforcing the nonlinear orthogonality condition $\bar{\nu}^T\bar{x} = 0$. Several variations of this scheme are possible, depending on the precise form of the LCP to be solved. For LCP(3.11), the recipe might be:

1. Make the constant vector $[\ c^T \quad b^T\ ]^T$ non-negative: if c_j or b_i is negative, multiply that equation by -1.

2. Choose as basic variables those $\bar{\nu}_j$ and $\bar{u}_i$ which have a coefficient of +1. For each equation without such a basic variable, add a different artificial variable w_k.

3. Minimise the infeasibility function $\sum w_k$ by means of the Simplex algorithm. To satisfy complementarity, the rule for selecting the pivot column in the Simplex algorithm is modified by a restricted basis entry: if $\bar{\nu}_j$ is already a basic variable, then $\bar{x}_j$ cannot be made basic; and if $\bar{u}_i$ is already a basic variable, then $\bar{\lambda}_i$ cannot be made basic. If the scheme terminates with $\sum w_k = 0$, then the LCP is solved; otherwise it is infeasible.

THE WOLFE-MARKOWITZ ALGORITHM

The solving of the investment portfolio problem proposed by Markowitz, see Boot[7], is equivalent to the solving of a parametric QP. He suggested the association of a parameter $\gamma \geq 0$ with the linear term of the objective function z

$$z(x) = \frac{1}{2}\, x^T D\, x + \gamma\, c^T x\ , \tag{3.26}$$

so that a sequence of solutions could be obtained as γ is varied monotonically. Wolfe's algorithm, in its long form, brings the systematic calculation of the Simplex method to Markowitz's original scheme. By embedding the QP – which is obtained when $\gamma = 1$ – in such a parametric system, Wolfe offered a resolution of the convergence issue which arises when **D** is positive semidefinite.

This algorithm – we shall follow Dantzig in calling it the Wolfe-Markowitz algorithm – can be used for solving LCPs. A simple variant has been exploited for elastoplastic structural analysis, Lloyd Smith[7]. The parameter γ is a load factor which proportionately varies the applied loads. Typically, the LCP which describes such problems has the form

$$\begin{bmatrix} 0 & A^T \\ A & E \end{bmatrix} \begin{bmatrix} \bar{x} \\ \bar{\lambda} \end{bmatrix} + \begin{bmatrix} \bar{\nu} \\ 0 \end{bmatrix} + \begin{bmatrix} c\gamma \\ b\gamma \end{bmatrix} = \begin{bmatrix} c_o \\ b_o \end{bmatrix} \tag{3.27}$$

$$\bar{x} \geq 0 \qquad \bar{\nu}^T\, \bar{x} = 0 \qquad \bar{\nu} \geq 0$$

This is a *parametric linear complementarity problem* (PLCP). It has the form of LCP(3.11) with **D** = **0**, $\bar{\mathbf{u}}$ = **0** and $\bar{\lambda}$ unrestricted, but with the addition of a column vector parametric in γ.

The algorithm achieves a sequence of basic solutions as γ increases monotonically if we simply adopt the objective of maximising γ. Such a sequence of basic solutions follows the evolution of plastic deformation in the structure, and the sequence terminates when γ attains the value of the plastic limit load.

For the PLCP(3.27) where the constant vectors $\mathbf{b}_o$ and $\mathbf{c}_o$ are non-negative, the Wolfe-Markowitz type of algorithm is:

1. Perform a sequence of m pivot operations on the matrix **E**. The initially basic variables are then the unrestricted variables $\bar{\lambda}_i$ (i = 1,2,...,m), and the non-negative variables $\bar{r}_j$ (j = 1,2,...,n).

2. Select the parametric column $[\ \mathbf{c}^T \ \ \mathbf{b}^T\ ]^T$ as pivot column, then γ enters the basis.

3. Select the pivot row from the upper set of n constraint rows by the usual Simplex rule, then some non-negative variable, say $\bar{r}_j$, leaves the basis. All the unrestricted variables remain basic.

4. Carry out a pivot operation to obtain a new tableau.

5. Enforce complementarity through a restricted basis entry. The only variable that can enter the basis is the complement (in this case $\bar{x}_j$) of the variable (in this case $\bar{r}_j$) which has just been removed. The pivot column, say column s, is thus chosen automatically.

6. Select the pivot row, say row r, by the usual rule: $\theta_r = (\bar{c}_{or}/\bar{a}_{rs}) = \min (\bar{c}_{oi}/\bar{a}_{is})$ from the n upper constraint rows and for pivot column elements $\bar{a}_{is} > 0$. If $\bar{a}_{is} \leq 0$ for all i, STOP - no basic solution can be found for greater values of γ.

7. Return to 4 and continue.

REFERENCES

1. Cottle, R. W., Symmetric dual quadratic programs, Q. Appl. Math. **21** (1963) 237-243.

2. Dorn, W. S., Duality in quadratic programming, Q. Appl. Math. **18** (1960) 155-162.

3. Lloyd Smith, D., Plastic limit analysis and synthesis by linear programming, PhD Thesis, University of London 1974.

4. Beale, E. M. L., Mathematical Programming in Practice, Pitman 1968.

5. Fletcher, R., A general quadratic programming algorithm, J. Inst. Math. Appl. **7** (1971) 76–91.

6. Wolfe, P., The simplex method for quadratic programming, Econometrica **27** (1959) 382–398.

7. Lloyd Smith, D., The Wolfe–Markowitz algorithm for nonholonomic elastoplastic analysis, Eng. Structs. **1** (1978) 8–16.

CHAPTER 4

STATICS AND KINEMATICS

D. Lloyd Smith
Imperial College, London, U.K.

ABSTRACT

The statics and kinematics of simple framed structures are presented in the setting of small displacements and deformations. Firstly, the similarity of mathematical form between structural mechanics and other engineering problems is expressed in terms of network theory, and this approach is used to emphasise the need for a systematic description of the structural problem. Two, quite different, but equally systematic ways of setting out the static and kinematic laws are suggested: the mesh and nodal descriptions. A guiding adjoint formalism – static-kinematic duality – is seen to pervade the mathematical representation of statics and kinematics, and this contributes to the considerable symmetry in formulations of structural problems. For improved data generation, the nodal description is presented in system coordinates, a procedure for assembling a mesh description for frames with rectangular meshes is suggested, and the transformation of nodal into mesh equilibrium equations is mentioned.

INTRODUCTION

The mechanics of deformable structures is ususally founded upon three types of mechanical laws.

Firstly, Newton's law for a system of mass particles in motion provides for the balance of momenta. In a structural system, this balance gives expression to the laws of *kinetics* which relate the internal forces in the structural elements to the applied loads through the changes in momenta of the masses in motion. When the course of motion of the structure evolves in such a manner that the changes in momenta are sensibly zero, the laws of kinetics become the laws of *statics* or of *equilibrium*. It must be emphasised that such laws are required to describe the equilibrium of the structure for each deformed configuration in the course of its motion. Generally, therefore, the relations of equilibrium will depend on the current total displacements of each of the structural elements. However, while the current total displacements remain "small", it is common practice to employ equilibrium equations which are based upon the originally undisplaced configuration of the structure: such equations simply relate applied loads to internal forces in the structural elements without the intervention of displacements.

Secondly, there are laws of *kinematics* which describe the allowable configurations of the deformable structure at each instant in the course of its motion. Such laws relate the deformations in each of the structural elements to the current total displacements of the whole system. Implicit in the kinematic relations is the notion of *compatibility* – that is, the structural elements remain properly connected within the whole system throughout the evolution of the entire deformation process. The laws of kinematics are therefore founded on simple Euclidean geometry. In general, they provide equations which are rather nonlinear; but, while the current total displacements and consequent element deformations remain "small", the resulting kinematic equations which connect them become linear.

Thirdly, there are the *constitutive relations* which describe the manner of response of the construction material, at each instant of time, to a class of allowable disturbances. Typically, such relations express the strain rates induced in the material by imposed stress rates; and this response may further depend on the current total stress and strain state of the material, or even on the entire course of its evolution to this state. For structures modelled by finite elements, the constitutive relations are synthesised for the whole element in terms of the rates of change associated with element forces and deformations.

The laws of statics and kinematics are independent of constitutive relations. In this chapter we shall study, albeit rather briefly, the nature of static and kinematic laws for framed structures in the régime of "small" displacements; the results will therefore provide a systematic and proper foundation for the mechanics of structures constructed from any material.

THE NETWORK REPRESENTATION OF A FRAME

A structure can be envisaged as a particular kind of network, in the same way that the pipework for a town's water supply or the distribution of cables for a factory's electrical power supply may be considered as networks. A network is an abstract, but unifying, concept.

Firstly, a network comprises *elements* or *branches*, each of which may be represented as a line on a diagram, and the terminals of such an element may be connected to other elements at points called *nodes*. Closed circuits formed by a succession of elements are called *meshes*. The diagram is referred to as the *graph* of the network; and if the direction of each element is indicated by an arrow, the resulting diagram is called the *directed graph* of the network. The associated graph conveniently specifies the manner in which the elements are connected together within the network – its *topological* properties.

Secondly, a network needs an appropriate *algebra*. Associated with each element are *flow variables*; they represent physical entities that pass through the element – like stress resultants, fluid flowrate or electrical current. Also associated with each element are *potential variables*; they are measured across the element terminals – like relative displacements or deformations, fluid head or pressure difference, and electrical potential difference or electo-motive force. The laws of physics which relate such variables are described in terms of an appropriate algebra which enables the distribution of the values of all variables to be calculated throughout the network. For a structural system of connected elements, it is the linear algebra of real vector spaces – the algebra of vectors and matrices – that is usually employed. Both Seely[1] and MacFarlane[2] give most readable accounts of the theory of networks.

THE DISCRETE FRAME ELEMENT

Each branch of the graph model represents a structural element or member of finite length – the finite element. Consider a rectilinear element of length L dissected from a planar frame. There is a set of *member end forces* or *element nodal forces* Q_{Mi}' (i = 1,2,...,S = 6) which represent the forces of interaction or of interconnection with the structural system from which the element has been removed. They are shown in Figure 4.1(a), the prime denoting the use of a local coordinate system.

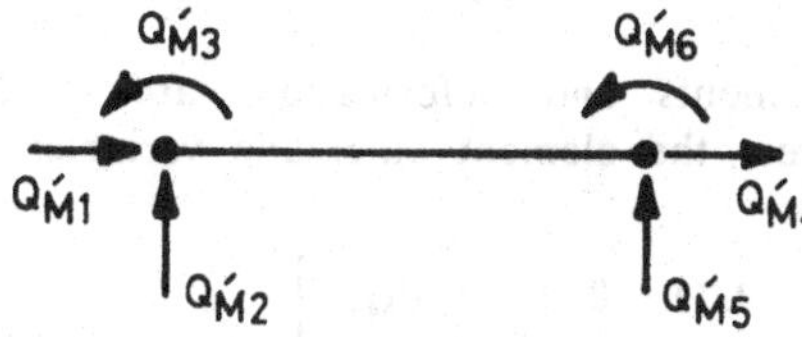

Figure 4.1(a)

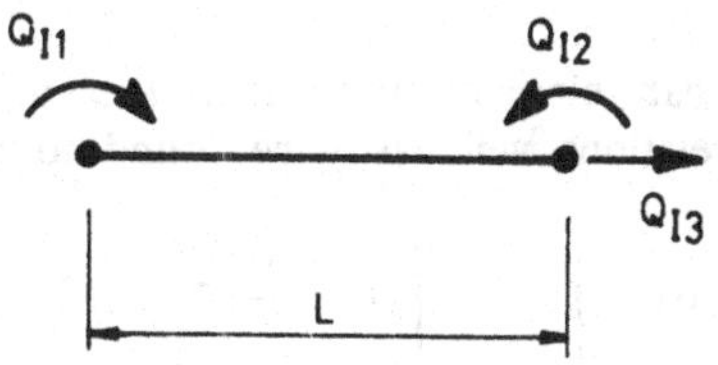

Figure 4.1(b)

For a planar element, there are three independent conditions of equilibrium, and so the six forces Q_{Mi}' can always be expressed in terms of a particular set of three *independent member forces* Q_{Ij} (j = 1,2,..,s = 3). One possible set is suggested in Figure 4.1(b): an alternative set is that of the three member end forces at the right hand end of the element. There are clearly several other equally valid sets.

The element forces $Q_{Mi'}$ or Q_{Ij} are applied at the element nodes. Together with any loads that may be distributed over the length of the element, these forces induce *stress resultants* or *generalised stresses* $Q_k(x)$ ($k = 1,2,\ldots,s$) to flow along the element. For the plane frame element, $Q_1(x)$ is the axial force, $Q_2(x)$ is the transverse shear force and $Q_3(x)$ is the bending moment at coordinate x on the longitudinal axis of the element. Once the element forces are established, the stress resultants can be calculated: attention must therefore be focussed on the former.

There is also a set of *member end displacements* or *element nodal displacements* $q_{Mi'}$ ($i = 1,2,\ldots,S = 6$), and again the prime indicates their reference to the same local coordinate system of the element. The $q_{Mi'}$ are shown in Figure 4.2(a), and the positive sense of each conforms with that chosen for the corresponding force $Q_{Mi'}$. Six degrees of freedom for the planar element are implied by the member end displacements: they may be decomposed into three rigid body motions in the plane, together with three *independent member deformations* q_{Ij} ($j = 1,2,\ldots,s = 3$). Each q_{Ij} shown in Figure 4.2(b) is chosen to conform in type and positive sense with the corresponding force Q_{Ij}. The straight line connecting the ends of the deformed element is called the *chord*. Again, other sets of independent member deformations are possible.

At each coordinate x on the longitudinal axis of the element there are *generalised strains* $q_k(x)$ ($k = 1,2,\ldots,s$). For the plane frame element, $q_1(x)$, $q_2(x)$, $q_3(x)$ are respectively the axial and shear strains and the curvature. Once they are established, the independent element deformations q_{Ij} can be calculated by integration.

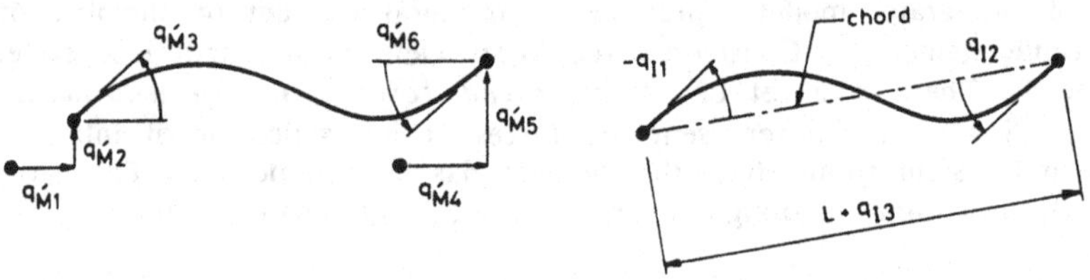

Figure 4.2(a) *Figure 4.2(b)*

For a planar element in which the displacements and deformations are small, the following relations will easily be found to connect the element kinematic variables.

$$\begin{bmatrix} q_{I1} \\ q_{I2} \\ q_{I3} \end{bmatrix} = \begin{bmatrix} 0 & -L^{-1} & -1 & 0 & L^{-1} & 0 \\ 0 & L^{-1} & 0 & 0 & -L^{-1} & 1 \\ -1 & 0 & 0 & 1 & 0 & 0 \end{bmatrix} \begin{bmatrix} q_{M1'} \\ q_{M2'} \\ q_{M3'} \\ q_{M4'} \\ q_{M5'} \\ q_{M6'} \end{bmatrix} \qquad (4.1)$$

If equilibrium of the element is based upon its undisplaced configuration, then, in the absence of element loads, the following relations will be found to connect the element static variables.

$$\begin{bmatrix} Q_{M1}' \\ Q_{M2}' \\ Q_{M3}' \\ Q_{M4}' \\ Q_{M5}' \\ Q_{M6}' \end{bmatrix} = \begin{bmatrix} 0 & 0 & -1 \\ -L^{-1} & L^{-1} & 0 \\ -1 & 0 & 0 \\ 0 & 0 & 1 \\ L^{-1} & -L^{-1} & 0 \\ 0 & 1 & 0 \end{bmatrix} \begin{bmatrix} Q_{I1} \\ Q_{I2} \\ Q_{I3} \end{bmatrix} \qquad (4.2)$$

Similar relations may be written for a spatial frame element for which there are twelve member end forces and twelve member end displacements ($S = 12$), and for which there are six independent member forces and six independent member deformations ($s = 6$).

STATIC-KINEMATIC DUALITY

Relations (4.1) and (4.2) for the collection of structural elements have the form

$$q_I = D\, q_M' , \qquad Q_M' = D^T Q_I \qquad (4.3\text{-}4)$$

where superscript T denotes the transpose of a matrix. The kinematic relations (4.3) are *independent* of the static relations (4.4), yet they are connected through the adjoint operation expressed by transposition of the matrix **D**.

This *adjoint relationship* is an example of what Munro[3] has called *static-kinematic duality* (SKD): it is a relationship that will recur throughout this presentation of the statics and kinematics of structures in the régime of small displacements. In what was probably its first formal statement, Maxwell[4] referred to a statically determinate or isostatic framework of pin-jointed bars:

> If *p* be the tension in bar *A* due to a tension-unity (unit loads directed towards each other) between joints *B* and *C*, then an extension-unity (unit extension) taking place in bar *A* will bring joints *B* and *C* nearer by a distance *p*.

Unlike Maxwell's more well-known theorem of reciprocal displacements, which was given in the same celebrated communication, SKD is entirely *independent* of material properties.

Subsequently, Jenkins[5] rediscovered static-kinematic transformations like (4.3-4) in connection with the analysis of thin cylindrical shells. He referred to them as *contragredient transformations*, the name given by J. J. Sylvester, one of the originators of the algebra of matrices and determinants.

THE MESH DESCRIPTION

The network mesh and node laws associated with the name of Kirchhoff may be translated directly into structural terms: Spillers[6] gives a simple but effective presentation of these ideas. The *mesh* form of describing statics and kinematics for the entire structure will be presented first.

Static Indeterminacy

When the equilibrium conditions are not sufficient to determine *uniquely* the forces in all elements of a framed structure, it is said to be statically indeterminate or hyperstatic. It is then necessary to determine the number of *independent* forces in the structure whose determination will allow all member end forces to be established: this is the *static indeterminacy number* α.

It is the presence of cycles or meshes (closed circuits) in the graph model that leads to such indeterminacy. Let M be the number of branches or frame elements, N the number of nodes, wherein all foundation nodes are collected into (and counted as) a single node, and C the number of *independent* cycles or meshes; then

$$C = M - N + 1 = M - N_f \tag{4.5}$$

where N_f is the number of free (non-foundation) nodes. C is sometimes called the *cyclomatic number*. The number of independent forces in each cycle is $s_c = 6$ for a spatial frame and $s_c = 3$ for a planar frame or a grid. Hence the static indeterminacy number α is

$$\alpha = s_c\, C - r \tag{4.6}$$

where r is the number of mechanical articulations or releases.

Statics

The static or equilibrium conditions of a structure can be expressed in the mesh form

$$\mathbf{Q}_I = \mathbf{B}\,\mathbf{p} + \mathbf{B}_o\,\mathbf{F} \tag{4.7}$$

where $\mathbf{Q}_I$ collects the independent member forces for each member in the structure, **p** is a list of the α selected independent *mesh* or *hyperstatic forces* (sometimes unfortunately called redundants), and **F** is a list of applied loads. Matrix **B** may be called a *mesh matrix*. The forces $\mathbf{Q}_I$ are thus separated by (4.7) into two sets: one set is self-equilibrating – the complementary solution –, the other equilibrates the external loading – the particular solution.

The elementary approach is to select α releases of internal forces in the frame members, and then to apply unit self-equilibrating force pairs (biactions) **p** at these releases, whence the matrix **B** can be calculated. Such an approach has little scope for automation. However, for a planar frame constructed with meshes of the same regular – for instance, rectangular – form, a considerable improvement in automatic data generation is possible.

Consider a single rectangular mesh or cell: three hyperstatic forces **p** may be associated with such a cell, and a possible basic set of bending moment diagrams is suggested in Figure 4.3. These diagrams are both self-equilibrating and linearly

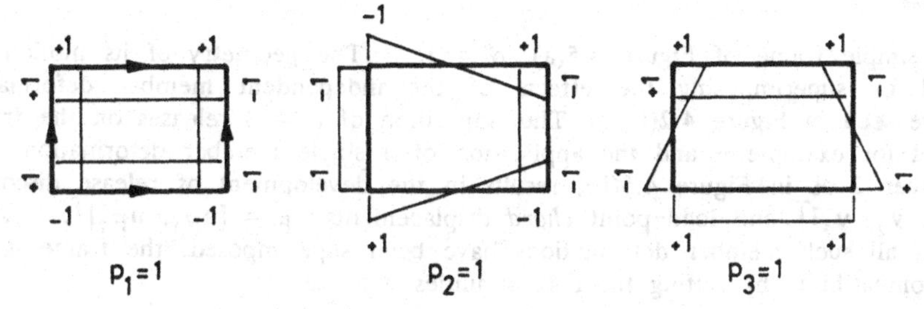

Figure 4.3

independent, and the ordinates have been scaled to convenient values. The axial forces associated with such bending moment basic functions may easily be worked out, although they will depend on the dimensions of the mesh.

For a more general plane frame with rectangular meshes, such as that of Figure 4.4, each cell may be assigned the same set of three bending moment basic functions. Of course, the set of multipliers p_{ij} $(i = 1,2,..,s_c = 3)$ for cell j will be different from the set p_{ik} for cell k, but the **B** matrix can be generated automatically from the data of Figure 4.3, and is sparse. The $\alpha = 12$ columns of the **B** matrix for the frame of Figure 4.4 are linearly independent and form a vector basis for representing all possible self-equilibrating components of Q_I.

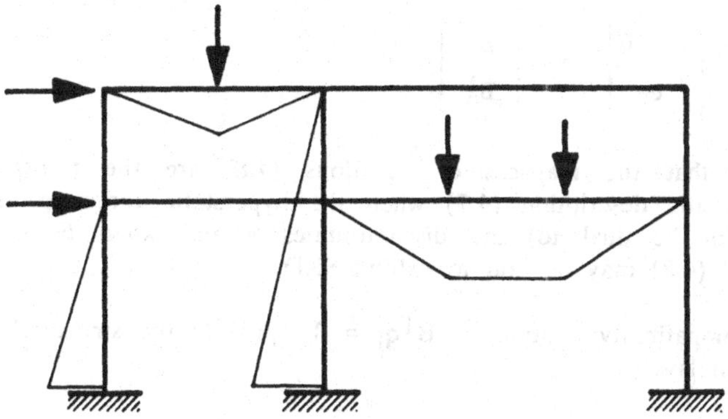

Figure 4.4

The load equilibrating bending moment diagrams shown in Figure 4.4 indicate that each load may be equilibrated separately in the most convenient manner: thus, horizontal loads may be supported on the columns as cantilevers, and vertical loads on the beams as though simply-supported.

Kinematics

For the simple frame of Figure 4.5(a), $\alpha = 3$. The geometry of its motion may be described by superimposing the effects of the independent member deformations $\mathbf{q}_I$ which are seen in Figure 4.2(b). The imposition of $\alpha = 3$ releases on the frame - a single cut for example - and the application of a single member deformation, say q_{I1}, for member 2 as in Figure 4.5(b), results in the development of release discontinuities $\mathbf{v} = [v_1, v_2, v_3]^T$ and load-point *chord* displacements $\mathbf{u}_F = [u_{F1}, u_{F2}]^T$. After the effects of all such member deformations have been superimposed, the frame is restored to full compatibility by setting the discontinuities $\mathbf{v}$ to zero.

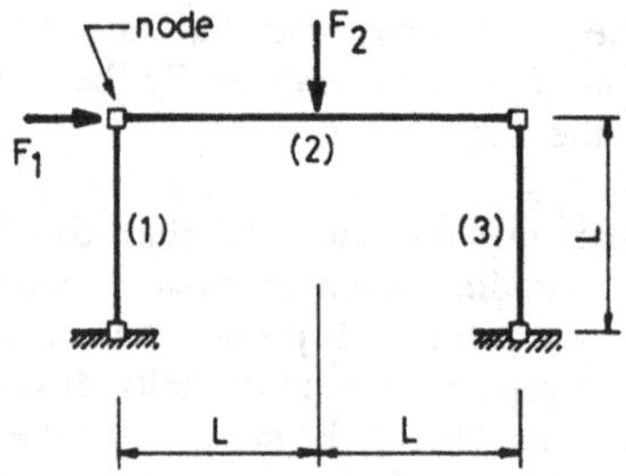

Figure 4.5(a)

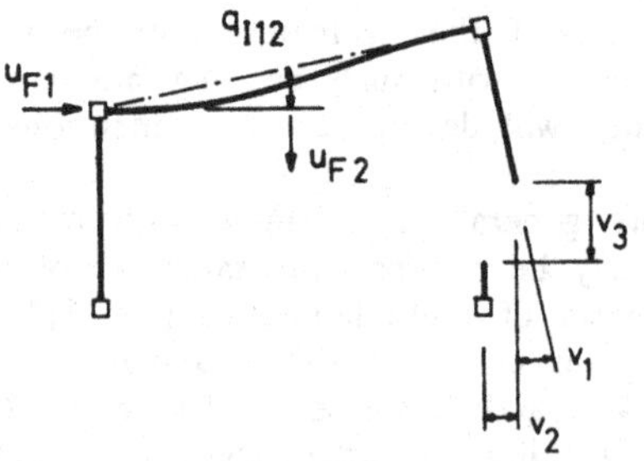

Figure 4.5(b)

In general, therefore, kinematics is expressed in mesh form by relations of the kind

$$\begin{bmatrix} \mathbf{v} = \mathbf{0} \\ \mathbf{u}_F \end{bmatrix} = \begin{bmatrix} \mathbf{B}^T \\ \mathbf{B}_o^T \end{bmatrix} \mathbf{q}_I \qquad (4.8)$$

It will be found that the matrices of equations (4.8) are the transpose of those appearing in the static description (4.7) when the hyperstatic forces $\mathbf{p}$ are chosen to correspond with (or be dual to) the discontinuities $\mathbf{v}$ and when $\mathbf{F}$ is dual to $\mathbf{u}_F$. Relations (4.7) and (4.8) may be said to exhibit SKD.

In (4.8), the α compatibility relations $\mathbf{B}^T\mathbf{q}_I = \mathbf{0}$ provide the structural manifestation of Kirchhoff's mesh law:

> The vectorial sum of the element deformations around each of the C independent cycles or meshes is zero.

The kinematic description (4.8) can be inferred from the static one (4.7). It can easily be shown [7] that this remains true when the mesh matrices $\mathbf{B}$, $\mathbf{B}_o$ are constructed in the manner suggested by Figures 4.3 and 4.4. The parameters $\mathbf{v}$ then have no direct physical interpretation, but their vanishing still enforces full compatibility.

Incorporation of Releases

It is clear that the employment of the simple procedure suggested by Figures 4.3 and 4.4 requires that all planar stress resultants be generally sustainable everywhere in the framed structure. Suppose, however, that one or more of the column bases in Figure 4.4 is pinned. Then the same simple static-kinematic description may be retained, provided that some additional conditions are imposed. We may write

$$Q_R = T_R^T Q_I, \qquad q_I = q_{Id} + T_R q_R, \qquad (4.9\text{-}10)$$

where Q_R records the bending moments implied by Figures 4.3 and 4.4 at the column bases where bending moment releases are required, and q_R are the dual rotations that may then exist at those bases, Figure 4.6. Clearly, q_{Id} contains the independent member deformations due to all sources other than the release motions q_R. By introducing these results into (4.7) and (4.8), we achieve the following mesh description:

$$\begin{bmatrix} Q_I \\ Q_R \end{bmatrix} = \begin{bmatrix} B & B_o \\ T_R^T B & T_R^T B_o \end{bmatrix} \begin{bmatrix} p \\ F \end{bmatrix} \quad (4.11) \qquad \begin{bmatrix} v = 0 \\ u_F \end{bmatrix} = \begin{bmatrix} B^T & B^T T_R \\ B_o^T & B_o^T T_R \end{bmatrix} \begin{bmatrix} q_{Id} \\ q_R \end{bmatrix} \quad (4.12)$$

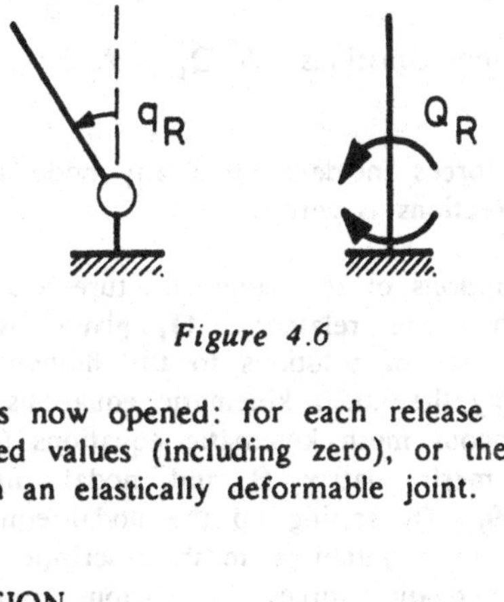

Figure 4.6

A range of possibilities is now opened: for each release station k, either Q_{Rk} or q_{Rk} may be assigned prescribed values (including zero), or they may be connected by some constitutive relation, as in an elastically deformable joint.

THE NODAL DESCRIPTION

Kinematic Indeterminacy

Relations 4.1(a) indicate that the member deformations q_I are determined completely by the member end or nodal displacements q_M' of that element. The member end displacements for the complete assembly of elements may be expressed entirely in terms of a certain number of *independent* displacements of the structure's node points. The number of these displacements is the *kinematic indeterminacy number* β of the structural system: β may be said to record the number of degrees of freedom in the nodal matrix in network theory, it may be remarked that this same matrix is also represented occasionally by the symbol **C** – the connection matrix.

Statics

For the discrete model of the structural system shown in Figure 4.7, which has $\beta = 2$, equilibrium between the independent member forces $\mathbf{Q}_I$ (internal forces) and the external loads **F** can be expressed in terms of two, more generally β, ***independent nodal forces*** or ***nodal constraint forces*** $\mathbf{U} = [U_1, U_2]^T$: these forces act at the positions and in the directions of the β independent nodal displacements or degrees of freedom. By means of SKD, we may infer from (4.14) the following laws

$$\mathbf{U}_I = \mathbf{A}^T \mathbf{Q}_I \quad (4.15) \qquad\qquad \mathbf{U}_o = \mathbf{A}_o^T \mathbf{F} \quad (4.16)$$

which may easily be interpreted. In (4.15), the $\mathbf{U}_I$ will be found to represent loads which, when applied in coincidence with the degrees of freedom, are in equilibrium with the independent member forces $\mathbf{Q}_I$. In (4.16), the $\mathbf{U}_o$ represent *equivalent nodal loads* – loads that, when applied in coincidence with the degrees of freedom, are statically equivalent to the actual applied loads **F**. Equilibrium is obtained in the original structure when the nodal constraint forces $\mathbf{U} = \mathbf{U}_I - \mathbf{U}_o$ vanish.

$$\mathbf{U} = \mathbf{0} = \begin{bmatrix} \mathbf{A}^T & \mathbf{A}_o^T \end{bmatrix} \begin{bmatrix} \mathbf{Q}_I \\ -\mathbf{F} \end{bmatrix} \qquad (4.17)$$

In system (4.17), the β equilibrium equations $\mathbf{A}^T\mathbf{Q}_I - \mathbf{A}_o^T\mathbf{F} = \mathbf{0}$ constitute Kirchhoff's node law for structural networks:

> The vectorial sum of the forces incident upon any node and measured in any of its independent freedom directions is zero.

Although mesh and nodal descriptions of the same structure are apparently unconnected, there is a sense in which they are related: $\mathbf{Q}_I$ given by the mesh equilibrium equations (4.7) forms the entire set of solutions to the homogeneous nodal equilibrium equations (4.17), and $\mathbf{q}_I$ given by the nodal kinematic equations (4.14) forms the entire set of solutions to the homogeneous mesh kinematic equations (4.8). From these last two relations, for any valid mesh matrix **B** and nodal matrix **A**, we find that $\mathbf{B}^T\mathbf{A} = \mathbf{0}$, and hence $\mathbf{A}^T\mathbf{B} = \mathbf{0}$. By setting up the nodal equilibrium equations (4.17) and solving them for $\mathbf{Q}_I$, one may obtain a mesh description of equilibrium without having to pre-select the hyperstatic forces. Various numerical procedures for performing this nodal-to-mesh transformation are discussed by Kaneko, Lawo and Thierauf[8].

This representation of the nodal description – that is, in terms of the independent member forces $\mathbf{Q}_I$ and independent member deformations $\mathbf{q}_I$ – lacks scope for the automatic generation of data. For this purpose, it is necessary to perform the description in terms of the member end forces $\mathbf{Q}_M$ and member end displacements $\mathbf{q}_M$ in a universal coordinate system – one which is common to all elements in the system.

possible displaced configurations of the chosen finite element model of the system. Let s_m be the number of independent member deformations q_{Ijm} $(j = 1,2,..,s_m)$ for element m, Figure 4.2(b), and M the number of elements. Then

$$\beta = \sum_{m=1}^{M} s_m - \alpha \tag{4.13}$$

gives the number of independent degrees of freedom. Unlike α, which is a topological invariant, β depends on the number of elements employed in the modelling.

<u>Kinematics</u>

For the frame shown in Figure 4.7, $\alpha = 2$. If a simple plastic limit analysis is contemplated, for example, the only independent member deformations would be the rotations q_I at the $c = 4$ potential plastic hinge positions. Then (4.13) gives $\beta = 2$, and the geometry of the collapse motion may be expressed in terms of any two *independent nodal displacements, degrees of freedom* or *hyperkinematic displacements* $\mathbf{u} = [u_1, u_2]^T$. This motion may be represented by the general form:

$$\begin{bmatrix} q_I \\ u_F \end{bmatrix} = \begin{bmatrix} \mathbf{A} \\ \mathbf{A}_o \end{bmatrix} \mathbf{u} \tag{4.14}$$

Matrix **A** is a *nodal matrix*. Concerning notation, while **A** is commonly used for the

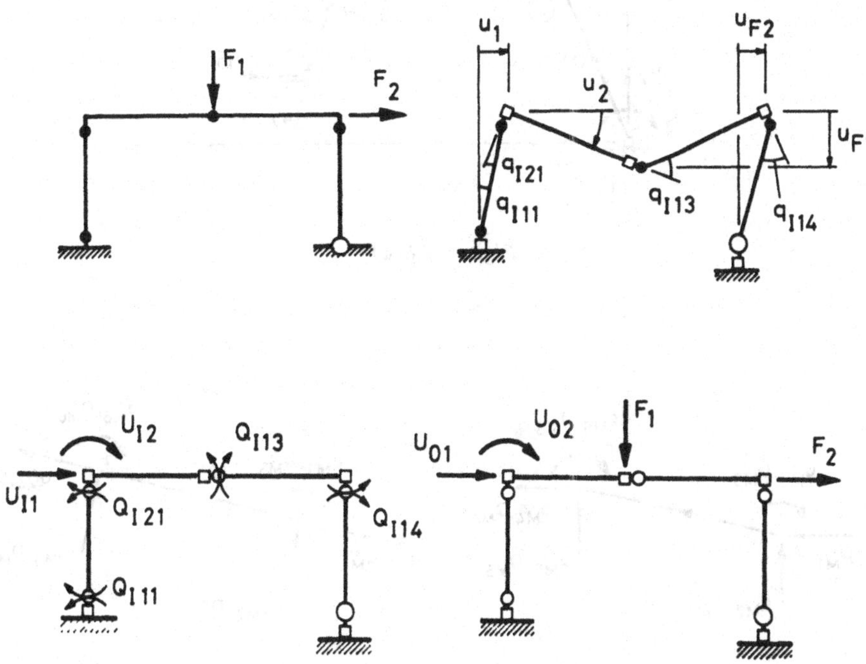

Figure 4.7

THE NODAL DESCRIPTION IN SYSTEM COORDINATES

Statics and Kinematics

The simple nodal description can be put into a form which is more suitable for the automatic assembly of the whole system of element relations: the price to be paid is a considerable augmentation of the static and kinematic descriptions.

Firstly, all foundation and other structural supports – that is, displacement constraints – may be removed. Secondly, each element is assigned its full complement of independent member deformations, $s_m = 3$ for a planar element and $s_m = 6$ for a spatial element: any constraint on these deformations – for instance, axial inextensibility – must be imposed as an additional condition. As a result, every node has three degrees of freedom for planar motion, or six for spatial motion. The planar frame of Figure 4.8(a), for example, then has twelve independent nodal displacements or degrees of freedom represented by $\mathbf{u} = [u_1, u_2, \ldots, u_{12}]^T$ which are referred to global or

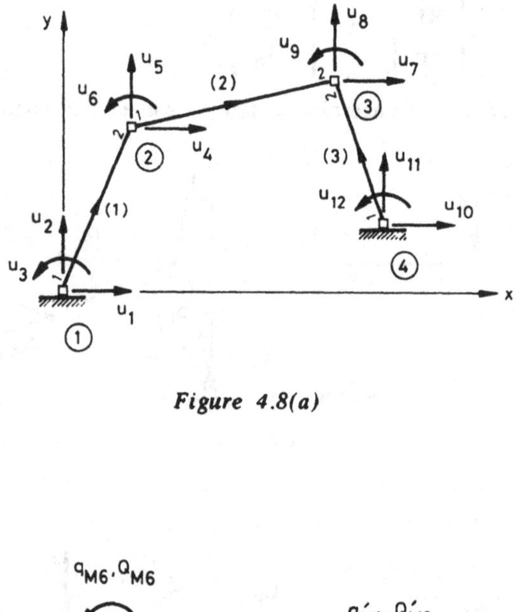

Figure 4.8(a)

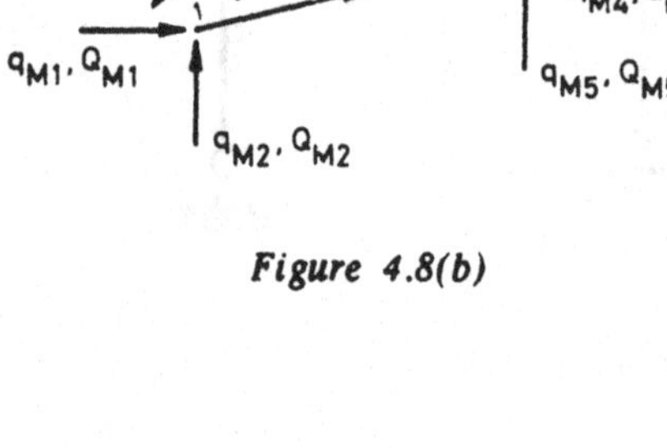

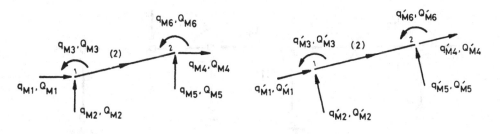

Figure 4.8(b)

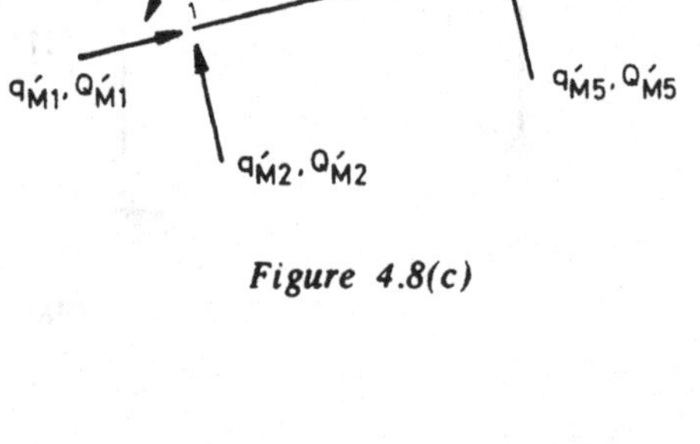

Figure 4.8(c)

system coordinates. The nodal description of kinematics in system coordinates has the form

$$U_o = A^T Q_M \qquad q_M = A\,u \qquad (4.18\text{-}19)$$

where Q_M and q_M are respectively the member end forces and displacements for all elements referred to global or system coordinates, Figure 4.8(b), and U_o are the applied nodal loads in the same coordinates. The nodal matrix **A** in (4.19) now has the simple form

$$\begin{bmatrix} q_{M11} \\ q_{M21} \\ q_{M12} \\ q_{M22} \\ q_{M13} \\ q_{M23} \end{bmatrix} = \begin{bmatrix} I & 0 & 0 & 0 \\ 0 & I & 0 & 0 \\ 0 & I & 0 & 0 \\ 0 & 0 & I & 0 \\ 0 & 0 & 0 & I \\ 0 & 0 & I & 0 \end{bmatrix} \begin{bmatrix} u_1 \\ u_2 \\ u_3 \\ u_4 \end{bmatrix} \qquad (4.20)$$

where **I** is a 3x3 identity matrix and **0** a 3x3 null matrix, q_{Mij} is the set of three member end displacements at end i of element j in system coordinates and u_k is the set of three independent nodal displacements (degrees of freedom) at node k.

Constitutive properties of elements are conveniently referred to local or element coordinates, Figure 4.8(c), so the following transformations are required between the two coordinate systems

$$Q_M = L^T Q_M' \qquad q_M' = L\, q_M \qquad (4.21\text{-}22)$$

Here **L** is a block diagonal matrix of direction cosines, and the transformation is orthogonal, i.e. $L^{-1} = L^T$ or $L^T L = I$.

Then (4.21-22), together with (4.18-19) and (4.3-4), give the desired results

$$U_o = [DLA]^T Q_I \qquad q_I = [DLA]\, u \qquad (4.23\text{-}24)$$

The matrix product **[DLA]** is symbolic: once the matrix product **DL** has been calculated for each unconnected member, then, if it is needed, **[DLA]** can be directly assembled for the whole connected structure without formal matrix multiplication due to the special nature of the augmented nodal matrix **A**, as exemplified by (4.20).

So far, all the structure nodes have remained entirely free of constraint. When the formulation is completed by the addition of appropriate constitutive relations connecting Q_I and q_I for all elements, the necessary support or displacement boundary conditions may be imposed on the nodal displacements **u**.

Incorporation of Releases

The additional freedom exhibited by a structure in which mechanical releases are incorporated into some of the members may be taken into reckoning through the release motions q_R which appear in (4.10). If (4.9) is appended to (4.23) and (4.10) is substituted into (4.24), the following is obtained:

$$\begin{bmatrix} U_o \\ -Q_R \end{bmatrix} = \begin{bmatrix} [DLA]^T \\ -T_R^T \end{bmatrix} Q_I \qquad q_{Id} = \begin{bmatrix} [DLA] & -T_R \end{bmatrix} \begin{bmatrix} u \\ q_R \end{bmatrix} \qquad (4.25\text{-}6)$$

The advantages of the nodal description in system coordinates are thereby retained. Again, Q_I and q_{Id} are to be related through appropriate constitutive laws for the members of the structure; and, at each release station k, Q_{Rk} or q_{Rk} may take prescribed values (including zero), or they may also be related through a constitutive law.

REFERENCES

1. Seely, S., Dynamic Systems Analysis, Reinhold 1964.

2. MacFarlane, A. G. J., Dynamical System Models, Harrap 1970.

3. Munro, J. and Smith, D. Lloyd., Linear programming in plastic analysis and synthesis, Proc. Int. Symposium on Computer-aided Design, University of Warwick 1972.

4. Maxwell, J. Clerk, On the calculation of the equilibrium and stiffness of frames, Philosophical Magazine, 7*th* Series **27** (1964) 294.

5. Jenkins, R. S., Theory and Design of Cylindrical Shell Structures, Ove Arup & Partners, London 1947.

6. Spillers, W. R., Automated Structural Analysis, Pergamon 1972.

7. Lloyd Smith D., Plastic Limit Analysis and Synthesis of Structures by Linear Programming, PhD Thesis, University of London 1974.

8. Kaneko, I., Lawo, M. and Thierauf, G., On computational procedures for the force method, Int. J. for Numerical Methods in Engineering, **18** (1982) 1469–1495.

CHAPTER 5

PLASTIC LIMIT ANALYSIS

D. Lloyd Smith
Imperial College, London, U.K.

ABSTRACT

Systematic mesh and nodal descriptions of the laws of statics and kinematics for the limiting state of plastic collapse in a structural system are set out. Then the constitutive relations appropriate to this condition are presented in such a way as to emphasise their inherent complementarity. The mixing together of these three independent ingredients – statics, kinematics and material constitution – gives rise to the vectorial formulation which governs plastic collapse: it is identified as a linear complementarity problem. From it are derived the dual linear programs which give expression to the variational principles associated with upper and lower bounds on the collapse load factor.

INTRODUCTION

The modern systems methods for plastic limit analysis have their genesis in the very first systematic methods developed for manual calculation by Neal and Symonds - the method of inequalities[1] of 1950 and the method of combination of mechanisms[2] of 1952. Of course, the notion of utilising the plastic ductility of structural materials in improving the design of engineering structures goes back, at least, to Kazinczy and to 1914; a brief review of some of the literature of this first period of theoretical and practical development is given by Baker, Horne and Heyman[3] and by Neal[4]. From 1950, mathematical programming has played a significant rôle in plastic solid mechanics, and particularly in plastic limit analysis; a historical perspective on this phase is provided by Maier and Munro[5].

Limit Analysis is concerned with establishing the strength of a structure - its capacity for the supporting of loads. It is not at all concerned with deformation: it cannot therefore provide the load carrying capacity for a structure with elements that have a limited ductility or deformability, nor for a structure which becomes unstable because of the displacements induced by plastic deformation.

PLASTIC COLLAPSE MECHANISMS

Plastic collapse occurs when a structure is converted into a mechanism by the provision of a suitable number and disposition of plastic zones. If the material in these zones exhibits perfect plasticity, a further *small* change in configuration of the structure can be induced without any change in the applied loads, and the mechanism motion that ensues is that of a set of contiguous *rigid* bodies articulated through deformations in the plastic zones.

For plane frames, the typical structural element is that shown in Figure 5.1. In the simplest type of plastic modelling, it is assumed that the plastic zones, if they exist at all, are concentrated at points, or critical sections, adjacent to the element terminals. Within the element, at each critical section, there may consequently occur three

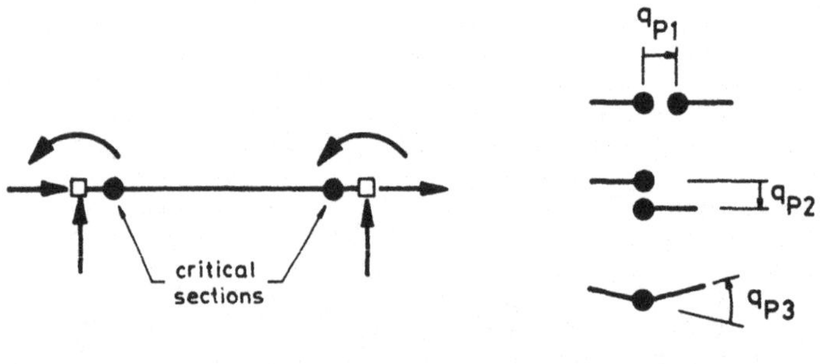

Figure 5.1

possible plastic strain resultants $q_P = [q_{P1}, q_{P2}, q_{P3}]^T$, generated by the corresponding stress resultants $Q = [Q_1, Q_2, Q_3]^T$. Such internal deformations within the element contribute to the relative displacements between the element nodes. They are represented by the three independent member deformations q_I of Figure 4.2(b) which themselves may be calculated from the six member end displacements q_M' (in local coordinates) of Figure 4.2(a).

Suppose that a structure has, at time t, just reached the state of plastic collapse. Let the total independent member deformations of all the structural elements be

$$q_I = q_{IE} + q_{IP}, \tag{5.1}$$

where q_{IE} are the member deformations due to elastic strains distributed continuously along the element, and where q_{IP} are the member deformations due to discrete plastic strain resultants q_P at the critical sections of each element. Then, at a small time increment Δt later, the deformations become

$$q_I + \dot{q}_I\Delta t = q_{IE} + \dot{q}_{IE}\Delta t + q_{IP} + \dot{q}_{IP}\Delta t. \tag{5.2}$$

The deformations of (5.1) and (5.2) must necessarily constitute compatible configurations of the structure at incipient plastic collapse and just subsequent to plastic collapse respectively, as suggested in Figure 5.2(a). Referring to the mesh description of compatibility, we may write

$$v = B^T q_I = 0, \qquad v + \dot{v}\Delta t = B^T[q_I + \dot{q}_I\Delta t] = 0. \tag{5.3-4}$$

By subtracting these two sets of equations, noting that Δt is small but otherwise quite arbitrary, we obtain the compatibility conditions which govern the motion of the structure in the initial phase of plastic collapse

$$\dot{v} = B^T\dot{q}_I = B^T[\dot{q}_{IE} + \dot{q}_{IP}] = 0. \tag{5.5}$$

These equations suggest that the difference between the two configurations of Figure 5.2(a) is itself a compatible configuration – the plastic collapse mechanism of Figure 5.2(b). What characterises the simple plastic collapse mechanism is that the structural

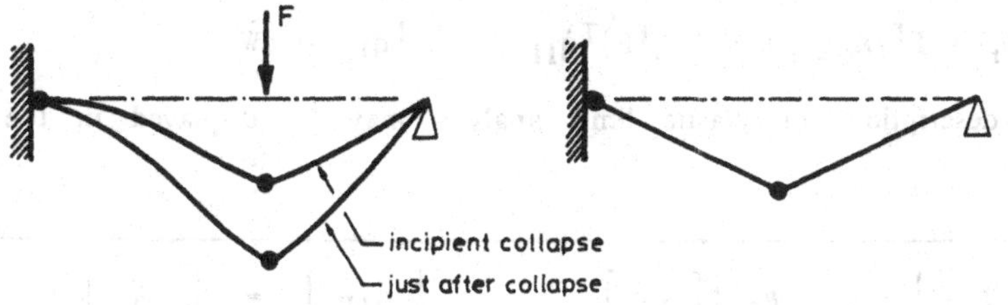

Figure 5.2(a) Figure 5.2(b)

elements tend to move as a system of rigid bodies which is articulated through the presence of deformations in the plastic zones. This means that, in the initial phase of plastic collapse, the increments of elastic deformation $\dot{q}_{IE}$ are zero. Therefore

$$\dot{v} = B^T\dot{q}_{IP} = 0 \tag{5.6}$$

expresses the compatibility requirement for any feasible plastic collapse mechanism.

STATICS AND KINEMATICS

For plastic limit analysis it is usual to express the set of applied loads **F** in terms of a single variable parameter, the *load factor* s. Then $F = F_o s$, where F_o is a column vector of constant loads. In the mesh description detailed in Chapter 4, the second component of the independent member forces Q_I in the equilibrium equations (4.7) may be written in the form $B_oF = B_oF_os = b_os$, where b_o is a column vector. From (4.8) we have the load point displacement rates $\dot{u}_F$. The workrate of the applied loads upon the possible collapse mechanism is then

$$F^T\dot{u}_F = F^T[B_o{}^T\dot{q}_{IP}] = [B_oF]^T\dot{q}_{IP} = sb_o{}^T\dot{q}_{IP} = s\dot{W} \tag{5.7}$$

where $\dot{W}$ is the workrate at unit load factor. We may write the mesh description for plastic limit analysis in the following manner

$$Q_I = \begin{bmatrix} B & b_o \end{bmatrix} \begin{bmatrix} p \\ s \end{bmatrix} \tag{5.8}$$

$$\begin{bmatrix} \dot{v} = 0 \\ \dot{W} \end{bmatrix} = \begin{bmatrix} B^T \\ b_o^T \end{bmatrix} \dot{q}_{IP} \tag{5.9}$$

MESH DESCRIPTION

The nodal equilibrium equations (4.15) may be rewritten so that the second component now becomes $A_o^TF = A_o^TF_os = a_os$, where a_o is a column vector. Then the load point displacement rates $\dot{u}_F$ are obtained from (4.12), and the workrate of the applied loads on the possible collapse mechanism is therefore

$$F^T\dot{u}_F = F^T[A_o\dot{q}_{IP}] = [A_o{}^TF]^T\dot{q}_{IP} = sa_o{}^T\dot{q}_{IP} = s\dot{W}. \tag{5.10}$$

The nodal description for plastic limit analysis may be displayed in the following manner.

$$U = 0 = \begin{bmatrix} A^T & a_o \end{bmatrix} \begin{bmatrix} Q_I \\ -s \end{bmatrix} \tag{5.11}$$

$$\begin{bmatrix} \dot{q}_{IP} \\ \dot{W} \end{bmatrix} = \begin{bmatrix} A \\ a_o^T \end{bmatrix} \dot{u} \tag{5.12}$$

NODAL DESCRIPTION

It is also possible to present the nodal description in system coordinates if member end forces Q_M' and displacement rates $\dot{q}_M'$ are employed in the local coordinates appropriate to each element in the structure. We shall consider the applied nodal loads as sU_o. Then (4.21) and (4.22) give the augmented nodal description in which the nodal loads sU_o and the nodal displacement rates $\dot{u}$ of the structure are measured in system coordinates, and **L** is the matrix of direction cosines which relates the local coordinate system of each element to the global or system coordinates.

$$sU_o = [L\ A]^T\ Q_M' \qquad (5.13)$$

$$\dot{q}_{MP}' = [L\ A]\ \dot{u} \qquad (5.14)$$

$$Q_M' = D^T\ Q_I \qquad (5.15)$$

$$\dot{q}_{IP} = D\ \dot{q}_{MP}' \qquad (5.16)$$

NODAL DESCRIPTION IN SYSTEM COORDINATES

Relations (5.15) and (5.16), formerly given as (4.4) and (4.3), allow reference to be made to independent member forces Q_I and independent plastic deformation rates $\dot{q}_{IP}$ for the whole structure. For the individual element of Figure 5.1, the matrix D^T is recorded in equations (4.2) of Chapter 4. The workrate of the applied nodal loads on any possible collapse mechanism may now simply be written $sU_o{}^T\dot{u} = s\dot{W}$.

CONSTITUTIVE RELATIONS

At each critical section, there are three stress resultants, the axial force Q_1, the transverse shear force Q_2 and the bending moment Q_3. They may be represented as a vector in a three dimensional stress resultant space; then, for those combinations of the stress resultants that induce plastic yielding, the vector traces out a surface – the yield surface. Such a surface may be simplified by its being made piecewise linear, as indicated in Figure 5.3(a) where yielding is supposed to be governed by Q_1 and Q_3

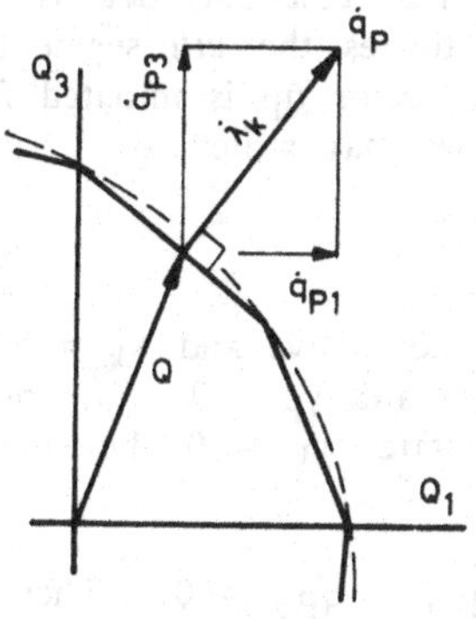

Figure 5.3(a)

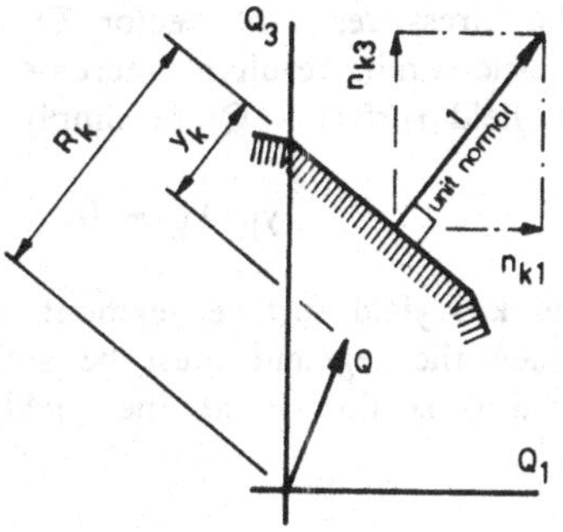

Figure 5.3(b)

only. Permissible combinations of stress resultants are those for which the vector **Q** lies within the closed set of points contained by the yield surface; these constraints on **Q** are known as the *yield conditions*. One of the linearised segments of the yield surface, say the *kth* (k = 1,2,...,y), is shown in Figure 5.3(b). With respect to this segment, the yield condition is expressed by the two relations

$$[n_{k1}Q_1 + n_{k3}Q_3] + y_k = R_k, \qquad y_k \geq 0. \qquad (5.17\text{-}18)$$

In (5.17), the n_{kj} are the direction cosines of the normal to segment k, and R_k, the orthogonal distance from the origin to segment k, is a measure of the *plastic capacity* of the critical section. That **Q** should lie within the feasible region is expressed by (5.18) through the slack variable y_k, the *plastic potential*. If preferred, y_k may be replaced by a non-positive plastic potential ψ_k with a change of the positive sign in (5.17).

A plastic material that does not harden or soften with the development of plastic deformation is said to be *perfectly plastic*. For such a material, the yield surface retains its initial position in stress resultant space throughout the entire deformation process. Plastic limit analysis is only concerned with perfectly plastic materials.

The components $\dot{q}p_1$ and $\dot{q}p_3$ of the plastic strain resultant increment vector $\dot{\mathbf{q}}\mathbf{p}$ are assumed to develop in proportion to each other in the following manner. If the vector $\dot{\mathbf{q}}\mathbf{p}$ is superimposed upon the space of the stress resultants, it will be directed along the *outward* normal to the yield surface; this is the *normal* or *associated flow rule*, so called because the law governing plastic flow is directly associated with the yield surface. From Figures 5.3(a) and 5.3(b) the flow rule is seen to be represented by

$$\dot{q}p_1 = n_{k1}\dot{\lambda}_k, \qquad \dot{q}p_3 = n_{k3}\dot{\lambda}_k, \qquad \dot{\lambda}_k \geq 0, \qquad (5.19\text{-}20\text{-}21)$$

In these relations, $\dot{\lambda}_k$ is the *plastic multiplier* for the *kth* segment of the yield surface; it gives the length of the vector $\dot{\mathbf{q}}\mathbf{p}$, and its non-negativity reflects the requirement that $\dot{\mathbf{q}}\mathbf{p}$ is directed outwardly along the normal.

Finally, the stress resultants **Q** at a critical section must be *associated* with or related to the plastic strain resultant increments $\dot{\mathbf{q}}\mathbf{p}$. The essential idea is represented in Figure 5.3(a); if the stress resultant vector **Q** activates the *kth* segment of the yield surface, then the plastic strain resultant increment vector $\dot{\mathbf{q}}\mathbf{p}$ is initiated from the point of activation on the yield surface. Quite simply, we may write

$$y_k \, \dot{\lambda}_k = 0. \qquad (5.22)$$

If $y_k > 0$, then the *kth* yield surface segment is not active and $\dot{\lambda}_k = 0$; on the other hand, if $\dot{\lambda}_k > 0$, then the segment must be active and $y_k = 0$. Of course, it is also possible that the critical section is at the yield state, $y_k = 0$, but that there is no plastic flow, $\dot{\lambda} = 0$.

A special case of these relations occurs when $\dot{q}p_1 = \dot{q}p_2 = 0$. Then we encounter the familiar plastic hinge. The yield surface comprises two horizontal planes at distances $R_1 = M_p^+$ and $R_2 = M_p^-$ above and below the origin, thus identifying the plastic moments of resistance for bending of either sense. From $Q_3 + y_1 = R_1$ and

$-Q_3 + y_2 = R_2$ we have the plastic potentials, in terms of which the yield conditions $y_1 \geq 0$ and $y_2 \geq 0$, or alternatively $Q_3 \leq M_p^+$ and $Q_3 \geq -M_p^-$, are expressed. The normal flow rule then requires that $\dot{q}_{P3} = \dot{\lambda}_1 - \dot{\lambda}_2$ be the relative angular velocity in the plastic hinge.

These constitutive relations must now be written for all the *y* segments forming the yield surface at a particular critical section, and for all the *c* critical sections in the structural system. They are essentially in the form suggested by Maier[6], a form for which he was able to propose many interesting mathematical programming applications.

$$\begin{bmatrix} 0 & N^T \\ N & 0 \end{bmatrix} \begin{bmatrix} \dot{\lambda} \\ Q \end{bmatrix} + \begin{bmatrix} y \\ 0 \end{bmatrix} = \begin{bmatrix} R \\ \dot{q}_P \end{bmatrix} \qquad \begin{matrix} (5.23) \\ (5.24) \end{matrix}$$

$$y \geq 0 \quad (5.25) \qquad y^T \dot{\lambda} = 0 \quad (5.26) \qquad \dot{\lambda} \geq 0 \quad (5.27)$$

Notice that (5.26) is a condition of *orthogonality* between the vectors **y** and $\dot{\lambda}$ which, through the non-negativity of these vectors, ensures that the relation (5.22) of association between y_k and $\dot{\lambda}_k$ is satisfied for all segments k at each critical section. Relations (5.25) to (5.27) form the familiar triad of complementarity conditions.

It will be observed that the plasticity relations are expressed in terms of stress resultants Q and strain resultant increments $\dot{q}_P$ at the critical sections. The elements, however, are described in terms of the independent member forces Q_I and deformation rates $\dot{q}_{IP}$, or, alternatively, in terms of the member end forces Q_M' and velocities $\dot{q}_{MP}'$. It is therefore necessary to set up the following transformations

$$\dot{q}_{IP} = T\, \dot{q}_P \qquad\qquad Q = T^T Q_I \qquad\qquad (5.28\text{-}29)$$

While these relations are written here for the whole structural system, they are incorporated into the governing relations on an element-by-element basis. For example, considering critical sections at both left (L) and right (R) ends of the element of length L in Figure 5.1, the relations implied by (5.29) are

$$\begin{bmatrix} Q_{1L} \\ Q_{2L} \\ Q_{3L} \\ \hline Q_{1R} \\ Q_{2R} \\ Q_{3R} \end{bmatrix} = \begin{bmatrix} 0 & 0 & 1 \\ -L^{-1} & L^{-1} & 0 \\ 1 & 0 & 0 \\ \hline 0 & 0 & 1 \\ -L^{-1} & L^{-1} & 0 \\ 0 & 1 & 0 \end{bmatrix} \begin{bmatrix} Q_{I1} \\ Q_{I2} \\ Q_{I3} \end{bmatrix} \qquad (5.30)$$

GOVERNING RELATIONS

We seek to establish the load factor s_c at which the state of plastic collapse is induced in a structure. If the load factor is held steady at $s = (1 - \varepsilon)s_c$ for some small and positive ε, the structure exhibits a *static state* in which its displaced configuration, involving both elastic and plastic deformations, is stationary. On the other hand, if the load factor is steady at $s = (1 + \varepsilon)s_c$, the structure presents a *dynamic state* in which there occurs an accelerated motion of the elements compatible with the mechanism of plastic collapse, the accelerations depending upon the magnitude and distribution of the structural masses. When the steady load factor is $s = s_c$, the structure has what we might choose to call a *neutral state*, or a state of balance, in which the collapse mechanism acceleration is zero.

It is possible that, in the neutral state, the velocity of the collapse mechanism is also zero, as would occur when the load factor is *slowly* increased to the value s_c. However, it is entirely consistent with the neutral state that the structure may have impressed upon it any constant velocity field or distribution, provided that it is compatible with the plastic collapse mechanism and observes the sense or direction of the mechanism motion implied by the loads. When a structure is in the neutral state, an arbitrary, but small and appropriately distributed, impulse would propel the structure into a motion of the collapse mechanism with arbitrary but *constant* velocity; and yet it would sensibly leave unchanged the applied loads.

The state of simple or static plastic collapse is therefore a neutral state in which the collapse mechanism velocity is constant and has the appropriate sense; but, mathematically, we may consider it otherwise arbitrary. This means that the workrate $\dot{W}$ expended on the collapse mechanism is also arbitrary, although it should not be negative if the mechanism is to have the appropriate sense. Setting $\dot{W} = 1$ is a convenient way of *normalising* the mechanism velocity. Then the governing relations are obtained by combining statics, kinematics and the plasticity relations.

For the mesh formulation, we obtain (5.31).

$$\left[\begin{array}{c|cc} 0 & [T\,N]^T b_o & [T\,N]^T B \\ \hline b_o^T T\,N & 0 & 0 \\ B^T\,T\,N & 0 & 0 \end{array}\right] \left[\begin{array}{c} \dot{\lambda} \\ \hline s \\ p \end{array}\right] + \left[\begin{array}{c} y \\ \hline 0 \\ 0 \end{array}\right] = \left[\begin{array}{c} R \\ \hline 1 \\ 0 \end{array}\right]$$

$$y \geq 0 \qquad y^T \dot{\lambda} = 0 \qquad \dot{\lambda} \geq 0 \qquad (5.31)$$

For the nodal formulation, the governing relations are those of (5.32).

$$\left[\begin{array}{cc|cc} 0 & 0 & a_o & -A^T \\ 0 & 0 & 0 & [T\ N]^T \\ \hline a_o^T & 0 & 0 & 0 \\ -A & T\ N & 0 & 0 \end{array}\right] \left[\begin{array}{c} \dot{u} \\ \dot{\lambda} \\ \hline s \\ Q_I \end{array}\right] + \left[\begin{array}{c} 0 \\ y \\ \hline 0 \\ 0 \end{array}\right] = \left[\begin{array}{c} 0 \\ R \\ \hline 1 \\ 0 \end{array}\right]$$

$$y \geq 0 \qquad y^T \dot{\lambda} = 0 \qquad \dot{\lambda} \geq 0 \tag{5.32}$$

The augmented nodal formulation in system coordinates is that of (5.33).

$$\left[\begin{array}{cc|cc} 0 & 0 & U_o & -[DLA]^T \\ 0 & 0 & 0 & [T\ N]^T \\ \hline U_o^T & 0 & 0 & 0 \\ -[DLA] & [T\ N] & 0 & 0 \end{array}\right] \left[\begin{array}{c} \dot{u} \\ \dot{\lambda} \\ \hline s \\ Q_I \end{array}\right] + \left[\begin{array}{c} 0 \\ y \\ \hline 0 \\ 0 \end{array}\right] = \left[\begin{array}{c} 0 \\ R \\ \hline 1 \\ 0 \end{array}\right]$$

$$y \geq 0 \qquad y^T \dot{\lambda} = 0 \qquad \dot{\lambda} \geq 0 \tag{5.33}$$

where **A** has a simplified form, similar to (4.20), and the various matrices can be directly assembled from the contributions made by individual members or elements. However, matrix **A** in the augmented nodal description is usually assembled with all structure nodes absolutely unconstrained or free to move in all directions. The appropriate support or boundary conditions must therefore be imposed: homogeneous conditions on components of the system nodal velocities $\dot{u}$ – that is, $\dot{u}_k = 0$ – are easily applied by deleting the appropriate rows and corresponding columns of the matrix in (5.33).

The governing relations (5.31), (5.32) and (5.33) are each examples of a linear complementarity problem (LCP). However the form of this LCP is special. It is symmetric – due to SKD and to constitutive reciprocity or the normal flow rule – and the static variables and kinematic variables are uncoupled.

LINEAR PROGRAMS FOR PLASTIC LIMIT ANALYSIS

The special structure of each of the LCPs governing plastic limit analysis is almost identical to that of the LCP(2.16) which has been discussed in Chapter 2. There it will be seen that LCP(2.16) constitutes the necessary and sufficient conditions – the Karush–Kuhn–Tucker conditions – for the optimal solutions of the dual linear programs LP(2.27) and LP(2.28). There is an equivalence between the LCP and the dual linear programs.

For the mesh LCP (5.31), the associated dual LPs are given by (5.34) and (5.35).

Minimise z

$$\begin{bmatrix} R^T \end{bmatrix} \dot{\lambda} = z$$

$$\begin{bmatrix} b_o^T T N \\ B^T T N \end{bmatrix} \dot{\lambda} = \begin{bmatrix} 1 \\ 0 \end{bmatrix}$$

$$\dot{\lambda} \geq 0 \qquad (5.34)$$

KINEMATIC MESH LP

Maximise w

$$\begin{bmatrix} 1 & 0^T \end{bmatrix} \begin{bmatrix} s \\ p \end{bmatrix} = w$$

$$\begin{bmatrix} N^T T^T b_o & N^T T^T B \end{bmatrix} \begin{bmatrix} s \\ p \end{bmatrix} \leq R \qquad (5.35)$$

STATIC MESH LP

If the kinematic LP has m constraints in n variables, the static LP will have n constraints in m variables. For the mesh LPs (5.34) and (5.35), it will be seen that $m = (1 + \alpha)$ and $n = Y$, where α is the static indeterminacy number and Y is the total number of segments, representing all the yield surfaces for all the critical sections.

The nodal LCP(5.32) has the associated dual LPs set out in (5.36) and (5.37).

Minimise z

$$\begin{bmatrix} 0^T & R^T \end{bmatrix} \begin{bmatrix} \dot{u} \\ \dot{\lambda} \end{bmatrix} = z$$

$$\begin{bmatrix} a_o^T & 0^T \\ -A & T N \end{bmatrix} \begin{bmatrix} \dot{u} \\ \dot{\lambda} \end{bmatrix} \begin{matrix} = \\ = \end{matrix} \begin{bmatrix} 1 \\ 0 \end{bmatrix}$$

$$\dot{\lambda} \geq 0 \qquad (5.36)$$

KINEMATIC NODAL LP

Maximise w

$$\begin{bmatrix} 1 & 0^T \end{bmatrix} \begin{bmatrix} s \\ Q_I \end{bmatrix} = w$$

$$\begin{bmatrix} a_o & -A^T \\ 0 & N^T T^T \end{bmatrix} \begin{bmatrix} s \\ Q_I \end{bmatrix} \begin{matrix} = \\ \leq \end{matrix} \begin{bmatrix} 0 \\ R \end{bmatrix} \qquad (5.37)$$

STATIC NODAL LP

For the nodal LPs (5.36) and (5.37), it will be found that $m = (1 + \Sigma s_m)$, in which $m = 1,2,\ldots,M$ is the index of summation and M is the total number of members, and that $n = (\beta + Y)$. In the first of these relations, s_m is the number of independent member deformation rates in member m, with a maximum of three – $\dot{q}_{I1}, \dot{q}_{I2}, \dot{q}_{I3}$ – for a planar frame member. However, in limit analysis, not all of these need be present.

Finally, the nodal LCP (5.33), which is constructed in system coordinates, gives rise to the dual LPs of (5.38) and (5.39).

Minimise z

$$\begin{bmatrix} 0^T & R^T \end{bmatrix} \begin{bmatrix} \dot{u} \\ \dot{\lambda} \end{bmatrix} = z$$

$$\begin{bmatrix} U_o & 0^T \\ -[DLA] & [T\ N] \end{bmatrix} \begin{bmatrix} \dot{u} \\ \dot{\lambda} \end{bmatrix} \begin{matrix} = \\ = \end{matrix} \begin{bmatrix} 1 \\ 0 \end{bmatrix}$$

$$\dot{\lambda} \geq 0 \qquad (5.38)$$

KINEMATIC NODAL LP IN SYSTEM COORDINATES

Maximise w

$$\begin{bmatrix} 1 & 0^T \end{bmatrix} \begin{bmatrix} s \\ Q_I \end{bmatrix} = w$$

$$\begin{bmatrix} U_o^T & -[DLA]^T \\ 0 & [TN]^T \end{bmatrix} \begin{bmatrix} s \\ Q_I \end{bmatrix} \begin{matrix} = \\ \leq \end{matrix} \begin{bmatrix} 0 \\ R \end{bmatrix} \qquad (5.39)$$

STATIC NODAL LP IN SYSTEM COORDINATES

For the nodal LPs in system coordinates, $m = (1 + dM)$ and $n = (dN + Y)$. In these relations, d is the number of degrees of freedom at each node or member end – d = 3 for planar frames or for grids, d = 6 for spatial frames –, and N is the total number of nodes, including all foundation or constrained nodes.

In all cases, $m < n$; this implies that each kinematic LP has fewer constraints than the corresponding dual static LP.

LIMIT THEOREMS AND UNIQUENESS

As has been previously emphasised, the LCP which forms the governing relations is the vectorial representation of plastic limit analysis; it expresses or formulates the problem in terms of static and kinematic variable vectors through the three fundamental laws of structural mechanics – statics, kinematics and constitutive relations. This LCP is special in that the static and kinematic variables are uncoupled; it is equivalent to a dual pair of linear programs – one with static variables, the other with kinematic variables – which present solutions for the problem in terms of complementary variational principles.

Kinematic Linear Program

The variables appearing in the kinematic linear program relate to the velocity field of a possible collapse mechanism. The constraints impose the conditions that a potential plastic collapse mechanism must satisfy: compatibility, the normal flow rule and normalisation of the mechanism velocity – in which the essential ingredient is the sense of the velocity appropriate to the applied loading. The vector of variables, in satisfying the constraints, traces out the region of feasible solutions; it is occupied by by vectors – called *kinematically admissible solutions* – which represent all the possible mechanisms of plastic collapse.

Let T be the kinetic energy and V the potential energy of a system in motion. Newton's (vectorial) law of motion is equivalent to the conservation of mechanical energy – that is, $T + V = \text{constant}$ or $\dot{T} + \dot{V} = 0$. The increment of potential energy $\dot{V}$ in the structural system is $\dot{V} = \dot{U} + \dot{D} - s\dot{W}$, where $\dot{U}$ is the rate of change of elastic or strain energy, $\dot{D}$ is the rate of energy dissipation in the plastic zones and $s\dot{W}$ is the workrate of the applied loads. In the neutral state associated with static plastic collapse, $\dot{U} = 0$ because the elastic strains remain constant, and $\dot{T} = 0$ because the mechanism motion is unaccelerated. Conservation of mechanical energy in the state of plastic collapse therefore implies that

$$s\dot{W} = \dot{D} = \mathbf{R}^T \dot{\lambda}. \tag{5.40}$$

Consider any possible mechanism of plastic collapse. If we imagine the structure constrained so that it can only collapse in this mechanism, then the value $s = s_k$, given by (5.40), determines the level of loading required to mobilise the given mechanism. We might refer to s_k as a *kinematically admissible load factor*.

The usefulness of the chosen method of normalising the mechanism velocity is now quite apparent. With $\dot{W} = 1$, (5.40) gives $s_k = \mathbf{R}^T\dot{\lambda} = z$ – that is, the value of the objective function z for any feasible solution (possible collapse mechanism) of the kinematic LP is the load factor required for the mobilisation of that mechanism. We may state the following theorem which is directly implied by the kinematic LP.

> *The kinematic limit theorem*:
> Among all kinematically admissible solutions (possible collapse mechanisms), the one (or more) associated with plastic collapse minimises the kinematically admissible load factor s_k.

It follows that the load factor s_k required for the mobilisation of any kinematically admissible solution is an *upper bound* on the correct load factor s_c for the structure – that is, $s_c \leq s_k$.

One must always exercise care in the discussion of such bounds. The kinematic LP converges upon a limiting value s_c which depends upon the accuracy of the structural modelling – the modelling of statics, kinematics and the constitutive laws. If the approximating piecewise linear yield surface is contained within the actual yield surface, then the strength of the plastic zone will be underestimated, and it is possible that s_c will be underestimated too. While this value of s_c represents a conservative, and therefore practically appropriate, estimate of the actual collapse load factor, we cannot assert that any s_k determined with this modelling is an upper bound on the correct plastic collapse load factor.

Static Linear Program

The variables which appear in the static linear program determine the distribution of internal forces or stress resultants in the elements of a structure. The constraints impose such conditions as would be satisfied by the stress resultants when the structure is in the neutral state: equilibrium with the applied loads, and the yield conditions. The vector of variables, in satisfying the constraints, traces out a region of feasible solutions: it is occupied by vectors – called ***statically admissible solutions*** – which represent all the possible stress resultant distributions that could be associated with plastic collapse. Each set of stress resultants in a statically admissible solution is in equilibrium with the applied loads at load factor $s = s_s$. It may be useful to refer to s_s as a ***statically admissible load factor***, and we note that $w = s_s$ forms the objective function of the static LP.

We may state the following theorem which is given directly by the static LP.

> *Static limit theorem:*
> Among all statically admissible solutions (sets of possible stress resultants at plastic collapse) , the one (or more) associated with plastic collapse maximises the statically admissible load factor s_s.

It follows that the load factor s_s for any statically admissible solution is a ***lower bound*** on the correct load factor s_c for plastic collapse of the structure – that is, $s_s \leq s_c$.

Again, caution should pervade our use of the lower bound concept. Part of the kinematic modelling concerns our choice of the number and location of the critical sections or potential plastic zones. If our critical sections do not coincide with the actual plastic zones, then the static or kinematic LP will converge upon a value of s_c which overestimates the correct plastic collapse load factor. The same effect results from an approximating piecewiselinear yield surface which contains the actual yield surface. In such circumstances, we cannot say that any s_s determined from our modelling is a lower bound on the correct plastic collapse load factor.

It must be said that the lower bound aspect of the static limit theorem is a wonderful aid in structural design. It embodies the notion of safety – statically admissible solutions are safe ($s_s \leq s_c$), but kinematically admissible solutions are unsafe ($s_c \leq s_k$). Suppose that it is required to design a structure to support given applied loads with fixed load factor s. First, we devise ***any*** set of stress resultants which satisfies only the conditions of equilibrium. Then we ***design*** the elements by assigning them sufficient material that their yield conditions safely contain or support the devised stress resultants. The static limit theorem now assures us that the collapse load factor s_c for a structure so designed cannot be less than the design load factor s.

The safety inherent in this lower bound approach to practical design is often used with genuine intuitive understanding by experienced engineers; it may also fortuitously conceal gross mistakes in complicated structural analysis, provided that the incorrect stress resultants so determined are in equilibrium with the design loads.

Uniqueness of the Plastic Collapse Load Factor

The load factor s_c at plastic collapse, established by the static LP, is identical with the value for s_c determined by the kinematic LP. Uniqueness of the collapse load factor therefore follows from strong duality in linear programming; this insists that there is no duality gap.

> If the static LP has an optimal solution, then so does the kinematic LP, and Min z = Max $w = s_c$.

Since the static and kinematic LPs are equivalent to the corresponding LCP which forms the governing relations for plastic collapse, there is but one value $s = s_c$ which solves the LCP. The governing relations which comprise the LCP are of three types: relations of static admissibility which determine the feasible solutions of the static LP; relations of kinematic admissibilty which determine the feasible solutions of the kinematic LP; and the condition $y^T\dot{\lambda} = 0$ – the only constitutive condition that relates the static variables to the kinematic variables. Only *optimal* solutions of the static and kinematic LPs obey the last condition.

The stress resultants or element forces which appear in an optimal solution of the static LP are not necessarily unique. If there are alternative optima, they form a convex set of optimal solution vectors, and one of these vectors will represent the actual set of element forces which obtain at plastic collapse. Generally, the actual element forces will depend upon the previous history of elasto-plastic deformations, including initial strains and self-stresses, and hence upon the load path that terminates in plastic collapse. Such information is not required for plastic limit analysis in which the aim is to determine the capacity of a structure to carry loads.

The collapse mechanism revealed in an optimal solution of the kinematic LP is also not necessarily unique. If there are alternative optima, they form a convex set of optimal solution vectors which represents a collapse mechanism with many degrees of freedom.

COMPUTATION

Choice of linear program

If the LPs for plastic limit analysis are to be solved by the Simplex algorithm, then the computational effort will increase rapidly with M, the number of constraints, and will increase much less rapidly with N, the number of variables associated with the LP in standard form. The kinematic LP always has fewer constraints than the corresponding static LP; so we should choose to solve the kinematic LP.

For solution by the Simplex algorithm, the kinematic LP must be in standard form. The kinematic mesh LP(5.34) is already in standard form, and the other two LPs (5.36) and (5.38) may be brought into this form by conversion of their unrestricted variables to non-negative ones.

The kinematic mesh LP(5.34) is by far the least demanding of computational effort for framed structures, having $M = 1 + \alpha$ constraints. On the other hand, the augmented

The kinematic mesh LP(5.34) is by far the least demanding of computational effort for framed structures, having $M = 1 + \alpha$ constraints. On the other hand, the augmented kinematic nodal LP(5.38) in system coordinates – with $M = 1 + \mathrm{dM}$ constraints – requires a much higher computational effort, but it has the potential for automatic data generation. One possible compromise measure is to set up the data for an augmented nodal description, and then to transform it into a mesh description.

None of the static LPs is in standard form. All their variables are unrestricted and all, excepting the load factor s, must be replaced by non-negative variables. Although s is unrestricted, the LP process of maximisation will always find a positive value of s, provided that such is feasible. Furthermore, each static LP has Y yield conditions, expressed as inequalitites; they require the addition of Y slack variables.

Use of Simplex multipliers

For economy of effort, we choose to solve one of the kinematic LPs. Its optimal solution gives the unique plastic collapse load factor s_c and an associated mechanism of plastic collapse which is not necessarily unique. To obtain an optimal solution of the dual static LP, we do not have to solve it; the Simplex multipliers of the optimal solution to the kinematic LP give us directly a set of optimal static variables from which the element forces can be inferred. They, of course, are also not necessarily unique; but they are safe – the structure, designed to sustain these element forces, will not collapse at a load factor less than s_c.

Inefficiency of the full Simplex algorithm

One of the possible sources of inefficiency in the Simplex algorithm is that of initiating from an artificial basis – Phase I of the full algorithm – in order to find a first basic feasible solution. In the kinematic LP, the basic feasible solutions are possible collapse mechanisms, and, in many cases, the variables which would be basic in such feasible solutions are easily identified. Any such means of advancing the solution is usually of considerable benefit. A better solution process would take more account of the special structure of the kinematic or static LP so that the algorithm can operate on a more compact tableau, Teixeira de Freitas[7].

EXAMPLES

A Propped Cantilever

We shall consider first a trivial problem, one that can easily be solved by hand calculations. Our purpose is to clarify the setting up of the various matrices which are required by the foregoing linear programming formulations of plastic limit analysis. Let the propped cantilever shown in Figure 5.4 be considered as an assemblage of three members, those members being identified by the index m = (1),(2),(3). Each of the three potential plastic zones or critical sections – indexed by k = 1,2,3 – is able to develop only the rotational deformation rate $\dot{q}_{P3k}$ of a simple plastic hinge which has plastic moment of resistance M_p for bending moments of either sense. Member (1) therefore has two ($s_m = 2$) independent deformation rates $\dot{q}_{I1(1)}$, $\dot{q}_{I2(1)}$; member (2) has none; and member (3) just one such parameter $\dot{q}_{I1(3)}$.

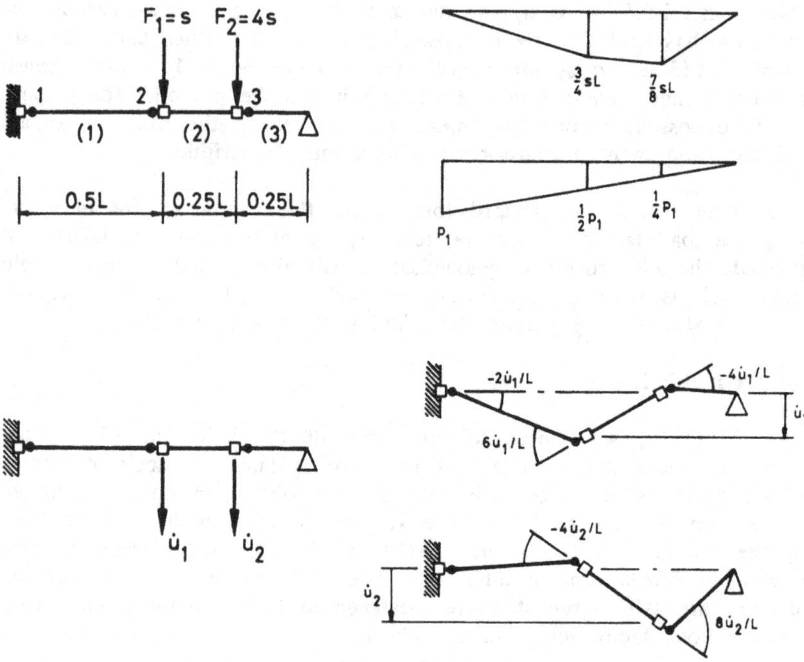

Figure 5.4

For this arrangement of members and critical sections, the matrices **T** and **N** and the vector $\mathbf{R}^T$ of plastic capacities have the following simple form:

$$\begin{bmatrix} \dot{q}_{I1(1)} \\ \dot{q}_{I2(1)} \\ \dot{q}_{I1(3)} \end{bmatrix} = \begin{bmatrix} 1 & 0 & 0 \\ 0 & 1 & 0 \\ 0 & 0 & 1 \end{bmatrix} \begin{bmatrix} \dot{q}_{P31} \\ \dot{q}_{P32} \\ \dot{q}_{P33} \end{bmatrix} \qquad \begin{bmatrix} \dot{q}_{P31} \\ \dot{q}_{P32} \\ \dot{q}_{P33} \\ \dot{D}/M_p \end{bmatrix} = \begin{bmatrix} 1 & 0 & 0 & -1 & 0 & 0 \\ 0 & 1 & 0 & 0 & -1 & 0 \\ 0 & 0 & 1 & 0 & 0 & -1 \\ 1 & 1 & 1 & 1 & 1 & 1 \end{bmatrix} \begin{bmatrix} \dot{\lambda}_1 \\ \dot{\lambda}_2 \\ \dot{\lambda}_3 \\ \dot{\lambda}_4 \\ \dot{\lambda}_5 \\ \dot{\lambda}_6 \end{bmatrix}$$

where the plastic multiplier rates $\dot{\lambda}_1$, $\dot{\lambda}_2$, $\dot{\lambda}_3$ give the plastic hinge rotation rates induced by positive bending moments at the three critical sections, and $\dot{\lambda}_4$, $\dot{\lambda}_5$, $\dot{\lambda}_6$ give the rotation rates at the same sections due to negative bending moments.

A mesh description for the structure is obtained from the load-equilibrating and hyperstatic bending moment diagrams given in Figure 5.4. The static indeterminacy number has the value $\alpha = 1$, and the single hyperstatic force p_1 is taken as the bending moment at the left hand end of the cantilever. Since the development of plastic deformation rates in the mechanism of plastic collapse is governed entirely by the bending moments Q_{3k} at the critical sections, only these values need be reported in the mesh matrices.

Alternatively, a nodal description for the structure may be constructed in the following manner. Since there are $s_m = 2$ independent member deformation rates for member (1) and $s_m = 1$ for member (3), the kinematic indeterminacy number for this modelling has the value $\beta = 2$; consequently, two hyperkinematic or independent nodal displacement rates $\dot{u}_1$ and $\dot{u}_2$ may be selected and the associated basic mechanisms or velocity profiles devised as shown in Figure 5.4.

The matrices b_0 and B of the mesh description and a_0^T and A of the nodal description are therefore assembled with the following form:

$$
\begin{bmatrix} Q_{11}(1) \\ Q_{12}(1) \\ Q_{11}(3) \end{bmatrix} = \begin{bmatrix} 0.000 \\ 0.750 \\ 0.875 \end{bmatrix} sL + \begin{bmatrix} 1.000 \\ 0.500 \\ 0.250 \end{bmatrix} P_1
$$

$$
\begin{bmatrix} \dot{w}/L \\ \dot{q}_{11}(1) \\ \dot{q}_{12}(1) \\ \dot{q}_{11}(3) \end{bmatrix} = \begin{bmatrix} 1 & 4 \\ -2 & 0 \\ 6 & -4 \\ -4 & 8 \end{bmatrix} \begin{bmatrix} \dot{u}_1/L \\ \dot{u}_2/L \end{bmatrix}
$$

When introduced into the appropriate kinematic LP (5.34) or (5.36), these data give the plastic limit state as characterised by

$$s_c = 1.4286 \, M_p/L, \quad \dot{\lambda}_3 = 1.1429/L, \quad \dot{\lambda}_4 = 0.2857/L, \quad \dot{u}_1 = 0.1429, \quad \dot{u}_2 = 0.2143$$

The appearance of $\dot{\lambda}_3$ and $\dot{\lambda}_4$ identifies a positive rotation rate at critical section 3 and a negative rotation rate at critical section 1 respectively; the plastic collapse mechanism is shown in Figure 5.5. In this same figure appears the bending moment diagram given by the Simplex multipliers for the plastic limit state.

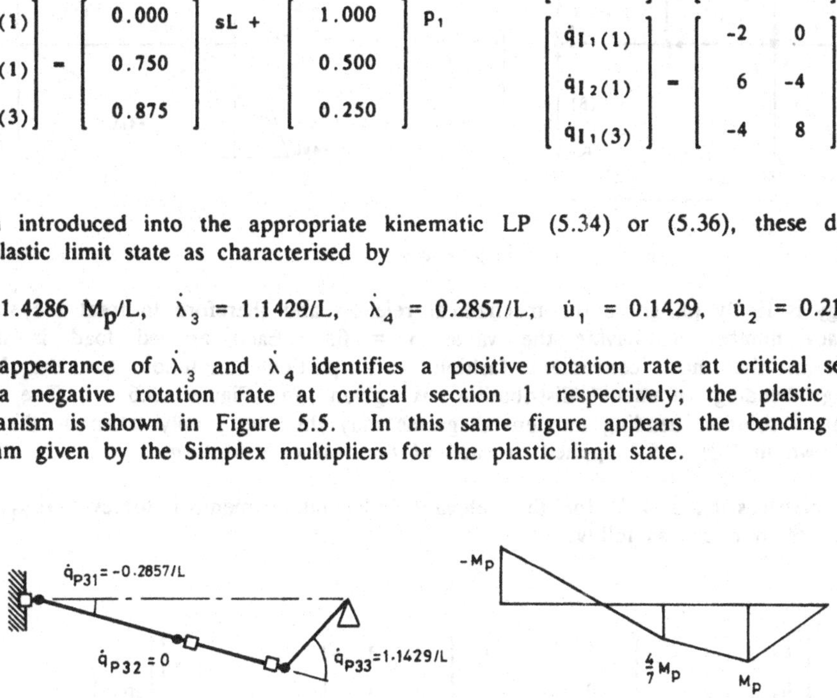

Figure 5.5

A Two-storey Frame with Hinged Column Bases

The second example illustrates the convenience of a cellular mesh description for a structure in which the geometrical layout exhibits a repetitive cellular form. The example also shows how this special mesh description can allow the incorporation of mechanical releases or articulations such as hinged column bases.

A two-storey planar frame is shown in Figure 5.6. Its geometrical form comprises two rectangular cells, and the sections R1 and R2 are the sites of bending moment releases.

The frame may be considered as an assemblage of eight rectilinear beam elements. Again, it is assumed that at any of the ten critical sections, numbered consecutively in the figure, a simple flexural plastic hinge may occur; for each such hinge, the plastic moment of resistance is M_p in response to bending moments of either sense.

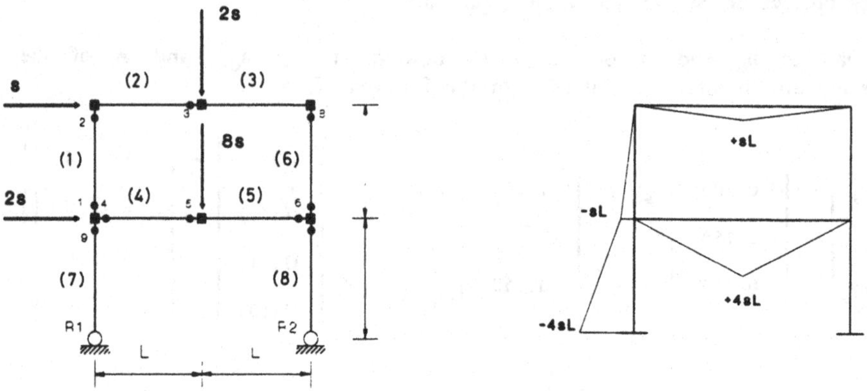

Figure 5.6

The strategy is firstly to ignore the mechanical releases and therefore to treat the static indeterminacy number as having the value $\alpha = 6$. Each applied load is then equilibrated in the most convenient fashion – a particular solution for the load equilibrating bending moment distribution is given in Figure 5.6. The six complementary solution bending moment diagrams may then be simply taken as the set of three shown in Figure 5.7 applied to each of the two cells in turn.

The mesh matrices $\mathbf{b_o}$ and $\mathbf{B}$ for the relevant independent member forces $Q_{Ij(m)}$ of member m are therefore as follows:

$$
\begin{bmatrix} Q_{I1(1)} \\ Q_{I2(1)} \\ Q_{I2(2)} \\ Q_{I1(4)} \\ Q_{I2(4)} \\ Q_{I2(5)} \\ Q_{I1(6)} \\ Q_{I2(6)} \\ Q_{I1(7)} \\ Q_{I2(7)} \\ Q_{I1(8)} \\ Q_{I2(8)} \end{bmatrix}
=
\begin{bmatrix} -1 \\ 0 \\ 1 \\ 0 \\ 4 \\ 0 \\ 0 \\ 0 \\ -4 \\ -1 \\ 0 \\ 0 \end{bmatrix} sL +
\begin{bmatrix} 1 & -1 & -1 & & & \\ 1 & -1 & 1 & & & \\ 1 & 0 & 1 & & & \\ -1 & 1 & 1 & 1 & -1 & 1 \\ -1 & 0 & 1 & 1 & 0 & 1 \\ -1 & -1 & 1 & 1 & 1 & 1 \\ -1 & 1 & 1 & & & \\ -1 & -1 & -1 & & & \\ & & & 1 & -1 & -1 \\ & & & 1 & -1 & 1 \\ & & & -1 & -1 & 1 \\ & & & -1 & -1 & -1 \end{bmatrix}
\begin{bmatrix} p_1 \\ p_2 \\ p_3 \\ p_4 \\ p_5 \\ p_6 \end{bmatrix}
$$

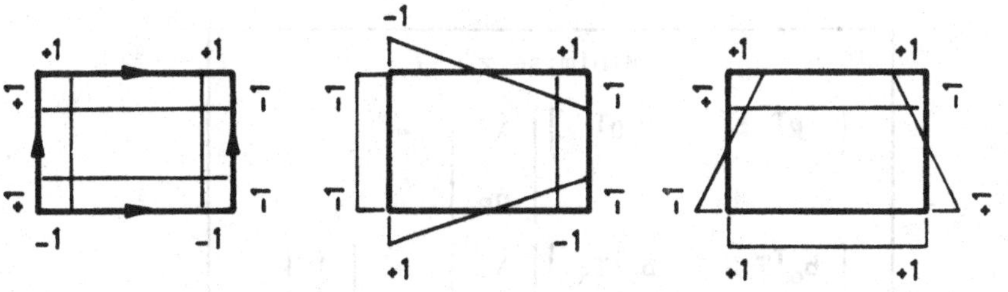

$$p_1 = 1 \text{ or } p_4 = 1 \qquad p_2 = 1 \text{ or } p_5 = 1 \qquad p_3 = 1 \text{ or } p_6 = 1$$

Figure 5.7

In addition, the relations $Q = T^T Q_I$ and $Q_R = T_R{}^T Q_I$, which transform the independent member forces $Q_{Ij(m)}$ into the relevant bending moments Q_{3k} at the plastic hinge sections and the appropriate bending moments Q_{R_1} and Q_{R_2} at the mechanical releases are:

$$
\begin{bmatrix}
Q_{31} \\
Q_{32} \\
Q_{33} \\
Q_{34} \\
Q_{35} \\
Q_{36} \\
Q_{37} \\
Q_{38} \\
Q_{39} \\
Q_{3,10} \\
Q_{R_1} \\
Q_{R_2}
\end{bmatrix}
=
\begin{bmatrix}
1 & 0 & 0 & 0 & 0 & 0 & 0 & 0 & 0 & 0 & 0 & 0 \\
0 & 1 & 0 & 0 & 0 & 0 & 0 & 0 & 0 & 0 & 0 & 0 \\
0 & 0 & 1 & 0 & 0 & 0 & 0 & 0 & 0 & 0 & 0 & 0 \\
0 & 0 & 0 & 1 & 0 & 0 & 0 & 0 & 0 & 0 & 0 & 0 \\
0 & 0 & 0 & 0 & 1 & 0 & 0 & 0 & 0 & 0 & 0 & 0 \\
0 & 0 & 0 & 0 & 0 & 1 & 0 & 0 & 0 & 0 & 0 & 0 \\
0 & 0 & 0 & 0 & 0 & 0 & 1 & 0 & 0 & 0 & 0 & 0 \\
0 & 0 & 0 & 0 & 0 & 0 & 0 & 1 & 0 & 0 & 0 & 0 \\
0 & 0 & 0 & 0 & 0 & 0 & 0 & 0 & 0 & 1 & 0 & 0 \\
0 & 0 & 0 & 0 & 0 & 0 & 0 & 0 & 0 & 0 & 0 & 1 \\
0 & 0 & 0 & 0 & 0 & 0 & 0 & 0 & 1 & 0 & 0 & 0 \\
0 & 0 & 0 & 0 & 0 & 0 & 0 & 0 & 0 & 0 & 1 & 0
\end{bmatrix}
\begin{bmatrix}
Q_{I_1}(1) \\
Q_{I_2}(1) \\
Q_{I_2}(2) \\
Q_{I_1}(4) \\
Q_{I_2}(4) \\
Q_{I_2}(5) \\
Q_{I_1}(6) \\
Q_{I_2}(6) \\
Q_{I_1}(7) \\
Q_{I_2}(7) \\
Q_{I_1}(8) \\
Q_{I_2}(8)
\end{bmatrix}
$$

The simple form of the mesh matrices given above is made possible by our willingness to impose the necessary conditions $Q_{R_1} = 0$ and $Q_{R_2} = 0$ as supplementary constraints in the static LP of plastic limit analysis. In fact, for the usual reasons of computational efficiency, we choose to solve the kinematic LP of plastic limit analysis; for the mesh description, incoporating mechanical releases, it may easily be shown to have the general form:

$$\text{Minimise } z$$

$$\begin{bmatrix} R^T & 0^T \end{bmatrix} \begin{bmatrix} \dot{\lambda} \\ \dot{q}_R \end{bmatrix} = z$$

$$\begin{bmatrix} b_o{}^T T\, N & b_o{}^T T_R \\ B^T\, T\, N & B^T\, T_R \end{bmatrix} \begin{bmatrix} \dot{\lambda} \\ \dot{q}_R \end{bmatrix} = \begin{bmatrix} 1 \\ 0 \end{bmatrix}$$

$$\dot{\lambda} > 0 \qquad (5.41)$$

KINEMATIC MESH LP WITH INCORPORATED RELEASES

In LP(5.41), the vector 0^T in the objective function derives from the right-hand side of the supplementary constraints $Q_R = 0$ which appear in the dual static LP; and these same dual constraints are associated with the primal variables $\dot{q}_R$, which are the rotation rates in the mechanical releases.

The solution of LP(5.41) for this problem produces the plastic collapse mechanism shown in Figure 5.8 at a load factor $s_c = 0.4545\ M_p/L$. The Simplex multipliers associated with this optimal solution identify the bending moment distribution shown in the same figure as a possible one for plastic collapse. This bending moment distribution is not uniquely determined at collapse since the plastic hinge at section 2 does not participate in the collapse mechanism and the bending moment there need not be at the full plastic value.

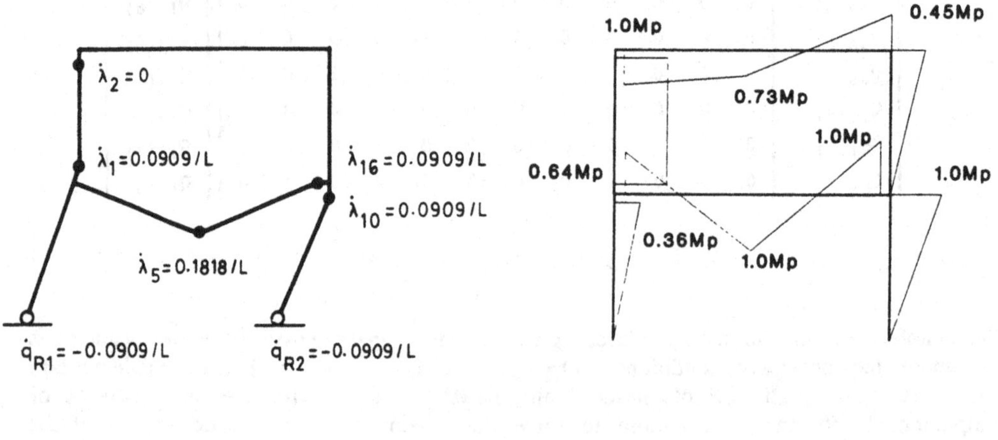

Figure 5.8

RESTRICTIONS

1. In plastic limit analysis, it is assumed that the structural material will exhibit perfect plasticity at the state of plastic collapse. When this state occurs, just prior to the impressing of a mechanism motion, the programme of loading which has brought the structure to the current state will have induced some *total* plastic deformations at the critical sections. If, at this level of total plastic deformation, any of the sections is suffering degradation of strength, or plastic softening, then the structure may be unable to mobilise the full load carrying capacity predicted by plastic limit analysis. When the structural material has only a very limited ductility, we may be unable to predict the strength of the structure with sufficient confidence unless we account for the total elastic and plastic deformations.

2. If the plasticity relations do not have an associated, standard or normal flow rule, then the governing LCP is not symmetric and the usual complementary limit theorems do not exist. Constitutive laws which are required to represent a frictional component of material behaviour have, in general, non-standard flow rules. An excellent mathematical discussion of Coulomb friction in relation to soil mechanics is given by Salençon[8].

3. A structure may be subjected to many repetitions of a loading programme. Even if one application of the programme does not cause instantaneous plastic collapse, successive repetitions may produce total plastic deformations which constitute a ratchetting motion or incremental collapse. On the other hand, if the level of loading in the programme is sufficiently restricted, the structure may develop plastic deformations in the first few repetitions, but its subsequent response will be entirely elastic; the structure is said to *shake down*. These matters involve an interchange of elastic and plastic deformations and cannot therefore be addressed through plastic limit analysis.

4. A structure can fail – or lose its capacity to carry additional loading – at a load factor much less than that predicted by plastic limit analysis through the development of instability. For a static structure, stability is a quality associated with its equilibrium; the assessment of such a quality requires that the laws of equilibrium be accurately founded upon the geometry of the current displaced configuration of the structure. Equilibrium must now involve displacements, and, if the displacements of the current configuration are large, the laws of statics and of kinematics will be nonlinear.

<u>REFERENCES</u>

1. Neal, B. G. and Symonds, P. S., The calculation of collapse loads for framed structures, J. Inst. Civil Engineers, **35** (1950) 21.

2. Neal, B. G. and Symonds, P. S., The rapid calculation of the plastic collapse load for a framed structure, Proc. Inst. Civil Engineers, **1** (1952) 58.

3. Baker, J. F., Horne, M. R. and Heyman, J., The Steel Skeleton, vol. 2, Cambridge University Press 1956.

4. Neal, B. G., The Plastic Methods of Structural Analysis, *2nd* Edition, Chapman and Hall 1963.

5. Maier, G. and Munro, J., Mathematical programming applications to engineering plastic analysis, Applied Mech. Rev., **35** (1982) 1631–1643.

6. Maier, G., Linear flow-laws of elastoplasticity: a unified general approach, Rendic. Acad. Naz. Lincei, Series 8, **47** (1969) 266.

7. Teixeira de Freitas, J. A., An efficient Simplex method for the limit analysis of structures, Computers & Structures, **21** (1985) 1255–1265.

8. Salençon, J., Applications of the Theory of Plasticity in Soil Mechanics, Wiley 1977.

CHAPTER 6

PIECEWISE-LINEAR ELASTIC-PLASTIC STRESS-STRAIN RELATIONS

J. A. Teixeira de Freitas
Istituto Superior Tecnico, Lisbon, Portugal

Abstract: A matrix description for piecewise-linear models of inviscid, isothermal plastic behaviour is presented. A geometric interpretation of the plasticity conditions is used to illustrate the role of the intervening constitutive operators. A matrix description for incremental elastic-plastic stress-strain relations is presented and processed next through mathematical programming theory to illustrate its capacity for generating variational interpretations and statements of qualification regarding the existence, uniqueness and stability of elastoplastic solutions.

Introduction

The description of the material behaviour is one of the essential ingredients in the formulation of structural mechanics problems.

The selection of a model for the constitutive relations results from a balanced compromise between realism and simplicity. The model must be so chosen as to enable the analyst to simulate the features of the actual response he deems relevant. On the other hand the mathematical encodement of such a model must embody the simplicity required for developing useful methods of numerical implementation.

The linear elastic model, the foundations of which are due to Cauchy [1], is now widely used in structural engineering. The designs thus produced are safe but quite often overexpensive in consequence of not exploiting the inelastic behaviour exhibited by most structural materials.

The main objective of this paper is to present some models of inviscid, isothermal plastic behaviour.

Plasticity is herein treated from a phenomenological standpoint, as only the macroscopic aspects of the behaviour are to be modelled.

In the phenomenological theory of plasticity, which can be traced back to Tresca [2] and Guest [3], the material is assumed to be macroscopically homogeneous, isotropic and free of stress and strain in its virgin state. A review of the most relevant microscopic and thermodynamic aspects of plasticity has been recently presented by Dang [4].

A confrontation of the stress- and strain-based approaches to plasticity, in the manners of Drucker [5] and Ilyushin [6], respectively, can also be found in Dang [4], who recommends Zycznoski's bibliographic review [7] on the available literature on plasticity.

A stress-based approach is used in the formulation presented next. It rests heavily on previous works by Maier [8,9,10] and De Donato [11], who proposed a general and quite powerful matrix description of Koiter's theory of plasticity [12].

The ability of the classical theories of plasticity to describe realistically the plastic deformation of metals is nowadays generally accepted [13]. Certain recommendations [14] do however restrict their use in modelling the response of structural concrete.

Fundamental Phenomenological Aspects

The fundamental information on the mechanical properties of building materials is obtained through experiment. The macroscopic behaviour of the specimen thus tested is processed to set up empirical formula which are thenceforth used to model the constitutive relations of the structural materials.

The stress-strain diagrams obtained experimentally can be very sensitive to the testing conditions. There are however some fundamental characteristics that are found to typify certain classes of materials.

The stress-strain diagram represented in figure 1 exhibits the distinguishing phases in the response of the most commonly used structural materials.

In the first phase of the response, defined by line OAB, the material is able to regain completely its original dimensions upon removal of the perturbation responsible for the development of stress and strain. This phase in the response is termed elastic as it implies the absence of any permanent deformations.

If the relationship between stress and strain is essentially linear, the material is said to be linear elastic. If it displays a certain curvature, the material is said to be nonlinear elastic. Most structural materials exhibit a distinct, initial linear elastic behaviour.

In the second, inelastic phase in the response of the material, important phenomena are mobilized if straining is not monotonically increased.

If unstressing takes place, as depicted by line CD in figure 1, the response is essentially identical to that of the initial elastic phase. However a residual irreversible deformation sets in. This phase in the response of material is thus termed plastic, as it implies the presence of permanent deformations.

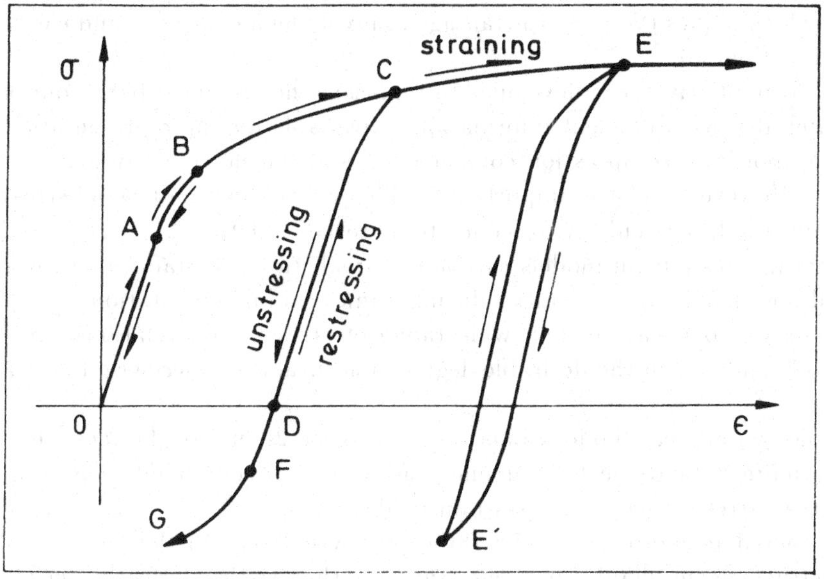

Figure 1

If restressing is implemented, the elastic phase is recovered and followed by the continuation of the original curve if further straining takes place, as depicted by line DCE in the graph presented in figure 1.

However, if further unstressing is implemented instead, as in path DFG, it is often found that the inelastic range is regained before exhausting fully the initial elastic capacity of the material, thus exposing the Bauschinger effect.

Yet another phenomena associated with inelastic materials is the hysteresis effect. When the material is stressed cyclically into the inelastic range, there is a tendency for small hysteresis loops EE' to develop, for instance along a line such as CD in figure 1.

Idealized Stress-Strain Diagrams

The analytical representation of the behaviour of structural materials requires the replacement of the experimentally determined stress-strain diagrams by simpler models, in which the fundamental phenomenological aspects are preserved. A review of the response of metals and alloys to different uniaxial and biaxial loading histories reported in the literature has been presented by Rees [15].

A wide range of stress-strain relations can be represented by the Ramberg-Osgood formula, which enjoys the major advantage of involving a continous and easily differentiable function [16].

Other formula have been developed to represent the nonlinear behaviour of structural materials [17]. Sufficient information is available for the representation of the uniaxial response of compression concrete [18] and tensile cold worked steel specimens [19]. However a host of important problems involving multi-axial stress states remains unsolved, particularly for concrete materials [20,21].

Nonlinear stress-strain models can be over-realistic for certain practical applications and pose particular difficulties during numerical implementation.

If these are to be avoided, a wide range of stress-strain relations can still be represented, and within the desirable degree of accuracy, by piecewise-linear approximations.

The piecewise-linear models, illustrated in figure 2, endure the inconveniency of requiring more data to be fully defined, as well as specific control procedures to identify the currently active stress-strain modes.

Represented in figure 3 are the three piecewise-linear models most commonly used in structural mechanics problems, namely the elastic, perfectly plastic model for mild steel, the elastic, linearly hardening model for cold worked steel, and the elastic, linearly softening model for concrete materials.

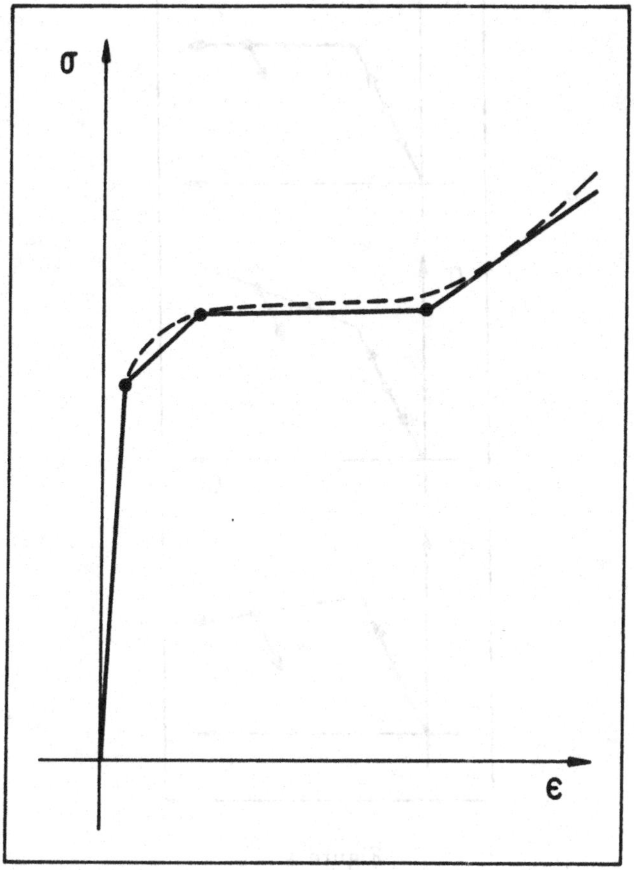

Figure 2

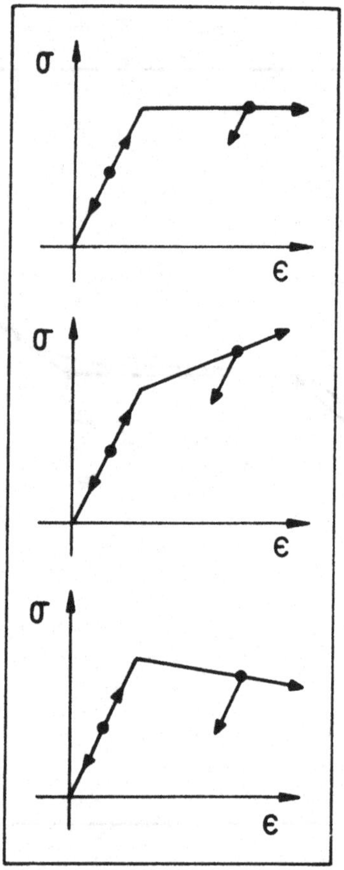

Figure 3

Piecewise-linear Stress-Strain Relations

Consider the piecewise stress-strain diagram represented in figure 4.

For simplicity of the presentation, it is assumed that the elastic phase is described by a single mode and that two modes suffice to represent the plastic range; σ^+ and σ^- denote the yield stress in tension and compression, respectively.

Each plastic mode is represented by the limiting stress, σ^+, and the associate

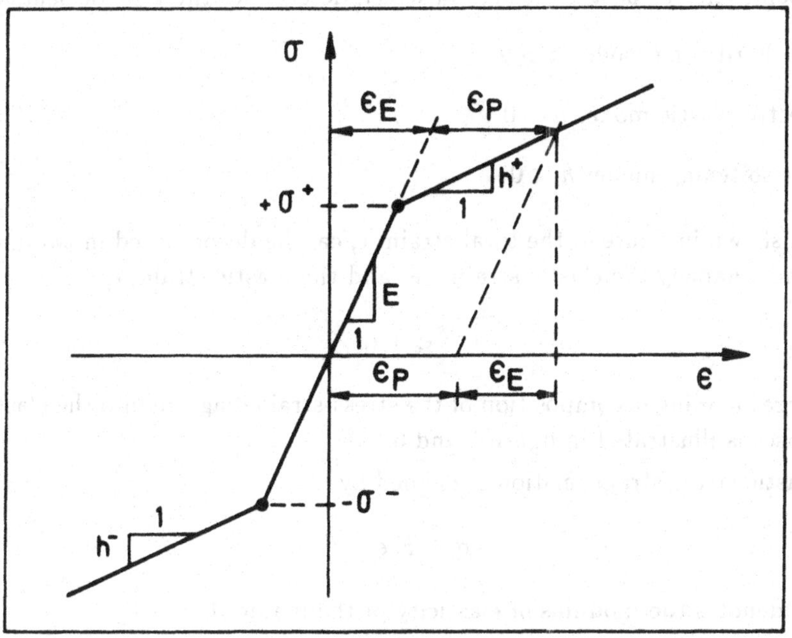

Figure 4

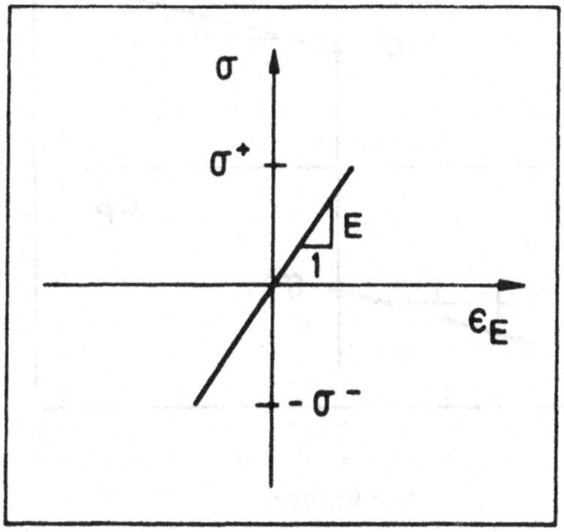

Figure 5

stiffness coefficient, h^+ or h^-, the sign of which determines the type of behaviour:

- linear hardening mode: $h > 0$
- perfectly plastic mode: $h = 0$
- linear softening mode: $h < 0$

As it is shown in figure 4, the total strain, ϵ, can be decomposed in two independent addends, namely the elastic strain, ϵ_E, and the plastic strain, ϵ_p:

$$\epsilon = \epsilon_e + \epsilon_p \tag{1}$$

The corresponding decomposition of the stress-strain diagram into the elastic and plastic phases is illustrated in figure 5 and 6.

The elastic stress-strain relation is defined by

$$\sigma = E\,\epsilon_e \tag{2}$$

wherein E denotes the modulus of elasticity of the material.

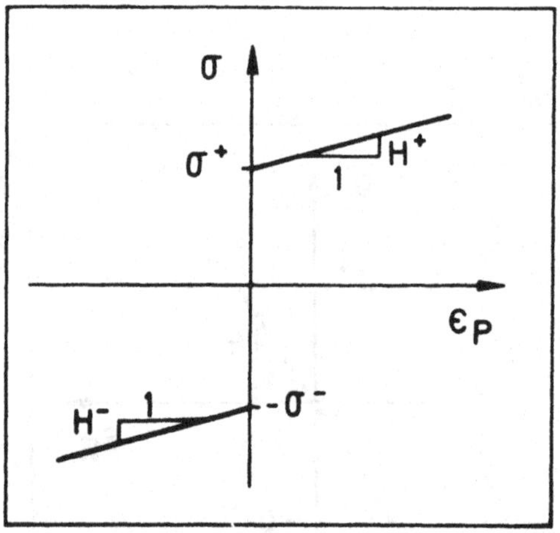

Figure 6

The plastic phase of the stress-strain relations can be expressed as follows:

$$\begin{aligned} -\sigma^- \leq \sigma \leq +\sigma^+ \quad &\text{if} \quad \epsilon_p = 0; \\ \sigma = h^+ \epsilon_p + \sigma^+ \quad &\text{if} \quad \epsilon_p > 0; \\ \sigma = h^- \epsilon_p - \sigma^- \quad &\text{if} \quad \epsilon_p < 0. \end{aligned} \tag{3}$$

Parameters h^+ and h^- denote the strain-hardening coefficients associated with the tension and compression yield modes, respectively. Simple geometric considerations show that they are defined thus:

$$h^\pm = \frac{EH^\pm}{E - H^\pm} \quad . \tag{4}$$

Let the plastic strain addend be represented by auxilliary variables, the plastic parameters ϵ^+ and ϵ^- associated with the tension and compression yield modes as shown in figure 7.

These parameters, which are organized in array

$$\epsilon_* = \begin{Bmatrix} \epsilon^+ \\ \epsilon^- \end{Bmatrix}, \tag{5}$$

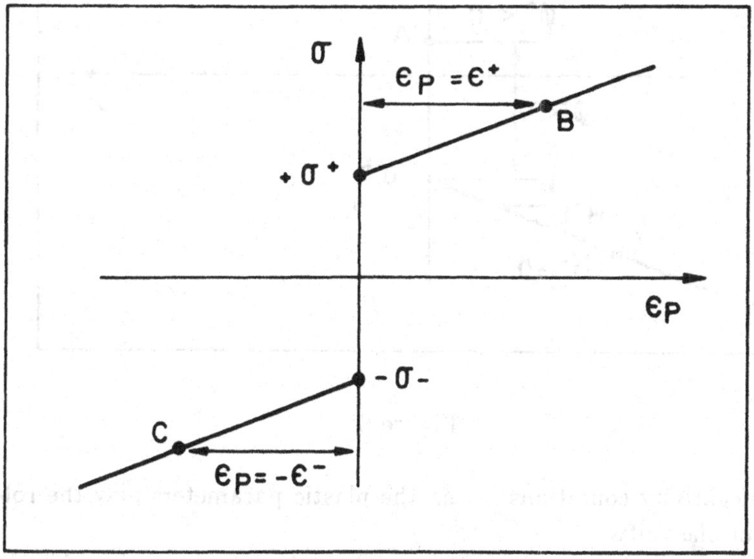

Figure 7

represent the accumulated plastic strain in fibers under tension and compression, respectively. By definition, they can not take negative values:

$$\epsilon_* \geq 0. \tag{6}$$

The accumulated plastic strain may now be expressed in form

$$\epsilon_p = \mathbf{V}\,\epsilon_*, \tag{7}$$

wherein

$$\mathbf{V} = \begin{bmatrix} +1 & -1 \end{bmatrix},$$

so that

$$\epsilon_p = -\epsilon^- \ , \ \epsilon^+ = 0 \ , \ \text{if} \ \ \epsilon_p < 0,$$

and

$$\epsilon_p = +\epsilon^+ \ , \ \epsilon^- = 0 \ , \ \text{if} \ \ \epsilon_p > 0,$$

as illustrated by points B and C in figure 8.

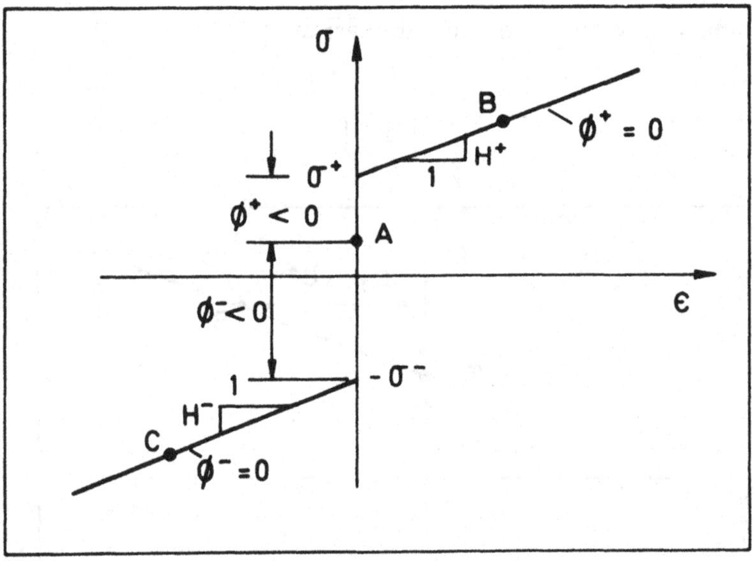

Figure 8

The non-negativity conditions (6) on the plastic parameters play the role of the flow rule (3) of plasticity.

An alternative description for conditions (3) can be found by using yield functions associated with the tension and compression yield modes, as shown in figure 8. These functions are defined by

$$\phi^+ = +\sigma - h^+ \epsilon^+ - \sigma^+ \quad , \quad \phi^- = -\sigma - h^- \epsilon^- - \sigma^-, \tag{8}$$

or, in matrix form,

$$\phi_* = \check{\mathbf{N}}\sigma - \mathbf{H}\,\epsilon_* - \sigma_*,$$

with

$$\phi_* = \begin{Bmatrix} \phi^+ \\ \phi^- \end{Bmatrix}, \quad \sigma_* = \begin{Bmatrix} \sigma^+ \\ \sigma^- \end{Bmatrix},$$

and

$$\mathbf{N} = \begin{bmatrix} +1 & -1 \end{bmatrix} \quad , \quad \mathbf{H} = \begin{bmatrix} h^+ & 0 \\ 0 & h^- \end{bmatrix}.$$

If, as required by condition (4), neither of the yield modes is active

$$\epsilon^+ = 0 \quad \epsilon^- = 0, \tag{9}$$

it suffices to require

$$\phi^+ < 0\,, \quad \phi^- < 0, \tag{10}$$

to recover condition (3), as illustrated by point A in figure 8.

Consider now point B in the same figure. If the tension yield mode is active

$$\epsilon^+ > 0\,, \quad \epsilon^- = 0, \tag{11}$$

the associate plastic function has to be set to zero

$$\phi^+ = 0\,, \quad \phi^- < 0, \tag{12}$$

in order to recover condition (3).

Similarly, conditions (3) and (4) for the activation of the compression yield mode can be expressed thus:

$$\epsilon^+ = 0\,, \quad \epsilon^- > 0, \tag{13}$$

$$\phi^+ < 0\,, \quad \phi^- = 0. \tag{14}$$

The non-positivity conditions (10), (12) and (14) in the plastic potentials (15)

$$\phi_* \leq \mathbf{0}, \tag{15}$$

play therefore the role of the yield rule (5) of plasticity.

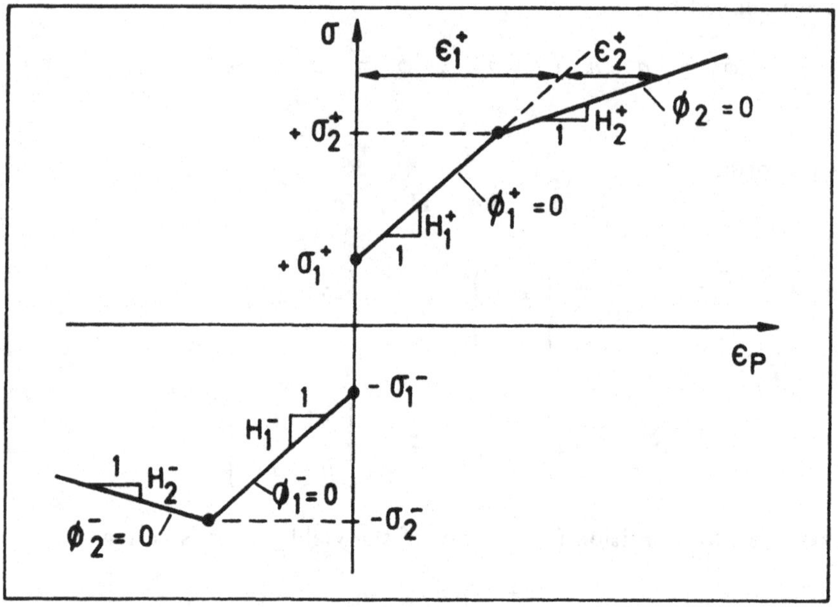

Figure 9

Moreover, results (9), (11) and (13) and (10), (12) and (14) show clearly that the plastic potentials and the correspondent plastic parameters are associated through the following complementarity conditions,

$$\phi^{+}\ \epsilon^{+} = 0\ ,\ \phi^{-}\ \epsilon^{-} = 0,$$

which can be written in form

$$\tilde{\phi}_{*}\ \epsilon_{*} = 0, \tag{16}$$

if conditions (6) and (15) are taken into account.

The aforepresented formulation can be easily generalized to describe multi-phase stress-strain diagrams.

In definition (7) the plastic parameters vector is now defined by

$$\epsilon_{*} = \left\{\ \epsilon_1^{+}\ \ \epsilon_2^{+}\ \ \ldots\ \ \epsilon_1^{-}\ \ \epsilon_2^{-}\ \ \ldots\ \right\},$$

with

$$\mathbf{V} = \begin{bmatrix} 1 & & & & -1 & & & \\ & 1 & & & & -1 & & \\ & & 1 & & & & -1 & \\ & & & 1 & & & & -1 \end{bmatrix}.$$

In definition (8) for the plastic functions vector

$$\phi_\star = \left\{ \phi_1^+ \quad \phi_2^+ \quad \ldots \quad \phi_1^- \quad \phi_2^- \quad \ldots \right\},$$

the associate yield stress vector is defined by

$$\sigma_\star = \left\{ \sigma_1^+ \quad \sigma_2^+ \quad \ldots \quad \sigma_1^- \quad \sigma_2^- \quad \ldots \right\},$$

with

$$\mathbf{N} = \begin{bmatrix} 1 & & & & -1 & & & \\ & 1 & & & & -1 & & \\ & & 1 & & & & -1 & \\ & & & 1 & & & & -1 \end{bmatrix},$$

and

$$\mathbf{H} = \begin{bmatrix} H_1^+ & & & & & & & \\ & H_2^+ & & & & & & \\ & & H_3^+ & & & & & \\ & & & \ldots & & & & \\ & & & & H_1^- & & & \\ & & & & & H_2^- & & \\ & & & & & & H_3^- & \\ & & & & & & & \ldots \end{bmatrix}.$$

Lagrangian Description of the Constitutive Relations

Assume now that n components σ_i, listed in the stress vector σ, are required to represent the state of stress developing in the test specimen, and let the associate strains ϵ_i be grouped in the strain vector ϵ.

Decomposition (1) of the strain field into elastic and plastic addends is now expressed in the following vector notation:

$$\boxed{\epsilon = \epsilon_e + \epsilon_p} \tag{17}$$

The elastic phase (2) of the constitutive relations generalizes into form (18a), wherein the stiffness matrix K collects the relevant elastic constants.

Lagrangian Description of Elasticity	
$\sigma = \mathbf{K}\ \epsilon_e \quad (a)$	$\epsilon_e = \mathbf{F}\ \sigma \quad (b)$
Stiffness	Flexibility

(18)

As matrix K is non-singular, the elastic phase can be represented in the alternative flexibility description (18), wherein

$$\mathbf{F} = \mathbf{K}^{-1}, \tag{19}$$

is the associate flexibility matrix.

Reciprocity in the interaction between elastic modes, which is typical of hookian materials, is represented by symmetric elastic causality operators:

$$\mathbf{K} = \tilde{\mathbf{K}}\ ,\ \ \mathbf{F} = \tilde{\mathbf{F}}. \tag{20}$$

Summarized in system (21) is the generalization of the plasticity relations (6), (7), (8), (15) and (16).

Lagrangian Description of Plasticity		
$\begin{bmatrix} -\mathbf{H} & \tilde{\mathbf{N}} \\ \mathbf{V} & \mathbf{0} \end{bmatrix} \begin{Bmatrix} \epsilon_\star \\ \sigma \end{Bmatrix} = \begin{Bmatrix} \phi_\star \\ \epsilon_p \end{Bmatrix} + \begin{Bmatrix} \sigma_\star \\ \mathbf{0} \end{Bmatrix} \begin{matrix} (a) \\ (b) \end{matrix}$		
$\phi_\star \leq \mathbf{0} \quad (c)$	$\tilde{\phi}\ \epsilon_\star = 0 \quad (d)$	$\epsilon_\star \geq \mathbf{0} \quad (e)$
Yield Rule	Complementarity	Flow Rule

(21)

System (21) corresponds essentially to Smith's encodement [22] of Maier's matrix description of plasticity for monotonic straining [9].

Definition (21a) shows that the columns of matrix $\mathbf{N}$ represent the gradients of the associate yield functions:

$$\nabla_\sigma \ (\phi_{i*}) = \mathbf{N}_i \tag{22}$$

Similarly, the columns of matrix $\mathbf{V}$ can be interpreted as the gradients of associated plastic potential functions, ψ_{i*}:

$$\nabla_\sigma \ (\psi_{i*}) = \mathbf{V}_i \tag{23}$$

When the plastic potential functions ψ_{i*} are so chosen as to identify with the corresponding yield functions ϕ_{i*}, as first postulated by Von Mises [23], the plastic response is said to exhibit associated flow laws and the following identification holds:

$$\mathbf{N} = \mathbf{V}. \tag{24}$$

As it is shown next, condition (24) implies that the plastic deformation vector is orthogonal to the currently active yield modes. Frictional materials present non-associated flow laws and the models which describe them [24-26] do not exhibit the normality condition (24).

The role of the plastic hardening matrix H in system (21) is to represent different forms of interaction between the plastic yield modes.

Reciprocity in the interaction between the yield modes is modelled by enforcing symmetry:

$$\mathbf{H} = \tilde{\mathbf{H}} \tag{25}$$

Non-softening {softening} behaviour is simulated by requiring matrix H to be non-negative definite {indefinite}.

As it has been shown by Maier [10] and De Donato [11], matrix H can be used to model a wide range of plastic hardening rules.

Geometric Interpretation of the Plasticity Conditions

Consider first the static phase (21a,c) in the response of plastic materials, which is characterized by variation of stress in the absence of plastic straining:

$$\epsilon_* = \mathbf{0}.$$

According to system (21a), the yield function associated with the i-th yield mode is defined by

$$\phi_{i*} = \tilde{\mathbf{N}}_i \, \sigma - \sigma_{i*}, \tag{26}$$

wherein $\mathbf{N}$ denotes the i-th column of matrix $\mathbf{N}$ and σ_{i*} is a non-negative constant.

The activation of the i-th yield mode, $\phi_{i*} = 0$, is represented in the σ-space by an hyperplane with equation

$$\phi_{i*} = \tilde{\mathbf{N}}_i\, \sigma - \sigma_{i*} = 0. \tag{27}$$

As it is shown in figure 10, this hyperplane bounds the half-space defined by the yield condition $\phi_{i*} \leq 0$.

Definition (22) shows that vector $\mathbf{N}_i$ represents the outward normal of the associate yield hyperplane. If the outward normals are so chosen as to represent unit vectors

$$\tilde{\mathbf{N}}_i \mathbf{N}_i = 1,$$

then, according to definition (26), the plastic capacity σ_{i*} identifies with the initial distance,

$$d_i = \sigma_{i*} \tag{28}$$

of the i-th yield hyperplane to the origin of the σ-space, as shown in figure 10.

The full set of hyperplanes (27) represent a polytope in the σ-space. This polytope, illustrated in figure 11, defines the initial yield locus, that is, the stress combinations capable of mobilizing one or more yield modes. Points contained by the yield locus satisfy condition (21c); they represent statically admissible states of stress.

Therefore and as first suggested by Hodge [27,28], description (24a,c) for the yield conditions can be used to represent piecewise linear discretizations of any of the commonly used yield criteria namely the criteria of Hencky-Von Mises [29-32] and Drucker-Prager [33-35].

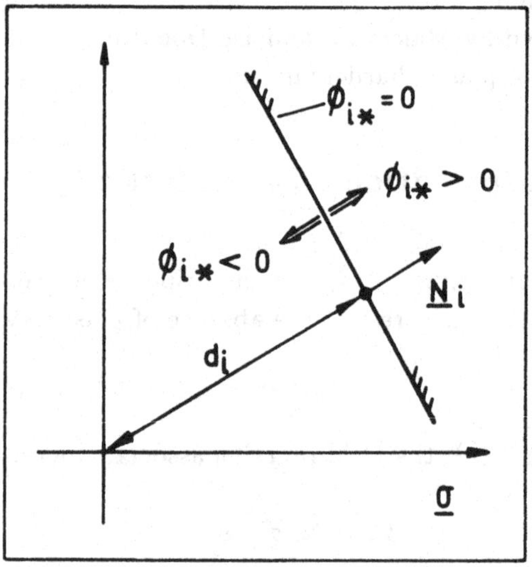

Figure 10

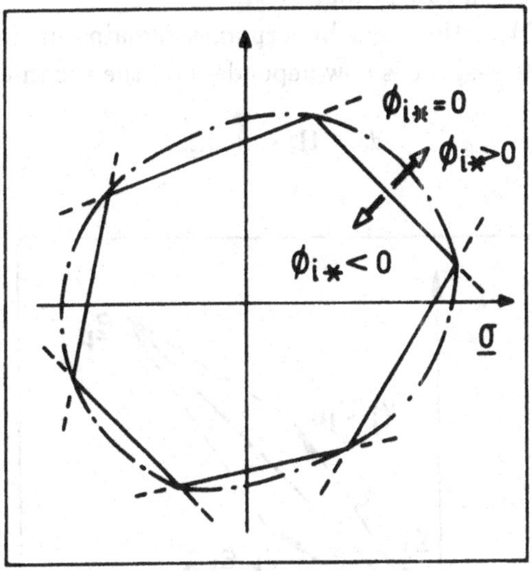

Figure 11

It can be easily shown that the yield locus is convex, thus satisfying a fundamental requirement for stable plastic responses, in the sense of Drucker [36-37] and Ilyushin [38].

Assume now that plastic straining is allowed to develop,

$$\epsilon_* \geq \mathbf{0}, \tag{29}$$

thus mobilizing the kinematic phase (21b,e) of the plastic response.

Let definition (21b) for the plastic strains be written in form

$$\epsilon_p = \sum_i \mathbf{V}_i \, \epsilon_{i*}, \tag{30}$$

wherein $\mathbf{V}_i$; denotes the i-th column of matrix $\mathbf{V}$.

The graphic representation of the kinematic admissibility conditions (29) and (30), given in figure 12, shows that the role of the flow rule (21c) is to enforce the plastic strain vector ϵ_p to be contained in the hypercone spanned by vectors $\mathbf{V}_i$. It has already been mentioned that these vectors can be interpreted as the outward normals (23) of associate plastic potential functions, as illustrated in figure 13.

When plastic straining sets in, definition (27) for the yield locus takes the following form

$$\phi_{i*} = \tilde{\mathbf{N}}_i \, \sigma - \mathbf{H}_i \, \epsilon_* - \sigma_{i*} = 0, \tag{31}$$

wherein $\mathbf{H}_i$ denotes the i-th row of matrix $\mathbf{H}$.

The orientation (22) of the yield hyperplanes remains unchanged but their distance to the origin of the σ-space is now dependent on the accumulated plastic strains:

$$d_i = \mathbf{H}_i \, \epsilon_* + \sigma_{i*}. \tag{32}$$

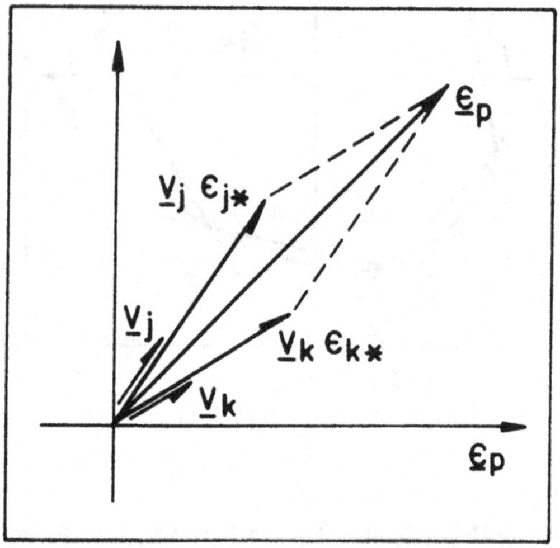

Figure 12

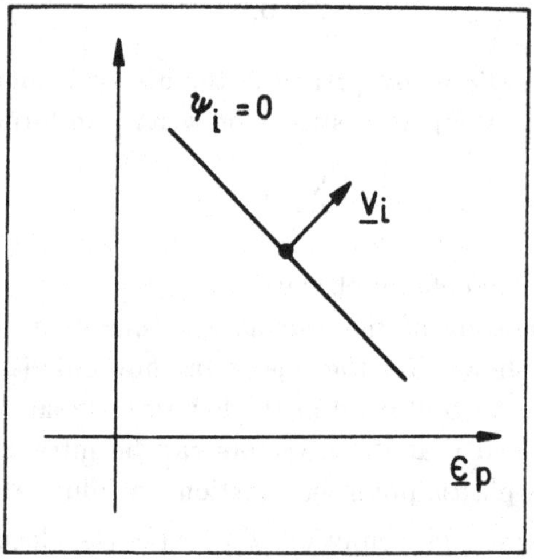

Figure 13

Assume for simplicity that matrix H is diagonal,

$$d_i = H_{ii}\, \epsilon_{i*} + \sigma_{i*},$$

as in Koiter's non-interactive hardening model [12].

As it is shown in figure 14, when the i-th hardening {softening} yield mode is mobilized, $\epsilon_{i*} > 0$, $H_{ii} > 0$ $\{ \epsilon_{i*} > 0,\ H_{ii} < 0 \}$, the associated yield hyperplane $\phi_{i*} = 0$ translates in the outward {inward} normal direction; yield hyperplanes associated with perfectly-plastic modes, $H_{ii} = 0$, are however unaffected by plastic straining.

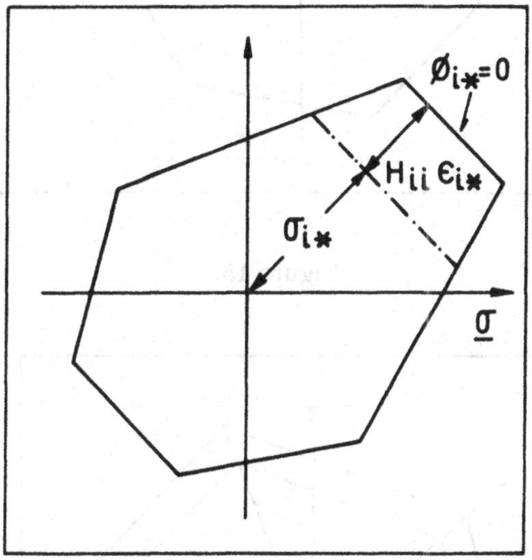

Figure 14

When the i-th yield mode is mobilized, $\phi_{i*} = 0$, complementary condition (21d) allows plastic straining to develop, $\epsilon_{i*} > 0$, according to the flow law (21b):

$$\epsilon_p = \mathbf{V}_i\, \epsilon_{i*}. \tag{33}$$

If the flow law is associated (24), the plastic deformation vector (33) is proportional to the outward normal to hyperplane $\phi_{i*} = 0$, as shown in figure 15.

Illustrated in the same figure is the simultaneous activation of several yield modes:

$$\phi_{j*} = \phi_{k*} = \ldots = 0,$$
$$\epsilon_{j*}\ ,\ \epsilon_{k*}, \ldots > 0.$$

The state of stress σ is now represented by a singular point of the yield polytope

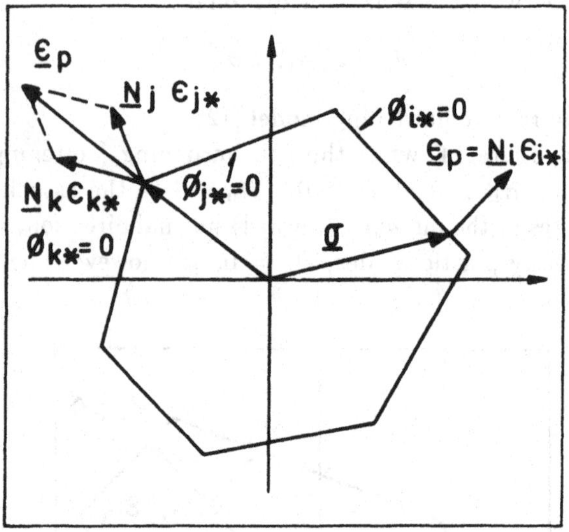

Figure 15

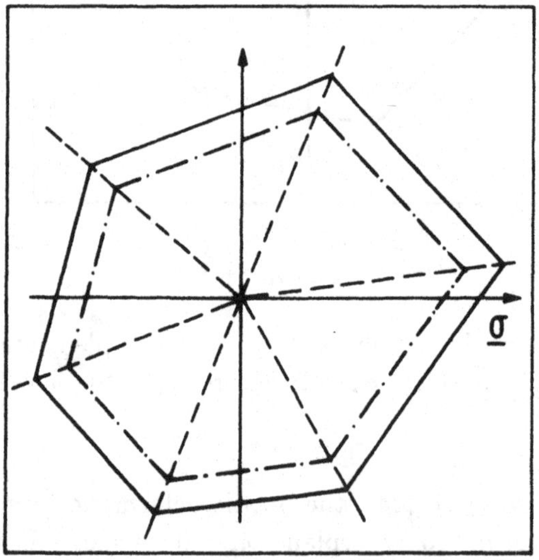

Figure 16

and the vector describing the associate state of strain,

$$\epsilon_p = \mathbf{V}_j \, \epsilon_{j*} + \mathbf{V}_k \, \epsilon_{k*} + \dots ,$$

is contained in the hypercone spanned by the gradient vectors $\mathbf{V}_j$ and $\mathbf{V}_k$, as suggested by Koiter [12] for materials with associated flow laws.

The movement the yield locus is subjected to during plastic straining is dictated by the hardening rule modelled by the entries of matrix $\mathbf{H}$.

The isotropic hardening rule [27,39-42] assumes that the yield locus expands during plastic flow, while preserving its shape and position with respect to the origin of the σ-space, as illustrated in figure 16.

The rule of kinematic hardening [24,43-46] is exemplified in figure 17. According to this rule, the yield locus translates without distortion in the direction of the strain flow.

Different rules of anisotropic hardening [47-52] are obtained by appropriate combinations of isotropic, kinematic and non-interacting hardening laws. The yield locus is now allowed to translate and change of shape simultaneously.

Reviews on most of the hardening rules proposed in the literature can be found in [4,53-55].

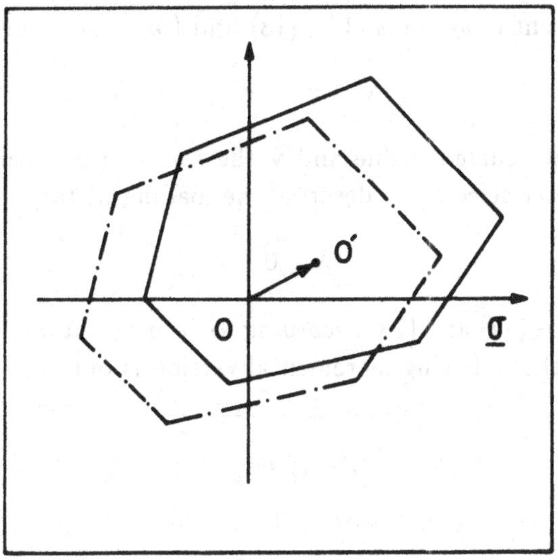

Figure 17

Incremental Description of Elastoplasticity

The several plasticity theories that have been proposed can be organized into two groups, according to whether the basic relationships connect stresses and strains or stresses and strain rates. The results generated by these theories, respectively known as deformation theories and flow theories, will only coincide when the development of yielding is regularly progressive.

The deformation approach presented above is uncapable of representing situations of plastic unstressing and allows for decreasing, i.e. reversible "plastic" strains. Plastic response is typically history dependent and must therefore be expressed in incremental form.

The activation laws and the flow laws are the basic ingredients of the incremental theories of plasticity. The activation laws define which yield modes are active at a given state of stress and strain and indicate at which points of the system plastic straining may develop. The flow laws distinguish among these points those which will in fact be further strained from the ones where plastic unstressing will take place during the incremental action to which that system will be subject.

The following presentation concentrates on the so called flow laws. The information they provide, insufficient per se, is assumed to be backed by a complete knowledge of the states of stress and strain prior to the increment.

Let variables present in systems (17), (18) and (21), say $\mathbf{v}$, be expressed in form

$$\mathbf{v} = \mathbf{v}^0 + \dot{\mathbf{v}}\,\lambda, \tag{34}$$

wherein $\mathbf{v^0}$ denotes its current value and $\dot{\mathbf{v}}$ the rate of variation with respect to a non-negative parameter selected to describe the loading history:

$$\lambda \geq 0. \tag{35}$$

Treating definitions (17) and (18) according to decomposition (34) and eliminating the load parameter, the following incremental version is obtained:

$$\boxed{\dot{\epsilon} = \dot{\epsilon}_e + \dot{\epsilon}_p} \tag{36}$$

(37)

Incremental Description of Elasticity	
$\dot{\epsilon}_e = \mathbf{F}\,\dot{\sigma}$ (a)	$\dot{\sigma} = \mathbf{K}\dot{\epsilon}_e$ (b)
Flexibility	Stiffness

After taking increments (34), the following description is found for the yield functions (21a)

$$\phi_* = \phi_*^0 + \dot{\phi}_* \lambda,$$

wherein

$$\dot{\phi}_* = -\mathbf{H}\, \dot{\epsilon}_* + \tilde{\mathbf{N}}\, \dot{\sigma}, \tag{38}$$

with

$$\phi_*^0 \leq \mathbf{0}, \tag{39}$$

and

$$\phi_*^0 + \dot{\phi}_*\, \lambda \leq \mathbf{0}, \tag{40}$$

representing now the yield rule (21c) of plasticity.

In the incremental description of the plastic flow law (21b),

$$\dot{\epsilon}_p = \mathbf{V}\, \dot{\epsilon}_*, \tag{41}$$

the plastic parameter rates are now required to be non-negative,

$$\dot{\epsilon}_* \geq \mathbf{0},$$

to ensure the irreversibility of plastic straining.

The situations the incremental description of plasticity must be able to represent are shown in figures 18 and 19.

As it is illustrated in figure 18, if the i-th yield mode is not active

$$\phi_{i*}^0 < 0 \ , \ \epsilon_{i*}^0 \geq 0,$$

the rate on the associated yield function can not be constrained in sign, but the current value of the plastic strain must be held unchanged:

$$\dot{\phi}_{i*} \neq 0 \ , \ \dot{\epsilon}_{i*} = 0.$$

According to the yield condition (39) and (40), the non-negative increment (35) on the loading parameter that exposes the activation of the yield mode, $\phi_{i*} = 0$, is defined by

$$\lambda = -\phi_{i*}^0 / \dot{\phi}_{i*}. \tag{42}$$

Assume now that the yield mode is active,

$$\phi_{i*}^0 = 0 \ , \ \epsilon_{i*}^0 \geq 0,$$

as illustrated in figure 19.

Plastic straining {unstressing} is implemented by requiring the rate on the associated plastic parameter {yield function} to be positive {negative}, while simultaneously constraining the current value of the yield function {plastic deformation} to remain unchanged:

$$\dot{\phi}_{i*} = 0 \quad , \quad \dot{\epsilon}_{i*} < 0,$$

$$\dot{\phi}_{i*} < 0 \quad , \quad \dot{\epsilon}_{i*} = 0.$$

In either case, the following conditions hold:

$$\dot{\phi}_{i*} \leq 0 \ , \dot{\phi}_{i*}\, \dot{\epsilon}_{i*} = 0 \ , \ \dot{\epsilon}_{i*} \geq 0 \tag{43}$$

The incremental description of plasticity (44) is obtained collecting information (43) regarding the currently active yield modes, as well as definitions (38) and (41) for the associated yield function and plastic deformation rates, respectively.

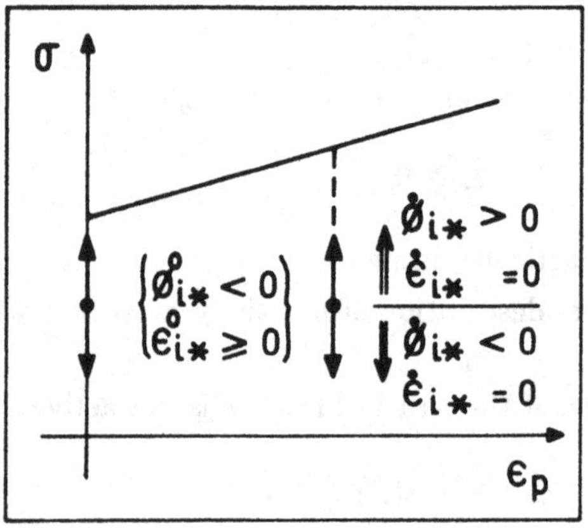

Figure 18

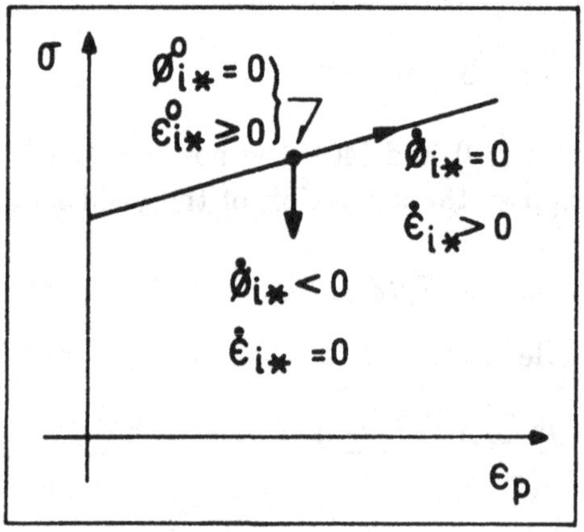

Figure 19

Incremental Description of Plasticity		
$$\begin{bmatrix} -\mathbf{H} & \tilde{\mathbf{N}} \\ \mathbf{V} & \mathbf{0} \end{bmatrix} \begin{Bmatrix} \dot{\epsilon}_* \\ \dot{\sigma} \end{Bmatrix} = \begin{Bmatrix} \dot{\phi}_* \\ \dot{\epsilon}_p \end{Bmatrix} \begin{matrix} (a) \\ (b) \end{matrix} \qquad (44)$$		
$\dot{\phi}_* \leq \mathbf{0}$ (c)	$\tilde{\dot{\varphi}}_* \dot{\epsilon}_* = 0$ (d)	$\dot{\epsilon}_* \geq \mathbf{0}$ (e)
Yield Rule	Complementarity	Flow Rule

The Incremental Governing System

To account for the elastic phase in the response, let the total strains (36) and the flexibility description (37a) of elasticity be substituted in the plastic flow condition (44b). The following governing system is found,

$$\begin{bmatrix} -\mathbf{H} & \tilde{\mathbf{N}} \\ \mathbf{N} & \mathbf{F} \end{bmatrix} \begin{Bmatrix} \dot{\epsilon}_* \\ \dot{\sigma} \end{Bmatrix} = \begin{Bmatrix} \dot{\phi}_* \\ \dot{\epsilon}_p \end{Bmatrix}, \quad \begin{matrix} (a) \\ (b) \end{matrix} \qquad (45)$$

$$\dot{\phi}_* \leq 0, \quad \dot{\epsilon}_* \geq \mathbf{0}, \quad \tilde{\dot{\phi}}_* \dot{\epsilon}_* = 0, \qquad (c-e)$$

where it has been assumed for simplicity that normality and reciprocity hold.

System (45) has a typical mathematical structure which is known in mathematical programming theory as a symmetric, linear complementarity problem. It has been shown [56,57] that a wide class of problems in elastic-plastic structural analysis is governed by similar systems.

When the total deformation rates $\dot{\epsilon}$ are assumed to be prescribed, it is found that by implementing the Karush-Kuhn-Tucker equivalence conditions [58,59], system (45) is dissociated into an equivalent pair of quadratic programs:

$$\text{Min } z = 1/2\, \tilde{\dot{\epsilon}}_* \mathbf{H}\, \dot{\epsilon}_* + 1/2\, \tilde{\dot{\sigma}}\, \mathbf{F}\, \dot{\sigma} - \tilde{\dot{\sigma}}\, \dot{\epsilon} \qquad (46)$$

$$\text{subject to} \quad -\mathbf{H}\, \dot{\epsilon}_* + \tilde{\mathbf{N}}\, \dot{\sigma} = \dot{\phi}_*, \; \dot{\phi}_* \leq 0 \qquad (47)$$

$$\text{Min } w = 1/2\, \tilde{\dot{\epsilon}}_* \mathbf{H}\, \dot{\epsilon}_* + 1/2\, \tilde{\dot{\sigma}}\, \mathbf{F}\, \dot{\sigma} \qquad (48)$$

$$\text{subject to} \quad \mathbf{N}\, \dot{\epsilon}_* + \mathbf{F}\, \dot{\sigma} = \dot{\epsilon}, \; \dot{\epsilon}_* \geq 0 \qquad (49)$$

Note that in the primal (46) and (47), the plastic parameter rates $\dot{\epsilon}_*$ are not required to be non-negative.

The above program can be read as follows,

(S1) - Among all statically admissible solutions (47) the actual one(s) make the potential co-energy increment (46) a minimum,

(S2) - Among all kinematically admissible solutions (49) the actual one(s) make the potential energy increment (48) a minimum,

thus establishing the incremental theorems of minimum potential complementary energy and minimum potential energy, respectively.

Existence, Uniqueness and Stability of Solutions

In the terminology of mathematical programming theory, a solution

$$\mathbf{v} = [\dot{\epsilon}_\star,\ \dot{\sigma}]$$

of program (46), (47) {(48)}, {(49) } is said to be optimal if it satisfies the constraints on static {kinematic} admissibility (47) {(49)} and minimizes the objective function (46) {(48)}.

Cottle's duality theorem [60] when specialized to programs (46) and (47) justifies the following statements on the existence of solutions:

(S3) - If $\mathbf{v}'$ and $\mathbf{v}''$ are respectively statically and kinematically admissible solutions, then $z + w > 0$.

(S4) - If $\mathbf{v}'$ {$\mathbf{v}''$} is an optimal solution of the primal {dual} program, there exists a $\dot{\epsilon}_\star$ {$\dot{\sigma}$} satisfying $\mathbf{H}\,\dot{\epsilon}_\star = \mathbf{H}\,\dot{\epsilon}'_\star$ {$\mathbf{F}\,\dot{\sigma} = \mathbf{F}\,\dot{\sigma}''$} such that $\mathbf{v}' = [\dot{\sigma}',\ \dot{\epsilon}_\star]$ {$\mathbf{v}'' = [\dot{\sigma},\ \dot{\epsilon}''_\star]$} is an optimal solution to the dual {primal} program and Min z = Max w.

(S5) - If kinematically {statically} admissible solutions do not exist but statically {kinematically} admissible solutions do exist, then z{w} is unbounded.

(S6) - If either a primal or a dual program has an optimal solution, then there is a solution $\mathbf{v}$ which is optimal for both primal and dual programs.

(S7) - If both statically and kinematically admissible solutions exist, then both primal and dual programs have optimal solutions.

The sufficient conditions stated below for the existence of unique optimal solutions result from the direct application of the uniqueness theorem [61] to programs (46) and (48).

(S8) - If v' $\{v''\}$ is a statically {kinematically} admissible solution, then:

As the elastic flexibility matrix $\mathbf{F}$ is positive definite, the static solution $\dot{\sigma}' = \dot{\sigma}''$ is the unique optimal solution: the kinematic solution may be multiple, $\dot{\epsilon}'_* \neq \dot{\epsilon}''_*$, at optimality.

(S9) - If the plastic hardening matrix $\mathbf{H}$ is positive definite, the kinematic solution $\dot{\epsilon}'_* = \dot{\epsilon}''_*$ is the unique optimal solution.

The composite form (50) of programs (46) and (48) requires the minimization of the difference between the primal and dual objective functions over the intersection of the corresponding constraint sets [60]:

$$
\begin{aligned}
&\text{Min } Q \;= \check{\dot{\epsilon}}_* \,\mathbf{H}\, \dot{\epsilon}_* + \check{\dot{\sigma}}\,\mathbf{F}\,\dot{\sigma} - \check{\dot{\sigma}}\,\epsilon && (a)\\
&\text{subject to: } \begin{bmatrix} -\mathbf{H} & \check{\mathbf{N}} \\ \mathbf{N} & \mathbf{F} \end{bmatrix} \begin{Bmatrix} \dot{\epsilon}_* \\ \dot{\sigma} \end{Bmatrix} = \begin{Bmatrix} \dot{\phi}_* \\ \dot{\epsilon} \end{Bmatrix} && \begin{matrix}(b)\\(c)\end{matrix} \qquad (50)\\
&\dot{\phi}_* \leq \mathbf{0}\,,\; \dot{\epsilon}_* \geq \mathbf{0} && (d,e)
\end{aligned}
$$

Therefore, if program (50) has optimal solutions, those solutions must be simultaneously statically and kinematically admissible. Furthermore and according to statement (S4), if program (50) has an optimal solution, then $Q = O$ at optimality.

The following statement on the conditions for the existence of multiple optimal solutions result from the direct application of the multiplicity theorem [61] to program (50).

(S10) - If $\mathbf{v}'$ is an optimal solution to program (50), then solution $\mathbf{v}'' = \mathbf{v}' + \alpha\,\delta\,\mathbf{v}$ such that

$$
\begin{bmatrix} -\mathbf{H} & \tilde{\mathbf{N}} \\ \mathbf{N} & \mathbf{F} \end{bmatrix} \begin{Bmatrix} \delta\,\dot{\epsilon}_* \\ \delta\,\sigma \end{Bmatrix} = \begin{Bmatrix} \delta\,\dot{\phi}_* \\ 0 \end{Bmatrix}
$$

$$
\alpha \begin{Bmatrix} \delta\,\dot{\epsilon}_* \\ \delta\,\dot{\phi}_* \end{Bmatrix} \begin{matrix} \geq \\ \leq \end{matrix} - \begin{Bmatrix} \dot{\epsilon}'_* \\ \dot{\phi}'_* \end{Bmatrix}
$$

$$
\left\{ \tilde{\dot{\epsilon}}'_* \;\; \tilde{\dot{\phi}}' \right\} \begin{Bmatrix} \delta\,\dot{\epsilon}_* \\ \delta\,\dot{\phi}_* \end{Bmatrix} = 0
$$

$$
\delta\,\tilde{\dot{\epsilon}}_*\;\delta\,\dot{\phi}_* = 0\,,\; \tilde{\dot{\epsilon}}\;\delta\,\dot{\sigma} = 0
$$

is also an optimal solution

Note that the conditions above reduce to

$$\mathbf{N}\ \delta\,\dot{\epsilon}_* = 0\ ,\ \dot{\phi}'_*\ \delta\,\dot{\epsilon}_* = 0\ ,\ \dot{\epsilon}'_* + \alpha\ \delta\,\dot{\epsilon}_* \geq 0$$

if the static solution is unique ($\delta\,\dot{\sigma} = \mathbf{0}$) and the material is elastic, perfectly plastic ($H = 0$).

Let the internal work rate be defined as

$$\dot{W} = \tilde{\dot{\sigma}}\,\dot{\epsilon} \tag{51}$$

Substituting condition (45b) in definition (51) and using results (45a) and (45c), the following alternative expression for the internal work rate is found

$$\dot{W} = \tilde{\mathbf{v}}\,\mathbf{S}\,\mathbf{v} \tag{52}$$

wherein

$$\mathbf{S} = \begin{bmatrix} \mathbf{H} & \mathbf{0} \\ \mathbf{0} & \mathbf{F} \end{bmatrix}$$

According to Drucker's criterion [37] the state of equilibrium is stable if the work rate is positive for any excursion $\dot{\epsilon}$ into a neighbouring configuration, the equilibrium of which is ensured by a variation $\dot{\sigma}$ on the state of stress. The equilibrium state is said to be critical {unstable} if there is at least one path for which $\dot{W} = 0$ $\{\dot{W} < 0\}$. According to result (52), Drucker's stability criterion can be read as follows:

(S11) - If matrix $\mathbf{S}$ is positive definite, the equilibrium state is stable.

(S12) - If matrix $\mathbf{S}$ is positive semi-definite, the equilibrium state is not unstable.

REFERENCES

1. Cauchy, A.: Recherches sur l'équilibre et le mouvement interieur des corps solides ou fluides, élastiques ou non élastiques, Bull. Soc. Philomatique, p 9-13, Paris, 1823

2. Tresca, H.: Mémoire sur l'écoulement des corps solides soumis à des fortes pressions, Compte-Rendu, 59, Paris, 1864

3. Guest, J.J.: On the strength of ductile materials under combined stress, Phil. Magazine, 50, 1900

4. Nguyen Dang Hung: Sur la plasticité et le calcul des états limites par éléments finis, Thése de Doctorat spécial, Univ. de Liège, 1985

5. Drucker, D.C.: Basic concepts in plasticity, Handbook of Engrg. Mech., W. Flügge, (Ed.),McGraw-Hill, 1962

6. Ilyushin, A.A.: On the foundations of the general mathematical theory of plasticity, Voprosy Teorii Plastichnosti, AN SSR, Moscow, 1961

7. Zyczkowsky, M.: Combined loadings in the theory of plasticity, Polish-Scientific Publ., 1981

8. Maier, G.: Some theorems for plastic strain rates and plastic strains, J. Mécanique, 8, 1969

9. Maier, G.: "Linear" flow-laws of elastoplasticity: a unified general approach, Accad. Naz. Lincei, VII, 47, 1969

10. Maier, G.: A matrix structural theory of piecewise linear elastoplasticity with interacting yield planes, Meccanica, 5, 1970

11. De Donato, O.: On piecewise-linear constitutive laws in plasticity, Tech. Rept. ISTC, 1974

12. Koiter, W.T.: Stress-strain relations, uniqueness and variational theorems for elastic-plastic materials with singular yield surfaces, Quart. Appl. Math., 11, 1953

13. Phillips, A. and C.W. Lee: Yield surfaces and loading surfaces. Experiments and recommendations, Int. J. Solids Struct., 15, 1979

14. Cohn, M.Z.(Ed.): Inelasticity and non-linearity in structural concrete. Study No.8, Solid. Mech. Div., University of Waterloo, 1973

15. Rees, D.W.A.: A review of stress-strain paths and constitutive relations in the plastic range, J. Strain Analysis for Engrg. Design, 16, 1981

16. Ramberg, W. and W.R. Osgood: Description of stress-strain curves by three parameters, NACA, TN 902, 1943

17. Margetton, J.: Tensile stress-strain characterization of nonlinear materials, J. Strain Analysis for Engrg. Design, 16, 1981

18. Sargin, M.: Stress-strain relationships for concrete and the analysis of structural concrete sections, Study No.4, SMD, Univ. of Waterloo, 1971

19. C.E.B.- F.I.P. - Model code for concrete structures, Comité Euro-International du Béton, 1978

20. Murray, D.W. and L. Chitnuyanondh: Concrete plasticity theory for biaxial stress analysis, J. Engrg. Mech. Div., Proc. ASCE, 105, 1979

21. Brazant, Z.P. and S.S. Kim: Plastic-fracturing theory for concrete, J. Engrg. Mech. Div., Proc. ASCE, 105, 1979

22. Smith, D.L.: Plastic limit analysis and synthesis of frames by linear programming, PhD Thesis, Univ. of London, 1974

23. Von Mises, R.: Mechanik der plastichen Formänderung von Kristallen,ZAMM, 8, 1928

24. Melan, L.: Zur Plastizität der raümlichen kontinuums, Ing. Archiv.,9, 1938

25. Mröz, Z.: On forms of constitutive laws for elastic-plastic solids, Arch. Mech. Stosow., I, 13, 1966

26. Schereyer, H.L.: A third-invariant plasticity theory for frictional materials, J. Struct. Mech., 11, 1983

27. Hodge, P.G.: The theory of piecewise-linear isotropic plasticity, IUTAM Colloquium, Madrid, 1955

28. Berman, J. and P.G. Hodge: A general theory of piecewise linear plasticity for initially anisotropic materials, Arch. Mech. Stosow., XI, 5, 1959

29. De Saint-Venant, B.: Mémoire sur l'établissement des équations différentielles des mouvements intérieures opérés dans les corps solides ductiles au-delá des limites oú l'élasticité pourrait les ramener à leur premier état. Compte-Rendu 70, Paris, 1870

30. Von Mises, R.: Mechanik der festen Körper in plastich deformablen Zustent, Nach. Math. Phys., KL, 1913

31. Hencky, H.: Zur Theorie plastischer Deformationem und der Hiedurch im Material hervargerufenem nachspannungen, ZAMM, 4, 1924

32. Hosford, W.F.: A generalized isotropic yield criterion, Trans ASME, E39, 3, 1972

33. Mohr, O.: Welche Umstände belingen die Elastizitatsrenze und den Bruch einess Materials, Verein Deutscher Ingenieure, Düsseldorf, 44, 45, 1900, 1901

34. Mohr, O.: Abhandlungen aus dem Gebiete der Technischen Mechanik, Auflage, Ernest und Solin, Berlin, 1914

35. Drucker, D.C. and W. Prager: Soil mechanics and plastic analysis or limit design, Aust. Appl. Math., 10, 1952

36. Drucker, D.C.: Some implications of workhardening and ideal plasticity, Quart. Appl. Maths, 7, 1950

37. Drucker, D.C.: A definition of stable inelastic materials, Trans ASME, J. Appl. Mech., 101, 1959

38. Ilyushin, A.A.: On the postulate of plasticity, Prik. Math. Melch., 25, 1961

39. Taylor, G.I. and H. Quinney: The plastic distortion of metals, Ph. Trans. Roy. Soc. London, A230, 1931

40. Schmidt, R.: Uber der Zusammenhang von Spannungen und Formanderungen im Verfestigungsgebiet, Ing.-Archiv, 3, 1932

41. Odqvist, F.K.G.: Die Verfestigung von flusseisenähnlichen Körpern, ZAMM, 13, 1933

42. Hill, R.: The mathematical theory of plasticity, Oxford, 1950

43. Prager, W.: Recent developments in the mathematical theory of plasticity, J. Appl. Phys., 20, 1949

44. Prager, W.: The theory of plasticity - a survey of recent achievements, Proc. Inst. Mech. Eng., 169, 41, 1955

45. Prager, W.: A new method of analysing stresses and strain in work-hardening plastic solids, J. Appl. Mech., Trans. ASME, 23, 4, 1956

46. Ziegler, Z.: A modification of Prager's hardening rule, Quart. Appl. Math, 17, 1, 1959

47. Prager, W.: Einfluss der Deformation auf die Fliessbedingung von Zahplastischen Körpen, ZAMM, 15, 1935

48. Baltov, A. and A. Sawczuk: A rule of anisotropic hardening, Acta Mechanica, 1, 2, 1964

49. Mröz, Z.: On the description of anisotropic work-hardening, J. Mech. Phys. Solids, 15, 1967

50. Krieg, R.D.: A practical two surface plasticity theory, J. Appl. Mech., Trans. ASME, 1975

51. Mröz, Z., H.P. Shrivastava and R.N. Dubey: A non-linear hardening model and its application to cyclic loading, Acta Mechanica, 25, 1976

52. Rees, D.W.A.: A theory of non-linear anisotropic hardening, Proc. Inst. Mech. Eng., C, Mech. Engrg. Sci, 197, 1983

53. Sawczuk, A.: A note on anisotropic hardening, Bull. Acad. Polonaise Sciences, Series des Sciences Techniques, 28, 1980

54. Rees, D.W.A.: Anisotropic hardening theory and the Bauschinger effect, J. Strain Analysis for Engrg. Design, 16, 1981

55. Mröz, Z.: Hardening and degradation rules for metals under monotonic and cyclic loading, Trans. ASME, J. Engrg. Materials and Techol., 105, 1983

56. Cohn, M.Z. and G. Maier (Eds.): Engineering plasticity by mathematical programming, Pergamon, New York, 1979

57. Maier, G. and J. Munro: Mathematical programming applications to engineering plasticity, Appl. Mech. Review, 35, 1982

58. Karush, W.: Minima of functions of several variables with inequalities as side conditions, M.S. Thesis, Dept. of Mathematics, Univ. of Chicago, 1939

59. Kuhn, H.W. and A.W. Tucker: Nonlinear programming, 2nd Berkeley Symp. Math. Statistics and Probability, Berkeley, 1951

60. Cottle, R.W.: Symmetric dual quadratic programs, Q. Appl. Maths., 21, 1963

61. Kunzi, H.P., W. Krelle and W. Oettli: Nonlinear Programming, Blaisdell, 1966

CHAPTER 7

COMPLEMENTARITY PROBLEMS AND UNILATERAL CONSTRAINTS

A. Borkowski
Polish Academy of Sciences, Warsaw, Poland

ABSTRACT

A simple formalism is proposed that leads to the unified treatment of many problems in Structural Mechanics. Such are the self-adjoint problems of statics of structures made of linear elastic, piecewise-linear elastic, elastic-plastic or rigid-plastic material. The linear part of the complete set of governing relations for each of those problems has symmetric matrix of coefficients with positive/negative definite submatrices along the diagonal. Hence, such a set can be replaced by a minimax problem, which in turn is equivalent to a pair of dual Quadratic Programming or Linear Programming problems. If inequalities are absent, then the dual problems are further reducable to the sets of equations corresponding to the Direct Stiffness (Displacement) Method and Flexibility (Force) Method. Otherwise no reduction is possible and either the dual problems must be solved directly (Direct Energy Approach) or the clasical methods should be applied in an iterative manner. In any case the proposed methodology allows us to establish easily the existence and uniqueness properties of the solution.

1. INTRODUCTION

The aim of the present paper is to present a certain simple formalism that allows one to construct in an easy and uniform way the complementary energy principles for a broad class of problems in Structural Mechanics. It will be shown that problems so different at first glance as the analysis of elastic unilaterally supported structures, the calculation of the response of cable-strut systems or the evaluation of the ultimate load have, in fact, an identical mathematical background. Namely, these problems can be shown to be reducible to certain Saddle Point Problems (SPP's) that, in turn, can be replaced by relevant pairs of dual Quadratic Programming Problems (QPP's).

In the author's opinion, the proposed approach is favourable for two reasons, at least. Firstly, it brings further unification into Structural Mechanics: the energy principles necessary for a consistent discretization of the structure, e.g. by the Finite Element Method, are derived almost automatically. Secondly, important questions regarding existence and uniqueness of solutions can be answered straightforwardly. Perhaps, one could mention the third reason as well: the solution of any of the problems dealt with can be obtained directly solving the QPP that corresponds to the proper energy principle. In certain cases such a direct energy approach to the numerical solution can be reasonable, though usually alternative methods based upon the iterative solution of nonlinear algebraic equations are more efficient.

The theory presented in the sequel has been developed in the late 1960's by G. Maier and his co-workers (e.g. [1] to [5]). There seems to be, however, some novelty in the way of presentation adopted here. We shall begin with a definition of a set of linear algebraic equations and inequalities of a certain internal structure. Then we shall define a potential of that set, demonstrate the saddle shape of such a function and proceed further to the equivalent dual QPP's. Thus the Complete Set of Governing Relations (CSGR) for each considered problem of Structural Mechanics will serve as a starting point for all further derivations.

Space limitations of the present book allow only a brief presentation of basic ideas. A detailed discussion of both theoretical and numerical aspects of the proposed method as well as numerous examples of applications can be found in the monograph [6].

2. SADDLE POINTS AND SYMMETRIC DUAL QPP'S

Consider a set of linear algebraic equations

$$\underline{A}\,\underline{x} + \underline{b} = \underline{0} \tag{2.1}$$

with a symmetric matrix of coefficients $\underline{A}$ of order n. The quadratic function

$$f(\underline{x}) = \tfrac{1}{2}\,\underline{\tilde{x}}\,\underline{A}\,\underline{x} + \underline{\tilde{b}}\,\underline{x} \tag{2.2}$$

is called the potential of this set because the gradient $\underline{\nabla} f$ of f generates the left-hand side of (2.1). Hence, the solution $\underline{x}_*$ of (2.1) is simultaneously the stationary point of f: $\underline{\nabla} f(\underline{x}_*) = \underline{0}$. If $\underline{A}$ is not only symmetric but also positive (negative) semidefinite then $\underline{x}_*$ corresponds to the minimum (maximum) of the potential (2.2).

Suppose that the set of equations has a little bit more complicated structure:

$$\begin{aligned} \underline{A}_{xx}\,\underline{x} + \underline{A}_{xy}\,\underline{y} + \underline{b}_x &= \underline{0}\,, \\ \underline{A}_{yx}\,\underline{x} + \underline{A}_{yy}\,\underline{y} + \underline{b}_y &= \underline{0} \end{aligned} \tag{2.3}$$

where the global matrix of coefficients is again symmetric ($\tilde{\underline{A}}_{xx} = \underline{A}_{xx}$, $\tilde{\underline{A}}_{yy} = \underline{A}_{yy}$, $\tilde{\underline{A}}_{xy} = \underline{A}_{yx}$), the submatrix $\underline{A}_{xx}$ is positive semidefinite and $\underline{A}_{yy}$ is negative semidefinite. Then the potential of (2.3) is the saddle function

$$f(\underline{x}, \underline{y}) = \tfrac{1}{2}\,\tilde{\underline{x}}\,\underline{A}_{xx}\,\underline{x} + \tfrac{1}{2}\,\tilde{\underline{y}}\,\underline{A}_{yy}\,\underline{y} + \tilde{\underline{x}}\,\underline{A}_{xy}\,\underline{y} + \tilde{\underline{b}}\ \underline{x} + \tilde{\underline{b}}\ \underline{y}\,. \tag{2.4}$$

It is convex with respect to $\underline{x}$ and concave with respect to $\underline{y}$. The solution $(\underline{x}_*, \underline{y}_*)$ of the SPP

$$\min_{\underline{x}}\ \max_{\underline{y}}\ f(\underline{x}, \underline{y}) \tag{2.5}$$

has to correspond to the vanishing gradients

$$\begin{aligned} \underline{\nabla} f_x &= \underline{A}_{xx}\,\underline{x} + \underline{A}_{xy}\,\underline{y} + \underline{b}_x = \underline{0}\,, \\ \underline{\nabla} f_y &= \underline{A}_{yx}\,\underline{x} + \underline{A}_{yy}\,\underline{y} + \underline{b}_y = \underline{0}\,. \end{aligned} \tag{2.6}$$

Hence, it is simultaneously the solution of (2.3).

Each SPP can be replaced by a pair of equivalent constrained extremum problems: a minimization problem and a maximization problem. In order to derive them we introduce the Legendre's transforms of f :

$$\begin{aligned} f' &= f - \tilde{\underline{y}}\,\underline{\nabla f}_y = \tfrac{1}{2}\,\tilde{\underline{x}}\,\underline{A}_{xx}\,\underline{x} - \tfrac{1}{2}\,\tilde{\underline{y}}\,\underline{A}_{yy}\,\underline{y} + \tilde{\underline{b}}_x\,\underline{x}\,, \\ f'' &= f - \tilde{\underline{x}}\,\underline{\nabla f}_x = -\tfrac{1}{2}\,\tilde{\underline{x}}\,\underline{A}_{xx}\,\underline{x} + \tfrac{1}{2}\,\tilde{\underline{y}}\,\underline{A}_{yy}\,\underline{y} + \tilde{\underline{b}}_y\,\underline{y}\,. \end{aligned} \tag{2.7}$$

The first one is convex with respect to both $\underline{x}$ and $\underline{y}$, whereas the second one is concave. It is clear that $f' = f$ provided $\underline{\nabla f}_y = \underline{0}$ and $f'' = f$ when $\underline{\nabla f}_x = \underline{0}$. Therefore, the solutions of the following constrained extremum problems

$$\min\ \{\ f'\ \mid\ \underline{\nabla f}_y = \underline{0}\ \}\,, \tag{2.8.a}$$

$$\max\ \{\ f''\ \mid\ \underline{\nabla f}_x = \underline{0}\ \} \tag{2.8.b}$$

must coincide with the solution of the original SPP (2.5). Since both f' and f'' are quadratic while the gradients $\underline{\nabla f}_x$ and $\underline{\nabla f}_y$ are linear with respect to $\underline{x}$ and $\underline{y}$, the equivalent problems (2.8) are the QPP's.

Since no inequalities have been involved so far, the equivalence between the set of equations (2.3), the SPP (2.5) and the QPP's (2.8) could be established in the frame of the classical Lagrangian approach. The situation changes if the saddle point is to be looked for among non-negative variables:

$$\min_{\underline{x} \geqq \underline{0}} \ \max_{\underline{y} \geqq \underline{0}} \ f(\underline{x}, \underline{y}) \tag{2.9}$$

In order to derive the necessary and sufficient conditions for $(\underline{x}_*, \underline{y}_*)$ to be the solution of (2.9), let us recall the conditions for the minimum of a convex function of a single non-negative variable. It is seen from Fig. 2.1 that three cases are to be considered: a) the unconstrained minimum is situated inside the admissible domain Ω ; b) it is located at the boundary of Ω ; c) it falls outside Ω . All these situations are covered by the following conditions: $x \geqq 0$, $df/dx \geqq 0$, $x\, df/dx = 0$.

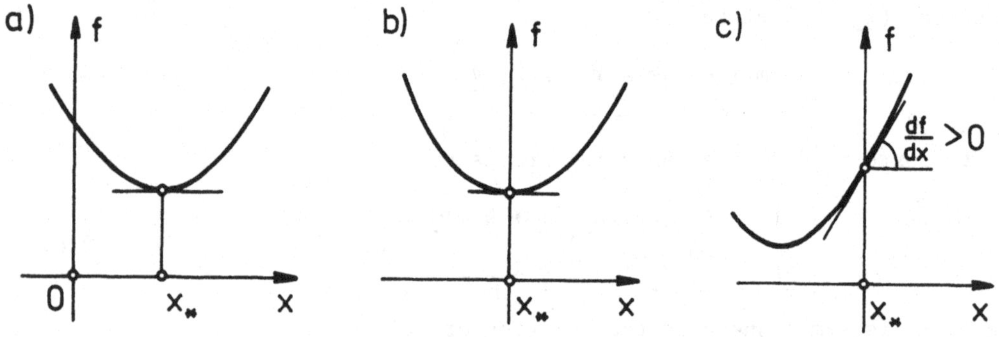

Fig. 2.1. Minima of convex functions of non-negative argument.

If we were looking for the maximum of a concave function, then the case b) would correspond to a negative value of derivative df/dx .. Hence, the solution x_* would have to satisfy the conditions: $x \geqq 0$, $df/dx \leqq 0$, $x\, df/dx = 0$. Combining these two cases together and generalizing them for multivariable functions we arrive at the necessary and sufficient conditions for $(\underline{x}_*, \underline{y}_*)$ to be the solution of (2.9):

$$\underline{\nabla f}_x \geqq \underline{0}, \quad \underline{\nabla f}_y \leqq \underline{0}, \tag{2.10.a}$$

$$\underline{x} \geqq \underline{0}\ , \quad \underline{y} \geqq \underline{0}, \tag{2.10.b}$$

$$\underline{\tilde{x}}\ \underline{\nabla f}_x = \underline{0}, \quad \underline{\tilde{y}}\ \underline{\nabla f}_y = \underline{0}. \tag{2.10.c}$$

Thus the set of equations and inequalities

$$\underline{A}_{xx}\,\underline{x} + \underline{A}_{xy}\,\underline{y} + \underline{b}_x \geq \underline{0}\ ,$$
$$\underline{A}_{yx}\,\underline{x} + \underline{A}_{yy}\,\underline{y} + \underline{b}_y \leq \underline{0}\ ,$$
$$\underline{x} \geq \underline{0}\ ,\quad \underline{y} \geq \underline{0}\ , \tag{2.11}$$
$$\tilde{\underline{x}}\,(\underline{A}_{xx}\,\underline{x} + \underline{A}_{xy}\,\underline{y} + \underline{b}_x) = 0\ ,$$
$$\tilde{\underline{y}}\,(\underline{A}_{yx}\,\underline{x} + \underline{A}_{yy}\,\underline{y} + \underline{b}_y) = 0$$

is equivalent to the original SPP (2.9).

Following the same arguments as in the case of free variables, we can show that (2.9) or (2.11) can be replaced by the pair of QPP's:

$$\min\ \{\ f'\ |\ \underline{\nabla} f_y \leq \underline{0},\ \ \underline{x} \geq \underline{0}\ \}, \tag{2.12.a}$$

$$\max\ \{\ f''|\ \underline{\nabla} f_x \geq \underline{0},\ \ \underline{y} \geq \underline{0}\ \} \tag{2.12.b}$$

These are the symmetric dual QPP's introduced by R. W. Cottle [7]. They are symmetric in the sense that the dual of the dual (2.12.b) is the primal (2.12.a) (which is not the case for general Nonlinear Programming Problems). Relations (2.10) are the Karush-Kuhn-Tucker conditions [8] for the QPP's (2.12).

Mixed free and non-negative variables, as well as mixed equality and inequality constraints, are treated in symmetric duality similarly as in conventional duality. Namely, each free primal variable generates a dual constraint of the equality type, while each non-negative primal variable yields an inequality. Conversely, each primal equality constraint generates a free dual variable and each inequality constraint of the primal problem corresponds to a non-negative variable of the dual problem.

Thus we have established the equivalence scheme: the set of equations (inequalities) <=> the SPP <=> the dual pair of QPP's. In Structural Mechanics the first part of that chain corresponds to the CSGR, the second - to the unconstrained energy principle, the third - to the pair of complementary energy principles.

After a certain mechanical problem has been formulated, it is important to check whether it has a solution for given data and whether that solution is unique. Such questions are much easier to answer at the level of QPP's than at the level of the original CSGR. Namely, there follows from the theory of duality that if one of the QPP's (2.8) or (2.12) has the solution ($\underline{x}_*$, $\underline{y}_*$) corresponding to the finite value of the minimized (maximized) function then it is simultaneously the solution of the dual problem. Moreover, there holds $f'(\underline{x}_*, \underline{y}_*) = f''(\underline{x}_*, \underline{y}_*)$. Thus, in order to establish the existence of the solution, one has to demonstrate for one of the dual QPP's that: a) there exists a point ($\underline{x}_*$, $\underline{y}_*$) satisfying the constraints, b) the relevant extremum is bounded.

Once its existence has been established, the uniqueness or non-uniqueness of the solution depends upon the properties of the minimized (maximized) function. Namely,the solution is unique when that function is strictly convex (concave). In the case of the quadratic forms f' , f'' considered here, that amounts to the requirement of positive (negative) definiteness of the submatrices $\underline{A}_{xx}$, $\underline{A}_{yy}$. Note that in any case those submatrices must be respectively positive and negative semidefinite in order to ensure the saddle shape of the potential f .

3. CONVENTIONAL PROBLEM OF ELASTIC ANALYSIS

First of all we apply the equivalence scheme introduced in the previous Section to the classical problem of static analysis of an elastic structure. Let the state of that structure be described by a finite number of parameters grouped into the four matrices:

- matrix of generalized displacements $\underline{u} \in E^n$,
- matrix of generalized strains $\underline{q} \in E^m$,
- matrix of generalized nodal forces $\underline{F} \in E^n$,
- matrix of generalized stresses $\underline{Q} \in E^m$.

Under assumption of small displacements the kinematic equation

$$\underline{q} = \underline{C}\,\underline{u} \tag{3.1}$$

and the equilibrium equation

$$\underline{F} = \tilde{\underline{C}}\,\underline{Q} \tag{3.2}$$

are mutually adjoint. The compatibility matrix $\underline{C}$ has the dimensions m x n with $m \geq n$. In a properly discretized and geometrically stable structure the rank of $\underline{C}$ is equal to the number of degrees of freedom n. We confine our attention to hyperstatic structures ($r = m - n > 0$), since the analysis of isostatic systems ($r = 0$) is rather trivial.

For a structure made of linearly elastic material the following constitutive relations hold:

$$\underline{Q} = \underline{E}\,\underline{q} \,, \tag{3.3.a}$$

$$\underline{q} = \underline{E}^{-1}\underline{Q} \,. \tag{3.3.b}$$

Here $\underline{E}$ is a positive definite elasticity matrix of rank m .

The conventional problem of static analysis is stated as follows: given $\underline{E}$, $\underline{C}$ and $\underline{F}$ find the response $\underline{u}_*$, $\underline{q}_*$, $\underline{Q}_*$. The CSGR for such a problem includes the relations (3.1), (3.2) and (3.3). Since no inequalities are present, the equivalence scheme (2.3) <=> (2.5) <=> (2.8) might be applicable. In order to check whether it is, we compare the governing equations against the pattern (2.3). The complete set considered can be written in the form of table (3.4), where the symbols in brackets indicate the dimensions of the submatrices.

We see now that the global matrix of coefficients of the set (3.4) is symmetric. Let the kinematic variables $\underline{q}$, $\underline{u}$ play the role of $\underline{x}$ while the static variable $\underline{Q}$ is of the type $\underline{y}$. This division is indicated in (3.4) by the dashed line. The only non-zero submatrix that appears in $\underline{A}_{xx}$ is the positive definite elasticity matrix $\underline{E}$. Hence $\underline{A}_{xx}$ is also positive definite. The submatrix $\underline{A}_{yy} = \underline{0}$ can be formally treated as negative semidefinite. Thus all the requirements of the scheme (2.3) are met by the CSGR (3.4).

$$
\begin{array}{rc|cc|c|c|cl}
 & & \tilde{\underline{q}} & \tilde{\underline{u}} & \tilde{\underline{Q}} & 1 & & \\
\hline
\underline{\nabla f}_q = & & \underline{E} & \underline{0} & -\underline{I} & \underline{0} & = \underline{0} & (m) \\
\underline{\nabla f}_u = & & \underline{0} & \underline{0} & \tilde{\underline{C}} & -\underline{F} & = \underline{0} & (n) \\
\hline
\underline{\nabla f}_Q = & & -\underline{I} & \underline{C} & \underline{0} & \underline{0} & = \underline{0} & (m) \\
\hline
 & & (m) & (n) & (m) & (1) & &
\end{array}
\qquad (3.4)
$$

According to (2.4) the potential of (3.4) reads:

$$f(\underline{q},\ \underline{u},\ \underline{Q}) = \tfrac{1}{2}\,\tilde{\underline{q}}\,\underline{E}\,\underline{q} - \tilde{\underline{q}}\,\underline{I}\,\underline{Q} + \tilde{\underline{u}}\,\underline{C}\,\underline{Q} - \tilde{\underline{F}}\,\underline{u} \qquad (3.5)$$

The SPP equivalent to (3.4) is then

$$\min_{\underline{q},\ \underline{u}}\ \max_{\underline{Q}}\ f(\underline{q},\ \underline{u},\ \underline{Q}) \qquad (3.6)$$

Applying the formulae (2.7) we find the Legendre's transforms of f :

$$
\begin{aligned}
f' &= \tfrac{1}{2}\,\tilde{\underline{q}}\,\underline{E}\,\underline{q} - \tilde{\underline{F}}\,\underline{u}\ , \\
f'' &= -\tfrac{1}{2}\,\tilde{\underline{q}}\,\underline{E}\,\underline{q}
\end{aligned}
\qquad (3.7)
$$

Since $\underline{\nabla f}_x = \{\underline{\nabla f}_q\ ,\ \underline{\nabla f}_u\}$ and $\nabla f_y = \{\underline{\nabla f}_Q\}$,the dual QPP's defined in (2.8) assume the following form:

$$\min\{\tfrac{1}{2}\,\tilde{\underline{q}}\,\underline{E}\,\underline{q} - \tilde{\underline{F}}\,\underline{u}\ \mid\ -\underline{q} + \underline{C}\,\underline{u} = \underline{0}\}\ , \qquad (3.8.a)$$

$$\max\{-\tfrac{1}{2}\,\tilde{\underline{q}}\,\underline{E}\,\underline{q}\ \mid\ \underline{E}\,\underline{q} - \underline{Q} = \underline{0},\ \ \tilde{\underline{C}}\,\underline{Q} = \underline{F}\}. \qquad (3.8.b)$$

The primal problem (3.8.a) expresses the well known principle of the minimum total potential energy. The dual one (3.8.b) can be transformed further: the first of its constraints allows one to eliminate $\underline{q}$ from the unknowns. Having done that, we replace the maximization by the minimization changing the sign of the function. The final result is the principle of the minimum complementary energy:

$$\min\{\ \tfrac{1}{2}\,\tilde{\underline{Q}}\,\underline{E}\,\underline{Q}\ \mid\ \tilde{\underline{C}}\,\underline{Q} = \underline{F}\ \} \qquad (3.9)$$

Let us investigate now the existence and the uniqueness of solutions of the dual QPP's (3.8), (3.9). It is more convenient to do that using the simpler static principle (3.9) . Its single constraint is the set of equilibrium equations (3.2). For a hyperstatic structure that set is underdetermined: the number of equations n is less then the number m of unknown stresses . According to linear algebra such a set of equations has a non-unique solution

$$\underline{Q} = \underline{\tilde{C}}^{-} \underline{F} + \underline{\tilde{C}}^{0} \underline{Z} \tag{3.10}$$

that depends upon arbitrarily chosen parameters collected in the matrix of redundants $\underline{Z} \in E^{r}$. The matrices $\underline{C}^{-}$ and $\underline{C}^{0}$ appearing transposed in (3.10) are the generalized inverse and the kernel of $\underline{C}$, respectively. They fulfill the following conditions:

$$\underline{C}\,\underline{\tilde{C}}^{-} = \underline{I} , \qquad \underline{C}\,\underline{\tilde{C}}^{0} = \underline{0} \tag{3.11}$$

Thus there exists an admissible solution (3.10) of (3.9) for any given load $\underline{F}$. On the other hand, the function of complementary energy is strictly convex due to positive definiteness of $\underline{E}$ and it is bounded from below by zero. Therefore (3.9) is the correctly formulated convex QPP with the unique solution $\underline{Q}_{*}$ for any given load $\underline{F}$. The theory of duality tells us that in such a case the dual (3.8.a) has also the unique solution $\underline{q}_{*}$, $\underline{u}_{*}$.

All variables of the dual QPP's (3.8), (3.9) are free and there are no inequalities among the constraints of those problems. This property allows one to derive explicit conditions for the relevant extrema in the form of matrix equations. For example, we can solve the constraint of (3.8.a) with respect to $\underline{q}$ and then substitute the result into the expression of potential energy. This converts (3.8.a) into the unconstrained minimization problem

$$\min \{ \tfrac{1}{2} \underline{\tilde{u}} \underline{K} \underline{u} - \underline{\tilde{F}} \underline{u} \} \tag{3.12}$$

where

$$\underline{K} = \underline{\tilde{C}} \underline{E} \underline{C} \tag{3.13}$$

is the stiffness matrix of the structure. Obviously, the minimum (3.12) is attained at $\underline{u}_{*}$ that corresponds to the vanishing gradient of the minimized function. Hence, $\underline{u}_{*}$ has to satisfy the matrix equation

$$\underline{K} \underline{u} = \underline{F} \tag{3.14}$$

which is the canonical equation of the Direct Stiffness Method (Displacement Method). Similarly, solving the constraint of (3.9) with respect to $\underline{Q}$ and substituting the result (3.10) into the expression of complementary energy we can replace (3.9) by the unconstrained problem

$$\min \{ \tfrac{1}{2} \underline{\tilde{Z}} \underline{B} \underline{Z} + \underline{\tilde{D}} \underline{Z} \} \tag{3.15}$$

where

$$\underline{B} = \underline{C}^{0} \underline{E}^{-1} \underline{\tilde{C}}^{0} \tag{3.16}$$

is the flexibility matrix of the structure and

$$\underline{D} = \underline{C}^{0} \underline{E}^{-1} \underline{\tilde{C}}^{-} \underline{F} \tag{3.17}$$

depends upon the given load. The condition for the minimum (3.15) amounts to the canonical matrix equation

$$\underline{B}\,\underline{Z} + \underline{D} = \underline{0} \tag{3.18}$$

of the Flexibility Method (Force Method).

Thus, the complete set of $2m + n$ equations (3.4) is reducible either to n equations (3.14) or to r equations (3.18). It is much easier to solve one of those reduced sets of equations than to solve the equivalent QPP's (3.8), (3.9). Therefore, in the conventional elastic analysis Quadratic Programming finds no numerical application.

4. ELASTIC STRUCTURE ON UNILATERAL SUPPORTS

The versatility of the QP-approach appears in its full strength when mechanical problems include sign-constrained variables. The simplest and most common example encountered in practice of such a problem is the analysis of an elastic structure resting on unilateral supports.

Consider a cantilever beam shown in Fig. 4.1. There is a given gap u_o between its right end and the support. Depending upon the intensity of load, two states of the beam are possible:

a) there is no contact with the support (Fig. 4.1.a) - $u < u_o$, $R = 0$;
b) there is a contact with the support (Fig. 4.1.b) - $u = u_o$, $R \geq 0$.

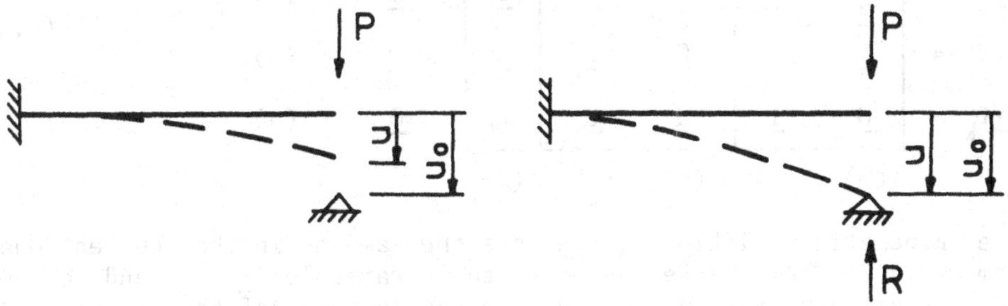

Fig. 4.1. Cantilever with a unilateral support.

Both states can be described by means of two linear inequalities and a single non-linear equation:

$$u - u_o \leq 0, \tag{4.1.a}$$

$$R \geq 0, \tag{4.1.b}$$

$$R(u - u_o) = 0 \tag{4.1.c}$$

In particular, there need be no gap: $u_o = 0$. We would say then that the beam is unilaterally supported at its right end.

Let us see how unilateral supports influence elastic response of the structure. To simplify the notation we assume that all generalized displacements are subject to the constraints of the type (4.1). This assumption is not restrictive since we can always assume large enough "gaps" for those displacements that are in fact free.

Introducing the matrix of reactions $\underline{R} \in E^n$ we can write the conditions (4.1) for the whole structure as

$$\underline{u} - \underline{u}_o \leq \underline{0}, \tag{4.2.a}$$

$$\underline{R} \geq \underline{0}, \tag{4.2.b}$$

$$\tilde{\underline{R}}\,(\underline{u} - \underline{u}_o) = 0. \tag{4.2.c}$$

These relations are to be incorporated into the CSGR together with the conventional equations (3.1), (3.2), (3.3). Since we have now non-negative variables (reactions) and the complementary slackness rule (4.2.c), the equivalence scheme (2.11) <=> (2.9) <=> (2.12) must be taken into account. To find out whether that scheme is applicable we try firstly to put the linear part of the governing relations into the symmetric form:

$$\begin{array}{rc|cc|cc|c|cl}
 & & \tilde{\underline{q}} & \tilde{\underline{u}} & \tilde{\underline{Q}} & \tilde{\underline{R}} & 1 & & \\
\hline
\underline{\nabla f}_q & = & \underline{E} & \underline{0} & -\underline{I} & \underline{0} & \underline{0} & = \underline{0} & (m) \\
\underline{\nabla f}_u & = & \underline{0} & \underline{0} & \tilde{\underline{C}} & \underline{I} & -\underline{F} & = \underline{0} & (n) \\
\hline
\underline{\nabla f}_Q & = & -\underline{I} & \underline{C} & \underline{0} & \underline{0} & \underline{0} & = \underline{0} & (m) \\
\underline{\nabla f}_R & = & \underline{0} & \underline{I} & \underline{0} & \underline{0} & -\underline{u}_o & \leq \underline{0} & (n) \\
\hline
 & & (m) & (n) & (m) & (n) & (1) & &
\end{array} \tag{4.3}$$

The kinematic variables $\underline{q}$, $\underline{u}$ are the same as in the conventional problem (3.4). There appears a new static variable $\underline{R} \geq \underline{0}$ and a new kinematic constraint (4.2.a) that enters the last row of the table (4.3). Note that the symmetry requires the identity matrix to appear in the second row of that table. This is correct since reactions have to be taken into account in the equilibrium equation:

$$\tilde{\underline{C}}\,\underline{Q} + \underline{R} = \underline{F} \tag{4.4}$$

Hence, we have now $\underline{x} = \{\underline{q},\ \underline{u}\}$ and $\underline{y} = \{\underline{Q},\ \underline{R}\}$. The submatrices $\underline{A}_{xx}$ and $\underline{A}_{yy}$ of the table (4.3) remain positive definite and negative semi-definite, respectively. Let us denote the potential (3.5) of the conventional elastic analysis problem by f_o. Then the potential of the CSGR in the presence of unilateral supports can be written as

$$f\,(\underline{q},\ \underline{u},\ \underline{Q},\ \underline{R}) = f_o + \tilde{\underline{u}}\,\underline{I}\,\underline{R} - \tilde{\underline{u}}_o\,\underline{R} \tag{4.5}$$

The equivalent SPP is

$$\min_{\underline{q},\,\underline{u}} \ \max_{\underline{Q},\,\underline{R} \geq \underline{0}} \ f(\underline{q},\,\underline{u},\,\underline{Q},\,\underline{R})\,. \tag{4.6}$$

Further application of the equivalence scheme leads to the complementary energy principles in the form of dual QPP's:

$$\min\{\tfrac{1}{2}\,\underline{\tilde{q}}\,\underline{E}\,\underline{q} - \underline{\tilde{F}}\,\underline{u} \mid -\underline{q} + \underline{C}\,\underline{u} = \underline{0},\ \ \underline{u} \leq \underline{u}_o\}, \tag{4.7.a}$$

$$\max\{-\tfrac{1}{2}\,\underline{\tilde{q}}\,\underline{E}\,\underline{q} + \underline{\tilde{u}}_o\,\underline{R} \mid \underline{E}\,\underline{q} - \underline{Q} = \underline{0},\ \ \underline{\tilde{C}}\,\underline{Q} + \underline{R} = \underline{F}\,,\ \ \underline{R} \geq \underline{0}\}. \tag{4.7.b}$$

The dual problem (4.7.b) can be transformed similarly to what was done in Section 3. Such a transformation yields the modified principle of the minimum complementary energy:

$$\min\{\tfrac{1}{2}\,\underline{\tilde{Q}}\,\underline{E}^{-1}\,\underline{Q} + \underline{\tilde{u}}_o\,\underline{R} \mid \underline{\tilde{C}}\,\underline{Q} + \underline{R} = \underline{F}\,,\ \ \underline{R} \geq \underline{0}\}. \tag{4.8}$$

Despite similarities in the overall structure of the CSGR's and of the complementary energy principles, the problem considered now differs substantially from that of the conventional elastic analysis presented in Section 3. First of all, it is no longer linear: a substantial non-linearity comes through the equation (4.2.c). Then, its solution may not exist for certain $\underline{F}$. That is clearly seen from the constraints of (4.8): there must exist such a stress state that equilibrates $\underline{F}$ together with non-negative reactions. One can imagine easily an example of a unilaterally supported beam (Fig. 4.2) that is in equilibrium only under non-negative load. On the other hand, if the solution exists then it is unique due to the convexity of the quadratic energy forms in the QPP's (4.7), (4.8).

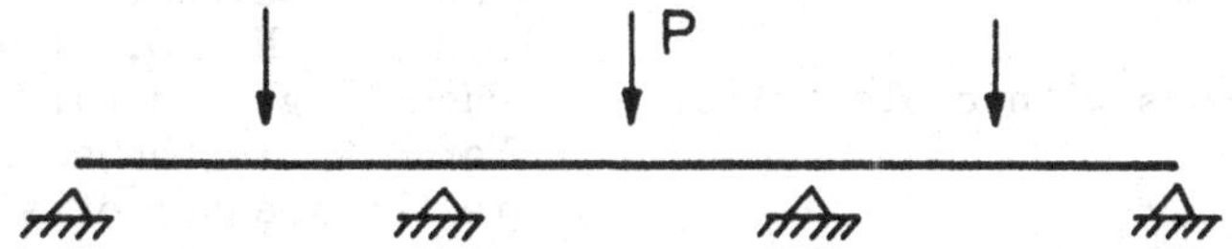

Fig. 4.2. Continuous beam resting on unilateral supports.

The next characteristic feature of the elastic analysis in the presence of unilateral supports is that such a problem can not be reduced to a single set of equations of the type (3.13) or (3.17). For example, eliminating strains from (4.7) we get the problem

$$\min\{\tfrac{1}{2}\,\underline{\tilde{u}}\,\underline{K}\,\underline{u} - \underline{\tilde{F}}\,\underline{u} \mid \underline{u} \leq \underline{u}_o\}, \tag{4.9}$$

which is not reducible any further. It does not mean, however, that the Direct Stiffness Method is useless for such problems. One can still use it in an iterative way: firstly all unilateral supports are assumed to be bilateral and the obtained ficticious structure is analysed in a conventional manner. Then the supports exhibiting negative reactions are released and the re-analysis is performed. Usually after few iterations one comes up with a final result that satisfies all the conditions (4.2). Alas, there exists no convergence proof for such a heuristic procedure and in certain situations one can expect lengthy calculations. On the other hand, the rigorous energy principles (4.7), (4.8), when applied directly for numerical solution, lead to very reliable solvers.

5. CABLE-STRUT SYSTEMS

Structural members that are supposed to carry tension loads only appear frequently in skeletal structures. For example, the braces of trusses and frames are often cables or very slender bars (Fig. 5.1). We consider in the present Section such structures that are rigid enough to allow the geometrically linear approach but at the same time include tension-only members.

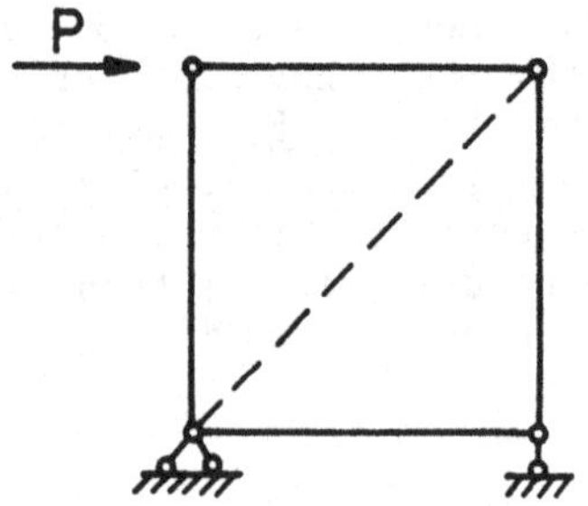

Fig. 5.1. Truss with cable brace.

In order to simplify things let us fix our attention to trusses. After the matrix model has been developed, it is easy to adapt it for any skeletal structure. We assume that the truss under consideration consists of two types of elements: a) ordinary struts able to transmit axial forces of arbitrary sign; b) tension-only members (cables). Such a member is shown in Fig. 5.2. It can be either in tension (Fig. 5.2.a): $N > 0$, $s = 0$; or slackened (Fig. 5.2.b): $N = 0$, $s \geq 0$. Here N is the axial force and s is the measure of slackness.

Denoting by q the total axial strain, e its elastic part, EA the stiffness in tension and L the length of element, we can write the constitutive law of such an element as follows:

$$q = e - s \ , \tag{5.1.a}$$

$$e = (L/EA)\, N \ , \tag{5.1.b}$$

$$N \geq 0, \tag{5.1.c}$$

$$s \geq 0, \tag{5.1.d}$$

$$s\, N = 0 \ . \tag{5.1.e}$$

Note that for an element in tension there is $N > 0$ which implies $s = 0$ according to (5.1.e). Then $q = e > 0$. On the other hand, for the slackened element there is $N = 0$ and $e = 0$, whereas $q = -s$ can be negative.

In order to have a single constitutive law for both ordinary struts and cables, we replace the sign constraint $N \geq 0$ in (5.1) by a more general condition $N \geq N_o$. Then the complementary slackness rule (5.1.e) must be replaced by $s(N - N_o) = 0$. The physical meaning of the above modification is obvious: the element remains linearly elastic until its axial force is greater then a given non-positive value N. When the axial force becomes equal to N_o, then an arbitrary shortening $-s$ of the element is possible without any additional effort. The situation $N < N_o$ is not feasible.

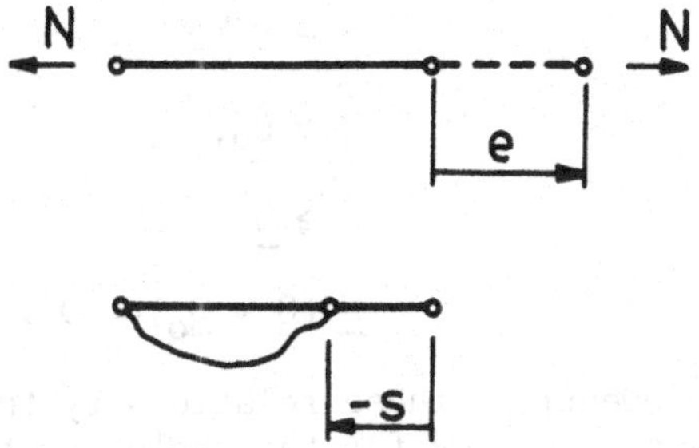

Fig. 5.2. Cable element.

Such a behaviour is shown graphically in Fig. 5.3.a. In particular, there might be $N_o = 0$ which corresponds to the cable member (Fig. 5.3.b). On the other hand, we can take a sufficiently large negative value of N_o if we want to exclude the slackened state. Thus ordinary struts can be modelled by the same law (5.1).

Coming to the matrix description of the structure, we collect the slackness parameters s_j of individual elements into the matrix $\underline{s} \in E^m$ and the physical constants N_{oj} into the matrix $\underline{Q}_o \in E^m$. The entries of the latter matrix are defined as follows:

$$Q_{oj} = \begin{cases} \text{zero, when the j-th element is a cable;} \\ \text{large negative number, when it is a strut.} \end{cases} \qquad (5.2)$$

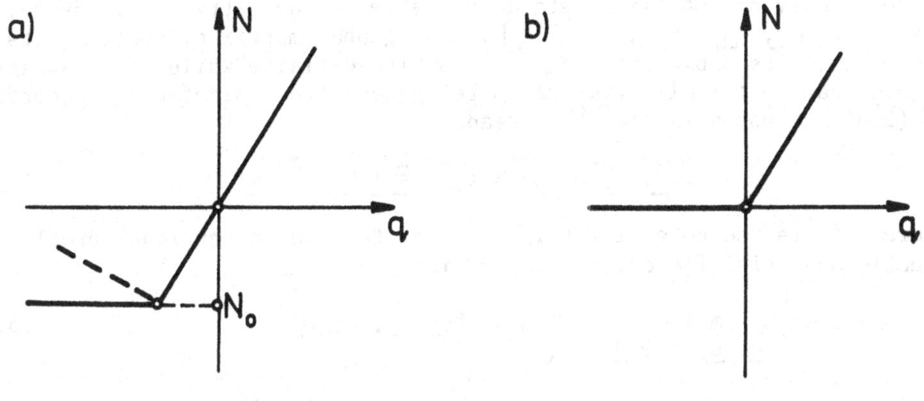

Fig. 5.3. Restricted resistance to compression.

One could introduce also certain intermediate values Q_{oj} for slender elements that are subject to buckling under compression. It should be kept in mind, however, that the post buckling behaviour can be of the geometrically softening type as shown by the dashed line in Fig. 5.3.a. For such elements the approach presented here would be inappropriate.

Taking into account the above considerations, we write the constitutive relations for the entire structure as follows:

$$\underline{q} = \underline{e} - \underline{s} \,, \tag{5.3.a}$$

$$\underline{e} = \underline{E}^{-1}\underline{Q} \,, \tag{5.3.b}$$

$$\underline{Q} \geq \underline{Q}_o , \tag{5.3.c}$$

$$\underline{s} \geq \underline{0} \,, \tag{5.3.d}$$

$$\tilde{\underline{s}}\,(\underline{Q} - \underline{Q}_o) = 0 \,. \tag{5.3.e}$$

Supplementing those relations by the adjoint equations (3.1), (3.2) we obtain the CSGR for the cable-strut system. The linear part of that set can be presented in the form of table:

$$\begin{array}{rl|ccc|c|c|ll}
 & & \tilde{\underline{s}} & \tilde{\underline{e}} & \tilde{\underline{u}} & \tilde{\underline{Q}} & 1 & & \\
\hline
\underline{\nabla f}_s & = & \underline{0} & \underline{0} & \underline{0} & \underline{I} & -\underline{Q}_o & \geq \underline{0} & (m) \\
\underline{\nabla f}_e & = & \underline{0} & \underline{E} & \underline{0} & -\underline{I} & \underline{0} & = \underline{0} & (m) \\
\underline{\nabla f}_u & = & \underline{0} & \underline{0} & \underline{0} & \tilde{\underline{C}} & -\underline{F} & = \underline{0} & (n) \\
\hline
\underline{\nabla f}_Q & = & \underline{I} & -\underline{I} & \underline{C} & \underline{0} & \underline{0} & = \underline{0} & (m) \\
\hline
 & & (m) & (m) & (n) & (m) & (1) & &
\end{array} \tag{5.4}$$

This time the kinematic variables include $\underline{e}$, $\underline{u}$ and $\underline{s} \geq \underline{0}$, while there is only one matrix of static variables - the matrix $\underline{Q}$. Hence, we take $\underline{x} = \{\underline{s}, \underline{e}, \underline{u}\}$, $\underline{y} = \{\underline{Q}\}$. The global matrix of coefficients is symmetric, its submatrix $\underline{A}_{xx}$ is positive definite while the submatrix $\underline{A}_{yy} = \underline{0}$ can be formally treated as being negative semidefinite. According to (2.4) the potential function reads

$$f(\underline{s}, \underline{e}, \underline{u}, \underline{Q}) = f_o + \tilde{\underline{s}}\,\underline{I}\,\underline{Q} - \tilde{\underline{Q}}_o\,\underline{s} \tag{5.5}$$

where f_o is the potential (3.5) derived for the conventional problem of elastic analysis. The equivalent SPP is

$$\min_{\underline{e},\, \underline{u},\, \underline{s} \geq \underline{0}} \; \max_{\underline{Q}} \; f(\underline{s}, \underline{e}, \underline{u}, \underline{Q}) \tag{5.6}$$

It is easy to check using the patterns (2.8), (2.12) that the equivalent dual QPP's have the form:

$$\min\{\tfrac{1}{2}\underline{\tilde{e}}\,\underline{E}\,\underline{e}-\underline{\tilde{F}}\,\underline{u}-\underline{\tilde{Q}}_o\,\underline{s}\;|\;\underline{C}\,\underline{u}+\underline{s}-\underline{e}=\underline{0},\quad \underline{s}\geq\underline{0}\}\qquad(5.7.a)$$

$$\max\{-\tfrac{1}{2}\underline{\tilde{e}}\,\underline{E}\,\underline{e}\;|\;\underline{E}\,\underline{e}-\underline{Q}=\underline{0},\quad \underline{\tilde{C}}\,\underline{Q}=\underline{F},\quad \underline{Q}\geq\underline{Q}_o\}\qquad(5.7.b)$$

The latter QPP can be transformed into the modified principle of minimum complementary energy:

$$\min\{\tfrac{1}{2}\underline{\tilde{Q}}\,\underline{E}^{-1}\underline{Q}\;|\;\underline{\tilde{C}}\,\underline{Q}=\underline{F},\quad \underline{Q}\geq\underline{Q}_o\}\qquad(5.8)$$

From the mathematical point of view the problem considered now is very similar to that of an elastic structure resting on unilateral supports. It is also non-linear (due to equation (5.3.e)) and its solution is unique provided it exists. A possible lack of solution is clearly seen in Fig. 5.1: the truss braced by means of a cable can sustain $P \geq 0$ only.

Since the constraints of the QPP's (5.7), (5.8) include inequalities, the analysis of a cable-strut system is not reducible to the solution of equations of the type (3.13) or (3.17). For example, eliminating elastic strains $\underline{e}$ from (5.7.a) we end up with a constrained extremum problem:

$$\min\{\tfrac{1}{2}\underline{\tilde{u}}\,\underline{K}\,\underline{u}+\tfrac{1}{2}\underline{\tilde{s}}\,\underline{E}\,\underline{s}-\underline{\tilde{s}}\,\underline{E}\,\underline{C}\,\underline{u}-\underline{\tilde{F}}\,\underline{u}-\underline{\tilde{Q}}_o\,\underline{s}\;|\;\underline{s}\geq\underline{0}\}\qquad(5.9)$$

Note that if all elements of the structure were ordinary struts, then all the entries of $\underline{Q}_o$ would have large negative values. That would lead to the solution $\underline{s}_* = \underline{0}$ since any $s_j > 0$ increases considerably the value of the minimized function. Obviously, under $\underline{s} = 0$ the QPP (5.9) reduces to the conventional form (3.11) that generates the canonical equations of the Direct Stiffness Method.

Similarly, as in the case of unilateral supports, the cable-strut systems can be analysed through iterative application of (3.13) or (3.17). At the beginning, all cables are replaced by ordinary members. Those of them that turn out to be in compression are then removed in the subsequent iteration. For simple structures it is easy to predict the signs of individual member forces but for more complicated systems the convergence of the above mentioned iterative procedure can be slow. Then the direct application of the energy principles (5.7), (5.8) is preferable.

6. PLASTIC LIMIT ANALYSIS

Evaluation of the safety factor against plastic collapse is perhaps the method of non-linear analysis most commonly used in engineering practice . Since the theory of Plastic Limit Analysis and its relations to Mathematical Programming are dealt with in other chapters of the present book, we restrict ourselves to a brief demonstration of the advantages that the proposed methodology brings in that field.

Exploiting the fact that the ultimate load does not depend upon elastic properties of the material, we introduce a simple rigid-perfectly-plastic model of constitutive behaviour:

$$\underline{N}\,\underline{Q} \leq \underline{R}\,, \tag{6.1.a}$$

$$\underline{\dot{q}} = \underline{N}\,\underline{\dot{\lambda}}\,, \tag{6.1.b}$$

$$\underline{\dot{\lambda}} \geq \underline{0}\,, \tag{6.1.c}$$

$$\underline{\tilde{\dot{\lambda}}}\,(\underline{N}\,\underline{Q} - \underline{R}) = 0. \tag{6.1.d}$$

Here $\underline{N}$ is the $(m \times \ell)$-matrix of gradients of the piecewise-linear yield surface, $\underline{R} \in E^{\ell}$ is the matrix of plastic moduli and $\underline{\dot{\lambda}} \in E^{\ell}$ is the matrix of plastic multipliers.

The set of inequalities (6.1.a) defines the convex polyhedron of admissible stresses, the equation (6.1.b) together with the sign constraint (6.1.c) expresses the associated flow rule, whereas the non-linear equation (6.1.d) excludes the consideration of plastic multipliers corresponding to the passive yield planes.

Let the load increase monotonically according to the relation

$$\underline{F} = s\,\underline{F}_o \tag{6.2}$$

where $\underline{F}_o$ denotes a certain reference loading and s is the load factor. Our aim is to find the ultimate value s_* of that factor together with the stress state $\underline{Q}_*$ corresponding to the incipient plastic collapse and the mechanism $\underline{\dot{q}}_*$, $\underline{\dot{u}}_*$ of that collapse.

In order to obtain the CSGR of the limit analysis we must supplement the constitutive law (6.1) by the kinematic equation

$$-\,\underline{\dot{q}} + \underline{C}\,\underline{\dot{u}} = \underline{0} \tag{6.3}$$

and by the equilibrium equation

$$\underline{\tilde{C}}\,\underline{Q} - s\,\underline{F}_o = \underline{0}\,. \tag{6.4}$$

We take now $\underline{x} = \{\underline{\dot{\lambda}}\,,\ \underline{\dot{u}}\}$ and $\underline{y} = \{\underline{Q},\ s\}$. After the plastic multipliers have been computed, the strain rates in the collapse mechanism can be found from (6.1) . Therefore $\underline{\dot{q}}$ is not included in $\underline{x}$.

The linear part of the CSGR constitutes the following table:

$$\begin{array}{rc|ccc|c|cc}
 & \underline{\tilde{\dot{\lambda}}} & \underline{\tilde{u}} & \underline{\tilde{Q}} & s & 1 & & \\
\hline
\underline{\nabla f}_{\lambda} = & \underline{0} & \underline{0} & -\underline{\tilde{N}} & \underline{0} & \underline{R} & \geqslant \underline{0} & (\ell) \\
\underline{\nabla f}_{u} = & \underline{0} & \underline{0} & \underline{\tilde{C}} & -\underline{F}_o & \underline{0} & = \underline{0} & (n) \\
\hline
\underline{\nabla f}_{Q} = & -\underline{N} & \underline{C} & \underline{0} & \underline{0} & \underline{0} & = \underline{0} & (m) \\
\nabla f_{s} = & \underline{0} & -\underline{\tilde{F}}_o & \underline{0} & 0 & 1 & = 0 & (1) \\
\hline
 & (\ell) & (n) & (m) & (1) & (1) & &
\end{array} \tag{6.5}$$

The first row of it corresponds to the stress admissibility condition

(6.1) , the second - to the equilibrium equation (6.4) and the third - to the kinematic equation (6.3) with $\dot{\underline{q}}$ replaced by its expression through $\dot{\underline{\lambda}}$ according to (6.1) . Since we have four unknowns $\dot{\underline{\lambda}}$, $\dot{\underline{u}}$, $\underline{Q}$ and s , there must be the fourth row of the table corresponding to the derivative of potential with respect to s . The first four entries of that row are defined by the requirement that the global matrix of coefficients must be symmetric: they follow from the transposition of the fourth column. The fifth entry is an arbitrary constant, e.g. unity, and there must be an equality sign attached to that row since s is treated as a free variable. Hence, the fourth row contains the equation

$$\dot{W} = \tilde{\underline{F}}_o \dot{\underline{u}} = 1 \tag{6.6}$$

that should be present in the CSGR. Its physical meaning is simple: one has to normalize the external power W because otherwise the motion of the collapse mechanism would be unrestricted. This shows one of the advantages of the systematic approach presented here: even if certain necessary relations have been omitted when constructing the CSGR, they come out automatically through the symmetry condition.

Both $\underline{A}_{xx}$ and $\underline{A}_{yy}$ are zero matrices now, which does not prevent us regarding them formally as being positive and negative semidefinite, respectively. Thus all the prerequisites for the application of the equivalence scheme are fulfilled. The potential function contains only bilinear and linear terms:

$$f(\dot{\underline{\lambda}}, \dot{\underline{u}}, \underline{Q}, s) = -\dot{\tilde{\underline{\lambda}}}\,\tilde{\underline{N}}\,\underline{Q} + \dot{\tilde{\underline{u}}}\,\tilde{\underline{C}}\,\underline{Q} - s\,\tilde{\underline{F}}_o\dot{\underline{u}} + \tilde{\underline{R}}\,\dot{\underline{\lambda}} + s\,. \tag{6.7}$$

The solution ($\dot{\underline{\lambda}}_*$, $\dot{\underline{u}}_*$, $\underline{Q}_*$, s_*) corresponds to the saddle point

$$\min_{\dot{\underline{u}},\ \dot{\underline{\lambda}} \ge \underline{0}}\ \max_{\underline{Q},\ s}\ f(\dot{\underline{\lambda}}, \dot{\underline{u}}, \underline{Q}, s), \tag{6.8}$$

or to the solutions of the following dual Linear Programming Problems (LPP's):

$$\min \{ \tilde{\underline{R}}\,\dot{\underline{\lambda}} \mid \underline{C}\,\dot{\underline{u}} - \underline{N}\,\dot{\underline{\lambda}} = \underline{0},\ \tilde{\underline{F}}_o\dot{\underline{u}} = 1,\ \dot{\underline{\lambda}} \ge \underline{0} \}, \tag{6.9.a}$$

$$\max \{ s \mid \tilde{\underline{C}}\,\underline{Q} - s\,\underline{F}_o = \underline{0},\ \tilde{\underline{N}}\,\underline{Q} \le \underline{R} \}. \tag{6.9.b}$$

The problem (6.9.b) is recognizable immediately as the static theorem of limit analysis. Its dual (6.9.a) represents the kinematic theorem though in a slightly modified form. Usually the latter is formulated as follows: the limit load factor s_* is the smallest of factors

$$s_\kappa = \dot{D} / \dot{W} \tag{6.10}$$

computed for all kinematically admissible collapse mechanisms. Here $\dot{D}$ stands for the dissipated power and $\dot{W}$ denotes the external power. It can be easily shown that

$$\dot{D} = \tilde{\underline{Q}}\,\dot{\underline{q}} = \tilde{\underline{R}}\,\dot{\underline{\lambda}}\,. \tag{6.11}$$

On the other hand, there follows from (6.10) that $s_{\kappa} = \dot{D}$, when $\dot{W} = 1$. Thus in problem (6.9.a) we are in fact minimizing the kinematic load factor.

The static problem (6.9.b) has at least one admissible solution for any given $\underline{F}_o$, namely, the trivial solution $\underline{Q}_* = 0$, $s_* = 0$. The maximum of s is unique and finite. The duality theory tells us that the kinematic problem (6.9.a) has also a solution $\dot{\underline{\lambda}}_*$, $\dot{\underline{u}}_*$ that corresponds to the minimum of dissipated power. However, neither s_* nor $\underline{\lambda}_*$ and $\dot{\underline{u}}_*$ need necessarily be unique. This follows from the semi-definiteness of the submatrices $\underline{A}_{xx}$ and $\underline{A}_{yy}$ of table (6.5). The linear functions f′, f′′ appearing in the dual problems (6.9) can represent the hyperplanes parallel to certain faces of the admissible domains. In such a case any point belonging to that face is a solution of the relevant LPP.

Physically the non-uniqueness of the stress states and collapse mechanisms comes from the simplified nature of the rigid-perfectly-plastic model. For example, a collapse mechanism of the hyperstatic structure is often incomplete: a certain part of the structure may col-lap-se while the rest of it remains statically indeterminate. Then the distribution of internal forces in the latter part can not be determined uniquely unless the elastic properties of the material are known.

7. CONCLUDING REMARKS

The equivalence scheme CSGR <=> SPP <=> dual QPP's can be applied to any problem of Structural Mechanics that posseses symmetry (self-adjointness) of the linear part of the CSGR and piecewise-linearity of the constitutive law. Many problems of that kind arise in the theory of plasticity: the incremental elastic-plastic analysis, the shakedown problem, the optimum plastic design. They are covered by other chapters of the present book.

Assuming that the conceptual advantages of the proposed methodology have been already demonstrated, let us add a few comments regarding numerical applications. As mentioned already in the Introduction, the efficiency of the available QP-solvers is less when compared to that of the solvers of linear algebraic equations. Hence, whenever the problem has large dimensions and when it is at the same time only moderately non-linear, an ad hoc modification of any available F.E.M.-code with regard to the non-negative variables and complementary slackness rules is reasonable. On the other hand, direct solution of the dual QPP's or LPP's is rational for strongly non-linear problems of a moderate size. There is little doubt, for example, that one should use easily available and fast Simplex codes for the evaluation of ultimate load. Problems involving elastic properties can be efficiently solved , e.g., by the QPSOL code developed by P.E. Gill et al [9] or by the ZQPCVX procedure written by M. J. D. Powell [10].

For educational purposes, as well as for tests of new algorithms and solution procedures, one would like to have a flexible solver of all mathematical problems related to the proposed methodology. Such is the package MATRIX developed by the author and his co-workers [11]. It is a user friendly conversational system implemented on IBM PC computer. Besides the usual operations of matrix calculus and linear algebra (multiplication, inversion, solution of equations, eigenvalues and eigenvectors, etc.) the package solves the LP- and QP-problems. Low cost and sufficient efficiency make of it a convenient tool for research and education.

REFERENCES

1. Maier, G.: Quadratic programming and theory of elastic-perfectly plastic structures, Meccanica, 3 (1968), 265-273.
2 Maier, G.: A matrix structural theory of piecewise linear elastoplasticity with interacting yield planes, Meccanica, 5 (1970), 54-66.
3. Capurso, M. and G. Maier: Incremental elastoplastic analysis and quadratic optimization, Meccanica, 5 (1970), 107-116.
4. De Donato, O.: Sul calcolo della strutture nonlineari mediante programmazione quadratica, Rend. Ist. Lomb., Cl. Sci., Vol. 103, 1969.
5. Cohn, M. Z. and G. Maier, eds.: Engineering Plasticity by Mathematical Programming, Proc. NATO ASI, Waterloo 1977, Pergamon Press, New York 1979.
6. Borkowski, A.: Statics of Elastic and Elastoplastic Skeletal Structures, Elsevier-PWN, Warsaw 1987.
7. Cottle, R. W.: Symmetric dual quadratic programs, Quarterly of Applied Mathematics, 21 (1963) 237.
8. Mangasarian, O. L.: Nonlinear Programming, McGraw-Hill, New York 1969.
9. Gill, P. E., Murray, W. and M.A. Saunders: User's guide for SOL/QPSOL: a FORTRAN package for quadratic programming, Report SOL 83-7, Department of Operations Research, Stanford University, California 1983.
10. Powell, M. J. D.: Report of DAMTP 1983/NA 17, Department of Applied Mechanics, University of Cambridge, Cambridge 1983.
11. Borkowski, A., Siemiątkowska, B., Weigl, M.: User's guide for "MATRIX": a matrix interpretation system for linear algebra and mathematical programming, Laboratory of Adaptive Systems, Institute of Fundamental Technological Research, Warsaw 1985.

CHAPTER 8

ELASTIC-PLASTIC ANALYSIS OF STRUCTURAL CROSS-SECTIONS

J. A. Teixeira de Freitas
Istituto Superior Tecnico, Lisbon, Portugal

Abstract: **The structural cross-section is interpreted as an assembly of finite elements. General stress and strain functions are used to model multi-axial states and to generate consistent formulations for the fundamental conditions of equilibrium, compatibility and elastoplasticity. The characterization of limit states is encoded by linear {quadratic} programming problems for structural materials exhibiting unbounded {bounded} yield plateaux. An incremental procedure is used to trace the pre- and post-collapse response of the cross-section under analysis.**

Introduction

An essential requirement in the formulation of elastic-plastic analysis procedures of skeletal structures is the representation of the constitutive relations in terms of stress - and strain - resultants.

The first attempts to model the non-linear behaviour of symmetrical cross-sections are due to Saint-Venant [1] and Ewing [2].

The development of the plastic methods of structural analysis in the 1950's was accompanied by attempts to derive yield surface expressions for steel sections. The effect of axial force on the fully plastic moment was by then clearly understood [3]. However, the combined action of bending and twisting and the presence of shear proved extremely difficult to formulate, particularly for unsymmetrical sections. The approximate procedures suggested by Brown [4], Hill [5,6], Drucker [7] and Hodge [8] inspired most of the contributions published in the following years [9-17].

Limited ductility, creep, the effects of bond deterioration and cracking add further difficulties to the representation of the response of structural concrete sections. Selected contributions can be found in [18,19] and a detailed analysis of the literature on the problem of combined torsion, shear and bending of reinforced concrete elements is presented in [20]. Results of investigations into the load carrying capacity of reinforced concrete sections, composite concrete steel sections and prestressed concrete sections can be found in references [21-24], [25-27] and [19,28,29], respectively.

The behaviour of cross-sections is traditionaly represented by a set of integral equations relating stress- and strain-resultants. The stresses are required to satisfy equilibrium of forces and moments and the strains are customarily assumed to vary linearly for axial and flexural modes. Warping functions have however to be adopted in the representation of shear and torsion modes. Uniaxial stress-strain curves are used to quantify the stress-strain relationship.

Non-symmetrical sections or those subjected to combined stress-resultants lead to complicated boundary conditions and iterative or inverse numerical procedures have to be used to solve the governing integral equations.

Discretization techniques are beeing increasingly adopted in the analysis of the behaviour of cross-sections. The section is interpreted as an assembly of finite elements in which uniaxial states of stress are assumed so that the stress-strain curves for steel or concrete apply. Numerical integration over the width and depth of the cross-sectional mesh is then performed to evaluate the internal forces.

This procedure has been adopted by Abdel-Baset [30,31], Appleton [32,33] and Sacchi [34] in conjunction with mathematical programming techniques.

This approach is herein extended to include general stress and strain functions, which are used to derive consistent equilibrium, compatibility and constitutive relations, in which Static-Kinematic Duality is preserved. The determination of the stress-resultant limit states at selected locations on the nonlinear collapse locus is still encoded as a linear programming problem for structural materials exhibiting unbounded yield plateaux. A quadratic programming problem is however found to be associated with materials with limited ductility. An incremental mathematical programming procedure is used to trace the elastoplastic stress-strain resultant relationship representing the pre- and post-collapse response of the cross-section under analysis.

Equilibrium and Compatibility Conditions

Assume that a generic cross-section of a prismatic member is discretized into homogeneous finite elements, as illustrated in figure 1.

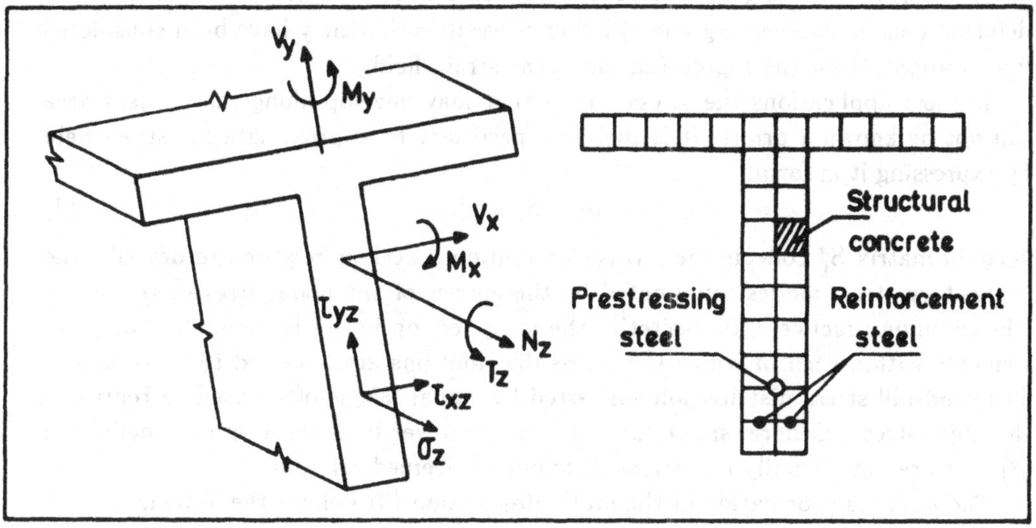

Figure 1

Let the components required to represent the state of strain developing in the i-th element be listed in the strain vector, ϵ_i.

The strain field can be expressed in form,

$$\epsilon_i = \mathbf{S}_i^t \, \mathbf{e}_i, \tag{1}$$

wherein matrix **S** collects the shape functions required to describe the element strain modes, defined by the entries of the deformation vector, $\mathbf{e}_i$. The strains are often assumed to vary linearly (plane-section assumption). Nonlinear shape functions can however be incorporated in definition (1) to simulate warping modes.

Let the stress vector, σ_i, collect the stresses associated with the strain components ϵ_i.

The stress-resultants $\mathbf{s}_i$ associated with the deformation parameters $\mathbf{e}_i$ are herein quantified by the dual transformation of the strain distribution definition (1), to yield,

$$\mathbf{s}_i = \int \tilde{\mathbf{S}}_i^{\epsilon} \, \sigma_i \, d\,A_i, \tag{2}$$

where A_i denotes the area of the element. Definition (2) ensures that the stress-resultants dissipate the same energy as the associate stress field:

$$\tilde{s}_i \, \mathbf{e}_i = \int \tilde{\sigma}_i \, \epsilon_i \, d\,A_i.$$

It can be easily verified that definition (2) generates the axial and shear forces or the bending and twisting moments and bimoments associated with the axial and shear deformations or the bending and twisting curvatures that may have been considered in definition (1) for the representation of the strain field.

In most applications the stress modes that may develop along the element area can not be known a priori. It is therefore necessary to approximate the stress field by expressing it in form,

$$\sigma_i - \mathbf{S}_i^{\sigma} \, \hat{\mathbf{s}}_i, \tag{3}$$

wherein matrix $\mathbf{S}_i^{\sigma}$ collects the stress distribution functions required to describe the elementary stress modes represented by the entries of the nodal stresses vector, $\hat{\mathbf{s}}_i$. The common practice is to discretize the cross section into a large number of small elements within each of which the stress distributions are assumed to be constant. The sand-hill stress distribution suggested by Nadai [35] is often used to represent the shear stress. General shape functions can however be incorporated in definition (3) to represent virtually any stress distribution deemed relevant.

The dual transformation of the static description (3) defines the kinematic variables associated with the nodal stresses $\hat{\mathbf{s}}_i$:

$$\hat{\mathbf{e}}_i = \int \tilde{\mathbf{S}}_i^{\sigma} \, \epsilon_i \, d\,A_i. \tag{4}$$

The above definition ensures that the nodal strain-resultants, $\hat{\mathbf{e}}_i$, dissipate the same energy as the associate strain components:

$$\tilde{\hat{\mathbf{e}}}_i \, \hat{\mathbf{s}}_i = \int \tilde{\epsilon}_i \, \sigma_i \, d\,A_i.$$

The element compatibility conditions are obtained substituting the strain field description (1) in definition (4) to yield,

$$\hat{\mathbf{e}}_i = \mathbf{C}_i \, \mathbf{e}_i, \tag{5}$$

wherein

$$\mathbf{C}_i = \int \tilde{\mathbf{S}}_i^\sigma \, \mathbf{S}_i^\epsilon \, dA_i. \tag{6}$$

The dual transformation of the compatibility condition (5) represents the element equilibrium conditions:

$$\mathbf{s}_i = \tilde{\mathbf{C}}_i \, \hat{\mathbf{s}}_i. \tag{7}$$

It is obtained substituting the stress field description (3) in definition (2) for the element stress-resultants.

In general, the entries of the element compatibility matrix (6) have to be determined using numerical integration procedures. Analytical solutions can however be obtained for simple, polyhedric elements, particularly when the strain and stress modes are represented by orthogonal shape functions.

Consider now the problem of assembling the elementary conditions (5) and (7) to obtain the systems implementing the equilibrium and compatibility conditions of the cross-section under analysis.

Let the parameters required to describe the section deformation modes be collected in the independent deformations vector, $\mathbf{e}$.

The deformations $\mathbf{e}_i$ developing in the elements into which the section has been discretized can be expressed in terms of the independent deformations through an incidence condition of the form:

$$\mathbf{e}_i = \mathbf{J}_i \, \mathbf{e}. \tag{8}$$

The dual transformation of the kinematic incidence condition (8) defines the static variables associated with the section independent deformations:

$$\mathbf{s} = \sum_i \tilde{\mathbf{J}}_i \, \mathbf{s}_i. \tag{9}$$

The independent stress-resultants $\mathbf{s}$ are therefore so defined as to ensure a global balance in the dissipation of energy throughout the section:

$$\tilde{\mathbf{s}} \, \mathbf{e} = \sum_i \tilde{\mathbf{s}}_i \, \mathbf{e}_i.$$

The compatibility and equilibrium conditions of the cross-section are obtained substituting conditions (8) in system (5) and the equilibrium condition (7) in definition (9). After collecting for all constituent elements,

$$\hat{\mathbf{s}} = \{\hat{\mathbf{s}}_i\} \ , \ \hat{\mathbf{e}} = \{\hat{\mathbf{e}}_i\},$$

the following result is obtained,

$$\hat{\mathbf{e}} = \mathbf{C} \, \mathbf{e}, \tag{10}$$

$$\mathbf{s} = \tilde{\mathbf{C}} \, \hat{\mathbf{s}}, \tag{11}$$

wherein $\mathbf{C} = \text{row}\,[(\mathbf{C}J)_i]$.

Equations (10) and (11) exhibit contragradient or dual transformations. Since the pairs of dual variables $\{\mathbf{s}, \hat{\mathbf{s}}\}$ and $\{\mathbf{e}, \hat{\mathbf{e}}\}$ are static and kinematic, the relationship implied in (10) and (11) is termed Static-Kinematic Duality (SKD) following Munro [36]. The virtual work equation,

$$D = W, \tag{12}$$

wherein

$$D = \tilde{\hat{\mathbf{s}}}\,\hat{\mathbf{e}}$$

represents the internal dissipation, and

$$W = \tilde{\mathbf{s}}\,\mathbf{e},$$

the external work, can be derived by performing the inner product of the equilibrium condition (11) with the compatibility condition (10). The virtual work equation which is often viewed as the axiomatic ingredient in linear structural mechanics [37] may therefore be interpreted as an energetic representation of SKD.

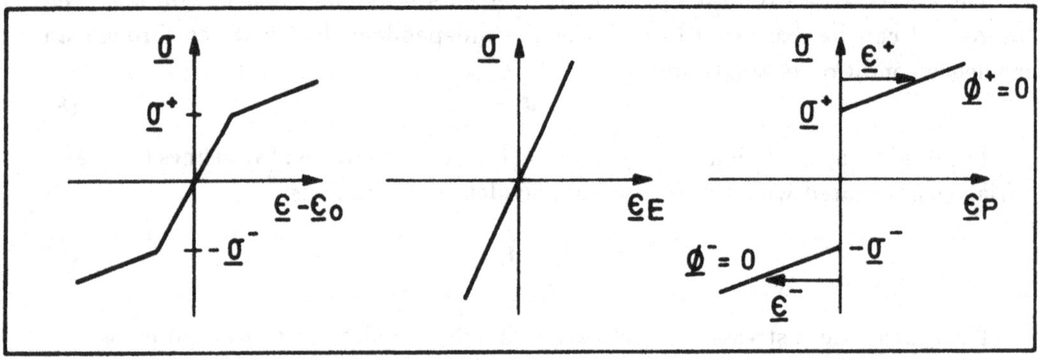

Figure 2

The analysis of the response of the cross-section is usually implemented assuming that the acting stress-resultants $\mathbf{s}$ vary proportionally to a load parameter, λ,

$$\mathbf{s} = \mathbf{C}_0\,\lambda, \tag{13}$$

yielding the following description for the equilibrium condition (11):

$$\mathbf{0} = \left[\,\tilde{\mathbf{C}} \quad \mathbf{C}_0\,\right] \left\{ \begin{array}{c} \hat{\mathbf{s}} \\ -\lambda \end{array} \right\}.$$

To preserve SKD, the compatibility condition (10) generalizes into form:

$$\left\{ \begin{array}{c} \hat{\mathbf{e}} \\ \bar{W} \end{array} \right\} = \left[\begin{array}{c} \mathbf{C} \\ \tilde{\mathbf{C}}_0 \end{array} \right] . \mathbf{e} \quad \begin{array}{c} (a) \\ (b) \end{array} \tag{14}$$

The virtual work equation (12) can be used to show that parameter $\bar{W}$ represents the external work dissipation rate:

$$\bar{W} = \tilde{\mathbf{s}}\, \mathbf{e}_{,}\lambda.$$

Elastoplastic Constitutive Relations

It is herein assumed that the structural materials constituting the finite elements into which the cross-section is discretized follow piecewise linear elastic-plastic laws relating stresses and strains.

Besides the residual strains ϵ_0 that may exist, two components may therefore be distinguished in the strain vector ϵ, namely the elastic and plastic addends ϵ_e and ϵ_p, respectively:

$$\epsilon = \epsilon_e + \epsilon_p + \epsilon_0. \tag{15}$$

The elastic strain developing in the i-th element can be expressed in form,

$$\epsilon_{ei} = \mathbf{F}'_i\, \sigma_i, \tag{16}$$

wherein $\mathbf{F}'_i$ denotes a flexibility matrix in which the elastic constants are organized.

To obtain the description of the elasticity conditions in terms of nodal stresses and strain-resultants,

$$\hat{\mathbf{e}}_{ei} = \mathbf{F}_i\, \hat{\mathbf{s}}_i, \tag{17}$$

it suffices to substitute the elastic causality condition (16) in definition (4), eliminating next the stress distribution using the modal description (3). The following expression is thus found for the element flexibility matrix:

$$\mathbf{F}_i = \int \tilde{\mathbf{S}}_i^{\sigma}\, \mathbf{F}'_i\, \mathbf{S}_i^{\sigma}\, d A_i. \tag{18}$$

Consider now the piecewise linear description of the plastic stress-strain relations (19) described in [38]:

$$\begin{aligned} \phi_{*i} &= \tilde{\mathbf{N}}_i' \, \sigma_i - \mathbf{H}_i' \, \epsilon_{*i} - \sigma_{*i}, && (a) \\ \epsilon_{pi} &= \mathbf{N}_i' \, \epsilon_{*i}, && (b) \\ \phi_{*i} &\leq \mathbf{0} \, , \; \epsilon_{*i} \geq \mathbf{0} \, , \; \tilde{\phi}_{*i}\epsilon_{*i} = 0. && (c-e) \end{aligned} \tag{19}$$

Summarized next is the description of plasticity in terms of nodal stresses and strain-resultants:

$$\begin{aligned} \boldsymbol{\phi}_{*i} &= \tilde{\mathbf{N}}_i \, \hat{\mathbf{s}}_i - \mathbf{H}_i \, \mathbf{e}_{*i} - \hat{\mathbf{s}}_{*i} && (a) \\ \hat{\mathbf{e}}_{pi} &= \mathbf{N}_i \, \mathbf{e}_{*i} && (b) \\ \boldsymbol{\phi}_{*i} &\leq \mathbf{0} \, , \; \mathbf{e}_{*i} \geq \mathbf{0} \, , \; \tilde{\boldsymbol{\phi}}_{*i}\mathbf{e}_{*i} = 0 && (c-e) \end{aligned} \tag{20}$$

Definition (20b) for the plastic component of the nodal strain-resultants, wherein

$$\mathbf{N}_i = \int \tilde{\mathbf{S}}_i^* \, \mathbf{N}_i' \, \mathbf{S}_i^* \, d\,A_i, \tag{21}$$

is obtained substituting the plastic flow condition (19b) in definition (4), after discretizing the plastic parameters field in the manner of the approximation (1):

$$\epsilon_{*i} = \mathbf{S}_i^* \, \mathbf{e}_{*i}. \tag{22}$$

The static variables associated with the plastic deformation parameters $\mathbf{e}_{*i}$ are defined by the dual transformation of definition (22):

$$\boldsymbol{\phi}_{*i} = \int \tilde{\mathbf{S}}_i^* \, \phi_{*i} \, d\,A_i. \tag{23}$$

As it is required by condition (20e), the above definition for the plastic yield resultant functions ensures that the local complementarity conditions (19e) are now satisfied globally over the element, since:

$$\tilde{\boldsymbol{\phi}}_{*i} \, \mathbf{e}_{*i} = \int \tilde{\phi}_{*i} \, \epsilon_{*i} \, d\,A_i.$$

Definition (20a) for the plastic yield resultant functions, wherein

$$\mathbf{H}_i = \int \tilde{\mathbf{S}}_i^* \, \mathbf{H}_i' \, \mathbf{S}_i^* \, d\,A_i, \tag{24}$$

represents the plastic hardening matrix, and

$$\hat{\mathbf{s}}_{*i} = \int \tilde{\mathbf{S}}_i^* \, \sigma_{*i} \, d\,A_i, \tag{25}$$

collects the plastic capacity resultants, is obtained substituting expressions (3), (19a) and (22) in definition (23).

When two or more stress modes combine to activate the plastic behaviour of the element, the entries of arrays (21), (24) and (25) become dependent on the position of the neutral axis. Consequently, the yield surface $\phi_{*i} = 0$ comes to be described by nonlinear functions on the nodal stresses and strain-resultants .

To recover a piecewise linear description of the plastic constitutive relations, the yield surface is replaced by an inscribed polyhedron, as illustrated in figure 3.

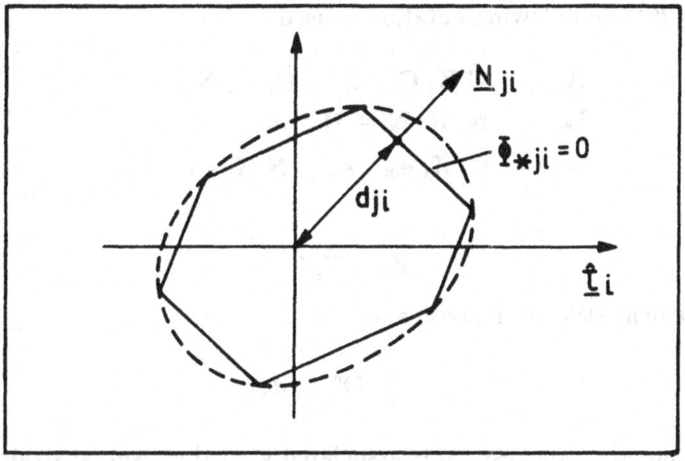

Figure 3

Description (20) can still be used to describe the plasticity conditions. The j-th column of matrix $\mathbf{N}_i$ represents now the outward normal of the j-th yield hiperplane $\phi_{*ij} = \mathbf{0}$, the distance of which to the origin of the s_i-place is given by,

$$d_{ji} = \mathbf{H}_{ji}\, \mathbf{e}_{*i} + \hat{s}_{*ji},$$

wherein $\mathbf{H}_{ji}$ denotes the j-th row of matrix $\mathbf{H}_i$. Conditions (20b,d) imply an associated flow law since the plastic deformation vector is assumed to be proportional to the outward normal to the currently active yield hyperplane:

$$\hat{\mathbf{e}}_{pi} = \mathbf{N}_{ji}\, e_{*ji}\ ,\ e_{*ji} \geq 0.$$

Governing System

After substituting the equilibruim and compatibility conditions (7) and (5), respectively, in description (15, 17, 20) for the elastic-plastic constitutive relations, the following system is found to govern the response of the finite elements into which the cross-section has been discretized:

$$\left\{\begin{array}{c} \mathbf{s} \\ -\boldsymbol{\phi}_{*} \end{array}\right\}_i = \left[\begin{array}{cc} \mathbf{A} & -\mathbf{B} \\ -\tilde{\mathbf{B}} & \mathbf{D} \end{array}\right]_i \left\{\begin{array}{c} \mathbf{e} \\ \mathbf{e}_{*} \end{array}\right\}_i + \left\{\begin{array}{c} -\mathbf{s}_0 \\ \mathbf{s}_{*} \end{array}\right\}, \quad \begin{array}{c} (a) \\ (b) \end{array} \qquad (26)$$

$$\boldsymbol{\phi}_{*i} \leq \mathbf{0}\ ,\ \mathbf{e}_{*i} \geq \mathbf{0}\ ,\ \tilde{\boldsymbol{\phi}}_{*i}\, \mathbf{e}_{*i} = 0 \qquad (c-e)$$

In system (26) the following notation is used,

$$\begin{aligned} \mathbf{A}_i &= \tilde{\mathbf{C}}\, \mathbf{K}_i\, \mathbf{C}_i\ ,\ \mathbf{B}_i = \tilde{\mathbf{C}}_i\, \mathbf{K}_i\, \mathbf{N}_i, \\ \mathbf{D}_i &= \tilde{\mathbf{N}}_i\, \mathbf{K}_i\, \mathbf{N}_i + \mathbf{H}_i, \\ \mathbf{s}_0 &= \tilde{\mathbf{C}}_i\, \mathbf{K}_i\, \hat{\mathbf{e}}_{0i}\ ,\ \mathbf{s}_{*i} = \tilde{\mathbf{N}}\, \mathbf{K}_i\, \hat{\mathbf{e}}_{0i} + \hat{\mathbf{s}}_{*i}, \end{aligned}$$

wherein

$$\mathbf{K}_i = \mathbf{F}_i^{-1}$$

denotes the element stiffness matrix, and

$$\hat{\mathbf{e}}_{0i} = \int \tilde{\mathbf{D}}_i^{\sigma}\, \epsilon_{0i}\, d\, A_i$$

represents the nodal strain-resultants associated with the residual strain field ϵ_{0i}.

The system governing the elastoplastic response of the cross-section is obtained substituting the kinematic incidence conditions (8) in system (26a,b) and the equilibrium conditions (26a) in definition (9) for the independent stress-resultants. Collecting for all constituent elements,

$$\boldsymbol{\phi}_{*} = \{\boldsymbol{\phi}_{*i}\}\ ,\ \mathbf{e}_{*} = \{\mathbf{e}_{*i}\}\ ,\ \mathbf{s}_{*} = \{\mathbf{s}_{*i}\}$$

the following system is obtained after inserting conditions (13) and (14b):

Elastoplastic Response
$$\left[\begin{array}{ccc} \mathbf{A} & -\mathbf{B} & -\mathbf{C}_0 \\ -\tilde{\mathbf{B}} & \mathbf{D} & \mathbf{0} \\ -\tilde{\mathbf{C}}_0 & \tilde{\mathbf{0}} & \mathbf{0} \end{array}\right] \left\{\begin{array}{c} \mathbf{e} \\ \mathbf{e}_{*} \\ \lambda \end{array}\right\} \begin{array}{c} = \\ \geq \\ = \end{array} \left\{\begin{array}{c} \mathbf{s}_0 \\ -\mathbf{s}_{*} \\ -\bar{W} \end{array}\right\} \quad \begin{array}{c} (a) \\ (b) \\ (c) \end{array}$$ $$\mathbf{e}_{*} \geq \mathbf{0} \quad (d)$$ $$\tilde{\mathbf{e}}_{*}\left[-\tilde{\mathbf{B}}\mathbf{e} + \mathbf{D}\mathbf{e}_{*} + \mathbf{s}_{*}\right] = 0 \quad (e)$$ (27)

To set up system (27), arrays

$$\begin{aligned} \mathbf{A} &= \textstyle\sum_i \tilde{\mathbf{J}}_i \mathbf{A}_i \mathbf{J}_i \quad , \quad \mathbf{s}_0 = \sum_i \tilde{\mathbf{J}}_i \mathbf{s}_{0i}, \\ B &= \mathrm{col}\ [(\mathbf{JB})_i] \quad , \quad \mathbf{D} = \mathrm{diag}\ [\mathbf{D}_i], \end{aligned}$$

are determined by direct allocation of the individual contribution of the constituting finite elements.

A similar procedure can be used to obtain system (28) governing the rigid-perfectly plastic response of the cross-section, wherein the following notation is used:

$$\hat{\mathbf{s}}_* = \{\hat{\mathbf{s}}_{*i}\}\ ,\ \mathbf{N} = \text{col}\ [\mathbf{N}_i]\ ,\ \mathbf{C} = \text{col}\ [(CJ)_i]$$

Plastic Response

$$\begin{bmatrix} \mathbf{0} & \mathbf{0} & -\tilde{\mathbf{C}} & \mathbf{C}_0 \\ \tilde{\mathbf{0}} & \mathbf{0} & \tilde{\mathbf{N}} & \mathbf{0} \\ -\mathbf{C} & \mathbf{N} & \mathbf{0} & \tilde{\mathbf{0}} \\ \tilde{\mathbf{C}}_0 & \tilde{\mathbf{0}} & \mathbf{0} & \mathbf{0} \end{bmatrix} \begin{Bmatrix} \dot{\mathbf{e}} \\ \dot{\mathbf{e}}_* \\ \hat{\mathbf{s}} \\ \lambda \end{Bmatrix} \begin{matrix} = \\ \leq \\ = \\ = \end{matrix} \begin{Bmatrix} \mathbf{0} \\ \hat{\mathbf{s}}_* \\ \mathbf{0} \\ 1 \end{Bmatrix} \quad \begin{matrix} (a) \\ (b) \\ (c) \\ (d) \end{matrix} \tag{28}$$

$$\dot{\mathbf{e}}_* \geq \mathbf{0} \quad (e)$$

$$\tilde{\dot{\mathbf{e}}}_*\left[-\tilde{\mathbf{N}}\hat{\mathbf{s}} - \hat{\mathbf{s}}_*\right] = 0 \quad (f)$$

In the above system, $\dot{\mathbf{e}}$ and $\dot{\mathbf{e}}_*$ represent deformation rates with respect to the unit external work:

$$\dot{\mathbf{e}} = \mathbf{e}/\dot{W}\ .\ \dot{\mathbf{e}}_* = \mathbf{e}_*/\dot{W}.$$

The effect of fixed loads, such as prestress forces acting on the cross-section, can be accounted for in the yield condition (28b), to yield

$$\hat{\mathbf{s}}_* := \hat{\mathbf{s}}_* - \tilde{\mathbf{N}}\ \hat{\mathbf{s}}_0,$$

with $\hat{\mathbf{s}}_0$ denoting any nodal stress distribution that equilibrates (11) the fixed loads $\mathbf{s}_0$.

Plastic Limit Analysis

System (28) represents an uncoupled, symmetric linear complementarity problem. There is an equivalence between this problem and the pair of primal-dual linear programs (29) and (30) which encode the static and kinematic theorems of plastic limit analysis [39], respectively:

$$\text{Max z} = \lambda\ .\ \text{subject to (28a,b)} \tag{29}$$

$$\text{Min w} = \tilde{\hat{\mathbf{s}}}_*\,\dot{e}\ .\ \text{subject to (28c-d)} \tag{30}$$

It can be shown [36,40] that at optimality the coefficients of the primal {dual} objective function represent the plastic parameter rates {plastic potencials} at collapse, and that the dual objective function identifies with the collapse load parameter.

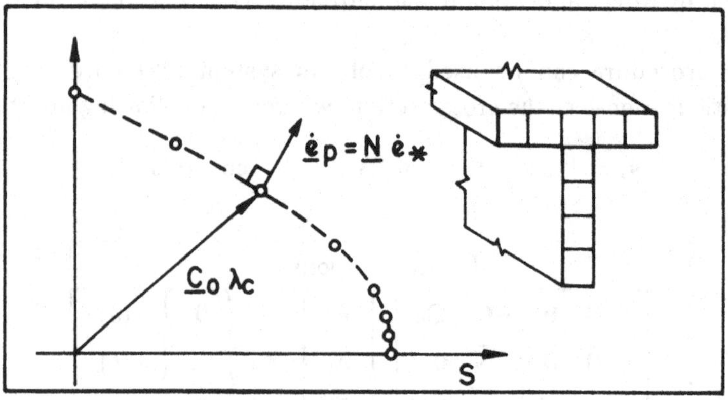

Figure 4

Either program can therefore be used to generate the points and outward normals required to describe the yield surface, as illustrated in figure 4. When the boundary conditions of the cross-section under analysis are simple and the collapse modes are known, solutions of program (29,30) become trivial and systems (28) can be used to obtain directly the yield surface.

Elastoplastic Deformation Analysis

Implementing the Karush-Kuhn-Tucker equivalence conditions, the coupled, symmetric linear complementarity problem (27) is dissociated into the following equivalent pair of primal-dual quadratic programs:

$$\text{Min } z = 1/2 \begin{Bmatrix} \tilde{e} \\ \tilde{e}_* \end{Bmatrix} \begin{bmatrix} A & B \\ -\tilde{B} & D \end{bmatrix} \begin{Bmatrix} e \\ e_* \end{Bmatrix} - \bar{W}\,\lambda,$$

subject to (27a,b);

$$\text{Min } w = 1/2 \begin{Bmatrix} \tilde{e} \\ \tilde{e}_* \end{Bmatrix} \begin{bmatrix} A & -B \\ -\tilde{B} & D \end{bmatrix} \begin{Bmatrix} \tilde{e} \\ e_* \end{Bmatrix} - \begin{Bmatrix} \tilde{e} \\ \tilde{e}_* \end{Bmatrix} \begin{Bmatrix} s_0 \\ -s_* \end{Bmatrix}$$

subject to (27c,d).

The above programs generate solutions associated with a prescribed work rate $\bar{W}$

and may therefore be used to implement analysis problem in which one deformation parameter or a weigthed combination of deformations $\mathbf{e}$ are controlled.

In general, if an elastoplastic analysis is to be performed for a given set of prescribed variables, say $\mathbf{v}$, the corresponding governing system is obtained simply by treating $\mathbf{v}$ as constant in system (27) and deleting therein its dual constraint.

This class of problems occurs frequently in the analysis of cross-sections with limited ductility.

The direct problem consists in determining the deformation induced by a prescribed loading $\mathbf{s}$ to check whether the bounds defined in the codes of practice are satisfied or not.

Load-controlled problems are implemented treating as constant, thus fixing the stress-resultants (13) acting upon the cross-section. The elastoplastic deformation analysis programs are thus found to reduce the following form:

$$\text{Min } z = 1/2 \begin{Bmatrix} \tilde{\mathbf{e}} \\ \tilde{\mathbf{e}}_* \end{Bmatrix} \begin{bmatrix} \mathbf{A} & -\mathbf{B} \\ -\tilde{\mathbf{B}} & \mathbf{D} \end{bmatrix} \begin{Bmatrix} \mathbf{e} \\ \mathbf{e}_* \end{Bmatrix}$$

subject to:

$$\begin{bmatrix} \mathbf{A} & -\mathbf{B} \\ -\tilde{\mathbf{B}} & \mathbf{D} \end{bmatrix} \begin{Bmatrix} \mathbf{e} \\ \mathbf{e}_* \end{Bmatrix} \begin{matrix} = \\ \geq \end{matrix} \begin{Bmatrix} \mathbf{s}_0 + \mathbf{s} \\ -\mathbf{s}_* \end{Bmatrix}$$

$$\text{Min } w = 1/2 \begin{Bmatrix} \tilde{\mathbf{e}} \\ \tilde{\mathbf{e}}_* \end{Bmatrix} \begin{bmatrix} \mathbf{A} & -\mathbf{B} \\ -\tilde{\mathbf{B}} & \mathbf{D} \end{bmatrix} \begin{Bmatrix} \mathbf{e} \\ \mathbf{e}_* \end{Bmatrix} - \begin{Bmatrix} \tilde{\mathbf{e}} \\ \tilde{\mathbf{e}}_* \end{Bmatrix} \begin{Bmatrix} \mathbf{s}_0 + \mathbf{s} \\ -\mathbf{s}_* \end{Bmatrix}$$

subject to:

$$\mathbf{e}_* \geq \mathbf{0}.$$

The inverse problem consists in prescribing the limit state on the deformations $\mathbf{e}$ and to determine next the associated load-carrying capacity, $\mathbf{s}$. Treating deformations $\mathbf{e}$ as prescribed variables the following programs are obtained:

$$\text{Min } z = 1/2\tilde{\mathbf{e}}_*\mathbf{D}\mathbf{e}_* \; : \; \mathbf{D}\mathbf{e}_* \geq -\mathbf{s}_* - \tilde{\mathbf{B}}\mathbf{e} \quad (a,b) \tag{31}$$

$$\text{Min } w = 1/2\tilde{\mathbf{e}}_*\mathbf{D}\mathbf{e}_* + \tilde{\mathbf{e}}_*(\mathbf{s}_* - \tilde{\mathbf{B}}\mathbf{e}) \; : \; \mathbf{e}_* \geq \mathbf{0} \quad (a,b) \tag{32}$$

After determining the plastic parameters $\mathbf{e}_*$, the elastic component of the nodal strain-resultants are computed from the compatibility conditions (5,8), to yield:

$$\hat{\mathbf{e}}_{Ei} = \mathbf{C}_i\,\mathbf{e}_i - \mathbf{N}_i\,\mathbf{e}_{*i}.$$

The nodal stresses in the elements are then determined through the stiffness version of the elasticity conditions (17) and fed into definitions (7,9) to obtain the stress-resultants, s.

This is therefore a general procedure to generate the stress-resultant combinations defining the interaction diagram that characterizes the limit states of structural concrete sections. When the boundary conditions of the cross-sections under analysis are simple, solution of either of programs (31) or (32) becomes irrelevant and only the latter part of the procedure has to be implemented.

Incremental Elastoplastic Analysis

The formulations for elastoplastic deformation analysis presented above were derived using the complementarity condition (27e) which assumes that plastic straining develops monotonically.

If plastic unstressing is to be accounted for, the plastic association conditions have to be expressed in incremental form:

$$\begin{aligned} &\check{\phi}_\star \, \Delta \, \mathbf{e}_\star = 0, && (a) \\ &\Delta \, \check{\phi}_\star \, \Delta \, \mathbf{e}_\star = 0. && (b) \end{aligned} \tag{33}$$

In the incremental elastoplastic analysis program,

$$\text{Max w, subject to (27a-d),} \tag{34}$$

the variables describing the plastic phase are dissociated into two groups, one describing the yield modes which are currently active,

$$\begin{aligned} \phi'_\star &= 0 \;, & \Delta\phi'_\star &= 0, \\ \mathbf{e}'_\star &\geq 0 \;, & \Delta\mathbf{e}'_\star &> 0, \end{aligned}$$

and the other those which are not:

$$\begin{aligned} \phi''_\star &< 0 \;, & \Delta\phi''_\star &\leq 0, \\ \mathbf{e}''_\star &\geq 0 \;, & \Delta\mathbf{e}''_\star &= 0. \end{aligned}$$

The association condition (33) is now implicitly satisfied.

The numerical implementation of program (34) to obtain the solution sequence tracing the elastoplastic response of the section consists essentially in sellecting the trajectory dictated by the currently active constraints and in determining the step lengths that expose the mobilization of a new yield mode or the deactivation of a currently active one [41]. The procedure is usually terminated when the load carrying capacity of the section is exhausted or when the limits prescribed on deformations variables are attained. The use of the external work rate W as a control variable, instead of the load parameter λ, is particularly useful from a numerical stand point when the post-collapse phase is to be investigated [41].

REFERENCES

1. Navier, C.L.M.H.: Résumé des leçons données à l'École des Ponts et Chaussées, 3rd Ed., Paris, 1864, with notes and appendices by B. de Saint-Venant.

2. Ewing, J.A.: The Strength of Materials, Cambridge, 1899.

3. Girkmann, K.: Bemessung von Rahmentragwerken unter Zugrundelegung eines ideal-plastichen Stahles, Sitzungsbenchte der Akademie der Wissenschaften in Wien, Abt. IIa, 140, 1931.

4. Brown, E.H.: Plastic asymmetrical bending of beams, Int. I. Mech. Sci., 9.

5. Hill, R. and M.P.L. Siebel: On combined bending and twisting of thin tubes in the plastic range, Phil. Magazine, 42, 1951.

6. Hill, R. and M.P.L. Siebel: On the plastic distorsion of solid bars by combined bending and twisting, J. Mech. Physics of Solids, 1, 1953.

7. Drucker, D.C.: The effect of shear on the plastic bending of beams, J. Appl. Mech., 23, Trans. ASME, 1956

8. Hodge, P.G.: Interaction diagrams for shear and bending of plastic beams, J. Appl. Mech., Trans ASME, 1957.

9. Gaydon, F.A. and H. Nuttal: On the combined bending and twisting of beams of various sections, J. Mech. Physics of Solids, 6, 1957

10. Neal, B.G.: The effect of shear and normal forces on the fully plastic moment of a beam of rectangular cross section, J. Appl. Mech., 28, Trans. ASME. 1961

11. Boulton, N.S.: Plastic twisting and bending of an I-beam in which the warp is restricted, Int. J. Mech. Sci., 4, 1962

12. Heyman, J.: The simple plastic bending of beams, Proc. Inst. Cor. Engrs.. 41, 1968.

13. Morris, G.A. and S.J. Fenves: Approximate yield surface equations, J. Engrg. Mech. Div., Proc ASCE, 95, 1969

14. Lescouarch, Y.: Resistence of a cross section submitted to various types of stresses, Construction Metallique, 14, 1977

15. Rafagofalan, K.S.: Combined torsion, bending and shear on L-beams, J. Struct. Div., Proc. ASCE, 106, 1980.

16. Szuladzinski, G.: Bending of beams with nonlinear material characteristics, J. Mech. Design, Trans ASME, 102, 1980

17. Benson, R.C.: Nonlinear bending and collapse of long, thin, open section beams and corrugated panels, J. Appl. Mech. Dev., University of Waterloo, 1973

18. Cohn, M.Z. (Ed.): Nonlinearity and continuity in prestress concrete, Univ. of Waterloo, 19, 1983

19. Chakraborty, M.: Ultimate strength of reinforced concrete rectangular beams under combined torsion - A critical review, I. Inst. Engineers (India) Civ. Engrg. Div., 60, 1980

20. Lachance, L.: Stress distribution in reinforced concrete sections subjected to biaxial bending, J. Am. Concrete Inst., 2, 1980

21. Sakai, K.: Moment-curvature relationship of reinforced concrete members subjected to combined bending and axial force, J. Am. Concrete Inst., 3, 1980.

22. Sharma, A.: Tests on concrete beams in biaxial bending, axial compression and torsion, Inst. Civ. Engineers, Proc, Research and Theory, 75, 1983

23. Mansur, M.A.: Combined bending and torsion in reinforced concrete beams with rectangular openings, Conc. Int. Design and Construction, 5, 1983

24. Rotter, J.M. and P. Ansourian: Cross section behaviour and ductility in composite beams, Inst. Civ. Engineers, Proc., Research and Theory, 67, 1979

25. Al-Noury, S.I.: Behaviour and design of reinforced and composite concrete sections, J. Struct. Div., Proc. ASCE, 108, 1982.

26. Lachance, L.: Ultimate strength of biaxially loaded composite sections, J.Div., Proc. ASCE, 108, 1982

27. Brondum-Nielsen, T.: Ultimate flexural capacity of partially or fully prestressed cracked arbitrarily concrete sections under axial load combined with biaxial bending, Conc. Int. Design and Construction, 5, 1983

28. Chushkewich, K.W.: Simplified cracked section analysis, J. Am. Concrete Inst., 80, 1983

29. Abdel-Baset, S.B.: Limit analysis of skeletal structures under combined stresses, Ph. D. Thesis, Univ. of Waterloo, 1976

30. Grierson, D.E. and S.B. Abdel-Baset: Plastic analysis under combined stresses, J. Engrg. Mech. Div., Proc. ASCE, 103, 1977.

31. Appleton, J.A.S.: Elastoplastic analysis of skeletal structures by mathematical programming, Ph. D. Thesis, Univ. of London, 1979.

32. Smith, D.L. and J.A.S. Appleton: A mathematical programming model for some problems in structural plasticity, 2nd Int. Conf, Applied Numerical Modelling, Madrid, 1978

33. Sacchi, P.L. and A. Tralli: A discrete model for the plastic analysis of thin walled beams of open cross section, J. Struct. Mech., 8, 1980.

34. Nadai, A.: Theory of Flow and Fracture of Solids, McGraw-Hill, 2nd edition,1950

35. Munro, J. and D.L. Smith: Linear programming duality in plastic analysis and synthesis, Int. Symp. Computer Aided Design, Univ. of Warwick, 1972.

36. Argyris, J.H. and S. Kelsey: Energy Theorems and Structural Analysis, Butterworth, 1960.

37. Freitas, J.A.T.: Piecewise linear elastic-plastic stress-strain relations, Mathematical Programming Methods in Structural Plasticity, CISM, Udine, 1986

38. Gvozdev, A.A.: The determination of the value of the collapse load for statically indeterminate systems undergoing plastic deformation,Proc.Conf. Plastic Deformations, Akad. Nauk. SSRC, Moscow,1936 (English Trans. in Int. J. Mech. Sci., 1)

39. Freitas, J.A.T.: An efficient simplex method for the limit analysis of structures. Computers and Struct., 21,1985

40. Freitas, J.A.T. and D.L. Smith: Plastic straining, unstressing and branching in large displacement perturbation analysis, Int. J. Num. Meth. Engrg., 20, 1984

41. Freitas, J.A.T. and J.P.B.M. Almeida: A nonlinear projection method for constrained optimization, Civ. Engrg. Syst., 1, 1984

CHAPTER 9

ELASTOPLASTIC ANALYSIS OF SKELETAL STRUCTURES

J. A. Teixeira de Freitas
Istituto Superior Tecnico, Lisbon, Portugal

Abstract: A unified finite element formulation for the elastoplastic analysis of skeletal structures is presented. Parametric expressions are used to describe not only the element displacement field but also the stress-resultant distribution, in both mesh and nodal representations. A finite element description, in terms of stress, strain and plastic multipliers, is incorporated to model the cross-sectional behaviour of the constitutive building elements. The ensuing formulation is encoded to perform both elastoplastic deformation and incremental analyses.

Introduction

Elastoplastic analysis of skeletal structures is today a well established field of structural mechanics.

The approaches that have been used in the solution of this problem can be organized in two categories.

One category embraces the iterative methods based on the initial stress and initial strain concepts suggested by Zienkiewicz [1] and Argyris [2]. It is characterized by a fundamentally elastic approach, modified to accommodate the essential features of plasticity.

The second category groups the finite methods inspired in Maier's approach [3] and are based on piecewise linear elastic-plastic laws and the application of mathematical programming theory [4]. Graph-theoretic descriptions have been employed by Smith and Munro [5] to generalize Maier's displacement based formulations.

The formulation presented next incorporates the four fundamental concepts that distinguish Smith's approach [6]:

- Structural discretization to allow representation by a finite algebraic mathematical system;
- Direct graph representation of structural networks to allow both mesh and nodal forms of the connectivity conditions to be employed in assembling the structural building elements;
- Static-kinematic duality (SKD) and constitutive reciprocity to induce symmetry by direct vectorial means in the governing mathematical system;
- Mathematical programming to complement the basic vectorial structural relations of the governing mathematical system with a discrete variational interpretation.

Smith's heuristic description of the mesh and nodal representations of equilibrium and compatibility is herein replaced by a general finite element formulation. Parametric expressions are used to describe not only the element displacement field but also the stress-resultant distribution. The variables associated with the discretization parameters are so identified as to preserve SKD [7].

Also abandoned is the plastic hinge hypotesis. The finite element description in terms of stress, strain and plastic multiplier models of the cross-sectional behaviour suggested in [8] is incorporated in the formulation presented herein and fed into the parametric representations of the element stress-resultant and plastic deformation fields. A similar model for the element constitutive relations has been proposed by Corradi [9]; besides the displacement field, only the plastic strain distribution is also represented parametrically, both over the cross-section and along the element lenght.

Description of Statics and Kinematics

Let the structure under analysis be discretized into finite elements. Assume that a typical element i, dissected from the structure is acted by end-forces $\mathbf{F}_i^k$ $(k = 1, 2)$ and by generalized forces $\mathbf{f}_i$ acting along the span, as illustrated in figure 1.

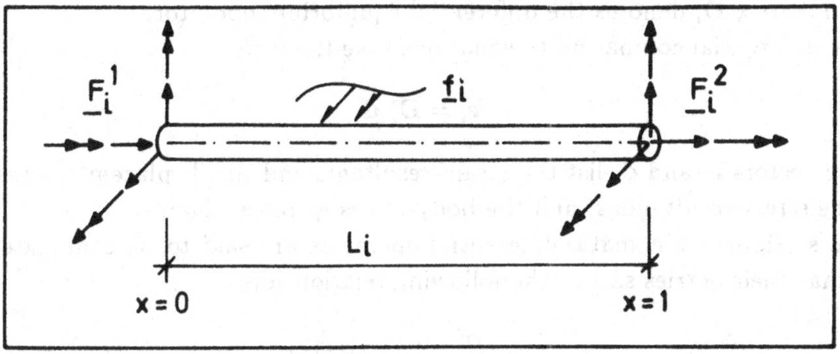

Figure 1

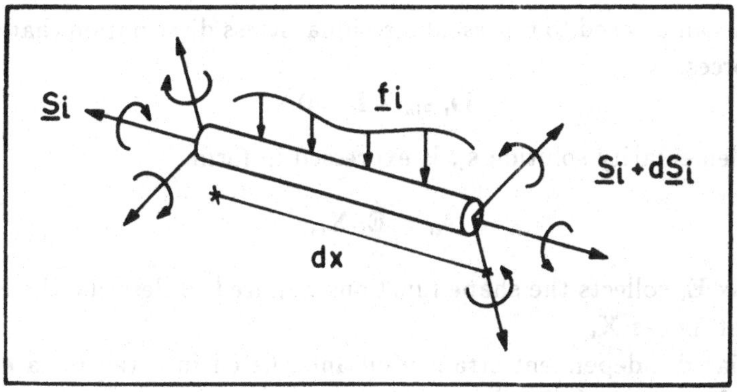

Figure 2

It is assumed that the acting forces are measured in local systems of reference, the origins of wich coincide with the baricenter of end-section $k = 1, 2$.

Let $\mathbf{s}_i$ denote the stress-resultants developing in a generic cross-section of the element. The end-forces can be defined by specifying the stress-resultant field at the boundaries of the element:

$$\mathbf{F}_i^k = \tilde{\mathbf{L}}_i^k \, \mathbf{s}_i^k \ , \quad k = 1, 2. \tag{1}$$

The equilibrium conditions of an infinitesimal segment of the element can be expressed in form

$$\mathbf{D}_i \, \mathbf{s}_i + \mathbf{f}_i = 0, \tag{2}$$

wherein matrix $\mathbf{D}_i$ denotes the differential equilibrium operator.

The differential compatibility equations take the form

$$\mathbf{e}_i = \mathbf{D}_i^* \, \mathbf{d}_i, \tag{3}$$

wherein vectors $\mathbf{e}_i$ and $\mathbf{d}_i$ list the strain-resultants and the displacements associated with the stress-resultants $\mathbf{s}_i$ and the body-forces $\mathbf{f}_i$, respectively.

The static and kinematic differential operators are said to be conjugate, in the sense that their entries satisfy the following relationship:

$$d_{ij}^* = \frac{\partial^k}{\partial \, x^k} = (-1)^{k+1} \, d_{ji}. \tag{4}$$

Let two components be distinguished in the representation of the stress-resultant field:

$$\mathbf{s}_i = \mathbf{s}_{ci} + \mathbf{s}_{0i}. \tag{5}$$

Addend $\mathbf{s}_{0i}$ can be used to represent a residual stress distribution that equilibrates the applied forces:

$$\mathbf{D}_i \, \mathbf{s}_{0i} + \mathbf{f}_i = 0. \tag{6}$$

The complementarity solution $\mathbf{s}_{ci}$ is expressed in form

$$\mathbf{s}_{ci} = \mathbf{E}_i \, \mathbf{X}_i, \tag{7}$$

wherein matrix $\mathbf{E}_i$ collects the shape functions required to describe the independent stress-resultant modes $\mathbf{X}_i$.

The associated independent strain-resultants, listed in array $\mathbf{u}_i$, are defined by the dual transformation of distribution (7),

$$\mathbf{u}_i = \int_0^1 \tilde{\mathbf{E}}_i \, \mathbf{e}_i \, dx, \tag{8}$$

thus ensuring an equivalent global dissipation of energy:

$$\tilde{\mathbf{u}}_i \, \mathbf{X}_i = \int_0^1 \tilde{\mathbf{e}}_i \, \mathbf{s}_{ei} \, dx.$$

Similarly to decomposition (5), let two addends be also distinguished in the representation of the displacement field:

$$\mathbf{d}_i = \mathbf{d}_{ei} + \mathbf{d}_{0i}. \tag{9}$$

The complementarity solution is expressed in form

$$\mathbf{d}_{ei} = \mathbf{C}_i \, \mathbf{q}_i, \tag{10}$$

wherein $\mathbf{q}_i$ denotes the element nodal displacements vector and the columns of matrix $\mathbf{c}_i$ collect the corresponding kinematic modes. The role of the particular solution $\mathbf{d}_{0i}$ is to correct the approximations induced by definition (9) for the displacement field.

The nodal description of the element compatibility conditions is obtained substituting results (3), (9) and (10) in definition (8) to yield,

$$\mathbf{u}_i = \mathbf{A}_i \, \mathbf{q}_i + \mathbf{u}_{0i}, \tag{11}$$

wherein

$$\mathbf{u}_{0i} = \int_0^1 \tilde{\mathbf{E}} \, (\mathbf{D}_i^* \, \mathbf{d}_{0i}) \, dx,$$

and

$$\mathbf{A}_i = \int_0^1 \tilde{\mathbf{E}}_i \, (\mathbf{D}_i^* \, \mathbf{C}_i) \, dx.$$

denotes the element compatibility matrix.

According to Static-Kinematic Duality [7], the dual transformation of the compatibility conditions (11),

$$\mathbf{Q}_i = \tilde{\mathbf{A}}_i \, \mathbf{X}_i + \mathbf{Q}_{0i}, \tag{12}$$

represents the member equilibrium conditions.

To derive definition (12) for the nodal forces $\mathbf{Q}_i$ associated with the nodal displacements $\mathbf{q}_i$, substitute results (5) and (7) in definition (2), premultiplied by the transpose of the nodal diplacements shape mode matrix, and integrate over the element the expression thus obtained:

$$\int_0^1 \tilde{\mathbf{C}}_i(\mathbf{D}_i \mathbf{E}_i) \, dx \, \mathbf{X}_i + \int_0^1 \tilde{\mathbf{C}}_i(\mathbf{D}_i \mathbf{s}_{0i} + \mathbf{f}_i) \, dx = 0.$$

To implement relationship (4), the term affecting the independent stress-resultants is integrated by parts, yielding the following definitions for the nodal forces present in the equilibrium condition (12):

$$\begin{aligned} \mathbf{Q}_i &= \left[\tilde{\mathbf{C}}_i \ \tilde{\mathbf{L}}_i \ \mathbf{s}_i \right]_0^1, \\ \mathbf{Q}_{0i} &= \int_0^1 \tilde{\mathbf{C}}_i \left(\mathbf{D}_i \, \mathbf{s}_{0i} - \mathbf{f}_i \right) dx - \left[\tilde{\mathbf{C}}_i \ \tilde{\mathbf{L}}_i \ \mathbf{s}_{0i} \right]_0^1 . \end{aligned}$$

The equilibrium and compatibility conditions of the structure under analysis can now be obtained processing the elementary relationships (11) and (12) through a standard assemblage procedure.

Let $\mathbf{q}$ list the independent nodal displacements of the structure, and

$$\mathbf{q}^* = \{ \ \mathbf{q}_i \ \} \tag{13}$$

collect the nodal displacement vector for all constituent finite elements.

They are related through a nodal incidence condition,

$$\mathbf{q}^* = \mathbf{J} \, \mathbf{q}, \tag{14}$$

wherein J is a boolean operator that implements the structure nodal connectivity properties, as well as its kinematic boundary conditions.

The structure compatibility conditions,

$$\mathbf{u} = \mathbf{A} \, \mathbf{q} + \mathbf{u}_0, \tag{15}$$

are obtained substituting (14) in conditions (11) extended to all structural elements in the manner of (14), yielding the following definition for the compatibility operator,

$$\mathbf{A} = \mathbf{A}^* \, \mathbf{J},$$

wherein $\mathbf{A}^*$ is a block diagonal matrix.

The nodal equilibrium condition,

$$\mathbf{Q} = \mathbf{0} = \tilde{\mathbf{A}} \, \mathbf{X} + \mathbf{Q}_0, \tag{16}$$

is obtained applying a similar assemblage procedure to the equilibrium conditions (12) of the disconnected elements:

$$\mathbf{Q} = \mathbf{0} = \tilde{\mathbf{J}} \, \mathbf{Q}^* .$$

Mesh Description of Statics and Kinematics

Let β represent the kinematic indeterminacy number of the structure, ie the dimension of array $\mathbf{q}$, and σ the number of entries in the independent stress-resultant vector $\mathbf{X}$; the dimensions of arrays $\mathbf{Q}$ and $\mathbf{u}$ are therefore β and σ, respectively.

Condition (16) shows clearly that only

$$\alpha = \sigma - \beta \tag{17}$$

of the elements of array $\mathbf{X}$ are linearly independent; α represents the static indeterminacy number of the structure.

This invariant of the structure is explored in the mesh description of Statics to express the stress-resultants as a direct function of the loading and of α indeterminate forces, collected in array $\mathbf{p}$:

$$\mathbf{X} = \mathbf{B}\,\mathbf{p} + \mathbf{X}_0. \tag{18}$$

The mesh description of Kinematics is defined by the dual transformation of the equilibrium conditions (18) to yield,

$$\mathbf{v} = \mathbf{0} = \tilde{\mathbf{B}}\,\mathbf{u} + \mathbf{v}_0, \tag{19}$$

wherein $\mathbf{v}$ represents the discontinuities associated with the hyperstatic forces $\mathbf{p}$. Condition (19) together with definition (17) show that only β elements of array $\mathbf{u}$ are linearly independent, as expected.

To derive results (18) and (19), let again the elementary stress-resultant and displacement fields be expressed in formats (5) and (9), respectively.

The particular solution for the stress-resultant distribution can still be expected to equilibrate the applied forces, as required by condition (6). However, the complementary solution is expressed in form

$$\mathbf{s}_{ci} = \mathbf{E}_i\,\mathbf{p}_i, \tag{20}$$

wherenow the columns of matrix $\mathbf{E}_i$ define linearly independent, self-equilibrated stress-resultant distributions, induced by hyperstatic forces $\mathbf{p}_i$ in meshes formed by one or more elements, as illustrated in figure 3 for the planar case.

In the representation (9) of the displacement field, the complementary solution is now associated with the independent strain-resultants,

$$\mathbf{d}_{ci} = \mathbf{C}_i\,\mathbf{u}_i, \tag{21}$$

and the particular solution is interpreted as a corrective addend comtemplating for instance, the rigid-body displacements.

The discontinuities associated with the hyperstatic forces are defined by the dual

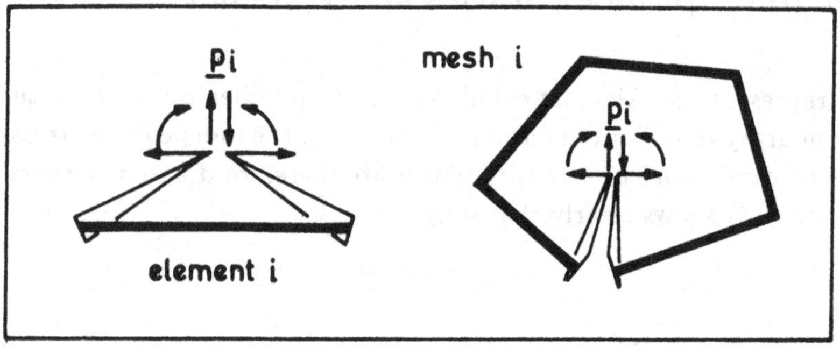

Figure 3

transformation of definition (20):

$$\mathbf{v}_i = \int_0^1 \tilde{\mathbf{E}}_i \, \mathbf{e}_i \, dx. \tag{22}$$

Substituting the above results (3), (9) and (21) the following expression is found for the mesh compatibility conditions,

$$\mathbf{v}_i = \tilde{\mathbf{B}}_i \, \mathbf{u}_i + \mathbf{v}_{0i}, \tag{23}$$

wherein

$$\tilde{\mathbf{B}}_i = \int_0^1 \tilde{\mathbf{E}}_i \, (\mathbf{D}_i^* \, \mathbf{C}_i) \, dx, \tag{24}$$

and

$$\mathbf{v}_{0i} = \int_0^1 \tilde{\mathbf{E}}_i \, (\mathbf{D}_i^* \, d_{0i}) \, dx.$$

To derive the mesh equilibrium conditions, substitute results (5) and (20) in definition (2), premultiply by the transpose of the strain-resultant shape mode matrix, and integrate over the element, to yield:

$$\int_0^1 \tilde{\mathbf{C}}_i (\mathbf{D}_i \, \mathbf{E}_i) dx \, \mathbf{p}_i + \int_0^1 \tilde{\mathbf{C}} (\mathbf{D}_i \, \mathbf{s}_{0i} + \mathbf{f}_i) dx = 0.$$

Integrating by parts the term affecting the hyperstatic forces and enforcing definitions (4) and (24), the following expression is found for the mesh equilibrium conditions

$$\mathbf{X}_i = \mathbf{B}_i \, \mathbf{p}_i + \mathbf{X}_{0i}, \tag{25}$$

wherein

$$\mathbf{X}_i = \left[\, \tilde{\mathbf{C}}_i \;\; \tilde{\mathbf{L}}_i \;\; \mathbf{s}_i \, \right]_0^1,$$

and

$$\mathbf{X}_{0i} = \left[\ \tilde{\mathbf{C}}_i \ \ \tilde{\mathbf{L}}_i \ \ \mathbf{s}_{0i} \ \right]_0^1 - \int_0^1 \tilde{\mathbf{C}}_i \left(\mathbf{D}_i \ \mathbf{s}_{0i} + \mathbf{f}_i\right) \, dx.$$

The mesh description of the equilibrium (18) and compatibility (19) conditions are recovered processing the elementary relations (23) and (25) through an assemblage procedure similar to the one used in the nodal description. The following definitions are found for the structure equilibrium matrix and for the particular solution of the structure discontinuities,

$$\mathbf{B} = \mathbf{B}^* \, \mathbf{J} \ , \ \ \mathbf{v}_0 = \tilde{\mathbf{J}} \ \mathbf{v}_0^*,$$

after relating the members hyperstatic forces with the structure indeterminate forces;

$$\mathbf{p}^* = \left\{ \ \mathbf{p}_i \ \right\} \quad , \quad \mathbf{p}^* = \mathbf{J} \ \mathbf{p}.$$

Elastoplastic Constitutive Relations

Assume that the material constituting the finite elements into which the structure is discretized, follows a piecewise-linear elastic-plastic law described by conditions (26) to (28), as suggested in [8]:

$$\mathbf{e}_i = \mathbf{e}_{ei} + \mathbf{e}_{pi}, \tag{26}$$

$$\mathbf{e}_{ei} = \mathbf{F}'_i \ \mathbf{s}_i, \tag{27}$$

$$\left\{ \begin{array}{c} \boldsymbol{\varphi}_* \\ \mathbf{e}_p \end{array} \right\}_i = \left[\begin{array}{cc} -\mathbf{H}' & \tilde{\mathbf{N}}' \\ \mathbf{V}' & \mathbf{0} \end{array} \right]_i \left\{ \begin{array}{c} \mathbf{e}_* \\ \mathbf{s} \end{array} \right\}_i - \left\{ \begin{array}{c} \mathbf{s}_* \\ \mathbf{0} \end{array} \right\}_i, \quad \begin{array}{c} (a) \\ (b) \end{array} \tag{28}$$

$$\boldsymbol{\varphi}_{*i} \leq \mathbf{0} \ , \ \ \boldsymbol{\varphi}_{*i} \ \mathbf{e}_{*i} = 0 \ , \ \ \mathbf{e}_{*i} \geq 0, \qquad (c-e)$$

To obtain the description of the elasticity conditions in terms of independent stress- and strain-resultants,

$$\mathbf{u}_{ei} = \mathbf{F}_i \ \mathbf{X}_i + \mathbf{u}_{e0i}, \tag{29}$$

the elastic causality condition (26) is substituted in definition (8) and the stress-resultant distribution is eliminated next using expressions (5) and (7), to yield,

$$\mathbf{F}_i = \int_0^1 \tilde{\mathbf{E}}_i \ \mathbf{F}'_i \ \mathbf{E}_i \ dx,$$

and

$$\mathbf{u}_{e0i} = \int_0^1 \tilde{\mathbf{E}}_i \ \tilde{\mathbf{F}}'_i \ \mathbf{s}_{0i} \ dx.$$

The stiffness version of the elastic constitutive relations (29) takes the form,

$$\mathbf{X}_i = \mathbf{K}_i\ \mathbf{u}_{ei} + \mathbf{X}_{e0i}, \tag{30}$$

wherenow

$$\mathbf{K}_i = \mathbf{F}_i^{-1},$$

and

$$\mathbf{X}_{e0i} = -\mathbf{K}_i\ \mathbf{u}_{e0i}.$$

A similar procedure can be used to derive the description of the plasticity conditions in terms of independent stress and strain-resultants:

$$\begin{Bmatrix} \phi_* \\ \mathbf{u}_p \end{Bmatrix}_i = \begin{bmatrix} -\mathbf{H} & \tilde{\mathbf{N}} \\ \mathbf{V} & \mathbf{0} \end{bmatrix}_i \begin{Bmatrix} \mathbf{u}_* \\ \mathbf{X} \end{Bmatrix}_i - \begin{Bmatrix} \mathbf{X}_* \\ \mathbf{0} \end{Bmatrix}_i, \qquad \begin{matrix} (a) \\ (b) \end{matrix} \tag{31}$$

$$\phi_{*i} \leq \mathbf{0}\,, \quad \tilde{\phi}_{*i}\ \mathbf{u}_{*i} = 0\,, \quad \mathbf{u}_{*i} \geq 0. \qquad (c-e)$$

Letting

$$\mathbf{e}_{*i} = \mathbf{C}_{*i}\ \mathbf{u}_{*i},$$

and

$$\phi_{*i} = \int_0^1 \tilde{\mathbf{C}}_{*i}\ \varphi_{*i}\ dx,$$

the following identifications are found:

$$\begin{aligned} \mathbf{H}_i &= \int_0^1 \tilde{\mathbf{C}}_{*i}\ \mathbf{H}'_i\ \mathbf{C}_{*i}\ dx, \\ \mathbf{N}_i &= \int_0^1 \tilde{\mathbf{E}}_i\ \mathbf{N}'_i\ \mathbf{C}_{*i}\ dx, \\ \mathbf{V}_i &= \int_0^1 \tilde{\mathbf{E}}_i\ \mathbf{V}'_i\ \mathbf{C}_{*i}\ dx, \\ \mathbf{X}_{*i} &= \int_0^1 \tilde{\mathbf{C}}_{*i}\ \left(\mathbf{s}_{*i} - \tilde{\mathbf{N}}'_i\ \mathbf{s}_{0i}\right)\ dx. \end{aligned}$$

Symmetry in the elastoplastic constitutive relations is consequent upon the following properties:

$$\mathbf{F}'_i = \tilde{\mathbf{F}}'_i\,,\ \mathbf{H}'_i = \tilde{\mathbf{H}}'_i\,,\ \mathbf{N}'_i = \mathbf{V}'_i.$$

The Equivalent Formulations of Elastoplastic Analysis

After combining the equilibrium and compatibility conditions (15,16) or (18,19) with the constitutive relations (29,31) or (30,31), the general description (32) is found for the system governing the response of elastoplastic structures:

$$\begin{bmatrix} \mathbf{D}_{11} & \mathbf{D}_{12} & \mathbf{M}_1 \\ \tilde{\mathbf{D}}_{12} & \mathbf{D}_{22} & \mathbf{M}_2 \\ \tilde{\mathbf{M}}_1 & \tilde{\mathbf{M}}_2 & -\mathbf{C} \end{bmatrix} \begin{Bmatrix} \mathbf{k}_1 \\ \mathbf{u}_* \\ \mathbf{s} \end{Bmatrix} = \begin{Bmatrix} \mathbf{d}_1 \\ \mathbf{d}_2 - \phi_* \\ \mathbf{c} \end{Bmatrix} \quad \begin{matrix} (a) \\ (b) \\ (c) \end{matrix} \qquad (32)$$

$$\phi_* \leq \mathbf{0}\,, \quad \tilde{\phi}_* \mathbf{u}_* = 0\,, \quad \mathbf{u}_* \geq \mathbf{0} \qquad (d-f)$$

The expressions for the matrices and vectors present in system (32) can be obtained using the identifications given in table 1 for the set of variables intervening in each of the four possible formulations.

Table 1	Nodal - Stiffness	Nodal - Flexibility	Mesh - Stiffness	Mesh - Flexibility
$\mathbf{k}_1$	$\mathbf{q}$	$\mathbf{q}$	$\mathbf{u}_e$	-
$\mathbf{s}$	-	$\mathbf{X}$	$\mathbf{p}$	$\mathbf{p}$
(32a)	β	β	σ	-
(32c)	ψ	ψ	ψ	ψ
(32c)	-	σ	α	α

Given in the same table are the dimensions of the main constraints (32a-c). Constraints (32b) define the plastic potentials associated with the ψ yield modes considered in the analysis of the structure. Constraints (32a) {(32c)}, when explicitly present, enforce the structure equilibrium {compatibility} conditions; in particular, in the nodal-stiffness {mesh-flexibility} formulation, matrix $\mathbf{D}$, {$\mathbf{C}$} represents the structure stiffness {flexibility} matrix. The results summarized in table 1 suggest that the force (mesh-flexibility) method should usually enjoy an advantage over the displacement (nodal-stiffness) method insofar as the size of the formulation and its numerical implementation are concerned, since in most applications the kinematic indeterminacy β of a structure far exceds its static redundancy α [10].

Elastoplastic Deformation Analysis

There is an equivalence between the symmetric, linear complementarity problem (32) and the pair of primal-dual quadratic programs (33,34), obtained by implementation of the Karush-Kuhn-Tucker equivalence conditions,

$$\text{Min } z \quad = \quad 1/2\,\tilde{\mathbf{k}}\mathbf{D}\mathbf{k} + 1/2\,\tilde{\mathbf{s}}\mathbf{C}\mathbf{s} + \tilde{\mathbf{s}}\mathbf{c} \;:\; \mathbf{D}\mathbf{k} + \mathbf{M}\mathbf{s} \left\{ \begin{matrix} = \\ \geq \end{matrix} \right\} d, \tag{33}$$

$$\text{Max } w \quad = \quad 1/2\,\tilde{\mathbf{k}}\mathbf{D}\mathbf{k} + 1/2\,\tilde{\mathbf{s}}\mathbf{C}\mathbf{s} - \tilde{\mathbf{k}}\mathbf{d} \;:\; \tilde{\mathbf{M}}\mathbf{k} - \mathbf{C}s = \mathbf{c},\; \mathbf{k}_2 \geq \mathbf{0}. \tag{34}$$

wherenow k_2 denotes the plastic parameters $\mathbf{u}_{\cdot}$.

The static and kinematic admissibility conditions (32a,b,c) and (32e,f) are separately implemented by programs (33) and (34). The identification of these programs with the Haar-Karman [11] and Kachanov-Hodge [12,13] theorems on the minimum potential co-energy and potential energy, respectively, was first suggested by Maier [14] for the nodal-stiffness formulation, and later extended by Smith [6] for both mesh and nodal description of elastic-perfectly plastic structures. The theorems of Friedrichs [15] and Kirchhoff [16] for elastic analysis are recovered by deleting in programs (33,34) the variables and constraints associated with the plastic phase $(\mathbf{u}_\star = \mathbf{0}, \phi_\star < \mathbf{0})$:

$$\text{Min } z \quad = \quad 1/2\,\tilde{\mathbf{k}}_1\mathbf{D}_{11}\mathbf{k}_1 + 1/2\,\tilde{\mathbf{s}}\mathbf{C}\mathbf{s} + \tilde{\mathbf{s}}\mathbf{c} \;:\; \mathbf{D}_{11}\mathbf{k}_1 + \mathbf{M}_1\mathbf{s}\; d_1,$$

$$\text{Max } w \quad = \quad 1/2\,\tilde{\mathbf{k}}_1\mathbf{D}_{11}\mathbf{k}1 + 1/2\,\tilde{\mathbf{s}}\mathbf{C}\mathbf{s} - \tilde{\mathbf{k}}_1\mathbf{d}_1 \;:\; \tilde{\mathbf{M}}_1\,\mathbf{k}_1 - \mathbf{C}\mathbf{s} = \mathbf{c}.$$

The general procedure for treating a set of variables in system (32) as prescribed is to interpret them as constants and remove next the associated set of constraints.

If the plastic deformations are treated as prescribed variables, program (33) and (34) are found to encode the variational theorems of Colonnetti [17,18] and Greenberg [19,14],

$$\text{Min } z \quad = \quad 1/2\,\tilde{\mathbf{k}}_1\mathbf{D}_{11}\mathbf{k}_1 + 1/2\,\tilde{\mathbf{s}}\mathbf{C}\mathbf{s} + \tilde{\mathbf{s}}\mathbf{c}' \;:\; \mathbf{D}_{11}\mathbf{k}_1 + \mathbf{M}_1\mathbf{s}\,\mathbf{d}'_1,$$

$$\text{Max } w \quad = \quad 1/2\,\tilde{\mathbf{k}}_1\mathbf{D}_{11}\mathbf{k}_1 + 1/2\,\tilde{\mathbf{s}}\mathbf{C}\mathbf{s} - \tilde{\mathbf{k}}_1\mathbf{d}'_1 \;:\; \tilde{\mathbf{M}}_1\,\mathbf{k}_1 - \mathbf{C}\,\mathbf{s} = \mathbf{c}'$$

wherenow

$$\mathbf{c}' = \mathbf{c} - \tilde{\mathbf{M}}_2\,\mathbf{k}_2,$$

and

$$\mathbf{d}'_1 = \mathbf{d}_1 - \mathbf{D}_{12}\,\mathbf{k}_2.$$

If variables $\mathbf{k}_1$ are instead treated as constants, the following programs are obtained,

$$\text{Min } z \quad = \quad 1/2\,\tilde{\mathbf{k}}_2\mathbf{D}_{22}\mathbf{k}_2 + 1/2\,\tilde{\mathbf{s}}\mathbf{C}\mathbf{s} + \tilde{\mathbf{s}}\mathbf{c}' \;:\; \mathbf{D}_{22}\mathbf{k}_2 + \mathbf{M}_2\mathbf{s} \geq d'_2,$$

$$\text{Max } w \quad = \quad 1/2\,\tilde{\mathbf{k}}_2\mathbf{D}_{22}\mathbf{k}_2 + 1/2\,\tilde{\mathbf{s}}\mathbf{C}\mathbf{s} - \tilde{\mathbf{k}}_2\mathbf{d}'_2 \;:\; \tilde{\mathbf{M}}_2\,\mathbf{k}_2 - \mathbf{C}\mathbf{s} = \mathbf{c}',$$

with

$$\mathbf{c}' = \mathbf{c} - \tilde{\mathbf{M}}_1\,\mathbf{k},$$

and

$$\mathbf{d}_2' = \mathbf{d}_2 - \tilde{\mathbf{D}}_{12}\,\mathbf{k}_1.$$

In the mesh-stiffness formulation, variable $\mathbf{k}_1$, identifies with the elastic deformations $\mathbf{u}_e$ which are uniquely associated with the stress-resultants $\mathbf{X}$ through the elastic causality conditions. The above programs may therefore be used to estimate the structure plastic deformations $\mathbf{k_2} = \mathbf{u}_*$ for a given stress distribution.

A similar result is obtained by specifying the static variables in programs (33,34), to yield,

$$\text{Min } z \;=\; 1/2\,\tilde{\mathbf{k}}\mathbf{D}\mathbf{k} \;:\; \mathbf{D}\mathbf{k} \left\{ \begin{array}{c} = \\ \geq \end{array} \right\} \mathrm{d}',$$

$$\text{Min } w \;-\; 1/2\,\tilde{\mathbf{k}}\mathbf{D}\,\mathbf{k} \;:\; \tilde{M}\;K = c' + C\;s\;,\;\mathbf{k}_2 \geq \mathbf{0},$$

wherenow

$$\mathbf{d}' = \mathbf{d} - \mathbf{M}\,\mathbf{s}$$

and

$$\mathbf{c}' = \mathbf{c} + \mathbf{C}\,\mathbf{s}$$

When variables $\mathbf{s}$ define the static field derived from plastic limit analysis, the above programs produce an estimate for the displacements developing in the structure at incipient collapse [20].

From the existing algorithms for solving quadratic programs and linear complementarity problems, those based in finite termination elimination techniques [21-23] have been particularly appealing for structural applications [4,24,25]. The use of more general gradient methods of non-linear programming has also been suggested [26,27].

Incremental Elastoplasticity Analysis

Let $\mathbf{v}$ denote the variables which are unrestricted in sign, that is, it collects $\mathbf{s}$ and $\mathbf{k}_1$, from system (32).

The right-hand side of (32) can be divided into terms which depend upon λ, being the parameter describing a proportional loading program, and load independent residuals $\mathbf{r}$.

System (32) can now be re-arranged and reduced to the following form:

$$\begin{bmatrix} \mathbf{E}_{11} & \mathbf{E}_{12} \\ \tilde{\mathbf{E}}_{12} & \mathbf{E}_{22} \end{bmatrix} \begin{Bmatrix} \mathbf{v} \\ \mathbf{u}_* \end{Bmatrix} + \begin{bmatrix} \mathbf{0} \\ \mathbf{I} \end{bmatrix} \phi_* = \begin{Bmatrix} \mathbf{e}_1 \\ \mathbf{e}_2 \end{Bmatrix} \lambda + \begin{Bmatrix} \mathbf{r}_1 \\ \mathbf{r}_2 \end{Bmatrix}, \quad \begin{matrix} (a) \\ (b) \end{matrix} \tag{35}$$

$$\phi_* \leq \mathbf{0}\ ,\ \tilde{\phi}_* = 0\ ,\ \mathbf{u}_* \geq \mathbf{0}. \qquad (c-e)$$

The parametric linear complementarity problem (35) may be solved by simple variants of the simplex algorithm [28-31]. Plastic unstressing is checked and implemented substituting the monotonic plastic straining condition (35d) by the appropriate incremental complementarity conditions:

$$\tilde{\phi}_*\, \Delta\, \mathbf{u}_* = \mathbf{0}\ ,\ \ \Delta\, \mathbf{u}_* = 0.$$

The incremental methods, by their very nature, allow for a direct control of those yield modes which are currently active. This essential feature is explored in the algorithms for incremental elastoplastic analysis suggested in [27,32] in which simple procedures are used to detect and control the phenomena of simple or multiple plastic straining and unstressing.

REFERENCES

1. Zienkiewicz, O.C., R. Valliapans and I.P. King: Elastoplastic solutions of engineering problems, initial stress, finite element approach, Int. J. Num. Meth. Engrg., 1, 1969

2. Argyris, J.H.: Methods of elastoplastic analysis, J. Appl. Math. Phys., 23, 1972

3. Maier, G.: Mathematical programming methods in structural analysis, Int. Conf. Variational Meth. Engrg. Univ. of Southampton, 1972

4. Cohn, M.Z. and G. Maier (Eds): Engineering Plasticity by Mathematical Programming, Pergamon, 1979

5. Smith, D.L. and J. Munro: On uniqueness in the elastoplastic analysis of frames, J. Struc. Mech., 6, 1978

6. Smith, D.L.: Plastic limit analysis and synthesis of frames by linear programming, Ph.D. Thesis, Univ. of London, 1974

7. Munro, J. and D.L. Smith: Linear programming duality in plastic analysis and synthesis, Int. Symp. Computer Aided Design, Univ. of Warwick,1972

8. Freitas, J.A.T.: Elastic-plastic analysis of structural cross-sections, Mathematical Programming Methods in Structural Plasticity, CISM, Udine, 1986

9. Corradi, L. and C. Poggi: A refined finite element model for the analysis of elastic-plastic frames,

10. Freitas, J.A.T. and J.P.B.M. Almeida: A direct flexibility approach to the force method of structural analysis, EPMESC, Macao, 1985

11. Haar, A. and T. von Karman: Zur Theorie der Spannungszustande in Plastichen und Sandartigen Medien, Nach der Wissenschaften zu Gottingen, 204-18, 1909

12. Kachanov, L.M.: Variational principles for elastoplastic solids, Prikl. Math. Mekh., 6, 1942

13. Hodge, P.G.: The mathematical theory of plasticity, Surveys in Applied Mathematics, Wiley, 1958

14. Maier, G.: A matrix structural theory of piecewise linear elastoplasticity with interacting yield planes, Meccanica, 5, 1970

15. Friedrichs, K.O.: Ein Verfahren der Variations rechnung das Minimum eines Integrals als das Maximum eines anderen Ausdruckes darzustellen, Nach. der Akademic der Wissenschaften in Gottigen, 13, 1929

16. Kirchhoff, G.: Uber das Gluchgenicht und die Bewegung einer Elastischen Schiebe, Creller J., 40, 1850

17. Colonnetti, G.: Elastic equilibrium in the presence of permanent set, O. Appl. Math., 7, 1950

18. Ceradini, G.: A maximum principle for the analysis of elastic plastic systems, Meccanica, 1, 1966

19. Greenberg, H.J.: On the variational principles of plasticity, Brown Univ., O.N.R. Report NR-041-032, 1949

20. Maier, G., D.E. Grierson and M.J. Best: Mathematical programming methods for deformation analysis at plastic collapse, Computers and Struc., 7, 1977

21. Wolfe, P.: - The simplex method for quadratic programming Econometrica, 27, 1959

22. Lemke, C.E.: Bimatrix equilibrium points and mathematical programming, Management Sci., 11, 1965

23. Cottle, R.W.: The principal pivoting method of quadratic programming, Linear Algebra and its Applications, 1, 1968

24. Smith, D.L. and J.A.S. Appleton: A mathematic programming model for some problems in plasticity, Proc. Int. Conf. Applied Numeric Modelling, E. Alarcon and C.A.Brebbia (Eds), Pentech, London, 1979

25. Franchi, A. and M.Z. Cohn: Computer analysis of elastic-plastic structures, Comp. Meth. Appl. Mech. Eng., 21, 1980

26. De Donato, O. and A. Franchi: A modified gradient method for finite element elastoplastic analysis by quadratic programming, Comp. Meth. Appl. Mech. Eng., 2, 1973

27. Freitas, J.A.T. and J.P.B.M. Almeida: A nonlinear projection method for constrained optimization, Civ. Engng. Syst., 1, 1984

28. De Donato, O. and G. Maier: Historical deformation analysis of elastoplastic structures as a parametric linear complementarity problem, Meccanica, 11, 1976

29. De Donato, O. and G. Maier: Local unloading in piecewise-linear plasticity, I.Engrg. Mech. Div. Proc. ASCE, 102, 1976

30. Smith, D.L.: The Wolfe-Markowitz algorithm for nonholonomic elastoplastic analysis, Engrg. Struct., 1, 1978

31. Maier, G., S. Giacomini and F. Paterlini: Combined elastoplastic and limit analysis via restricted bases linear programming, Comp. Meth. Appl. Mech. Eng., 19. 1979

32. Freitas, J.A.T. and D.L. Smith: Plastic straining, unstressing and branching in large displacement perturbation analysis, Int. J. Num. Meth. Engrg., 20, 1984

CHAPTER 10

A GRADIENT METHOD FOR ELASTIC-PLASTIC ANALYSIS OF STRUCTURES

J. A. Teixeira de Freitas
Istituto Superior Tecnico, Lisbon, Portugal

Abstract: **The systems governing the elastic, elastoplastic and plastic limit analysis of structures are encoded as mathematical programming problems. An iterative procedure based in the gradient method is presented and adapted for the solution of each of the aforementioned classes of structural analysis problems.**

Introduction

A large class of applications in elastic-plastic analysis of structures can be identified [1] as constrained optimization problems and encoded in the form:

$$\text{Max } z(\mathbf{x}) \ : \ \mathbf{f}\ (\mathbf{x}) = \mathbf{0} \ , \ \mathbf{g}\ (\mathbf{x}) \geq \mathbf{0}. \tag{1}$$

When the displacements and the deformations developing in the structure are assumed to be small and a piecewise-linear description is used to model the elastoplastic constitutive relations, it is found that the objective function is quadratic,

$$z\ (\mathbf{x}) = -1/2\ \tilde{\mathbf{x}}\ \mathbf{H}\ \mathbf{x} + \tilde{\mathbf{x}}\ \mathbf{h}_0, \tag{2}$$

and linear expressions are obtained for the intervening contraints:

$$\mathbf{f}\ (\mathbf{x}) = \mathbf{F}\ \mathbf{x} - \mathbf{f}_0 = \mathbf{0}, \tag{3}$$
$$\mathbf{g}\ (\mathbf{x}) = \mathbf{G}\ \mathbf{x} - \mathbf{g}_0 \geq 0. \tag{4}$$

The objective function (2) and the constraints (3,4) have a precise physical meaning and perform specific roles in each particular structural application.

In elastoplastic deformation analysis problems, constraints (3) and (4) represent the structure kinematic {static} admissibility conditions and the objective function (2) is found to relate with the system potential {co-} energy.

In plastic limit analysis problems, the hessian matrix $\mathbf{H}$ of the objective function (2) is the null matrix and program (1) becomes linear:

$$\text{Max } z(x) = \tilde{\mathbf{x}}\ \mathbf{h}_0 \ : \ \mathbf{f}\ (\mathbf{x}) = \mathbf{0} \ , \ \mathbf{g}\ (\mathbf{x}) \geq \mathbf{0}. \tag{5}$$

The inequality constraints (4) are typical of structural analysis problems in which plasticity is accounted for; they represent either the flow rule or the yield rule of plasticity.

These constraints are therefore absent in elastic analysis problems, which take the following general form:

$$\text{Min } z(\mathbf{x}) \ : \ \mathbf{f}\ (\mathbf{x}) = \mathbf{0}. \tag{6}$$

For particular formulations, such as those of the force and displacement methods of structural analysis, the equality constraints are not explicitly enforced. Program (6) is thus reduced to an unconstrained quadratic problem,

$$\text{Min } z(\mathbf{x}) = -1/2\ \tilde{\mathbf{x}}\ \mathbf{H}\ \mathbf{x} + \tilde{\mathbf{h}}_0 \mathbf{x}, \tag{7}$$

the solution of which is that of the equivalent systems of linear equations:

$$\mathbf{H}\ \mathbf{x} = \mathbf{h}_0. \tag{8}$$

The above mentioned programs are usually solved using finite-termination elimination techniques operating on the associated Karush-Kuhn-Tucker problem.

An alternative iterative solution procedure is discussed herein. It is essentially a gradient method [2] obtained by specialization of an algorithm developed in [3] for the solution of nonlinear programming problems.

A general description of the algorithm is given in the first part of the presentation. After introducing the theoretical basis of the method, a step-by-step account of the numerical implementation procedure is suggested next.

The second part of the presentation deals with applications of the algorithm to the problems of elastic analysis of structures. The program presented in [4] for the elastoplastic analysis of structures under prescribed static and kinematic fields can be processed in a similar manner. A nodal formulation is used throughout the text since few readers are equally familiar with the alternative, and often numerically more efficient, mesh approach.

General Description of the Algorithm

Let $\mathbf{x}_0$ represent a known feasible solution to program (1):

$$\mathbf{f}\,(\mathbf{x}_0) = \mathbf{0} \quad , \quad \mathbf{g}\,(\mathbf{x}_0) \geq \mathbf{0}.$$

Assume that solution $\mathbf{x}_0$ is non-optimal and consider the problem of determining a new configuration $\mathbf{x}$ for which an improvement of the current value of the objective function is obtained:

$$z\,(\mathbf{x}) > z\,(\mathbf{x}_0). \tag{9}$$

Let the trial solution be expressed in form,

$$\mathbf{x} = \mathbf{x}_0 + \mathbf{t}\; s, \tag{10}$$

wherein array $\mathbf{t}$ defines the selected trajectory and parameter s represents the adopted step increment:

$$s > 0. \tag{11}$$

Substitution of definition (10) in conditions (3) and (4) shows that the trial solution is also feasible provided that

$$\dot{\mathbf{f}}\,(\mathbf{t}) = \mathbf{0}, \tag{12}$$

and

$$\dot{\mathbf{g}}(\mathbf{t})\; s + \mathbf{g}(\mathbf{x}_0) \geq \mathbf{0}. \tag{13}$$

The following notation is used above:

$$\dot{\mathbf{f}}\,(\mathbf{t}) = \mathbf{F}\,\mathbf{t}, \tag{14}$$

$$\dot{\mathbf{g}}\,(\mathbf{t}) = \mathbf{G}\,\mathbf{t}, \tag{15}$$

and

$$\mathbf{g}\ (\mathbf{x}_0) = \mathbf{G}\ \mathbf{x}_0 - \mathbf{g}_0. \tag{16}$$

The procedure for the solution of program (1) consists in generating a sequence of configurations (10) for which a monotonic improvement (9) of the objective function is attained, while satisfying the admissibility conditions (12), (14) and (13), (15) (16).

Selection of the Trajectory

Assume that the inequality constraints (13), (15), (16) are inactive for the current solution point:

$$\mathbf{g}\ (\mathbf{x}_0) > \mathbf{0}.$$

It can be shown [3] that in these circumstances, a trajectory that complies with solutions (9) and (12), (14) is defined by

$$\mathbf{t} = \mathbf{h} - \mathbf{S}\ \tilde{\mathbf{F}}\ \mathbf{u}, \tag{17}$$

wherein $\mathbf{h}$ represents the gradient of the objective function at the current solution point:

$$\mathbf{h} = \mathbf{h}_0 - \mathbf{H}\ \mathbf{x}_0. \tag{18}$$

$\mathbf{S}$ is a scaling matrix and $\mathbf{u}$ is the solution of the following unconstrained optimization problem,

$$\text{Max w } = -1/2\ \tilde{\mathbf{u}}\ \mathbf{D}\ \mathbf{u} + \tilde{\mathbf{u}}\ \mathbf{d}, \tag{19}$$

with

$$\mathbf{D} = \mathbf{F}\ \mathbf{S}\ \tilde{\mathbf{F}}, \tag{20}$$

and

$$\mathbf{d} = \mathbf{F}\ \mathbf{h}, \tag{21}$$

which is equivalent to the following system of linear equations:

$$\mathbf{D}\ \mathbf{u} = \mathbf{d}. \tag{22}$$

Assume now that a subset of the inequality constraints (13), (15), (16) is active for the current solution point,

$$\mathbf{g}'\ (\mathbf{x}_0) = \mathbf{0},$$

and is to remain so for the trial solution (10):

$$\mathbf{g}'\ (\mathbf{x}) = \mathbf{0}.$$

Then, according to definitions (13), (15), (16), the trajectory must be so chosen as to further satisfy the following equality constraints:

$$\mathbf{G}'\,\mathbf{t} = \mathbf{0}. \tag{23}$$

Definition (17) can still be adopted provided the equality constraint operator **F** is extended to account for the additional conditions (23):

$$\mathbf{F} = \begin{bmatrix} \mathbf{F} \\ \mathbf{G}' \end{bmatrix} \tag{24}$$

Determination of the Step Increment

The increment (9) in the objective function

$$\Delta z = z\,(\mathbf{x}) - z\,(\mathbf{x}_0),$$

can easily be found to be defined by

$$\Delta z = -1/2\; a\; s^2 + b\; s, \tag{25}$$

wherein

$$b = \tilde{\mathbf{h}}\;\mathbf{t}, \tag{26}$$

and

$$a = \tilde{\mathbf{p}}\;\mathbf{t}, \tag{27}$$

with

$$\mathbf{p} = \mathbf{H}\,\mathbf{t}. \tag{28}$$

Definition (25) indicates that a stationary point is exposed by a step-length given by

$$s = b/a, \tag{29}$$

which represents a directional minimum whenever parameter a is positive; definitions (27), (28) show that this is consequent upon the positive definiteness of the hessian matrix **H**.

The step increment is also conditioned by the mobilization of one, or more, of the currently inactive constraints (13), (15), (16):

$$\begin{aligned} \dot{\mathbf{g}}''(\mathbf{t})\; s + \mathbf{g}''\,(\mathbf{x}_0) &\geq 0, \\ \mathbf{g}''\,(\mathbf{x}_0) &> 0. \end{aligned} \tag{30}$$

The activation of the m-th constraints present in subset (30) is exposed by condition

$$\dot{g}''_m\ s + g''_m = 0,$$

which generates the following bound for the step increment:

$$s = \min_m \left[-g''_m / \dot{g}''_m\right] > 0. \tag{31}$$

Summary of the Solution Procedure

The algorithm for solving the quadratic programming problem can be summarized as follows:

1. Select a starting feasible solution $\mathbf{x}_0$ and compute the corresponding value for the objective function using definition (2).
2. Set up the gradient and stipulation vectors using (18) and (16), respectively.
3. Use definitions (17) to (21) to determine the trajectory vector, $\mathbf{t}$.
4. For the currently active inequality constraints, compute the stipulation rates $\dot{\mathbf{g}}'\ (\mathbf{t})$ using definition (15); if it is found that $\dot{\mathbf{g}}' > \mathbf{0}$ demobilize this set of constraints and go to step 6.
5. Let $\dot{\mathbf{g}}'_m$ be the most negative entry of array $\dot{\mathbf{g}}$ and $\mathbf{G}'_m$ the associate row in the active inequality constraint matrix $\mathbf{G}'$. Extend the equality constraint matrix $\mathbf{F}$ according to definition (24) and return to step 3.
6. Compute parameter s_1 using definition (29).
7. Compute parameter s_2 using definition (31).
8. Set the step increment to $s = min\ (s_1, s_2)$; if $s \leq 0$, stop.
9. Update the program variables using definition (10).
10. Determine the increment Δz using definition (25) and update the function z.
11. Update the gradient vector using (18) and (28) to yield $\mathbf{h} = \mathbf{h} - \mathbf{p}\ s$.
12. Update the stipulation vector noticing that $\mathbf{g}' = \mathbf{0}$ and $g''_m = g''_m + \dot{g}''_m\ s$; return to step 3.

Elastic Analysis

The analysis of the elastic response of structure by the nodal-stiffness method [4] can be performed implementing an unconstrained quadratic programming problem

$$\text{Max } z = -1/2\, \tilde{\mathbf{q}}\, \mathbf{K}_*\, \mathbf{q} + \tilde{\mathbf{q}}\, \mathbf{Q}_0, \tag{32}$$

or the equivalent system of linear equations

$$\mathbf{K}_*\, \mathbf{q} = \mathbf{Q}_0, \tag{33}$$

wherein

$$\mathbf{K}_* = \tilde{\mathbf{A}}\, \mathbf{K}\, \mathbf{A}, \tag{34}$$

denotes the stiffness matrix of the structure.

System (33) can be solved using either direct or iterative methods [5,9] adapted to explore the special properties of matrix $\mathbf{K}_*$, namely symmetry and sparsity.

The procedure described above when applied to the solution of program (32) is found to identify with the steepest ascent method. It generates a sequence of orthogonal trajectory vectors which is known to converge poorly in the vicinity of the optimal solution. To improve the rate of convergence, this procedure should be coupled with a conjugate gradient approach [7,8].

Plastic Limit Analysis

As it is well known, the safe theorem of plastic limit analysis is encoded as a linear programming problem which can be expressed in form (1-4) letting [4]:

$$\mathbf{x} = \begin{Bmatrix} \mathbf{X} \\ \lambda \end{Bmatrix}, \quad \mathbf{h}_0 = \begin{Bmatrix} 0 \\ 1 \end{Bmatrix},$$

$$\mathbf{F} = \begin{bmatrix} -\tilde{\mathbf{A}} & \mathbf{a}_0 \end{bmatrix}, \quad \mathbf{f}_0 = \mathbf{Q}_0,$$

$$\mathbf{G} = \begin{bmatrix} -\tilde{\mathbf{N}} & \mathbf{0} \end{bmatrix}, \quad \mathbf{g}_0 = \mathbf{X}_*.$$

A distribution of stress-resultants $\mathbf{X}_0$ that equilibrates the permanent loads $\mathbf{Q}_0$ acting upon the structure can be used to define the initial feasible solution required in step 1 of the algorithm,

$$\mathbf{X} = \mathbf{X}_0 \;, \quad \lambda = 0,$$

wherein λ denotes the parameter associated with the live loads.

As the Hessian matrix is now the null matrix, $\mathbf{H} = \mathbf{0}$, the gradient (18) of the objective function is constant and criterion (29) for controlling the step increment is rendered irrelevant.

The numerical implementation effort concentrates now in step 3 of the algorithm, which requires the solution of either of problems (19) or (22).

Letting the scaling matrix be defined thus

$$\mathbf{S} = \begin{bmatrix} \mathbf{K}' & 0 \\ \tilde{0} & 0 \end{bmatrix},$$

the following expressions are found for definitions (13), (15), (16) and (17).

$$\mathbf{D} = \begin{bmatrix} \tilde{\mathbf{A}}\ \mathbf{K}'\ \mathbf{A} & -\tilde{\mathbf{A}}\ \mathbf{K}'\ \mathbf{N}' \\ -\tilde{\mathbf{N}}\ \mathbf{K}'\ \mathbf{A} & \tilde{\mathbf{N}}'\ \mathbf{K}'\ \mathbf{N}' \end{bmatrix}, \quad \mathbf{d} = \left\{ \begin{matrix} \mathbf{a}_0 \\ \mathbf{0} \end{matrix} \right\}, \tag{35}$$

wherein $\mathbf{N}'$ denotes the normality matrix associated with the currently active yield modes (4).

The above results show that the solution of problems (19) or (22) can be interpreted as quasi-elastic structural analysis problem (32) or (33). The scaling matrix $\mathbf{K}'$ may in fact be identified as the stiffness matrix relating the stress- and strain-resultants developing in the elements into which the structure is discretized ($\mathbf{K}' = \mathbf{K}$) or simplified to collect only its main diagonal $K'_{ij} = \delta_{ij}\ K_{ij}$. The operations implied by definitions (35) are implemented using the standard finite element techniques for direct allocation of the individual contributions of the structure building elements. As stated before for the elastic analysis problem, the direct and/or iterative methods selected to solve problems (19) or (22) should be designed to explore the special properties enjoyed by the "elastic-plastic" stiffness matrix $\mathbf{D}$.

Elastoplastic Deformation Analysis

The deformation analysis of elastoplastic structures by the nodal-stiffness method [4] can be performed implementing the following quadratic programming problem,

$$\text{Max } z = -1/2 \left\{ \begin{matrix} \tilde{\mathbf{q}} \\ \mathbf{u}_* \end{matrix} \right\} \begin{bmatrix} \mathbf{K}_* & \mathbf{P} \\ \tilde{\mathbf{P}} & \mathbf{H}_* \end{bmatrix} \left\{ \begin{matrix} \mathbf{q} \\ \mathbf{u}_* \end{matrix} \right\} - \left\{ \begin{matrix} \tilde{\mathbf{q}} \\ \mathbf{u}_* \end{matrix} \right\} \left\{ \begin{matrix} \mathbf{Q}_0 \\ -\mathbf{X}_* \end{matrix} \right\}, \quad \mathbf{u}_* \geq \mathbf{0},$$

wherein

$$\mathbf{P} = -\tilde{\mathbf{A}}\ \mathbf{K}\ \mathbf{N},$$

and

$$\mathbf{H}_* = \tilde{\mathbf{N}}\ \mathbf{K}\ \mathbf{N} + \mathbf{H}.$$

The above program is expressed in format (1-4) by delecting the equality constraint set (3), and letting

$$\mathbf{G} = \begin{bmatrix} \mathbf{0} & \mathbf{I} \end{bmatrix} \ , \ \mathbf{g}_0 = \mathbf{0} \tag{36}$$

A starting feasible solution is obtained setting the plastic multipliers $\mathbf{u}_*$ to zero. The independent displacements $\mathbf{q}$ can be treated similarly or identified with the (known) solution of the associated elastic analysis problem (34).

The equality constraint matrix (24) takes now the form,

$$\mathbf{F} = \begin{bmatrix} \mathbf{0} & \mathbf{J} \end{bmatrix},$$

and is obtained collecting the rows of matriz $\mathbf{G}$ defining the currently null-plastic multipliers; $\mathbf{J}$ is a boolian matrix with one unit element per row, all the remaining being zero.

Decomposing the gradient vector (11) into two terms, one associated with displacements $\mathbf{q}$ and the other with plastic parameters $\mathbf{u}_*$,

$$\mathbf{h} = \begin{Bmatrix} \mathbf{h}_1 \\ \mathbf{h}_2 \end{Bmatrix},$$

the following description is found for system (22),

$$\mathbf{u} = \tilde{\mathbf{J}} \, \mathbf{h}_2,$$

if the identify matrix is adopted for scaling matrix.

As the solution of step 3 becomes trivial, the computational effort required to implement the algorithm described above concentrates now on the determination of the gradient vector $\mathbf{h}$. The operations required to enforce steps 2, 6 and 11 can be minimized using direct allocation techniques to set up locally the entries of the intervening operators.

REFERENCES

1. Cohn, M.Z. and G. Maier (Ed.): Engineering Plasticity by Mathematical Programming, Pergammon, 1979

2. Freitas, J.A.T. and J.P.B.M. Almeida: A nonlinear projection method for constrained optimization, Civ. Engrg. Syst., 1, 1984

3. Freitas, J.A.T.: Elastoplastic analysis of skeletal structures, Mathematical Programming Methods in Structural Plasticity, CISM, Udine, 1986

4. Jennings, A.: Matrix Computation for Engineers and Scientists, Wiley, 1977

5. Burden, R.L. and J.D. Faires: Numerical Analysis, PWS, 1985

6. Hestenes, M.R. and E. Steifez: Methods of conjugate gradients for solving linear systems, Nat. Sur. Rept. nr.1659, 1952

7. Ginsburg, T.: The conjugate gradient methods, Numer. Math., 5, 1963

CHAPTER 11

PLASTIC SHAKEDOWN ANALYSIS

Nguyen Dang Hung and P. Morelle
University of Liège, Liège, Belgium

1. INTRODUCTION

Structures of mechanical engineering for instance power plants, reactors, pressure vessels, etc. or civil engineering for instance, frames, grids, bridge decks etc. are exposed to variable loading particularly cycliC or repeated loading. In these situations, classical limit analysis which assumes proportional loading are out of question because the results deduced present no security step-by-step elastic-plastic calculation is somewhat very costly in computing times. The most efficient way to handle the problem is to apply the shakedown theory. This theory is based on experimental facts obtained from realistic structures or laboratory specimen [1-4]. It offers a direct method as like as limit analysis to perform the analysis of the problem.

To clarify the phenomena of shakedown one may summarize the state of the problem in the following general terms.

General variable loadings $\bar{f}_k(t)$ may be represented by loading domain within its individual load can vary independently under any form. Let $\bar{f}^o_k$ be a reference applied forces variable between two bounds $\bar{f}^o_{k-}$ and $\bar{f}^o_{k+}$ so that one may describe the general loading domain P (figure 1) by using :

a loading factor α

$$\bar{f}_k = \alpha\, \bar{f}^o_k,\ k \quad [1,n] \qquad \bar{f}^o_{k-} \le \bar{f}^o_k \le \bar{f}^o_{k+} \tag{1}$$

where n is the number of independent loads.

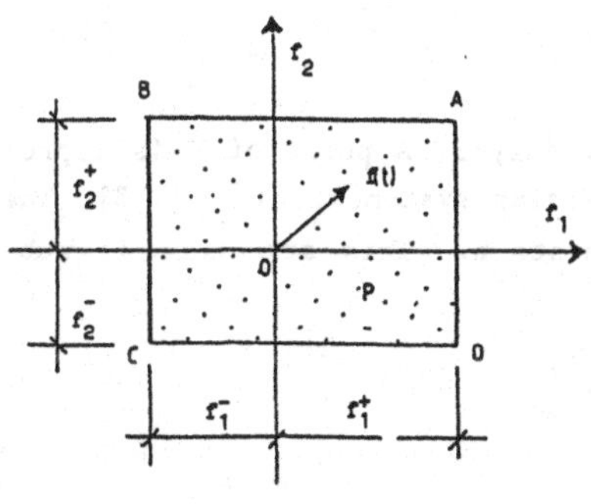

FIGURE 1. Loading domain

The principal purpose of the shakedown theory is to determine a limit shakedown domain defined by a limit factor α^a such that inside this one observes a stabilization of plastic deformations everywhere in the structure (figure 2.a.). This behaviour is called elastic shakedown or adaptation.

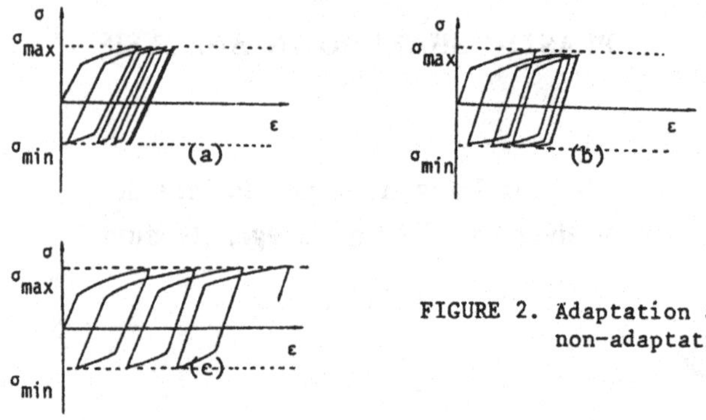

FIGURE 2. Adaptation and non-adaptation

Beyond this limit domain two dangerous situations may happen :

- accumulation of plastic deformations leading to incremental collapse or racheting (figure 2.b.)
- alternating plastic deformations leading to low cycle fatigue and fracture of the structure after an important member of cycles (figure 2.c.).

The shakedown theory is introduced firstly by BLEICH [5] and MELAN [6] during the thirthies.
At present time, the literature about this question is considerable. One can mention mainly the contributions of MELAN [6], SYMONDS [7], KOITER [8], NEAL [9] (kinematical aspects), PRAGER [10], ROSENBLUM [11], GOKHFELD [12], KONIG [13,14] (thermal aspects), HO HWA-SHAN [15], CORRADI and MAIER [16, 17], POLIZZOTTO [18] (dynamic aspects). In this paper, after recalling some classical theorems, we try to present a brief survey of the numerical formulations and results performed by the school of Liège (NGUYEN DANG HUNG, PALGEN, DE SAXCE, MORELLE) [18-25]).

2. DEFINITIONS AND ASSUMPTIONS

Let us consider a solid body with its boundary S. A point of V is represented by the three coordinates of a cartesian system x_i, i [1,3]. The unit normal vector of S has three components n_i. This structure is subjected to time-dependent sollicitations :

- body forces $\overline{f}_i$ in V
- imposed surfaces forces $\overline{t}_i$ on part S_σ of the overall boundary S
- imposed displacement $\overline{u}_i$ on the remaining part $S_u = S - S_\sigma$

The strain is supposed to be small and the loading varies quasi-statically between given limits so that the dynamic effects are negligible. The material is assumed elastic-perfectly plastic and the plastic strains are derived from the yield function f by the normality law :

$$\dot{\epsilon}^p_{ij} = \dot{\lambda} \frac{\partial f}{\partial \sigma_{ij}} , \quad \dot{\lambda} \geq 0 \tag{2}$$

For non yield function like the Tresca yield criterion the operator (2) must be replaced by a sub-derivative.

The criterion $f \leq 0$ (3)
defined a convex elastic domain so that the Drucker inequality : σ^o_{ij}

$$(\sigma_{ij} - \sigma^o_{ij}) \, \dot{\epsilon}^p_{ij} \geq 0 \tag{4}$$

is verified for any plastically admissible stress σ^o_{ij} satisfying (3).
It is useful to define the following quantities :

σ^s_{ij} : any statically admissible stress field satisfying the equilibrium equations of solid continuum :

$$\frac{\partial \sigma^s_{ij}}{\partial x_j} + \overline{f}_i = 0 \quad \text{in V}$$
$$n_j \, \sigma^s_{ij} = t^s_i = \overline{t}_i \quad \text{on } S_\sigma \tag{5}$$

σ^l_{ij} : any licit stress field which satisfies both (5) and (3). So that a licit stress field is both statically admissible and plastically admissible.

ϵ^k_{ij} : any kinematically admissible strain field which satisfies the compatibility equations :

$$\epsilon^k_{ij} = \frac{1}{2} \left(\frac{\partial u^k_i}{\partial x_j} + \frac{\partial u^k_j}{\partial x_i}\right) \quad \text{in V}$$
$$u^k_i = \overline{u}_i \quad \text{on } S_u \tag{6}$$

ϵ^l_{ij} : any licit strain field which verifies (6) and furnishes an external work positive :

$$\int \overline{f}_i \, u^l_i \, dV + \int \overline{t}_i \, u^l_i \, dS_\sigma > 0 \tag{7}$$

ϵ_{ij} (x,t) the exact time dependent strain field satisfying (6) with the exact displacement field u_i. This field is splitted into two parts :

$$\epsilon_{ij} (x,t) = \epsilon^e_{ij} (x,t) + \epsilon^p_{ij} (x,t) \tag{8}$$

where the elastic part of strain is calculated from the classical generalized Hooke's law :

$$\epsilon^{e}_{ij}\ (x,t) = A_{ijkl}\ \sigma_{kl}\ (x,t) \tag{9}$$

and the plastic part of strain rate obeys to the normality law (2).
$\sigma_{ij}(x,t)$ the exact time dependent stress field satisfying (5), (3) and (9).
$\sigma^{E}_{ij}(x,t)$ the fictitious time dependent stress field satisfying the equilibrium equation (5) and the Hooke's law (9) with the corresponding strain field $\epsilon^{E}_{ij}(x,t)$. This fictitious elastic strain field satisfies obviously compatibility condition (6) with the corresponding fictitious displacement field $u^{E}_{i}(x,t)$.
One defines also : $\rho_{ij}(x,t) = \sigma_{ij}(x,t) - \sigma^{E}_{ij}(x,t)$ (10)
the residual stress field that is in self-equilibrium stress field with no external forces :

$$\frac{\partial \rho_{ij}}{\partial x_j} = 0 \quad \text{in } V$$
$$n_j\ \rho_{ij} = 0 \quad \text{in } S_\sigma \tag{11}$$

$\eta_{ij}(x,t) = \epsilon_{ij}(x,t) - \epsilon^{E}_{ij}(x,t) =$ the residual strain field which relates to the residual displacement field :

$$v_i\ (x,t) = u_i\ (x,t) - u^{E}_{i}\ (x,t) \tag{12}$$

by the following compatibility relation :

$$\eta_i = -\left(\frac{\partial v_i}{\partial x_j} + \frac{\partial v_j}{\partial x_i}\right) \quad \text{in } V \tag{13}$$
$$v_i = 0 \quad \text{on } S_u$$

It may be seen that if

$$\eta_{ij} = \eta^{e}_{ij} + \eta^{p}_{ij} \tag{14}$$

one has :

$$\eta^{p}_{ij} = \epsilon^{p}_{ij} \tag{15}$$
$$\eta^{e}_{ij} = \epsilon_{ij} - \epsilon^{E}_{ij} - \epsilon^{p}_{ij} \tag{16}$$

By same analytical calculation one may show without any difficulty that the following word equation is obtained from any couple of stress field σ^{s}_{ij} , ϵ^{k}_{ij}

$$\int \sigma^{s}_{ij}\ \epsilon^{k}_{ij}\ dV = \int \bar{f}_i\ u^{k}_{i}\ dV + \int \bar{t}_i\ u^{k}_{i}\ dS_\sigma + \int t^{s}_{i}\ \bar{u}_i\ dS_u \tag{17}$$

Now using (11) and (6), one deduces :

$$\int \rho_{ij}(x,t)\ \eta_{ij}(x,t)\ dV = 0 \tag{18}$$

On the other hand, if $\sigma^{Eo}_{ij}(x,t)$ is the fictitious elastic stress response corresponding to the reference loading $\bar{f}^{o}_{i}$, $\bar{t}^{o}_{i}$ by using (1) and (10) one may write down :

$$\sigma^{E}_{ij}(x,t) = \alpha\, \sigma^{Eo}_{ij}(x,t)$$
$$\rho_{ij} + \alpha\, \sigma^{Eo}_{ij} = \sigma_{ij} \tag{19}$$

The above quantities allow to define now the dissipation power function :

$$\dot{D}\,(\dot{\epsilon}^{P}_{ij}) = \sup_{\sigma_{ij}} (\sigma_{ij}\, \dot{\epsilon}^{P}_{ij}) \tag{20}$$

for any stress field satisfies the yield condition (2).

Basing on (2) and (4) one may show that $\dot{D}\,(\dot{\epsilon}^{P}_{ij})$ is a convex and homogeneous function of order 1 :

$$\dot{D}\,(\dot{\gamma}\, \dot{\epsilon}^{P}_{ij}) = \dot{\gamma}\, \dot{D}\,(\dot{\epsilon}^{P}_{ij}) \tag{21}$$

In the case of Von Mises yield criterion, one has :

$$f = \sigma_e/\sigma_p - 1 = \frac{1}{k}\sqrt{s_{ij}\, s_{ij}/2} - 1 \tag{22}$$

where σ_e is the effective stress, σ_p is the elastic limit stress, $k = \sigma_p/\sqrt{3}$ the yield shear stress and $s_{ij} = \sigma_{ij} - \sigma_m\, \delta_{ij}$ the deviator stresses.

Therefore, the dissipation power may be written under the following form :

$$\dot{D}\,(\dot{\epsilon}^{P}_{ij}) = \sigma_p \sqrt{2/3\, \dot{\epsilon}^{P}_{ij}\, \dot{\epsilon}^{P}_{ij}} = k\sqrt{2\, \dot{\epsilon}^{P}_{ij}\, \dot{\epsilon}^{P}_{ij}} \tag{23}$$

For piecewise linear yield condition, one may show that :

$$\dot{D}\,(\dot{\epsilon}^{P}_{ij}) = \sigma_p\, ||\dot{\epsilon}^{P}_{ij}|| \tag{24}$$

where $||\dot{\epsilon}^{P}_{ij}||$ is a certain norm of the plastic strain rate.

By the fact that $\dot{\lambda} f$ is always nul, the following Legendre transformation :

$$\dot{D}\,(\dot{\epsilon}^{P}_{ij}) = \sigma_{ij}\, \dot{\epsilon}^{P}_{ij} - \dot{\lambda} f \tag{25}$$

allows to obtain the inverse relation for stresses

$$\sigma_{ij} = \frac{\partial\, \dot{D}(\dot{\epsilon}^{P}_{ij})}{\partial \dot{\epsilon}^{P}_{ij}} \tag{26}$$

Now the shakedown situation may be defined in terms of dissipation power in the following. Under a certain history of cyclic loading, a structure will shakedown if and only if everywhere the following condition is fulfilled.

$$\lim_{t \to t_\infty} \int_o^t dt \int_V \dot{D} (\dot{\epsilon}^p_{ij}) \, dV < \infty \tag{27}$$

It may be shown [26] that the shakedown condition (27) occurs if and only if

$$\lim_{t \to \infty} \dot{\epsilon}^p_{ij} = 0 \tag{28}$$

3. GENERAL THEOREMS

3.1. MELAN's theorem

Shakedown occurs if and only if there exists a permanent residual stress field $\bar{\rho}^s_{ij}$ such that :

$$f(\sigma^E_{ij}(t) + \bar{\rho}^s_{ij}) < 0 \tag{29}$$

Indeed using (15), (18) with the strain rate instead of the strain one may find :

$$\int \rho^s_{ij} \, \dot{\epsilon}^p_{ij} \, dV = - \, d/dt \, (1/2 \int \rho^s_{ij} \, H_{ijkl} \, \rho^s_{kl} \, dV) \tag{30}$$

if shakedown occurs, condition (28) is happened and because of the definite positivity of the Hooke's matrix after a certain time $\rho_{ij}(t) \to \bar{\rho}^s_{ij} = cste$.

Now on the contrary if $\bar{\rho}^s_{ij}$ is found and (29) is verified, there exists a factor $\gamma > 1$ such that :

$$f \, [\gamma \, (\sigma^E_{ij}(t) + \bar{\rho}^s_{ij})] \le 0, \quad \gamma > 1 \tag{31}$$

From (4) with $\sigma^o_{ij} = \gamma \, (\sigma^E_{ij} + \bar{\rho}^s_{ij})$, one may deduce :

$$\sigma_{ij} \, \dot{\epsilon}^p_{ij} \le \frac{\gamma}{\gamma - 1} \, (\rho^s_{ij} - \bar{\rho}^s_{ij}) \, \dot{\epsilon}^p_{ij} = - \frac{\gamma}{\gamma - 1} \, \dot{\pi} \tag{32}$$

with

$$\pi = 1/2 \int (\rho^s_{ij} - \bar{\rho}^s_{ij}) \, H_{ijkl} \, (\rho^s_{kl} - \bar{\rho}^s_{kl}) \, dV > 0 \tag{33}$$

Integrating the history of loading, one has :

$$\int_o^{t\infty} dt \int \sigma_{ij} \, \epsilon^p_{ij} \, dV \le \frac{\gamma}{\gamma - 1} \, [\pi(0) - \pi(t_\infty)] < \infty$$

So by the definition (27) shakedown occurs. The MELAN's theorem introduces the permanent admissible residual stress field $\bar{\rho}^s_{ij}$ which will play a major role of the shakedown analysis under statical expert.

Under kinematical aspect the following Koiter's theorem introduces the admissible cycle of plastic strain $\Delta\epsilon^{pk}_{ij}$ defined in the following :

$$\Delta\epsilon^{pk}_{ij} = \oint \dot{\epsilon}^{pk}_{ij}\, dt$$

$$\Delta\epsilon^{pk}_{ij} = 1/2\, \left(\frac{\partial \Delta u^k_i}{\partial x_j} + \frac{\partial \Delta u^k_j}{\partial x_i}\right) \tag{34}$$

$$\Delta u^k_i = 0 \text{ on } S_u$$

where the integration may be performed over the cycle of loading.

From (14) and (15) the residual admissible cycle of strain corresponding to (34) is :

$$\Delta\, \eta^k_{ij} = \oint \dot{\eta}^k_{ij}\, dt = \Delta\eta^{ek}_{ij} + \Delta\epsilon^{pk}_{ij} \tag{35}$$

with

$$\Delta\eta^{ek}_{ij} = \oint \dot{\eta}^{ek}_{ij}\, dt \tag{36}$$

Let us consider now the problem of a body with no external loading but with imposed plastic strain rate $\dot{\epsilon}^{pk}_{ij}$. For this elastic problem the solution is uniquely determined and the corresponding statically admissible residual stress rate $\dot{\rho}^k_{ij}$ is associated to the residual elastic strain rate $\dot{\eta}^{ek}_{ij}$ by the Hooke's law (9) :

$$\dot{\eta}^{ek}_{ij} = H_{ijkl}\, \dot{\rho}^k_{kl}, \quad \Delta\eta^{ek}_{ij} = H_{ijkl}\, \Delta\rho^k_{kl}$$

$$\Delta\rho^k_{ij} = \oint \dot{\rho}^k_{ij}\, dt \tag{37}$$

By the fact that $\Delta\epsilon^{pk}_{ij}$ satisfies (34), using the virtual work equation (18), one has :

$$\int \Delta\rho^k_{ij}\, \Delta\eta^k_{ij}\, dV = \int \Delta\rho^k_{ij}\, H_{ijkl}\, \Delta\rho^k_{ij} \tag{38}$$

Again, the virtual work equation (18) may be utilized for the couple of residual stress $\Delta\rho^k_{ij}$ and residual strain $\Delta\eta^k_{ij}$ which are respectively statically admissible and kinematically admissible with no external loading or imposed displacement, to deduce that $\Delta\rho^k_{ij}$ must be nul. Otherwise, any residual stress $\rho^k_{ij}(t)$ associated to admissible cycle of plastic strain (34) must be periodic.

$$\rho^k_{ij}\,(t) = \rho^k_{ij}\,(t+T) \tag{39}$$

where T is the period of the loading. This discussion seems to be important for the understanding of the new notion of admissible cycle of plastic strain introduced by Koiter. One is now ready to state the Koiter theorem.

3.2. Koiter's theorem

Shakedown occurs if the following inequality is satisfied :

$$\oint dt \, [\int \bar{f}_i \dot{u}_i^k \, dV + \oint \bar{t}_i \dot{u}_i^k \, dS_\sigma] \le \int dt \int \dot{D} \, (\dot{\epsilon}_{ij}^{pk}) dV \tag{40}$$

for any admissible cycle of plastic strain.

Shakedown cannot occur if on the contrary the following inequality holds :

$$\oint dt \, [\int \bar{f}_i \dot{u}_i^k \, dV + \int \bar{t} \dot{u}_i^k \, dS_\sigma > \oint dt \int \dot{D} \, (\dot{\epsilon}_{ij}^{pk}) \, dV \tag{41}$$

The prove of the first proposition of Koiter's theorem is based on the previous Melan's theorem. Indeed, if shakedown occurs there exists a certain permanent residual stress field $\bar{\rho}_{ij}^s$ which satisfies (29). Let σ_{ij}^k the stress field associated to the plastic strain rate field $\dot{\epsilon}_{ij}^{pk}$ by means of the gradient law (26). The Drucker inequality takes now the form :

$$[\sigma_{ij}^k - (\dot{\sigma}_{ij}^E + \bar{\rho}_{ij}^s)] \, \dot{\epsilon}_{ij}^{pk} \ge 0 \tag{42}$$

where σ_{ij}^E the fictitious elastic response corresponding to external applied forces $\bar{f}_i$, $\bar{t}_i$.

After performing the integration of inequality (42) over the cycle and the body, one has :

$$\oint dt \int \dot{D} \, (\dot{\epsilon}_{ij}^{pk}) \, dV \ge \oint dt \int \sigma_{ij}^E \, \dot{\epsilon}_{ij}^{pk} \, dV + \int \bar{\rho}_{ij}^s \oint \dot{\epsilon}_{ij}^{pk} \, dt \, dV \tag{43}$$

The last term of the second member of this inequality is vanished according to (11), (18) and (34).

Now let $\dot{\eta}_{ij}^k$ the uniquely determined residual strain rate response of the elastic problem of imposed plastic strain rate $\dot{\epsilon}_{ij}^{pk}$ and no external forces. The elastic part of this field is related to the residual stress rate by relation (37) and the plastic part by the relation (15).
It appears that the first term of the second member of inequality (43) may be transformed in the following way.

$$\oint dt \int \sigma_{ij}^E \dot{\epsilon}_{ij}^{pk} dV = \oint dt \int \sigma_{ij}^E \dot{\eta}_{ij}^k - \oint dt \int \epsilon_{ij}^E \dot{\rho}_{ij}^k dV \tag{44}$$

By the fact that ϵ_{ij}^E is compatible with no imposed displacement on S_u and $\dot{\rho}_{ij}^k$ is in equilibrium with no external forces, the last term of (44) vanishes. As to the first term of (44), it may be transformed according to equation (17) :

$$\oint dt \int \sigma_{ij}^E \dot{\eta}_{ij}^k = \oint dt \int \bar{f}_i \dot{u}_i^k dV + \oint dt \int \bar{t}_i \dot{u}_i^k dS_\sigma$$

After replacing this expression in (43), one obtains finally the inequality (40).

The second proposition of the Koiter's theorem is obvious because if it is not true, a contradiction with the first one.

3.3. Characterization of the adaptation behaviour

It appears that the shakedown situation better called adaptation behaviour of structure is bounded by a limit characterized by existence of both circumstanes after some history of repeated loading, a permanent residual stress field statically admissible and a cycle of plastic strain kinematically admissible in Koiter's sense. However, the previous discussion has pointed out that a cycle of admissible strain leads to a periodic residual stress field. Fortunately, there is here no contradiction because that the simplest periodic residual stress field is a permanent one and the frontier between the adaptation and the contrary behaviour, the inadaptation, has to be understood in a continuity sense.

On the other hand, it is useful to insist on the fact that adaptation is a limit evolution of the body after some history of repeated loading but not a state fixed in the times. As shown by the Koiter's theorem, such a situation exists if loading doesn't give more work that those dissipated in the body during the cycle.
The two following theorems, the lower bounds theorem and the upper bounds theorem allow to realize the approaches of this limit behaviour by respectively statical and kinematical way. Before presenting these practical theorems it is useful to recall here some results of convex analysis.

3.4. Theorems of convex analysis

Shakedown will occur for a loading domain P (figure 1) convex and compact if and only if it occurs on the frontier $\hat{P}$ of P. The prove of this classical theorem is given in reference [13,18]. A corollary of this theorem may be stated in following terms :
shakedown will occur for a loading domain P if it occurs for a cycle of loading passing on the vertexes of the frontier $\hat{P}$ of P.
This theorem allows to examine only the cycles of loading represented by the vertexes of P instead of all other cycles of loading within the domain P.

3.5. Lower bounds theorem a statical theorem.

The load factor α^s corresponding to any Melan's admissible residual stress field constitutes a lower bound of the adaptation limit factor α^a

Let σ^s_{ij}, $\bar{\rho}^s_{ij}$ the statically admissible stress fields in Melan's sense.
Let σ_{ij}, $\bar{\rho}_{ij}$, $\Delta\epsilon^p_{ij}$ the exact stress field and plastic cycle corresponding to the limit adaptation behaviour.

In basing on (19) and (34), one has :

$$\oint dt \int \sigma^s_{ij}\, \dot{\epsilon}_{ij} dV = \int \bar{\rho}^s_{ij}\, \Delta\epsilon^p_{ij}\, dV + \alpha^s \oint dt \int \sigma^{Eo}_{ij}\, \dot{\epsilon}^p_{ij}\, dV \tag{45}$$

The first term of the second member is vanished because of identity (18).

On the other hand for the exact fields, one has :

$$\oint dt \int \sigma_{ij}\, \dot{\epsilon}_{ij}\, dV = \oint dt \int \dot{D}\, (\dot{\epsilon}_{ij})\, dV = \alpha^a \oint dt \int \sigma^{Eo}_{ij}\, \dot{\epsilon}^p_{ij}\, dV \tag{46}$$

Now by comparison of (45) and (46), one obtains :

$$\oint dt \int (\sigma_{ij} - \sigma^s_{ij})\, \dot{\epsilon}^p_{ij}\, dV = (\alpha^s - \alpha^a) \oint dt \int \sigma^{Eo}_{ij}\, \dot{\epsilon}^p_{ij}\, dV$$

Finally according to (4) and (8) one deduces :

$$\alpha^s \le \alpha^a \tag{47}$$

3.6. Upper bounds theorem of kinematical theorem. The load factor α^k corresponding to any Koiter's admissible plastic strain field constitutes an upper bound of the adaptation limit factor α^a

Let $\Delta\epsilon^{pk}_{ij}$ the admissible plastic strain field defined by (34) and σ^k_{ij} the associated stress field deduced from the gradient law (26).

By the same reasons as in 3.5., one may write down :

$$\oint dt \int \sigma^k_{ij} \, \dot{\epsilon}^{pk}_{ij} \, dV = \oint dt \int \dot{D} \, (\dot{\epsilon}^{pk}_{ij}) \, dV = \alpha^k \oint dt \int \sigma^{Eo}_{ij} \, \dot{\epsilon}^p_{ij} \, dV \tag{48}$$

$$\oint dt \int \sigma_{ij} \, \dot{\epsilon}_{ij} \, dV = \alpha^a \oint dt \int \sigma^{Eo} \, \dot{\epsilon}^p_{ij} \, dV \tag{49}$$

Again according to (4), (8) and by comparison of (48) and (49), one obtains finally :

$$\alpha^a \leq \alpha^k \tag{50}$$

4. MATHEMATICAL PROGRAMMING FORMULATION

4.1. General nonlinear statical formulation

This formulation is a direct application of the lower bound theorem :

$$\begin{aligned} &\text{subjected to } \alpha^s = \text{ maximize } \alpha^- \, (\bar{\rho}^s_{ij}) \\ &\frac{\partial \bar{\rho}^s_{ij}}{\partial x_j} = 0 \qquad \text{on } V \\ &n_j \, \bar{\rho}^s_{ij} = 0 \qquad \text{on } S_\sigma \\ &f \, (\bar{\rho}^s_{ij} + \bar{\alpha} \, \sigma^{Eo}_{ij}) \leq 0 \end{aligned} \tag{51}$$

The problem (51) constitutes a particular form of nonlinear mathematical programming problem (NLP).

4.2. Piecewise linear statical formulation

In the case of piecewise linear yield domain bounded by plans in stress space :

$$f \, (\sigma_{ij}) \rightarrow f_m = N_{mij} \, \sigma_{ij} - h_m \leq 0 \tag{52}$$

the problem (51) may be simplified if one considers the envelope of the elastic fictitious responses :

$$a_m = \max_t N_{mij}\, \sigma^{Eo}_{ij}(t) = \max_{P} N_{mij}\, \sigma^{Eo}_{ij}(\hat{P})$$

where $\hat{P}$ defined in section 3.4. is the frontier of the loading domain.

Practically, the envelope a_m are obtained by taking the elastic responses corresponding to the vertexes of loading domain.

The problem (51) becomes a piecewise linear programming problem (PWP) :

$$\begin{aligned} \alpha^s &= \text{maximize } \alpha^- \, (\bar{\rho}^s_{ij}) \\ \text{subjected to} \quad &: \frac{\partial \bar{\rho}^s_{ij}}{\partial x_j} = 0 \quad \text{in } V \\ & n_j\, \bar{\rho}^s_{ij} = 0 \quad \text{on } S_\sigma \\ & \alpha^- a_m + N_{mij}\, \bar{\rho}^s_{ij} - h_m \le 0 \text{ in } V \end{aligned} \tag{53}$$

4.3. General nonlinear kinematical formulation

A formal kinematical formulation may be deduced directly from the upper bounds theorem :

$$\begin{aligned} \alpha^k &= \text{minimize } \alpha^+ = \oint dt \int \dot{D}\, (\dot{\epsilon}^{pk}_{ij})\, dV \\ \text{subjected to} : &\oint dt \int \sigma^{Eo}_{ij}\, \dot{\epsilon}^{pk}_{ij}\, dV = 1 \\ &\Delta\epsilon^{pt}_{ij} = \oint \dot{\epsilon}^{pk}_{ij}\, dt \\ &\Delta\epsilon^{pk}_{ij} = 1/2 \left(\frac{\partial \Delta u^k_i}{\partial x_j} + \frac{\partial \Delta u^k_j}{\partial x_i}\right) \quad \text{in } V \\ &\Delta u^k_i = 0 \quad \text{on } S_u \end{aligned} \tag{54}$$

Unfortunately, this formulation is not practical to handle because it implies both the plastic strain rate and the plastic strain and also it needs some integrations on time.

It appears that a modified formulation may be performed in the following way.

According to the theorem of convex analysis 3.4., one may replace any cycle of loading by particular severe cycle of loading $\hat{P}$ passing uniquely on the vertexes n of the loading domain.

Let τ_n the necessary times to reach the vertex n :

$$\max_n \tau_n = \tau \tag{55}$$

where τ the period of the loading.

The elastic response of each such cycle may be written under the following form :

$$\sigma_{ij}^{Eo}(x,t) = \sum_n \delta_n(t)\, \sigma_{ij}^{Eo}(\tau_n) \tag{56}$$

where :

$$\delta_n = 1 \qquad \text{if } t = \tau_n$$
$$\delta_n = 0 \qquad \text{if } t \neq \tau_n$$

$\sigma_{ij}^{Eo}(\tau_n)$ is the elastic response corresponding to the loading represented by the vertex n.

The plastic strain rate has some value only when t coïncides with τ_n :

$$\dot\epsilon_{ij}^{pk}(x,t) = \begin{vmatrix} 0 \text{ if } t \neq \tau_n \\ \dot\epsilon_{ij}^{p}(x,\tau_n) \text{ if } t = \tau_n \end{vmatrix} \tag{57}$$

Substituting (56) and (57) in (54) one obtains :

$$\int dt \int \sigma_{ij}^{Eo}\, \dot\epsilon_{ij}^{p}\, dV = \sum_n \int \sigma_{ij}^{Eo}(x,\tau_n)\, \Delta\epsilon_{ij}^{pn}\, dV$$

with $$\Delta\epsilon_{ij}^{pn} = \dot\epsilon_{ij}^{p}(x,\tau_n)\, \bar\tau_n \tag{58}$$

and $\bar\tau_n$ is the interval of time :

$$\bar\tau_n = \tau_n - \tau_{n-1},\ \tau_o = 0$$

By the fact that $\dot D(\dot\epsilon_{ij}^{pk})$ is homogeneous of order one, one deduces also :

$$\oint dt \int \dot D(\dot\epsilon_{ij}^{pk})\, dV = \sum_n \int \dot D(\Delta\epsilon_{ij}^{pn})\, dV$$

So that the problem (54) becomes :

$$\alpha^k = \underset{1}{\text{minimize}}\ \alpha^+ = \sum \int \dot D(\Delta\epsilon_{ij}^{pn})\, dV$$

subjected to :

$$\sum_n \int \sigma_{ij}^{Eo}(x,n)\, \Delta\epsilon_{ij}^{pk}\, dV = 1 \tag{59}$$

$$\Delta\epsilon_{ij}^{pk} = \sum_n \Delta\epsilon_{ij}^{pn}$$

$$\Delta\epsilon_{ij}^{pn} = 1/2 \left(\frac{\partial \Delta u_i^n}{\partial x_i} + \frac{\partial \Delta u_i^n}{\partial x_j}\right) \text{ in V}$$

It is clear that the problem (59) independent of times, constitutes again a particular form of nonlinear mathematical programming technique (NLP).

4.4. Piecewise linear kinematical formulation (PWL)

In the case of piecewise linear yield condition (52) the times independent form of kinematical formulation is obtained more easily according to the form (24) of the dissipation :

$$\oint dt \int \dot{D}\,(\dot{\epsilon}^{pk}_{ij})\, dV = \oint dt \int \sigma_{ij}\, \dot{\epsilon}^{pk}_{ij}\, dV = \int h_m\, \Delta\lambda^k_m\, dV \qquad (60)$$

where :

$$\Delta\lambda^k_m = \oint \dot{\lambda}^k_m\, dt$$

$$\dot{\epsilon}^{pk}_{ij} = N_{mij}\, \dot{\lambda}^k_m$$

On the other hand, let a^{Eo}_m the envelope of the elastic response corresponding to the loading represented by the vertexes of the loading domain P :

$$a^{Eo}_m = \sup_{\tau_n} \; [N_{mij}\, \sigma^{Eo}_{ij}\, (\tau_n)]$$

one may see that :

$$\oint dt \int \sigma^{Eo}_{ij}\, \dot{\epsilon}^{pk}\, dV \le \int a^{Eo}_m\, \Delta\lambda^k_m\, dV$$

Thus, finally the problem (54) is reduced to :

$$\alpha^k = \text{minimize inf } \alpha^+ = \int h_m\, \Delta\lambda^k_m\, dV$$

$$\text{subjected to : } \int a^{Eo}_m\, \Delta\lambda^k_m\, dV = 1$$

$$\Delta\epsilon^{pk}_{ij} = N_{mij}\, \Delta\lambda^k_m \qquad (61)$$

$$\Delta\epsilon^{pk}_{ij} = 1/2\, \left(\frac{\partial \Delta u_i}{\partial x_j} + \frac{\partial \Delta u_j}{\partial x_i}\right) \text{ in } V$$

$$\Delta u_i = 0 \qquad \text{on } S_u$$

The duality between the problems (50) and (53) is largely discussed in the literature [26,27]. The duality between the problem (51) and (59) is treated in the thesis [22,24] and in the report [25].

4.5. Remarks

a. In the case of proportional or simple loading, $\bar{f}^o_{m-} = \bar{f}^o_{m+}$ and the loading domain is reduced to a point and shakedown analysis coïncides with classical. limit analysis.

b. For shakedown analysis both statical and kinematical approaches need the exact fictitious elastic solution.

   In general, almost for complex structures, the corresponding exact elastic solutions have to be replaced by an approximate one. This elastic errors may cause some perturbation in the calculation of the shakedown multiplier α^a and sometimes the nature of the bounds can not be preserved (see table 1 of section 6.1.).

c. In shakedown analysis of complex structures, the plastic yield condition cannot be satisfied everywhere in the solid body but only on a finite number of points. Consequently, the generated stress field is not entirely plastic admissible. This plastic error may lead to another perturbation of the nature of the solution.

d. The same remark may be stated for the geometric-plastic compatibility condition in kinematical approaches.

5. FINITE ELEMENT FORMULATION

5.1. Statical formulation

Both nonlinear (51), and linear (53) formulation of the lower bound approach require the generation of elastic stress σ^{Eo}_{ij} and the permanent residual stress $\bar{\rho}^s_{ij}$. In the present statical formulation, these two fields will be obtained by the same discretization of the structure into a mesh of equilibrium finite elements [18,25].For the simplicity of the presentation, it is assumed in the following that body applied forces $\bar{f}_i$ are absent.

5.1.1. Finite element formulation for elastic response

The elastic stress field in element is chosen such that equilibrium equation (5) must be satisfied exactly a priori

$$\{\sigma^{Ee}\} = [S^e (x)] \{s^{Ee}\} \tag{62}$$

where $\{s^{Ee}\}$ is a system of parameters letter called generalized stresses and $[S^e]$ a matrix of polynomial functions. The stress vector on the boundary Γ^e of this element is perfectly defined in function of the discrete variable $\{s^{E^e}\}$.

$$\{\dot{t}^e\} = [S^e(\Gamma)] \; \{s^{Ee}\} \tag{62}$$

where $[S^e \; (\Gamma)]$ depends on the geometry of the element.
A system of generalized forces $\{F^e\}$ better called connectors must be chosen on the boundary Γ^e such that the stress vector (62) on each edge of which would be perfectly defined to guarantee the continuous transmission of stress between each element and its neighborhood :

$$\{F^e\} = [C^e] \quad \{s^{Ee}\} \tag{63}$$

where $[C^e]$ is a connection matrix which relates the internal forces $\{s^{Ee}\}$ to the external forces $\{d^{Ee}\}$.The dual qualities of $\{s^{Ee}\}$ called generalized deformation $\{d^{Ee}\}$ is deduced from the work equation on the boundary Γ^e :

$$\int \{t\}^T \, \bar{u} \; d\Gamma^e = \{s^{Ee}\}^T \int [S^e(\Gamma)]^T \, \{\bar{u}\} \; d\Gamma^e = \{s^{Ee}\}^T \, \{d^{Ee}\}$$

Thus :

$$[d^{Ee}] = \int [S^e(t)]^T \{\bar{u}\} \; d\Gamma^e \tag{64}$$

Relation (64) explains clearly the nature of equilibrium finite element discretization where deformation is known in a weak manner and the whole boundary of element is considered belong to Γ_u.

The compatibility between generalized deformation $\{d^{Ee}\}$ and generalized displacement $\{u^e\}$ is obtained by writing the equation of conservation of energy :

$$\{s^{Ee}\}^T \, \{d^{Ee}\} = \{F^e\}^T \, \{u^e\}$$

Therefore, according to (63) :

$$\{d_e^{Ee}\} = [C^e]^T \, \{u^e\} \tag{65}$$

Now, utilizing the principle of complementary virtual work for elastic solid :

$$\delta \{F^e\}^T\{u^e\} - \delta \{s^{Ee}\}^T \{d^{Ee}\} - \delta \{s^{Ee}\}^T [\int [S^e]^T [S^e] \, dv^e] \, s^{Ee}$$

one can find :

$$\{d_e^{Ee}\} = [E^e] \{s^{Ee}\} \tag{66}$$

where $[E^e]$ is a generalized Hooke's matrix :

$$[E^e] = \int [S^e]^T [A][S^e] \, dv^e \tag{67}$$

The latter matrix is always definite and positive so that its inversion exists. Finally, after elimination of $\{s^{Ee}\}$ in (63) by means of (65) and (66), one obtains the classical form of displacement method :

$$\{F^e\} = [K^e] \{u^e\} \tag{68}$$

where $[K^e]$ is the stiffness matrix of equilibrium formulation :

$$[K^e] = [C^e] [E^e]^{-1}[C^e]^T \tag{69}$$

The assemblage of finite elements is realized in a classical way. Let $[L_j^e]$ and $[R_j^e]$ [22] the localization matrix (boolean matrix) relates respectively the local quantities $\{u^e\}$, $\{s^e\}$ to the global qualities $\{u\}$, $\{s\}$:

$$\{u^e\} = [L^e] \{u\}, \ \{s^e\} = \{H^e\} \{s\} \tag{70}$$

It is not difficult to deduce the assembled characteristic of the whole structure :

$$\{F\} = [K] \{u\}, \ \{F\} = [C] \{s\}, \ \{d\} = [C]^T \{u\} \tag{71}$$

where :

$$\{F\} = \sum_e [L^e] \{F^e\}$$

$$[K] = \sum_e [L^e]^T [K^e][L^e]$$

$$[C] = \sum_e [L^e]^T [C^e] [H^e]$$

5.1.2. Finite element formulation for residual stress

Using the same assumpting (61) for the residual stress $\bar{\rho}^e_{ij}$ so the equilibrium equations (11) are satisfied :

$$\{\rho^e\} = [S^e(x)]\ \{r^e\} \tag{72}$$

and the same global connection matrix C may be obtained from the elastic formulation :

$$0 = [C]\ \{r\} \tag{73}$$

Relation (73) indicates that residual stresses are self-equilibrium stress and don't correspond to any external forces.

Let {r} the dependent part of the sequence of parameter {r} and r the dependent one, relation (73) may be partitioned in the following way :

$$[C_D\ :\ C_I] \begin{bmatrix} r_D \\ r_I \end{bmatrix} = 0$$

One may deduce :

$$\{r\} = [G]\ \{r_I\}\ ,\ \{r_D\} = -\ [C_D]^{-1}[C_I]\ \{r_I\} \tag{74}$$

where $\qquad [G] = \begin{bmatrix} -[C_D]^{-1}[C_D] \\ I \end{bmatrix}$

and [I] is a unity matrix.

As the discretizations of ρ_{ij} and σ^E_{ij} are the same, the previous assembled quantities may be used here :

$$\{r^e\} = [U^e]\ \{r\},\ [r^e] = [G^e]\ \{r_D\},\ [G^e] = [U^e][G] \tag{75}$$

Finally, the discretized form of statical approach (51) is simply :

$$\begin{aligned} &\alpha^s = \text{maximize } \alpha^-\ (\{r_D\}) \\ &\text{subjected to } f\ (\{s^e\}^n) \leq 0 \\ &\text{with } \{s^e\}^n = \alpha^-\ \{s^{Eo}\}^n + [G^e]\{r_D\} \end{aligned} \tag{76}$$

In the nonlinear mathematical programming form of statical approach (76) the yield condition must be considered for a finite number of point in each element e and each vertex n of the loading domain. Therefore the number of inequalities could be very important. An approximate method is proposed in

[18,19] to replace the local plastic criterion by a criterion in the mean:

$$<f> (s^e) = \frac{1}{V^e} \int f\ (\sigma^e)\ dV^e$$

so that only one inequality exists for each element. Such approach is performed for plane stress problem described in section 6.1.

5.1.3. Statical discretized PWL form

Under the same conditions, if piecewise linear yield condition (52) is used, the problem (53) may be reduced to the following sample discretized form :

$$\begin{aligned} &\alpha^s = \text{maximize } \alpha^- \{r_D\}) \\ &\text{subjected to} \\ &\alpha^- \{a\}^e + (N)(G^e)\{r\} - \{h^e\} \leq 0 \end{aligned} \tag{77}$$

5.2. Kinematical formulation

In the following we will avoid to distinguish the element quantities (index e) and the general quantities (without index) because the problem of assemblage of displacement finite elements is wellknown in the literature.

5.2.1. Kinematical discretized NLP form

The finite element discretization is based on the modified formulation (59) where assumptions are made on the special admissible plastic strain field and its associated displacement field :

$$\begin{aligned} &\Delta \{u_n\} = (M(x))\ \{d_n\} \\ &\Delta \{\epsilon^p_n\} = \partial(M(x))\ \{d_n\} \end{aligned} \tag{78}$$

where $\{d_n\}$ is the generalized deformation corresponding to vertex n of the loading domain.

(M(x)) is a polynomial interpolation matrix of the compatbile finite element approach

The total admissible plastic strain corresponding to the severe cycle of loading $\hat{P}$:

$$\Delta\{\epsilon^P\} = \sum_n \Delta\ \{\epsilon^P_n\} = \partial(M(x))\ \{d\}$$

with :

$$\{d\} = \sum_n \{d_n\}$$

which may be associated to a fictitious displacement field :

$$\Delta(u) = \sum_n \Delta(u_n) = \{M(x)\}\ \{d\} \tag{80}$$

Let q be the nodal displacements which assures the compatibility between the finite elements and C the corresponding connection matrix between internal generalized strains {d} and external displacement {q} :

$$\{q\}^T = \{C\}\ \ \{d\}\ ,\ \{d\} = \{C\}^T\ \{q\}, \tag{81}$$

it appears clearly that the problem (59) has the following discretized form :

$$\begin{aligned} \alpha^k &= \text{minimize } \alpha^+ = \sum_n \int \dot{D}\ (d_n)\ dV \\ &\text{subjected to : } \sum_n \{s_n^{Eo}\}\{d_n\} = 1 \\ &\qquad\qquad\quad \sum_n \{d_n\} = \{C\}^T\ \{q\} \end{aligned} \tag{82}$$

where the generalized elastic envelope $\{s_n^{Eo}\}$ is :

$$\{s_n^{Eo}\} = \int \partial\ \{M\}^T\ \{\sigma^{Eo}(x,n)\}\ dV$$

and in case of VON MISES yield condition, the density of dissipative power is :

$$\dot{D}\ (d_n) = \sqrt{\tfrac{2}{3}}\ \sigma_p \int [\{d_n\}^T\ \partial\{M\}^T\ \partial\{M\}\ \{d_n\}]^{1/2}\ dV$$

5.2.2. Kinematical discretized PWL form

Besides assumptions (79), (80) and the connection (81) it is necessary to discretize the plastic multiplier $\Delta(\lambda)$ such that :

$$\Delta(\lambda) = \{Z\ (x)\}\ \{d\}^p \tag{83}$$

By the fact that $\Delta\lambda_m$ are almost non negative the interpolation {Z} must possess some special forms. The details of this discussion may be found in the recent Ph. D. Thesis [24].

In these conditions, the internal compatibility relation becomes :

$$\partial \{M\} \{C\}^T \{q\} = \{N\} \{Z\} \{d\}^P \qquad (84)$$

It is not difficult to see that the problem (60) may be transformed to the following discretized problem :

$$\begin{aligned} &\alpha^k = \text{minimize } \alpha^+ (\{q\}, \{d\}^P) = \{p\}^T \{d\}^P \\ &\text{subjected to :} \\ &\{s^{Eo}\}^T \{d\}^P = 1 \\ &\partial\{M\} \{C\} \{q\} = \{N\} \{Z\} \{d\}^P \\ &\{d\}^P \geq 0 \end{aligned} \qquad (85)$$

where

$$\{p\} = \int \{Z\}^T \{h\} \, dV$$

$$\{s^{Eo}\} = \int \{Z\} \{a\}^{Eo} \, dV$$

and the compatibility condition (84) has to be fixed on a finite number of points in the structure.

6. NUMERICAL RESULTS

Because of lack of place, we restrict to mention here only two typical numerical exemples. The reader who is interested to the question may find other useful exemples in the works of the Belgian group mainly given by NGUYEN DANG HUNG, P. MORELLE and G. DE SAXCE [18-24].

6.1. Tension specimen with excentric hole

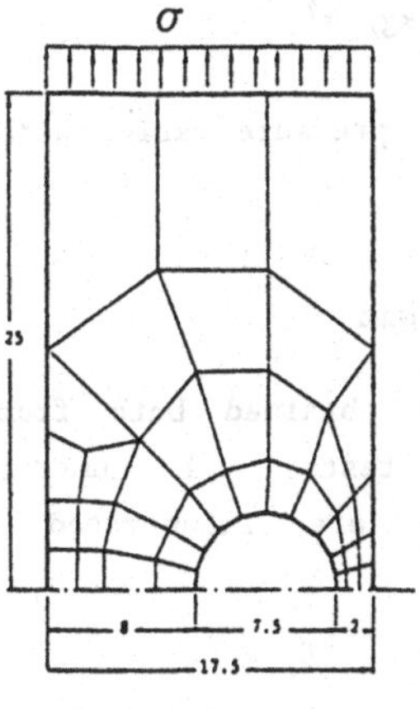

FIGURE 3. Specimen with excentric hole

The first exemple is a plane stress problem (figure 3) where loading is a tension variable from o to σ. The nonlinear yield condition is the classical Von Mises criteria :

$$f(\sigma_{ij}) = \frac{1}{\sigma_p} (\sigma_x + \sigma_y - \sigma_x \sigma_y + 3\tau_{xy}^2)^{1/2} - 1 \leq 0$$

and the piecewise linear yield condition is a polyedral domain of 14 facets in the ellipsoïd of VON MISES [22,24]. The equilibrium quadrilateral element composed of 4 constant stress triangles(5 internal parameters, 8 degrees of freedom) is used for statical shakedown analysis and the isoparametric compatible quadrangle of 8 nodal displacements is used for kinematical approaches. The elastic responses are obtained with these same elements with the Poisson's ratio ν equal to 0.3. The figure 3 shows also the mesh of 37 quadrilateral finite elements. The table 1 gives the comparison of different approaches obtained in Liège.

TABLE 1. Comparison of the shakedown multipliers obtained from different formulations.

Number of elements	$\alpha^a\ \sigma/\sigma_p$	Nature of approach	Formulation	Reference
37	0.413	NLP	(76)	[18,19]
37	0.400	NLP	(76)	[24]
64	0.411	NLP	(76)	[24]
37	0.377	PWL	(77)	[24]
64	0.376	PWL	(77)	[24]
37	0.521	NLP	(82)	[24]
37	0.388	PWL	(85)	[24]
64	0.413	PWL	(85)	[24]

The first line corresponds to an earlier result obtained by NGUYEN DANG HUNG by using a criterion in the mean performed for each quadrangle [18, 19, 22]. It appears that the quality of the elastic responses have an influence on the convergence of the shakedown multiplier.

6.2. Torispherical pressure vessel head

The geometric dimensions of this structure is shown by figure 4 and the material characteristics are the following :

σ_p = 18 kg/mm²

ν = 0.3

E = 21.000 kg/mm²

The internal pressure varies within the range :

$$0 \leq p(t) \leq p_{max}$$

The results obtained both from laboratory tests and numerical calculations are illustrated by table 2.

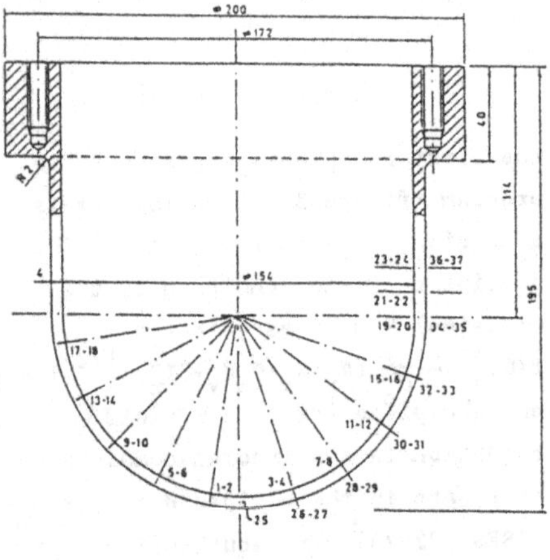

FIGURE 4. Torispherical vessel head

TABLE 2. Shakedown limit load p^a of the torispherical vessel head

	Experimental results (kg/cm²)	Numerical results (kg/cm²)
shakedown load	160-180	172
ultimate load	210	173

The numerical shakedown load is obtained by the NLP formulation (76) using a series of conical shell and cylindrical shell elements [22,24].

7. CONCLUSIONS

After recalling the basic assumptions and theoretical results of the shakedown theory, the paper tries to present carefully the dual aspects of the shakedown analysis both under continuous forms and discretized form. It appears that lower bound and upper bound approach may be performed in a same finite element software using mathematical programming solvers. These direct calculations of the shakedown load are generally cheaper in C.P.U. than the classical step-by-step procedure.

REFERENCES

[1] GRUNING, M. : Die Tragfähigkeit con Balken aus Stahl mit Berücksichtigung des plastischen Verformungsvermögens, Bautechnik, (1930).

[2] GURALNICK, S.A., SURANDA SINGH, ERBER T. : Plastic collapse, shakedown and hysteresis, Jl. Struct. Eng., 110, n° 9, pp. 2103-2119, (1984).

[3] MASSONNET, Ch. : Essais d'adaptation et de stabilisation plastiques sur des poutrelles laminées (Mémoires de l'AIPC, vol. 13, p. 239 (1953).

[4] MASSONNET, Ch. et SAVE, M. : Plastic Analysis and Design. Vol. 1 : Beams and Frames. Blaisdell, Publ. Co, (1965), éd. française, Centre Belgo-Luxembourgeois d'Information de l'Acier, Bruxelles, (1961).

[5] BLEICH, H. : Uber die Bemessung statisch unbestimmter stahltragwerke unter Berücksichtigung des elastisch-plastischen Verhaltens des Baustoffes, Bauingenieur, 19/20, (1932), p. 261

[6] MELAN, E. : Theorie statisch unbestimmter Systeme. Prelim. Publ., 2nd Cong. Int. Assoc. Bridge and Struct. Eng., Berlin, 43 (1936).

[7] SYMONDS, P.S. : Shakedown in Continuous Media, J. Appl. Mech., 18 (1951), p. 85.

[8] KOITER, W.T. : Some Remarks on Plastic Shakedown Theorems, Proc. 8th Int. Cong. Appl. Mech., 1, 220, Istanbul (1952).

[9] NEAL, B.G. : The Plastic Method of Structural Analysis, London, Chapman and Hall, 1956.

[10] PRAGER, W. : Shakedown in Elastic-Plastic Media Subjected to Cycles of Load and Temperature, Symp. sulla plasticita nella scienza delle costruzioni, Bologna (1957).

[11] ROSENBLUM, V.I. : Adaptation des corps élasto-plastiques soumis à une température non uniforme (en russe). Izv. Ak. N., URSS, OTN n° 7 (1957).

[12] GOKHFELD, D.A. : Carrying Capacity of Structures under Variable Thermal Condition (en russe). Mashinostroyeniye, Moscou (1970).

[13] KONIG, J.A. : Engineering Applications of Shakedown Theory, Lecture Notes, CISM, Udine (1977).

[14] HO HWA-SHAN : Shakedown in Elastic-Plastic Systems under Dynamic Loading, J. Appl. Mech. v. 39 (1972), pp. 416-421.

[15] CORRADI, L. and MAIER, G. : Inadaptation Theorems in the Dynamics of Elastic-Work Hardening Structures. Ing. Arch., 43 (1973), pp. 44-57.

[16] CORRADI, L. and MAIER, G. : Dynamic Non-Shakedown Theorem for Elastic Perfectly Plastic Continua. J. Mech. Phys. Solids, 22 (1974), pp. 401-413.

[17] POLIZZOTTO, C. : Adaptation of Rigid-Plastic Continua Under Dynamic Loadings. Tech. Rep. SISTA-77-OMS-2, Facolta di Architettura di Palermo, Palermo (1977).

[18] NGUYEN DANG HUNG and KONIG, J.A. : A Finite Element Formulation for Shakedown Problems Using a Yield Criterion of the Mean. Comp. Meth. Appl. Mech. Eng., vol. 8, n° 2, pp. 179-192 (1976).

[19] NGUYEN DANG HUNG and PALGEN, L. : Shakedown Analysis by Displacement Method and Equilibrium Finite Element. Trans. of the CSME, vol. 6, n° 1, pp. 34-40, (1980-81).

[20] NGUYEN DANG HUNG and MORELLE, P. : Numerical Shakedown Analysis of Plates and Shells of Revolution. New and Future Developments in Commercial Finite Element Methods, édité par J. Robinson, pp. 422-435 (1980).

[21] MORELLE, P. and NGUYEN DANG HUNG : Etude numérique de l'adaptation plastique des plaques et des coques de révolution par les éléments finis d'équilibre. J. de Méc. Théorique et appliquée, vol. 2, n° 4, pp. 567-599 (1983).

[22] NGUYEN DANG HUNG : Sur la plasticité et le calcul des états limites par éléments finis. Thèse de doctorat spéciale, University of Liège, (1984).

[23] DE SAXCE, G. : Sur quelques problèmes de mécanique des solides considérés comme matériaux à potentiels convexes. Ph. D. Thesis, University of Liège (1986).

[24] MORELLE, P. : Analyse duale de l'adaptation plastique des structures par la méthode des éléments et la programmation mathématique, Ph. D. Thesis, University of Liège (1989).

[25] SAVE, M. DE SAXCE, G. BORKOWSKI, A. : Computation of Shakedown Loads Feasibility Study. Final Report, Contract FA 1-0100-B (GDF/DSF), Commission of the European Communities, (1987).

[26] MAIER, G. : Shakedown Theory in Perfect Elastoplasticity with Associated and Non-Associated Flow-Laws, A Finite Element Linear Programming Approach, Meccanica, 4 (1969), p. 250.

[27] MAIER, G. : Shakedown of Plastic Structures with Unstable Parts. ASCE, J. Eng. Mech. Div. vol. 98 (1972), pp. 1322.

CHAPTER 12

OPTIMAL PLASTIC DESIGN AND THE DEVELOPMENT OF PRACTICAL SOFTWARE

Nguyen Dang Hung and P. Morelle
University of Liège, Liège, Belgium

1 INTRODUCTION

The purpose of this work is to present a simple, practical and automatic procedure for the rigid-plastic analysis and design of frame structures under both proportional and repeated loadings. The applicability of linear programming techniques to this kind of problem was pointed out some twenty years ago. The principle of the method is widely exposed in the literature[1–17]. In this paper we restrict ourselves to some particular aspects of the automatic calculations and practical implementations. These are:

1) the efficient choice between the statical and kinematical formulations leading to a minimum number of variables. This affects considerably the size of the problem:

2) the automatic construction of the characteristic matrices, mainly the independent equilibrium-compatibility matrix;

3) the direct calculations of dual variables which allows the necessary results to be obtained by a simple procedure;

4) the realistic verifications of the stability conditions imposed by European norms for steel structures.

These aspects are taken into account during the realization of a multipurpose computer program named CEPAO written in FORTRAN IV, available in the Department

TABLE 1

Linear programming formulations of plastic analysis and design. The dimensions are calculated in the initial form

		Approach	Formulation	No. of variables	No. of equalities	No. of inequalities	Total No. of variables	Total No. of constraints	Dimension
Analysis	Proportional loading	Static	$\max \lambda \left\| \begin{array}{l} \lambda_e - \mathbf{C}\mathbf{m} = 0 \\ \mathbf{m} - \mathbf{m}_p \le 0 \end{array} \right.$	$2n_r + 1$	n_m	$2n_r$	$4n_r + 1$	$n_m + 2n_r$	$(4n_r + 1) \times (n_m + 2n_r + 1)$
		Kinematic	$\min \lambda = \mathbf{m}_p^T\|\dot{\mathbf{r}}\| \left\| \begin{array}{l} \dot{\mathbf{r}} - \mathbf{C}^T\dot{\mathbf{w}} = 0 \\ \mathbf{e}^T\dot{\mathbf{w}} = 1 \end{array} \right.$	$2(n_r + n_m)$	$n_r + 1$	0	$2(n_r + n_m)$	$n_r + 1$	$2(n_r + n_m) \times (n_r + 2)$
	Variable repeated loading	Static	$\max \lambda_S \left\| \begin{array}{l} \mathbf{C}\rho = 0 \\ \lambda_S \mathbf{m}_E + \rho - \mathbf{m}_p \le 0 \end{array} \right.$	$2n_r + 1$	n_m	$2n_r$	$4n_r + 1$	$n_m + 2n_r$	$(4n_r + 1) \times (n_m + 2n_r + 1)$
		Kinematic	$\min \lambda_S = \mathbf{m}_p^T\|\dot{\mathbf{r}}\| \left\| \begin{array}{l} \dot{\mathbf{r}} - \mathbf{C}^T\dot{\mathbf{w}} = 0 \\ \mathbf{m}_L^T\dot{\mathbf{r}} = 1 \end{array} \right.$	$2(n_r + n_m)$	$n_r + 1$	0	$2(n_r + n_m)$	$n_r + 1$	$2(n_r + n_m) \times (n_r + 2)$
Design	Proportional loading	Static	$\min Z = \mathbf{m}_p^T \mathbf{l} \left\| \begin{array}{l} \mathbf{e} - \mathbf{C}\mathbf{m} = 0 \\ \mathbf{m} - \mathbf{D}^T\mathbf{m}_p \le 0 \end{array} \right.$	$2n_r + n_p$	n_m	$2n_r$	$4n_r + n_p$	$2n_r + n_m$	$(4n_r + n_p) \times (n_m + 2n_r + 1)$
		Kinematic	$\min Z = \mathbf{m}_p^T \mathbf{l} \| \mathbf{b} - \mathbf{A}\mathbf{m}_p \le 0$	n_p	0	n_q	$n_p + n_q$	n_q	$(n_p + n_q) \times (n_q + 1)$
	Variable repeated loading	Static	$\min Z_S = \mathbf{m}_p^T \mathbf{l} \left\| \begin{array}{l} \mathbf{C}\rho = 0 \\ \mathbf{m}_E + \rho - \mathbf{D}^T\mathbf{m}_p \le 0 \end{array} \right.$	$2n_r + n_p$	n_m	$2n_r$	$4n_r + n_p$	$n_m + 2n_r$	$(4n_r + n_p) \times (n_m + 2n_r + 1)$
		Kinematic	$\min Z_S = \mathbf{m}_p^T \mathbf{l} \| \bar{\mathbf{b}} - \mathbf{A}\mathbf{m}_p \le 0$	n_p	0	$n_r + n_q$	$n_p + n_q + n_r$	$n_q + n_r$	$(n_p + n_q + n_r) \times (n_q + n_p + 1)$

of Structural Mechanics and Stability of Constructions, University of Liège. This package, incorporating facilities such as interactive input data, graphic display of output results, is running now on IBM, VAX, and some mini-computers.

2 LINEAR PROGRAMMING FORMS OF PLASTIC ANALYSIS AND DESIGN

The object of plastic analysis is to determine the collapse load of a structure when its plastic capacities are known. In optimal plastic design one has to minimize some merit function (usually the total weight) when the loads and the geometric dimensions of the structure are specified. The following assumptions are generally made:

—the strains are small;
—the loading is quasi-static;
—the material behaviour is rigid perfectly plastic;
—the flexural action is predominant. Shear, torsion, and axial forces are negligible or are taken into account only *a posteriori*;
—the geometric effects of stability intervene only after the plastic collapse state;
—the plastic hinges are concentrated at critical sections;
—the weight of a member is proportional to its plastic capacity.

Under these assumptions, the plastic analysis and design problems can be identified as linear programming problems.

As shown by Cohn and his collaborators in some synthesis papers[6,7] there are four common plastic analysis and design problems with fixed and variable loadings. By the choice of variables, each problem may be formulated by either static or kinematic approaches. Table 1 summarizes the situation. In column 2 one can see the linear programming form of each case. The notation used in this table is:

$\mathbf{b} = \{b_i\}$ external work associated with the i-th limit equilibrium equation;
$\bar{\mathbf{b}} = \{\bar{b}_i\}$ work associated with the envelope elastic moments in the i-th possible possible collapse mode;
$\mathbf{C} = \{c_{ij}\}$ independent equilibrium matrix (compatibility matrix $\mathbf{C}^T$);
$\mathbf{D} = \{d_{kj}\}$ technological matrix:
$d_{kj} = 1$; k-th design variable governs section j;
$d_{kj} = 0$; k-th design variable does not govern section j;
$\mathbf{G} = \{g_{kj}\}$ coefficient of the k-th design variable in the i-th limit equilibrium equation;
$\mathbf{e} = \{e_i\}$ reduced applied forces—load term of i-th independent equilibrium equation;
$\mathbf{l} = \{l_k\}$ vector of length of the design member. Alternatively, l_k is the length over which the k-th design variable has constant value;
$\mathbf{m} = [M_j]$ current moment at critical section j;
$\mathbf{m}_p = [M_{pj}]$ plastic moment capacity of the critical section j;
$\mathbf{m}_E = [M_{Ej}]$ elastic envelope moment at section j;
n_r number of critical section;
n_p number of design variables;
n_m number of independent mechanisms;
n_b number of member elements;
$Z = \mathbf{l}^T\mathbf{m}_p$ merit function proportional to the total weight of the structure.

$\dot{\mathbf{r}} = \{\theta_j\}$ relative rotation rate at section j;
$\dot{\mathbf{w}} = \{W_i\}$ displacement rate of independent mechnaism i;
λ collapse load under proportional loading;
λ_s shakedown collapse load;
$\boldsymbol{\rho} = \{\rho_j\}$ residual moment at section j.

Columns 3 to 8 of Table I give respectively the number of variables, the number of equalities, the number of inequalities, the total number of variables, the total number of constraints and the dimensions of the problem in the initial form. The detailed calculations of these numbers are considered in Section 4 of this paper.

It appears that the four following forms are suitable for computer solution because their limited dimensions:

Limit analysis (Analysis under proportional loading): Kinematically admissible approach:

$$\text{Min } \lambda = \mathbf{m}_p^T |\dot{\mathbf{r}}| \left| \begin{array}{l} \dot{\mathbf{r}} - \mathbf{C}^T \dot{\mathbf{w}} = 0 \\ \mathbf{e}^T \dot{\mathbf{w}} = 1 \end{array} \right. \tag{1}$$

Limit design (Optimal plastic design under proportional loading): statically admissible approach:

$$\text{Min } Z = \mathbf{m}_p^T \mathbf{l} \left| \begin{array}{l} \mathbf{e} - \mathbf{Cm} = 0 \\ \mathbf{m}^T - \mathbf{D}^T \mathbf{m}_p \leq 0 \end{array} \right. \tag{2}$$

Shakedown analysis (Analysis under variable repeated loading): Kinematically admissible approach:

$$\text{Min } \lambda_s = \mathbf{m}_p^T |\dot{\mathbf{r}}| \left| \begin{array}{l} \dot{\mathbf{r}} - \mathbf{C}^T \dot{\mathbf{w}} = 0 \\ \mathbf{m}_E^T |\dot{\mathbf{r}}| = 1 \end{array} \right. \tag{3}$$

Shakedown design (Optimal plastic design under variable repeated loading): Statically admissible approach:

$$\text{Min } Z_s = \mathbf{m}_p^T \mathbf{l} \left| \begin{array}{l} \mathbf{C}\boldsymbol{\rho} = 0 \\ \mathbf{m}_E + \boldsymbol{\rho} - \mathbf{D}^T \mathbf{m}_p \leq 0 \end{array} \right. \tag{4}$$

The choice of these four approaches is also motivated by the fact that in these formulations the equilibrium-compatibility independent matrix C, hereafter called the rotation matrix, is the only structural characteristic matrix necessary to be constructed before the linear programming process. (The matrix D is a Boolean matrix easily performed by index calculation). Therefore in the automatic analysis and design of frame structures the central problem is the construction of the matrix C. We will summarize in the following how this is done.

3 AUTOMATIC ASSEMBLY OF THE ROTATION MATRIX

Let $\mathbf{u}^T = [u_1, u_2, \ldots, u_{n_b}]$ be the axial elongation vector of n_b members which form the plane frame structure, $\mathbf{v}^T = [v_1, v_2, \ldots, v_{n_b}]$ be the lateral deflection vector of these members, and $\mathbf{q}^T = [q_1, q_2, \ldots, q_{n_r}]$ be the displacement vector of the n_r critical sections (in general more than the total number of nodes) which constitute the initial and the final nodes of the member.

The following kinematical connection relation must exist between these quantities:

$$\mathbf{b} = \mathbf{B}\mathbf{q}; \tag{5}$$

$$\mathbf{d} = \mathbf{A}\mathbf{q} \tag{6}$$

where A, B are connection matrices of dimension $n_b \times 2 \times (n - n_3)$. n is the total number of nodes, n_3 is the total number of external constraints due to the fixation.

It is not difficult to deduce the following statical connection relation which constitutes the conjugate relations of Eqs. (5) and (6)

$$\mathbf{g} = \mathbf{A}^T\mathbf{n} + \mathbf{B}^T\mathbf{t} \tag{7}$$

where $\mathbf{g}^T = [g_1, g_2, \ldots, g_{n_r}]$ is the force vector conjugate to the displacement vector $\mathbf{q}$, $\mathbf{n}^T = [N_1, N_2, \ldots, N_{n_r}]$ is the vector of the member normal forces, and $\mathbf{t}^T = [T_1, T_2, \ldots, T_{n_r}]$ is the vector of the member shear forces.

Now, if the axial deformations of beams and columns are considered negligible compared to the lateral deflection, that is the lengths of all the members remain unchanged under loading, the relation (6) leads to the following:

$$\mathbf{A}\mathbf{q} = 0 \tag{8}$$

Relation (8) means that some relationships exist between the displacements and they are not independent: this suggests that $\mathbf{q}$ must be decomposed into two parts:

$$\mathbf{q}^T = [\mathbf{z}^T\bar{\mathbf{w}}] \tag{9}$$

where $\mathbf{z}$ is the n_b dependent components of displacement vector and $\bar{\mathbf{w}}$ is the $2(n - n_3) - n_b$ independent components of the displacement vector.

The constraint (8) may be written in the explicit form:

$$[\mathbf{A}_z \mathbf{A}_w]\begin{bmatrix}\mathbf{z}\\ \bar{\mathbf{w}}\end{bmatrix} = 0 \tag{10}$$

where $\mathbf{A}_z$ is a square matrix of dimension $n_b \times n_b$ and $\mathbf{A}_w$ is a rectangular matrix of dimension $n_b \times (2n - 2n_3 - n_b)$.

It is important to mention that the fixation must be performed during the construction of the matrix A. In these conditions the matrix $\mathbf{A}_z$ is invertible:

$$\mathbf{z} = -\mathbf{A}_z^{-1}\mathbf{A}_w\bar{\mathbf{w}} = \mathbf{G}\bar{\mathbf{w}},\ \mathbf{G} = -\mathbf{A}_z^{-1}\mathbf{A}_w \tag{11}$$

Similarly for Eq. (9), the dual decomposition of the force vector

$$\mathbf{g}^T = |\mathbf{h}^T\bar{\mathbf{e}}^T| \tag{12}$$

leads to the following form of relation (7)

$$\begin{bmatrix}\mathbf{h}\\ \bar{\mathbf{e}}\end{bmatrix} = \begin{bmatrix}\mathbf{A}_z^T & \mathbf{B}_z^T\\ \mathbf{A}_w^T & \mathbf{B}_w^T\end{bmatrix}\begin{bmatrix}\mathbf{n}\\ \mathbf{t}\end{bmatrix}$$

One may deduce from this that

$$\mathbf{n} = \mathbf{A}_z^{-T}[\mathbf{h} - \mathbf{B}_z^T\mathbf{t}] \tag{13}$$

$$\bar{\mathbf{e}} = -\mathbf{G}^T[\mathbf{h} - \mathbf{B}_z^T\mathbf{t} + \mathbf{B}_w^T\mathbf{t}] \tag{14}$$

The relation (13) gives the member normal forces when the member shear forces are known. The relation (14) will lead to the independent equilibrium equations relating the applied forces and the bending moment at all critical sections of the structure.

Let M_j be the bending moment at critical section j. The global bending moment vector of the whole structure is:

$$\mathbf{m}^T = |M_1, M_2, \ldots, M_{n_r}|$$

The shear force T_k of member k is related to its bending moments at the two ends by classical equilibrium relations:

$$T_k = \frac{M_i - M_j}{l_k} \tag{15}$$

where l_k is the length of this member. Accordingly, the global relationships of the whole structure may be written in matrix form as

$$\mathbf{t} = \mathbf{U}\mathbf{m} \tag{16}$$

where U is some matrix deduced from the relation (15) over all members. Replacing t from (14) by its value (16) gives

$$\mathbf{e}_1 = \bar{\mathbf{e}} + \mathbf{G}^T\mathbf{h} = \mathbf{C}_1\mathbf{m} \tag{17}$$

with $\mathbf{C}_1 = (\mathbf{G}^T\mathbf{B}_{=}^T + \mathbf{B}_w^T)\mathbf{U}$

Equation (17) represents the independent equilibrium equations of the structure. These equations correspond to the deflection mechanisms. It is necessary to add the equilibrium equations of type:

$$\mathbf{e}_2 = 0 = \mathbf{C}_2\mathbf{m}$$

corresponding to the rotation mechanisms to form the complete system of equilibrium equations:

$$\mathbf{e} = \mathbf{C}\mathbf{m} \tag{18}$$

$$\text{with } \mathbf{e}^T = [\mathbf{e}_1^T \quad 0]$$

$$\mathbf{C}^T = [\mathbf{C}_1^T \quad \mathbf{C}_2^T]$$

The detailed description of this procedure is given in Ref. 18 which also has some analytical examples.

4 FURTHER REDUCTION OF THE DIMENSION OF THE PROBLEM

4.1 *Limit Analysis*

The initial matrix form of the limit analysis recalled here:

$$\text{Min } \lambda = \mathbf{m}_p^T|\dot{\mathbf{r}}| \left| \begin{array}{l} \dot{\mathbf{r}} - \mathbf{C}^T\dot{\mathbf{w}} = 0 \\ \mathbf{e}^T\dot{\mathbf{w}} = 1 \end{array} \right. \tag{1}$$

is not suitable for the simplex solution technique of linear programming. All variables in the simplex process must be non-negative so that the actual form must be:

$$\text{Min } \lambda = \mathbf{m}_{p+}^T|\dot{\mathbf{r}}_+| + \mathbf{m}_{p-}^T|\dot{\mathbf{r}}| \left| \begin{array}{l} -\dot{\mathbf{r}}_+ + \dot{\mathbf{r}}_- + \mathbf{C}^T\dot{\mathbf{w}}_+ - \mathbf{C}^T\dot{\mathbf{w}}_- = 0 \\ \mathbf{e}^T(\dot{\mathbf{w}}_+ - \dot{\mathbf{w}}_-) = 1 \end{array} \right. \tag{19}$$

where all variables are decomposed into two non-negative components such that:

$$\dot{\mathbf{r}} = \dot{\mathbf{r}}_+ - \dot{\mathbf{r}}_-; \dot{\mathbf{w}} = \dot{\mathbf{w}}_+ - \dot{\mathbf{w}}_-$$

$$\text{with } \dot{\mathbf{r}}_+ \geq 0; \dot{\mathbf{r}}_- \geq 0; \dot{\mathbf{w}}_+ \geq 0; \dot{\mathbf{w}}_- \geq 0$$

The calculation of the size of the limit analysis problem (1) is based on the detailed form (19). This size is still large so a further appropriate reduction must be done. Let $\dot{w}^0$ be a constant vector to be found *a priori* such that the new system of displacement rate variables:

$$\dot{w}' = \dot{w} + \dot{w}_0 \geq 0$$

has only positive components.

Then the problem (19) is transformed into the following:

$$\text{Min}\ \lambda = m_{p+}^T|\dot{r}_+| + m_{p-}^T|\dot{r}_-| \left| \begin{array}{l} -\dot{r}_+ + \dot{r}_- + C^T\dot{w}' = C^T\dot{w}_0 \\ e^T\dot{w}' = 1 + e^T\dot{w}_0 \end{array} \right. \tag{20}$$

If we take the right hand sides of the system equalities (20) as initial values of the variable system:

$$x^T = [\dot{w}'^T\ r_+^T\ \ -\lambda\ r_-^T\ \ \eta]$$

where η is a convenient complementary variable, i.e.,

$$x_0^T = [0^T\ \ 0^T\ \ 0^T\ \ C^T\dot{w}_0\ \ 1 + e^T\dot{w}_0]$$

we are not sure that all components of x_0 are non-negative.

Some further preparation must be done again before the simplex process can commence. Let S be defined as the diagonal matrix such that:

$$S = \text{diag}[1 \times \text{sign of } C^T\dot{w}_0]$$
$$E = \text{unit matrix}$$

and let us consider the following new quantities:

$$\begin{aligned} \dot{r}_+ &= \tfrac{1}{2}[E + S]\dot{r}_+ + \tfrac{1}{2}(E - S)\dot{r}_- \\ \dot{r}_- &= \tfrac{1}{2}[E - S]\dot{r}_+ + \tfrac{1}{2}(E + S)\dot{r}_- \\ \bar{m}_{p+} &= \tfrac{1}{2}[E + S]m_{p+} + \tfrac{1}{2}(E - S)m_{p-} \\ \bar{m}_{p-} &= \tfrac{1}{2}[E - S]m_{p+} + \tfrac{1}{2}(E + S)m_{p-} \end{aligned}$$

Then problem (20) is transformed to the following simple linear programming problem.

$$\text{Min}\ \lambda = c^T x$$

subject to

$$Wx = b$$

with:

$$\begin{aligned} x^T &= [\dot{w}'^T\ \ \bar{r}_+\ \ \bar{r}_-\ \ \eta] \\ c^T &= [0^T\ \ \bar{m}_{p+}^T\ \ \bar{m}_{p-}^T\ \ 0^T] \\ b^T &= [0^T\ \ SC^T\ \ \dot{w}_0\ \ 1 + e^T\dot{w}_0] \end{aligned} \tag{21}$$

and:

$$W = \left[\begin{array}{c:c:cc} SC^T & -E & E & 0 \\ \hdashline e^T & 0 & 0 & 1 \end{array}\right],\ X_0 = \begin{bmatrix} E & 0 \\ 0 & 1 \end{bmatrix}$$

It can be seen that the basis matrix $\mathbf{X}$ of the simplex process is diagonal and the initial values of the variables x,

$$\mathbf{x}_0 = \mathbf{X}_0^{-1}\mathbf{b} = \mathbf{b}$$

must have non-negative values.

The form (21) is programmed in the CEPAO package.

4.2 *Optimal Plastic Design*

The initial matrix form of optimal plastic design (2) recalled here:

$$\text{Min } Z = \mathbf{m}_p^T\mathbf{l} \left| \begin{array}{l} \mathbf{e} - \mathbf{C}\mathbf{m} = 0 \\ \mathbf{m}^T - \mathbf{D}^T\mathbf{m}_p \leq 0 \end{array} \right. \tag{2}$$

It takes the following form in the simplex process:

$$\text{Min } Z = \mathbf{m}_p^T\mathbf{l} \left| \begin{array}{l} \mathbf{e} = \mathbf{C}(\mathbf{m}_+ - \mathbf{m}_-) \\ -\mathbf{D}^T\mathbf{m}_p + \mathbf{m}_+ - \mathbf{m}_- \leq 0 \\ -\mathbf{D}^T\mathbf{m}_p - \mathbf{m}_+ + \mathbf{m}_- \leq 0 \\ \mathbf{T}\mathbf{m}_p \leq 0 \end{array} \right. \tag{22}$$

where the last inequality represents additional technological conditions imposed by practical needs. The current moment vector is split into two vectors with non-negative components as required by the simplex process. To reduce the dimension of the problem it appears that the following change of variables is necessary:

$$\mathbf{m}' = \mathbf{m} + \bar{\mathbf{m}} \qquad M'_j = M_j + M_{pj}$$

Assuming that all sections of the structural members are symmetric:

$$M^+_{pj} = M^-_{pj} = M_{pj}$$

one can see that the plasticity conditions become:

$$0 \leq M'_j \leq 2M_{pj}$$

Therefore the two plasticity conditions of (22) are reduced to only one in terms of M'_j:

$$M'_j - 2\sum_i d_{ij}M_{pi} \leq 0$$

The problem (22) is reduced to the equivalent one:

$$\text{Min } Z = \mathbf{m}_p^T\mathbf{l} \left| \begin{array}{l} \mathbf{e} = \mathbf{C}\mathbf{m}' - \mathbf{C}\mathbf{D}^T\mathbf{m}_p \\ \mathbf{m}' - 2\mathbf{D}^T\mathbf{m}_p \leq 0 \\ \mathbf{T}\mathbf{m}_p \leq 0 \end{array} \right. \tag{23}$$

Problem (23) may be written in the following convenient form:

$$\text{Min } Z = \mathbf{c}^T\mathbf{x} \tag{24}$$

$$\text{subject to } \mathbf{W}\mathbf{x} = \mathbf{b}$$

with

$$\mathbf{c}^T = [\mathbf{0}^T \quad \mathbf{0}^T \quad \mathbf{0}^T \quad \mathbf{0}^T \quad \mathbf{0}^T \quad \mathbf{b}^T]$$

$$\mathbf{x}^T = [\mathbf{m}'^T \quad \mathbf{z}^T \quad \mathbf{p}^T \quad \mathbf{s}^t \quad \mathbf{q}^T \quad \mathbf{m}_p^T]$$

$$\mathbf{b}^T = [\mathbf{0}^T \quad \mathbf{0}^T \quad \mathbf{e}^T]$$

and

$$W = \begin{bmatrix} E & E \;\; 0 \;\; 0 & -2D^T \\ 0 & 0 \;\; E \;\; 0 & T \\ C & 0 \;\; 0 \;\; E & -CD^T \end{bmatrix}, \; X_0 = \begin{bmatrix} E & 0 & 0 \\ 0 & E & 0 \\ 0 & 0 & E \end{bmatrix}$$

The initial value of x:

$$x_0 = X_0^{-1} b = b$$

will have non-negative components if it is arranged such that the reduced force vector e has *a priori* non-negative components. p, s, z, and q are non-negative slack variables usually added in the simplex method. We shall see that they can play an important role in the calculation of the dual variables.

4.3 *Shakedown Plastic Analysis*

The partitioned form of this problem (3) using the kinematical approach is:

$$\text{Min } \lambda_s = m_{p+}^T |\dot{r}_+| + m_{p-}^T |\dot{r}_-| \tag{25}$$

subject to $\dot{r}_+ + \dot{r}_- + C^T\dot{w}_+ - C^T\dot{w}_- = 0$

$$m_{E+}^T \dot{r}_+ - m_{E-}^T \dot{r}_- = 1$$

where m_{E+}, m_{E-} are respectively the sequences of positive values and negative values of the envelope of the elastic responses of the considered loading domain. It is well known that if the loading is convex m_{E+} and m_{E-} may be chosen respectively as the maximum positive values and the norm of the minimum values of the elastic responses corresponding to every vertex of the loading domain.

By appropriate choice of $\dot{w}_0$ such that:

$$\dot{w}' = \dot{w} + \dot{w}_0 \geq 0$$

and by using the new quantities:

$$\bar{r}_+ = \tfrac{1}{2}(E + S)\dot{r}_+ + \tfrac{1}{2}(E - S)\dot{r}_-$$
$$\bar{r}_- = \tfrac{1}{2}(E - S)\dot{r}_+ + \tfrac{1}{2}(E + S)\dot{r}_-$$
$$\bar{m}_{p+} = \tfrac{1}{2}(E + S)m_{p+} + \tfrac{1}{2}(E - S)m_{p-}$$
$$\bar{m}_{p-} = \tfrac{1}{2}(E + S)m_{p-} + \tfrac{1}{2}(E - S)m_{p-}$$

with

$$S = \text{diag}[1 \times \text{sign of } C^T\dot{w}_0]$$
$$E = \text{unit matrix}$$

one may reduce the problem (25) to the following:

$$\text{Min } \lambda_s = c^T x \tag{26}$$

$$\text{subjected to: } Wx = b$$

where

$$x^T = [\dot{w}'^T \;\; \bar{r}_+ \;\; \bar{r}_- \;\; \eta]$$
$$c^T = [0^T \;\; \bar{m}_{p+} \;\; \bar{m}_{p-} \;\; 0]$$
$$b^T = [S \;\; C^T \;\; |\dot{w}_0| \;\; 1]$$

and

$$W = \begin{bmatrix} SC^T & -E & E & 0 \\ 0 & m_{E+}^T & m_{E-}^T & 1 \end{bmatrix}$$

It appears that the initial basic matrix X_0 is immediate:

$$X_0 = \begin{bmatrix} E & 0 \\ m_{E-}^T & 1 \end{bmatrix} \quad X_0^{-1} = \begin{bmatrix} E & 0 \\ -m_{E-}^T & 1 \end{bmatrix}$$

and the initial value x_0 necessary for the simplex process is:

$$x_0 = X_0^{-1} b = \begin{bmatrix} SC^T \dot{w}_0 \\ 1 - m_{E-}^T C^T \dot{w}_0 \end{bmatrix}$$

which has always positive components.

By comparison of the problems (21) and (26) it can be seen that there are no differences from the point of view of programming technique between the limit analysis problem and the shakedown analysis problem except for the choice of initial point and the basic matrix.

4.4 *Shakedown Plastic Design*

The partitioned form of problem (4) using the statical approach is:

$$\left.\begin{aligned} &\text{Min } Z_s = m_p^T 1^T \\ &\text{subject to:} \\ &C(\rho_+ - \rho_-) \\ &-Dm_p + \rho_+ - \rho_- \leq -m_{E+} \\ &-Dm_p + \rho_- \leq -m_{E-} \\ &Tm_p \leq 0 \end{aligned}\right\} \quad (27)$$

where m_{E+} and m_{E-} have the same meaning as in expression (25). An important but non-realistic assumption which must be mentioned here is that there is no dependence of the elastic responses upon the design variables m_p. Otherwise, one has to perform an iterative procedure to update the moments of inertia of the structural members whenever the member design plastic capacities m_p are obtained until there is no change in the response of the problem (27). The present CEPAO package is capable of performing this iteration automatically when the user expresses the desire.

Let us consider now the linear form (27) and try to reduce the dimension of this problem.

It may be verified without difficulties that if we perform the following transformation of variables:

$$m' = \rho + m_{E+} + Dm_p \geq 0$$

all components of the new variables m' are positive. The problem (27) is then reduced to the following interesting form:

$$\text{Min } Z_s = \mathbf{m}_p^T \mathbf{1}$$

subjected to

$$\left.\begin{aligned} &\mathbf{m}' + \mathbf{p}_+ - 2\mathbf{D}^T\mathbf{m}_p = 0 \\ &-\mathbf{p}_- + \mathbf{m}' + \mathbf{s} = (\mathbf{m}_{E+} + \mathbf{m}_{E-}) \\ &-\mathbf{C}\mathbf{m}' + \mathbf{C}\mathbf{D}^T\bar{\mathbf{m}}_p + \mathbf{q} = -\mathbf{C}\mathbf{m}_{E+} \\ &\mathbf{T}\mathbf{m}_p + \mathbf{z} = 0 \end{aligned}\right\} \tag{28}$$

where $\mathbf{p}_+$, $\mathbf{p}_-$, $\mathbf{q}$, $\mathbf{s}$, $\mathbf{z}$ are some positive slack variables necessary for the simplex technique.

The problem (28) may be summarized in the classical form:

$$\text{Min } Z_s = \mathbf{c}^T\mathbf{x} \tag{29}$$

subjected to

$$\mathbf{W}\mathbf{x} = \mathbf{b}$$

by putting

$$\begin{aligned} \mathbf{x}^T &= [\mathbf{m}' \quad \mathbf{p}_- \quad \mathbf{p}_+ \quad \mathbf{z} \quad \mathbf{s} \quad \mathbf{q} \quad \mathbf{m}_p] \\ c^T &= [0^T \quad 0^T \quad 0^T \quad 0^T \quad 0^T \quad 0^T \quad 1^T] \\ b^T &= [0^T \quad 0^T \quad (\mathbf{m}_{E+} + \mathbf{m}_{E-}) \quad C \quad m^+] \end{aligned}$$

and

$$\mathbf{W} = \begin{bmatrix} \mathbf{E} & 0 & \mathbf{E} & 0 & 0 & 0 & -2\mathbf{D}^T \\ 0 & 0 & 0 & \mathbf{E} & 0 & 0 & \mathbf{T} \\ \mathbf{E} & -\mathbf{E} & 0 & 0 & \mathbf{E} & 0 & 0 \\ \mathbf{C} & 0 & 0 & 0 & 0 & \mathbf{E} & -\mathbf{C}\mathbf{D}^T \end{bmatrix}, \quad \mathbf{X}_0 = \begin{bmatrix} \mathbf{E} & 0 & 0 & 0 \\ 0 & \mathbf{E} & 0 & 0 \\ 0 & 0 & \mathbf{E} & 0 \\ 0 & 0 & 0 & \mathbf{E} \end{bmatrix}$$

The initial basic matrix is simply a unit matrix and the initial variable $\mathbf{x}_0$ is:

$$\mathbf{x}_0 = \mathbf{X}_0^{-1}\mathbf{b} = \mathbf{b}$$

and is recognized to be always positive.

5 AUTOMATIC CALCULATION OF DUAL VARIABLES

We have seen in the previous sections that the choice of the kinematic formulation for the plastic analysis problems (limit analysis and shakedown analysis) and the static formulation for the optimization problems (optimal limit design and shakedown design) is motivated mainly by the limitation of the number of variables and constraints.

This choice involves that of the variables and the results given in the output are not complete if we do not try to handle the systematic calculation of the dual variables and the appropriate identification of these variables with physical quantities.

For example, for the limit analysis problem using the kinematically admissible formulation (1), the calculation of the dual variables enables us to get the current moments at every critical section in the ultimate state. By doing so we may avoid the expensive procedure of a second solution of the problem using a statically admissible formulation with moments at critical section as variables.

Let us consider a classical primal form of the linear programming problem:

$$\left.\begin{aligned} \text{Min } \mathbf{c}^T\mathbf{x} \\ \text{subjected to } \mathbf{W}\mathbf{x} = \mathbf{b} \\ \mathbf{x} \geq 0 \end{aligned}\right\} \tag{30}$$

and its canonical dual form:

$$\left.\begin{aligned} \text{Max } \mathbf{b}^T\mathbf{y} + \mathbf{0}^T\mathbf{h} \\ \mathbf{W}^T\mathbf{y} + \mathbf{h} = \mathbf{c} \\ \mathbf{y} \geq 0, \mathbf{h} \geq 0 \end{aligned}\right\} \tag{31}$$

In the canonical dual form, the additional variables $\mathbf{h}$ are called slack variables.

When one submits the problem (30) to a simplex process it will select iteratively trial subsets of the variable $\mathbf{x}$ named $\mathbf{x}_X$ and checks whether they are optimal, i.e., minimize the objective function. If they are optimal, the computations are terminated. If they are not, the method drops one variable from the current subset and adds a new one in such a manner that the objective function value is improved. The current vector $\mathbf{x}_X$ is the solution of the linear equation system:

$$\mathbf{X}\mathbf{x}_X = \mathbf{b}$$

where $\mathbf{X}$ is the current basic submatrix of the constraint matrix $\mathbf{W}$:

$$\mathbf{W} = [\mathbf{X} \quad \mathbf{Y}]$$

The test of the optimal situation is realized by the fact that the reduced costs:

$$\bar{\mathbf{c}} = \mathbf{c} - \mathbf{c}_X^T\mathbf{X}^{-1}\mathbf{W} \tag{32}$$

are non-negative.

As the solution of the dual problem (31) is given by:

$$\mathbf{y} = \mathbf{X}^{-T}\mathbf{c}_X$$

One may see that the reduced costs $\bar{\mathbf{c}}$ are identified to be exactly the slack variables $\mathbf{h}$ of the dual problem (31).

This important property of the simplex method constitutes an efficient tool for the computation of the structural dual quantities in the CEPAO package. For example, for limit analysis the dual form of the problem (21) is (31) with:

$$\mathbf{y}^T = [0 \quad -\mathbf{S}\mathbf{m} \quad \lambda]$$
$$\mathbf{h}^T = [0 \quad \mathbf{h}_+^T \quad \mathbf{h}_-^T \quad h_\lambda]$$

It can be seen that the dual form is nothing but the statically admissible form of the limit analysis problem shown in table 1. The detailed expression of the equality constraints of (31):

$$\begin{bmatrix} \mathbf{0}^T & \mathbf{S}\mathbf{C}^T & \mathbf{e}^T \\ \bar{\mathbf{m}}_{p+}^T & -\mathbf{E} & 0 \\ \bar{\mathbf{m}}_{p-}^T & \mathbf{E} & 0 \\ \mathbf{0}^T & 0 & 1 \end{bmatrix} \begin{bmatrix} 0 \\ -\mathbf{S}\mathbf{m} \\ \lambda \end{bmatrix} + \begin{bmatrix} 0 \\ \mathbf{h}_+ \\ \mathbf{h}_- \\ h_\lambda \end{bmatrix} = \begin{bmatrix} 0 \\ \bar{\mathbf{m}}_{p+} \\ \bar{\mathbf{m}}_{p-} \\ 0 \end{bmatrix}$$

shows clearly that the moments are:

$$\mathbf{m} = \tfrac{1}{2}\mathbf{S}[\overline{\mathbf{m}}_{p+} - \overline{\mathbf{m}}_{p-} + \mathbf{h}_{-} - \mathbf{h}_{+}]$$

where $\mathbf{h}_+$, $\mathbf{h}_-$ are the reduced costs of the simplex primal process already given in the output.

For the optimal limit design the dual problem (31) of the primal form (24) is identified with the following variables:

$$\mathbf{y}^T = [\dot{\mathbf{r}}'^T \quad \dot{\mathbf{r}}_s^T \quad -\dot{\mathbf{w}}^T]$$

$$\mathbf{h}^T = [\mathbf{h}_+^T \quad \mathbf{h}_-^T \quad \mathbf{h}_s^T \quad \mathbf{h}_w^T \quad \mathbf{h}_p^T]$$

and one may see without difficulty that the independent displacementates are identified with the reduced costs $\mathbf{h}_w$. Having obtained $\dot{\mathbf{w}}$, the relative rates of rotation $\dot{\mathbf{r}}$ are obtained directly by means of the first relation of (1).

A similar computation may be performed to obtain residual stresses in the shakedown analysis problem (26) and the collapse mechanism of the shakedown problem (29).

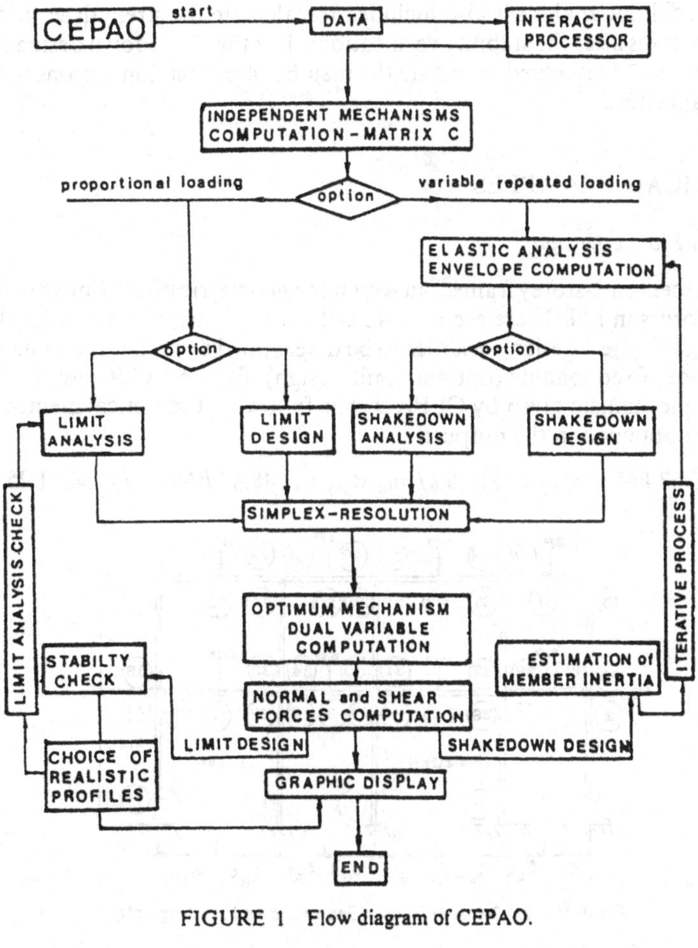

FIGURE 1 Flow diagram of CEPAO.

6 STABILITY VERIFICATIONS

In the design procedure (Problems (2) and (4)) the design variables which are the plastic capacities of the structural members are given in the output. These values represent theoretical, impractical ones. For practical purposes, an automatic choice of the manufacturer profiles is performed by furnishing to the computer a data vector containing a list of commercially available steel beams and columns. Whenever this choice is made, a systematic verification of the stability conditions imposed by the European Recommendations for Steel Constructions are carried out for all sections, beams and columns. So doing, the final structures proposed by the program may prevent premature local buckling, $P - \Delta$ effects, lateral buckling, imperfections, normal force and shear force effects. During this verification process the moments, normal forces and shear forces of the optimal state are taken as the current ones. The overall stability criteria and the limitations of the elasto-plastic deflections are not checked in the following examples. The detailed descriptions of the stability criteria are given in references [18] and [19].

A version, of the CEPAO program which includes interactive input data and graphic display is shown in Figure 1. It can be seen that an elastic analysis using the rotation matrix C is performed to obtain the elastic envelope before the shakedown computations. A control limit analysis is also included to calculate the strength security of the practical structures after the stability verifications. For the shakedown design problem an iterative procedure is added to update the member cross section inertias in terms of the plastic capacities.

7 NUMERICAL EXAMPLES

7.1 *Optimal Plastic Design*

Figure 2 represents a 2-storey frame[20] in which the geometric dimensions are in m and the applied forces in kN. There are $n_r = 42$ critical sections, $n_b = 26$ beam elements. Figure 3 indicates the 4 design variables to be determined such that the total weight is minimum under fixed loading (optimal limit design). Figure 4 illustrates the optimal FOULKES' mechanism given by CEPAO. The following theoretical member plastic capacities are obtained in the output:

$$M_{P_1} = 66.67\ kNm,\ M_{P_2} = 21.25\ kNm,\ M_{P_3} = 148.47\ kNm,\ M_{P_4} = 21.25\ kNm$$

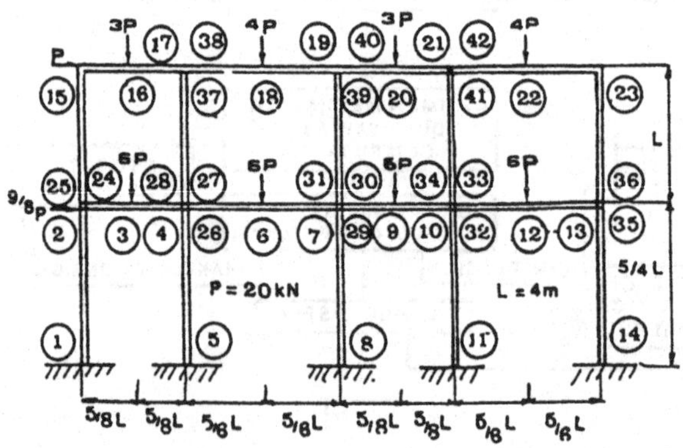

FIGURE 2 2-storey frame. Geometric and loading data.

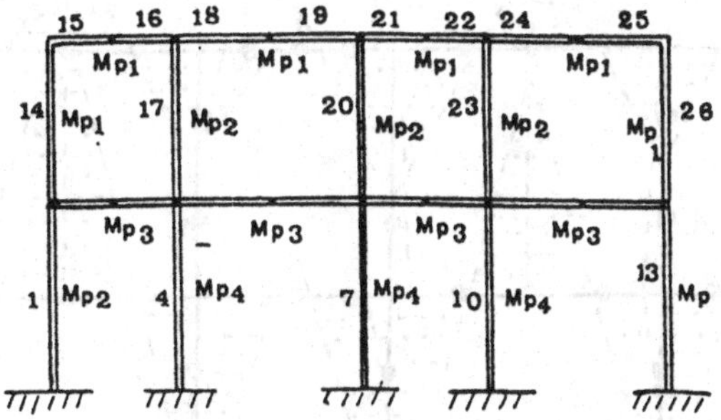

FIGURE 3 2-storey frame. Structural members and design variables.

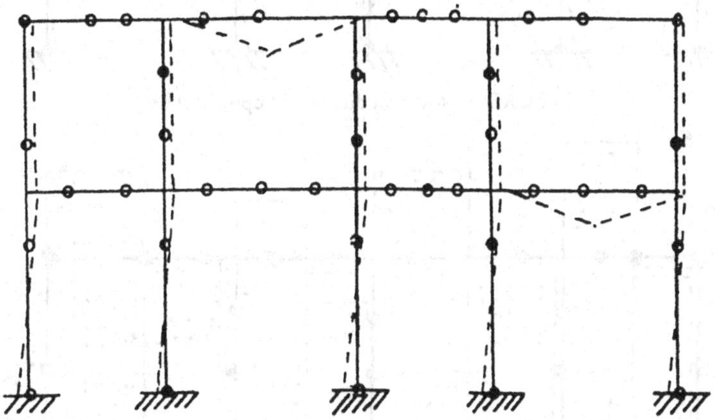

FIGURE 4 Foulkes' mechanism found by CEPAO.

Figures 5, 6 and 7 are respectively the moment, shear force and normal force diagrams stretched down by the graphic display processor. The corresponding numerical values are given respectively in Tables 2 and 3 where one may find also the relative rates of rotation of all critical sections.

If the data list of profiles is composed of current European beams IPE, HEA. HEB the first automatic choice performed by CEPAO is:

$$M_{p1} = 68.356 \text{ kNm : IPE 220}$$
$$M_{p2} = 24.904 \text{ kNm : HEB 100}$$
$$M_{p3} = 150.092 \text{ kNm : IPE 300}$$
$$M_{p4} = 24.904 \text{ kNm : HEB 100}$$

After all stability checkings the following final profiles are proposed (Figure 8):

$$\begin{aligned} M_{p1} &= 114.242 \text{ kNm : IPE 270} \\ M_{p2} &= 39.674 \text{ kNm : IPE 180} \\ M_{p3} &= 153.458 \text{ kNm : IPE 200} \\ M_{p4} &= 39.483 \text{ kNm : HEB 120} \end{aligned} \tag{33}$$

and the total weight of the structure is 2.544 tons.

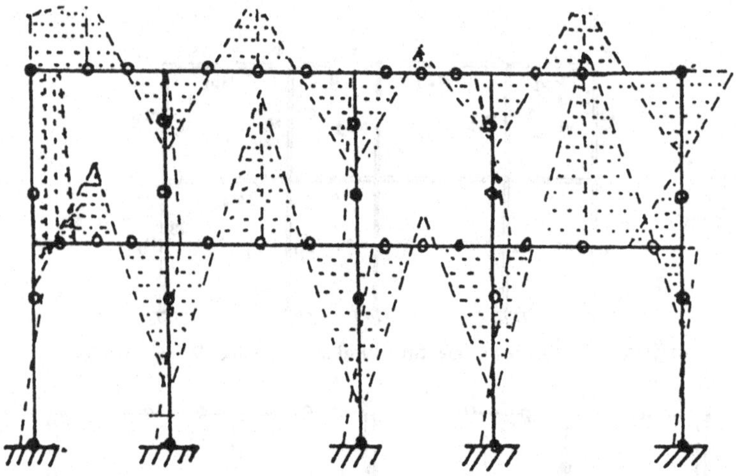

FIGURE 5 Moment diagram of optimal state.

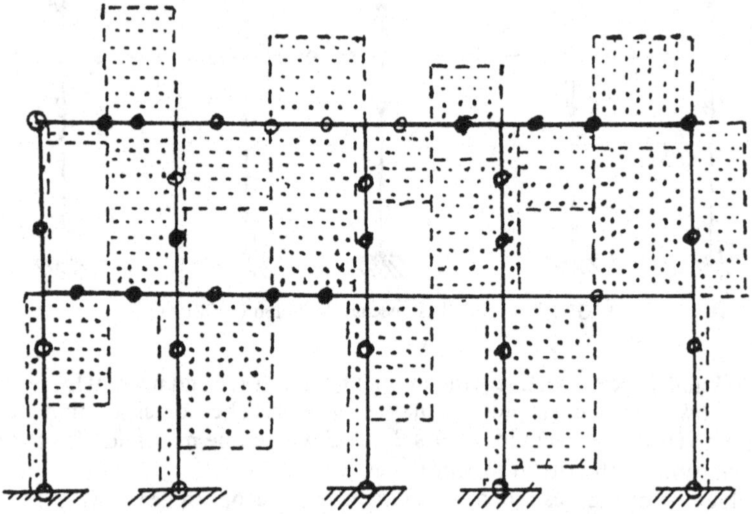

FIGURE 6 Shear force diagram of optimal state.

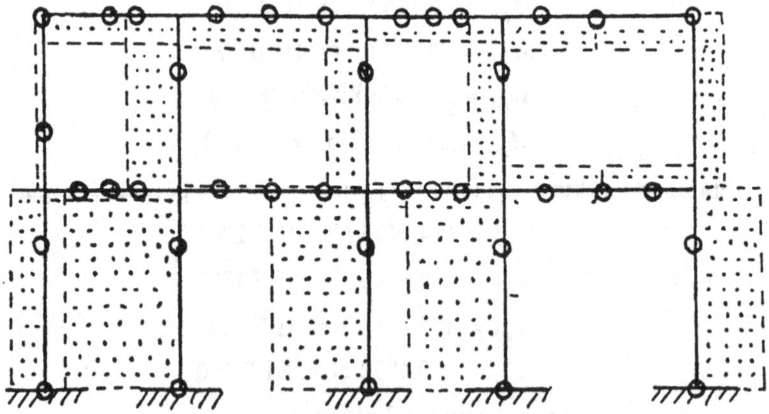

FIGURE 7 Normal force diagram of optimal state.

TABLE 2

Moment and relative rotation at critical section. Optimal state

Section	M_i (kNm)	Rotation θ_i	Section i	M_i (kNm)	Rotation θ_i	Section i	M_i (kNm)	Rotation θ_i
1	−21.25	−0.564	2	21.25	0.564	3	53.06	0
4	−148.47	0	5	−21.25	−0.564	6	121.1	0
7	−148.47	0	8	−21.25	−0.564	9	22.78	0
10	−148.47	−0.489	11	−21.25	−0.564	12	148.47	3
13	−66.67	0	14	21.25	0.564	15	50.0	0
16	66.67	0	17	−66.67	0	18	66.67	2.271
19	−66.67	−1.136	20	18.96	0	21	−45.42	0
22	66.67	0	23	−66.67	0	24	−45.42	0
25	66.67	0	26	21.25	0.564	27	−142.6	0
28	15.42	0	29	21.25	0.564	30	−106.0	0
31	−21.25	0	32	21.25	0.075	33	−148.5	−1.01
34	21.25	0.489	35	−21.25	−2.06	36	66.67	1.5
37	0	0	38	−66.67	−66.67	39	0	0
40	−66.67	0	41	−21.25	0	42	−66.67	0

TABLE 3

Shear and normal force of member

Member k	T_k (kN)	N_k (kN)	Member k	T_k (kN)	N_k (kN)	Member k	T_k (kN)	N_k (kN)
1	−8.50	−46.06	2	−39.39	−9.83	3	80.61	−9.83
4	−8.50	−253.1	5	−79.13	2.52	6	80.87	2.52
7	−8.50	−206.6	8	−51.50	5.71	9	68.5	5.71
10	−8.50	−223.3	11	−89.08	24.83	12	70.92	24.83
13	−8.50	−110.9	14	4.167	−6.667	15	−6.67	−24.17
16	53.33	−24.17	17	8.85	−93.33	18	−40.5	−28.02
19	40	−28.02	20	−5.31	−74.25	21	−34.25	−22.71
22	25.75	−22.71	23	10.62	−65.75	24	−40.0	−33.33
25	40	−33.33	26	−33.33	−40.00			

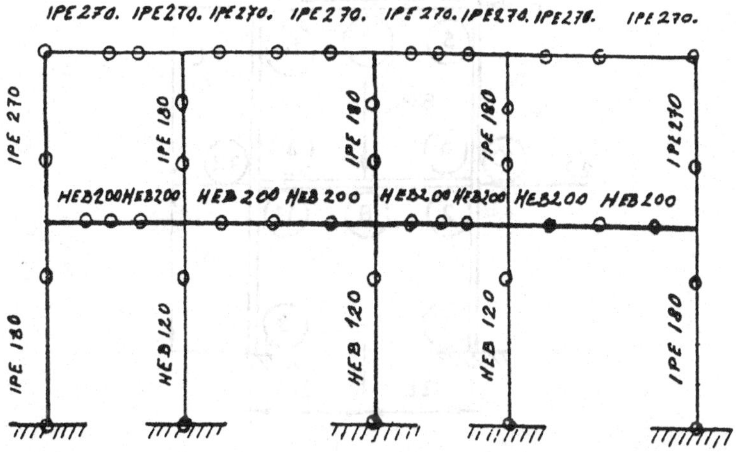

FIGURE 8 Optimal and stable structure proposed by CEPAO.

7.2 *Limit Analysis*

After reintroducing these realistic plastic capacities (33) into the limit analysis procedure, the limit loading factor $\lambda = 1.151$ is found. The corresponding collapse mechanism is illustrated by Figure 9.

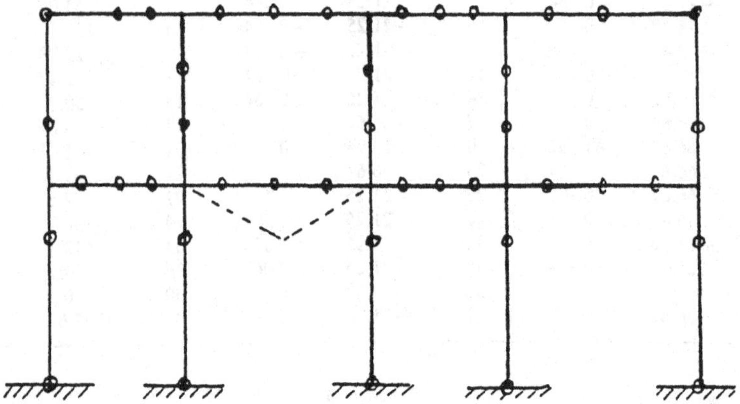

FIGURE 9 Limit analysis. Collapse mechanism of structure of Figure 8.

The moments, shear forces and normal forces of the collapse state are also automatically given by CEPAO but are not shown here.

7.3 *Shakedown Analysis*

Let us consider now the 2-storey frame (Figure 10) examined in reference (21), page 188.

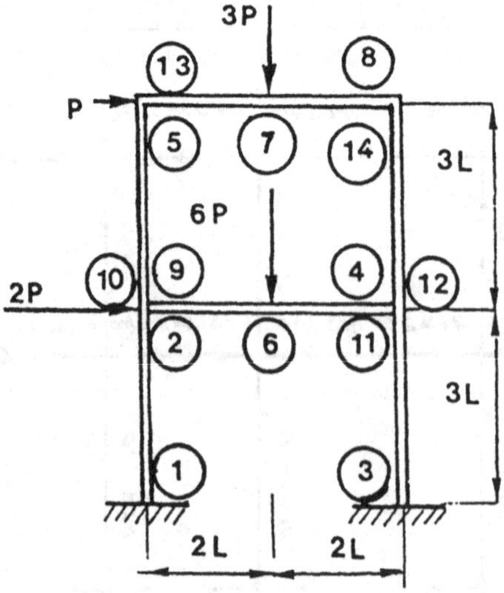

FIGURE 10 2-storey frame. Geometric and loading data.

Figure 11 shows the structural members and design variable numeration generated by CEPAO. For shakedown analysis the following were chosen: $M_{p1} = 2M_p$, $M_{p2} = M_{p3} = M_p$, $I_1 = 2.5I$, $I_2 = I_3 = I$.

Table 4 illustrates the comparison of results between CEPAO and reference (21). The difference is due to the fact that the elastic envelope found by CEPAO is calculated under assumptions that the normal force effects are negligible. It appears that the shakedown results depend strongly upon the elastic response. The following shakedown limit loads were found:

$$P_s = 0.508\ M_p/\bar{L} \qquad \text{(CEPAO)}$$

$$P_s = 0.509\ M_p/L \qquad \text{[reference (21)]}$$

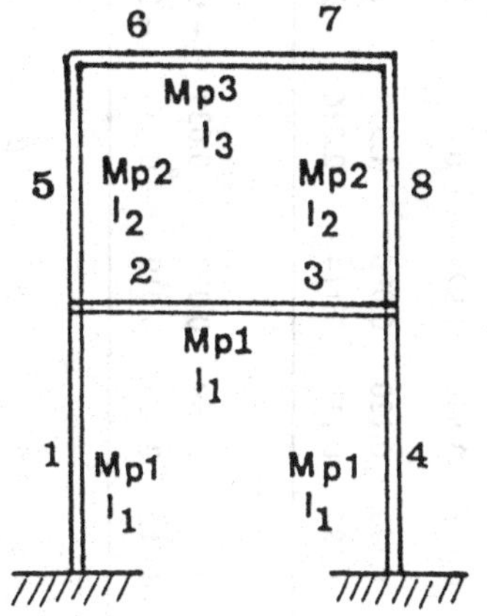

FIGURE 11 Structural members and design variables.

Table 4 shows also the residual moments of all critical sections and the collapse mechanism. It is important to mention that the considered loading domain is defined as:

$$P_2 \in [0, 2P]: P_5 \in [0, P]$$
$$P_6 \in [0, 6P]: P_7 \in [0, 3P] \tag{34}$$

where P_2 is understood as the force applied at section 2 (Figure 10)..

7.4 *Shakedown Design*

The same loading domain (34) is applied to the structure of Figure 10. The design variables M_{p1}, M_{p2}, M_{p3} are to be found such that the total weight is minimum.

The inertia of the members are assumed to be dependent to the plastic capacities in the following manner:

$$I_k/I = (M_{pk}/M_p)^{1.4}$$

An iterative procedure is necessary in this case because one has to update the relative moments of inertia of the cross section at each step. As shown by Table 5 four iterations

TABLE 4

Shakedown analysis. Comparison of CEPAO with Ref. (21)

Elastic envelope		Critical section 1	2	3	4	5, 13	6	7	8, 14	9	10	11	12
M_{E+}/PL	CEPAO	0.881	2.032	3.669	0.0	0.790	3.661	1.927	0.0	2.422	1.102	1.320	1.812
	Ref. (21)	0.840	2.025	3.621	0.0	0.831	3.630	1.929	0.0	2.388	1.214	0.306	1.883
M_{E-}/PL	CEPAO	2.948	1.762	0.160	4.881	1.170	0.120	0.096	1.959	2.459	0.7945	3.474	0.872
	Ref. (21)	2.934	1.680	0.153	4.872	1.185	0.114	0.114	2.016	2.484	0.773	3.399	0.104
Residual moment													
ρ/PL	CEPAO	−0.501	−0.075	0.135	0.481	0.045	0.139	0.021	−0.039	−0.203	0.128	0.560	0.08
	Ref. (21)	−0.507	−0.186	0.158	0.479	0.012	0.153	0.019	−0.026	−0.014	0.0	0.479	0.0
Collapse mechanism													
θ	CEPAO	-0.5θ	0.0	0.5θ	0.0	0.0	θ	$-\theta$	θ	0.0	0.0	0.0	0.0

TABLE 5

Iterative shakedown design of the structure of Figure 10

	Relative inertia				Critical section number												Design results			
Iteration	I_1/I	I_2/I	I_3/I		1	2	3	4	5, 13	6	7	8, 14	9	10	11	12	M_{p1}/PL	M_{p2}/PL	M_{p3}/PL	Z_s/PL^2
				M_{E+}/PL	0.881	2.034	3.67	0.0	0.79	3.66	1.93	0.0	2.42	1.10	0.32	1.81	3.962	1.348	2.511	57.701
1	2.5	1.0	1.0	M_{E-}/PL	2.95	1.76	0.16	4.88	1.17	0.12	0.096	1.96	2.46	0.79	3.47	0.084				
				ρ/PL	−1.01	−0.86	0.293	0.92	0.55	0.30	0.585	0.62	−0.316	−0.544	0.446	−0.473				
				M_{E+}/PL	0.96	1.96	3.75	0.0	0.79	3.72	2.23	0.0	2.42	0.76	0.25	1.47	3.967	1.172	2.662	57.355
2	2.5	0.457	1.320	M_{E-}/PL	2.91	1.92	0.126	4.79	0.87	0.09	0.01	1.67	2.38	0.76	3.63	0.05				
				ρ/PL	−1.05	−0.74	0.22	0.83	0.38	0.25	0.44	0.496	−0.33	−0.414	0.53	−0.296				
				M_{E+}/PL	0.98	1.94	3.76	0.0	0.79	3.73	2.32	0.0	2.42	0.67	0.22	1.37	3.964	1.121	2.71	57.210
3	2.5	0.454	1.431	M_{E-}/PL	2.90	1.95	0.112	4.77	0.77	0.08	0.096	1.56	2.35	0.75	3.67	0.04				
				ρ/PL	−1.06	−0.71	0.2	0.81	0.33	0.232	0.386	0.443	−0.343	−0.368	0.553	−0.254				
				M_{E+}/PL	0.98	1.93	3.78	0.0	0.79	3.74	2.35	0.0	2.42	0.64	0.21	1.34	3.963	1.104	2.72	57.160
4	2.5	0.426	1.468	M_{E-}/PL	2.89	1.96	0.107	4.76	0.74	0.08	0.094	1.53	2.34	0.75	3.68	0.04				
				ρ/PL	−1.07	−0.7	0.19	0.8	0.314	0.227	0.369	0.425	−0.348	−0.353	0.56	−0.241				

are sufficient if one accepts a tolerance on the merit function of

$$\frac{Z_s^{n+1} - Z_s^n}{Z_s^n} \leq 10^{-3}$$

The final distribution of residual moments (iteration number 4) are sketched in Figure 12. The final collapse mechanism corresponding to this optimum state is shown in Figure 13.

It is important to recall here that this collapse mechanism is that of the loading domain (34) (variable repeated loading) but not the applied loads outlined in Figure 10.

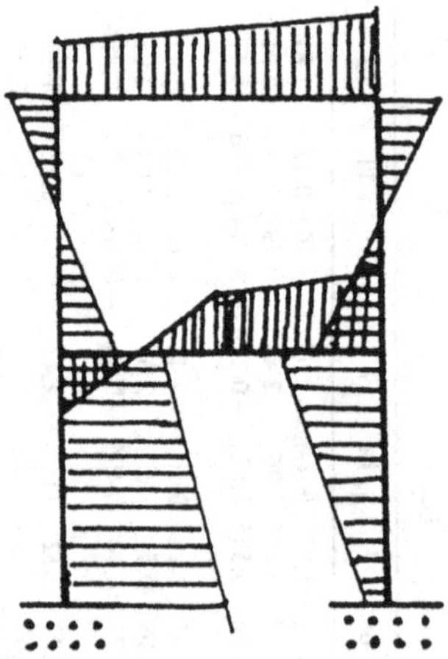

FIGURE 12 Final residual moment at optimum state.

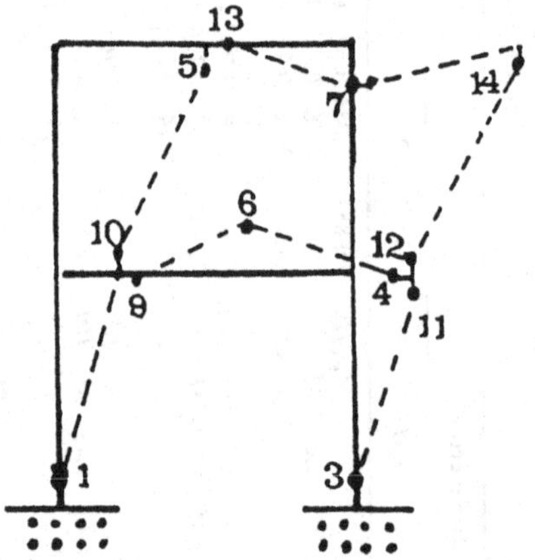

FIGURE 13 Final mechanism at optimum state.

8 CONCLUSIONS

Analysis and design of frame structures under proportional and variable repeated loadings are solved automatically in a unified computer program wherein the dimension of the computation is reduced considerably. For optimization problems the automatic choice of practical, available sections after checking stability conditions presents an efficient tool for practical purposes. An extension of this procedure to the optimization of joints seems to be advantageous as a future possibility. The facilities for data preparation and the automatic descriptions of the results, complete and direct, render plasticity as attractive a method for design as elasticity for engineer.

This is the "raison d'être" of the CEPAO program. The present version contains also a step by step method useful for elasto-plastic deflection checking in the plastic design of reinforced concrete structures. We hope to give in the near future a version in which the second order effects are taken into account.

REFERENCES

1. Ch. Massonnet and M. Save, *Calcul plastique des constructions. Vol. 1 Structures dépendent d'un paramètre.* 3ème édition, Nélissen B., Liège, (1976).
2. B. G. Neal, *The Plastic Methods of Structural Analysis.* Chapman and Hall, London, (1956).
3. P. G. Hodge, Jr., *Plastic Analysis of Structures.* McGraw-Hill, (1959).
4. D. B. Martin: Plasticity, Fundamentals and General Results. MIT Press, Cambridge, Massachusetts. Londres, 1975.
5. K. I. Majid, *Non-linear Structures.* Butterworths, London, (1978).
6. M. Z. Cohn, S. K. Ghosh, and S. R. Parimi, "A Unified Approach to Theory of Plastic Structures." *Solid Mech. Division, University of Waterloo, Canada, Report nr 26,* June (1973).
8. A. Charles, C. E. Lemke, and O. C. Zienkiewicz, "Virtual work. Linear Programming and Plastic Limit Analysis," *Proc. Royal Soc. A,* 251, (1959).
9. W. S. Dorn, "On the Plastic Collapse of Structures and Linear Programming," *Ph.D. Thesis,* Dept. of Mathematics, Carnegie Institute of Techn. Pittsburgh, Pa. May (1955).
10. W. S. Dorn and H. J. Greenberg, "Linear Programming and Plastic Limit Analysis of Structures." *Quart. J. Appl. Math.,* 15, (1957).
11. G. Ceradini and C. Gavarini, "Calcolo a rottura e programmazione lineare," *Giornale del Genio Civile,* 3, (1965).
12. D. E. Grierson and G. M. L. Gladwell, "Collapse Load Analysis Using Linear Programming." *J. Str. Div. ASCE.,* 97, (ST 5), May, (1971).
13. R. K. Livesley, "The Automatic Design of Structural Frames." *Quart. J. Mech. and Appl. Math.,* 8, (3), September (1956).
14. W. Prager, "Linear Programming and Structural Design," *Rand Research Memorandum KM-2021, ASTIA Document AP 150661,* December (1957).
15. G. Maier, R. Srininavan, and M. Save, "On Limit Design of Frames using Linear Programming." "*Proc. Int. Symp. on Computer-Aided Structural Design,* University of Warwick, Coventry, (1972).
16. G. Sacchi, M. Maier, and M. Save, "Limit Design of Frames for Movable Loads by Linear Programming," *IUTAM Symposium on Optimization in Structural Design.* Warsaw 1973, (ed. Sawczuk and Mroz), Springer, (1975).
17. G. Maier, "Mathematical Programming Methods in Structural Analysis." *Proc. Int. Symp. on Variational Methods in Engineering,* (ed. C. Brebbia and H. Tottenham), Southampton University Press, (1973).
18. Nguyen Dang Hung and G. De Saxce, "Analyse et dimensionnement plastique des structures à barres dans des conditions de stabilité," *Construction métallique,* (3), 15–28, (1981).
19. C. E. C. M., European Recommendations for Steel Constructions, *ECCS-EG 77-2E,* March (1978).
20. D. E. Grierson, "Elastic-Plastic Deformation Analysis using Linear Programming," *Solid Mech. Div., University of Waterloo, Report No., 88,* July (1971).
21. M. Z. Cohn, *Analysis and Design of Inelastic Structures. Vol. 2. Problems,* University of Waterloo Press, (1972).

CHAPTER 13

VARIATIONAL STATEMENTS AND MATHEMATICAL PROGRAMMING FORMULATIONS IN ELASTIC-PLASTIC ANALYSIS

L. Corradi
Politecnico di Milano, Milan, Italy

ABSTRACT

The basic results of classical (associated) rate plasticity theory are first summarized in this chapter and some variational (minimum) statements are recalled, which provide suitable bases for finite element discretization. Extensions to holonomic or piecewise holonomic representations of the elastic-plastic behavior are next discussed, under the assumption that the constitutive law can be expressed in a piecewise linear form. The discrete elastic-plastic problem for finite load increments is considered in this context and connections with Mathematical Programming formulations are underlined.

13.1.MATHEMATICAL DESCRIPTION OF ELASTIC-PLASTIC CONSTITUTIVE LAWS

13.1.1. Associated flow rule of classical plasticity

The constituive law known as flow theory of plasticity [1] is a relationship between stress rates $\dot{\sigma}_{ij}$ and strain rates $\dot{\varepsilon}_{ij}$, based on the following assumptions.

1. Additive decomposition of strains into an elastic (e_{ij}) and a plastic (p_{ij}) portion (given inelastic strains, such as thermal, are not considered for simplicity)

$$\dot{\varepsilon}_{ij} = \dot{e}_{ij} + \dot{p}_{ij} \tag{13.1}$$

where the elastic part is related to stresses by means of Hooke's law

$$\dot{\sigma}_{ij} = C_{ijhk}\,\dot{e}_{hk} \quad ; \quad \dot{e}_{ij} = A_{ijhk}\,\dot{\sigma}_{hk} \tag{13.2a,b}$$

(sum on repeated indeces is understood).

2. Existence of an instantaneous elastic domain, defined by the condition that a number (say Y) of yield functions be non-positive

$$\phi_\alpha(\sigma_{ij}, \text{plastic strain history}) \leq 0 \ , \ \alpha = 1,\ldots,Y \tag{13.3}$$

plastic strains may develop only if $\phi_\alpha = 0$ for some α.

3. Definition of a plastic flow rule, governing the expression of plastic strain rates. When associated flow rules are assumed, Eqs.(3) define a convex domain in stress space and plastic strain rates obey the so-called outward normality rule, which is expressed as follows

$$\dot{p}_{ij} = (\partial\phi_\alpha/\partial\sigma_{ij})\,\dot{\lambda}_\alpha \tag{13.4}$$

$$\dot{\lambda}_\alpha \geq 0 \quad \text{if} \quad \phi_\alpha = 0 \ \text{and} \ \dot{\phi}_\alpha = 0 \quad ; \quad \dot{\lambda}_\alpha = 0 \ \text{otherwise} \tag{13.5a,b}$$

or equivalently

$$\dot{\lambda}_\alpha \geq 0 \quad ; \quad \phi_\alpha\,\dot{\lambda}_\alpha = 0 \quad ; \quad \dot{\phi}_\alpha\,\dot{\lambda}_\alpha = 0 \tag{13.6a-c}$$

Perfectly plastic materials are defined by the condition that the elastic domain is fixed in stress space. In this case one has

$$\phi_\alpha = \phi_\alpha(\sigma_{ij}) \tag{13.7}$$

A typical example is given by Huber-Henky-Mises (H.H.M.) condition, which is expressed as follows

$$\phi = (3J_2)^{1/2} - \sigma_0 \qquad (Y = 1) \tag{13.8}$$

where J_2 is the quadratic invariant of the stress deviator tensor and σ_0 is the yield limit in uniaxial tension. Another example is provided by Tresca condition

$$\phi = |\tau_M| - \tau_0 \qquad (Y = 6) \tag{13.9}$$

where $|\tau_M|$ and τ_0 are the maximum shear stress and the shear yield limit, respectively. Both the above conditions depend on a single material datum.

If the yield function depends on plastic strain history, the instantaneous elastic range is modified because of yielding. With reference to the H.H.M. condition, the two hardening laws most frequently used are: (a) Isotropic hardening

$$\phi(\sigma_{ij},\kappa) = (3J_2)^{1/2} - \sigma_0(\kappa) \quad ; \quad \kappa = \int \sigma_{ij}\, \dot{p}_{ij} \tag{13.10a,b}$$

where κ is the dissipated energy in previous history, and: (ii) Kinematic hardening

$$\phi(\sigma_{ij}, p_{ij}) = (3J_2')^{1/2} - \sigma_0 \tag{13.11a}$$

where J_2' is the quadratic invariant of the deviatoric part of the shifted stress tensor

$$\sigma'_{ij} = \sigma_{ij} - c\, p_{ij} \qquad (c > 0) \tag{13.11b}$$

Attention will be first limited to so-called regular yield functions (Y=1, such as H.H.M. condition). If plastic strains may develop it must be $\phi = 0$ and, hence, $\dot{\phi} \leq 0$. Let $\dot{\phi}$ be assumed as a linear function of both $\dot{\sigma}_{ij}$ and $\dot{\lambda}$; such an expression can be written as follows

$$\dot{\phi} = (\partial\phi/\partial\sigma_{ij})\, \dot{\sigma}_{ij} - h\, \dot{\lambda} \tag{13.12}$$

where h is a hardening coefficient. For instance, for H.H.M. condition and isotropic and kinematic (c = const) hardening, respectively, one has

$$-h\dot{\lambda} = (\partial\phi/\partial\kappa)\, \dot{\kappa} = -(\partial\sigma_0/\partial\kappa)\, \sigma_{ij}(\partial\phi/\partial\sigma_{ij})\, \dot{\lambda}$$

$$-h\dot{\lambda} = (\partial\phi/\partial p_{ij})\, \dot{p}_{ij} = -c\, (\partial\phi/\partial\sigma_{ij})(\partial\phi/\partial\sigma_{ij})\, \dot{\lambda}$$

from which the relevant expressions for h are readily obtained.

If $\phi = 0$ at the current stress level, the associated flow rule of classical plasticity can be written as follows

$$\dot{\varepsilon}_{ij} = \dot{e}_{ij} + \dot{p}_{ij} \quad ; \quad \dot{\sigma}_{ij} = C_{ijhk}\, \dot{e}_{hk} \tag{13.13a,b}$$

$$\dot{p}_{ij} = (\partial\phi/\partial\sigma_{ij})\, \dot{\lambda} \quad ; \quad \dot{\phi} = (\partial\phi/\partial\sigma_{ij})\, \dot{\sigma}_{ij} - h\, \dot{\lambda} \tag{13.13c,d}$$

$$\dot{\lambda} \geq 0 \quad ; \quad \dot{\phi} \leq 0 \quad ; \quad \dot{\phi}\, \dot{\lambda} = 0 \tag{13.13e-g}$$

13.1.2. Explicit expressions for the constitutive laws

In this section expressions of the type $\dot{\sigma}_{ij} = \dot{\sigma}_{ij}(\dot{\varepsilon}_{hk})$ (direct constitutive law) and $\dot{\varepsilon}_{ij} = \dot{\varepsilon}_{ij}(\dot{\sigma}_{hk})$ (inverse constitutive law) are derived by eliminating $\dot{\lambda}$ from Eqs.(13).

(a) Inverse relation. From Eqs.(1),(2b) and (4) one obtains

$$\dot{\varepsilon}_{ij} = A_{ijhk}\, \dot{\sigma}_{hk} + (\partial\phi/\partial\sigma_{ij})\, \dot{\lambda} \tag{13.14}$$

if $h > 0$, the second addend in the expression Eq.(12) for $\dot{\phi}$ is non-positive. Hence, the condition $\dot{\phi}\, \dot{\lambda} = 0$ unambigously defines the following alternatives

(a1) If $(\partial\phi/\partial\sigma_{ij})\, \dot{\sigma}_{ij} > 0$, then $\dot{\lambda} = \frac{1}{h}(\partial\phi/\partial\sigma_{ij})\, \dot{\sigma}_{ij} > 0$ and

$$\dot{\varepsilon}_{ij} = A^{p}_{ijhk}\, \dot{\sigma}_{hk} \; ; \; A^{p}_{ijhk} = \frac{1}{h}(\partial\phi/\partial\sigma_{ij})(\partial\phi/\partial\sigma_{hk}) \tag{13.15a}$$

(a2) If $(\partial\phi/\partial\sigma_{ij})\, \dot{\sigma}_{ij} \leq 0$, then $\dot{\lambda} = 0$ and

$$\dot{\varepsilon}_{ij} = A_{ijhk}\, \dot{\sigma}_{hk} \tag{13.15b}$$

(b) Direct Relation. From Eqs.(2a),(1) and (4) one also obtains

$$\dot{\sigma}_{ij} = C_{ijhk}\, (\dot{\varepsilon}_{hk} - (\partial\phi/\partial\sigma_{hk})\dot{\lambda}\,) \tag{13.16}$$

which, when introduced in Eq.(12), yields

$$\dot{\phi} = C_{ijhk} \frac{\partial\phi}{\partial\sigma_{ij}}\, \dot{\varepsilon}_{hk} - (h + C_{ijhk} \frac{\partial\phi}{\partial\sigma_{ij}} \frac{\partial\phi}{\partial\sigma_{hk}}\,)\, \dot{\lambda} \tag{13.17}$$

non-positive when $\phi = 0$. If

$$h > h_{cr} \equiv - C_{ijhk}\, (\partial\phi/\partial\sigma_{ij})(\partial\phi/\partial\sigma_{hk}) \tag{13.18}$$

the second addend in Eq.(17) is non-positive and the following alternatives are unambigously defined (b1) If $C_{ijhk} \frac{\partial\phi}{\partial\sigma_{ij}} \dot{\varepsilon}_{hk} > 0$, then $\dot{\lambda} = \frac{1}{(h-h_{cr})} C_{lmpq} \frac{\partial\phi}{\partial\sigma_{lm}} \dot{\varepsilon}_{pq} > 0$ and

$$\dot{\sigma}_{ij} = C^P_{ijhk} \dot{\varepsilon}_{hk} \; ; \; C^P_{ijhk} = C_{ijhk} - \frac{1}{(h-h_{cr})} C_{ijst} C_{qrhk} \frac{\partial\phi}{\partial\sigma_{st}} \frac{\partial\phi}{\partial\sigma_{qr}} \tag{13.19a}$$

(b2) If $C_{ijhk} \frac{\partial\phi}{\partial\sigma_{ij}} \dot{\varepsilon}_{hk} \le 0$, then $\dot{\lambda} = 0$ and

$$\dot{\sigma}_{ij} = C_{ijhk} \dot{\varepsilon}_{hk} \tag{13.19b}$$

The following points are worth a comment.

1. The inverse law is uniquely defined (i.e., there always exists a unique $\dot{\varepsilon}_{ij}$ for any $\dot{\sigma}_{hk}$) if and only if $h > 0$ (hardening behavior). Analogously, the direct law is uniquely defined if and only if $h > h_{cr}$, Eq.(18). $h = h_{cr}$ implies critical softening behavior.

2. The elastic-plastic tensors A^P_{ijhk} and C^P_{ijhk} are defined whenever h and $h-h_{cr}$, respectively, are different from zero, even if negative. However, the (inverse or direct) laws are <u>uniquely</u> defined only if their values are positive. For instance, in the presence of softening $(0 > h > h_{cr})$ it is still possible to define A^P_{ijhk} ; however, both the alternatives Eqs.(15a) (loading, $\dot{\phi} = 0$) and (15b) (unloading, $\dot{\phi} < 0$) are consistent with $(\partial\phi/\partial\sigma_{ij}) \dot{\sigma}_{ij} < 0$, while no solution exists when this quantity is positive.

3. When both defined, tensors A^P_{ijhk} and C^P_{ijhk} are inverse to each other.

13.1.3. Drucker's postulate

Drucker's stability postulate defines a class of elastic-plastic materials for which the following inequalities hold

$$(\sigma_{ij} - \sigma^s_{ij}) \dot{p}_{ij} \ge 0 \quad \forall \; \sigma^s_{ij} \text{ such that } \phi(\sigma^s_{ij}) \le 0 \tag{13.20a}$$

$$\dot{\sigma}_{ij} \dot{p}_{ij} \ge 0 \tag{13.20b}$$

Materials obeying Drucker's postulate exhibit the following properties.

1. They comply with the associated flow rule of rate plasticity; i.e., Drucker's postulate ensures that the instantaneous elastic range

is a convex domain in stress space and that plastic strain rates obey the outward normality rule (proof is omitted).

2. The hardening coefficient h is non-negative. In fact, the condition $\dot{\phi}\,\dot{\lambda} = 0$ implies $h\,\dot{\lambda}^2 = \dot{\sigma}_{ij}\,(\partial\phi/\partial\sigma_{ij})\,\dot{\lambda} = \dot{\sigma}_{ij}\,\dot{p}_{ij}$, non-negative because of Eq.(20b).

3. Consider a stress state σ_{ij} such that $\phi(\sigma_{ij}) = 0$ and two pairs of stress-plastic strain rates from it (denoted by $(\ )^1$ and $(\ )^2$, respectively), each related by the constitutive relationship. Then, the following inequalities hold

$$(\dot{\sigma}^1_{ij} - \dot{\sigma}^2_{ij})(\dot{p}^1_{ij} - \dot{p}^2_{ij}) \geq 0 \tag{13.21a}$$

$$\dot{\sigma}^1_{ij}\,\dot{p}^1_{ij} + \dot{\sigma}^2_{ij}\,\dot{p}^2_{ij} - 2\,\dot{\sigma}^1_{ij}\,\dot{p}^2_{ij} \geq 0 \tag{13.21b}$$

In fact, one can write

$$\dot{\phi}^\alpha\dot{\lambda}^\beta = \dot{\sigma}^\alpha_{ij}\,\dot{p}^\beta_{ij} - h\,\dot{\lambda}^\alpha\dot{\lambda}^\beta \begin{cases} = 0 \text{ if } \alpha = \beta \\ \leq 0 \text{ if } \alpha \neq \beta \end{cases} \quad (\alpha,\beta = 1,2) \tag{13.22}$$

Proof of Eqs.(21) immediatly follows from direct substitution, account taken of the fact that $h \geq 0$ because of Drucker's postulate.

13.1.4. Piecewise linear yield functions

A particular case of non-regular yield condition is now considered. Let the instantaneous elastic range Eq.(3) be replaced by a polyhedron in stress space, defined by the Y linear inequalities

$$\phi_\alpha = n^\alpha_{ij}\,\sigma_{ij} - h_{\alpha\beta}\,\lambda_\beta - r_\alpha \leq 0 \quad ; \quad \alpha,\beta = 1,\dots,Y \tag{13.23}$$

where $n^\alpha_{ij} = \partial\phi_\alpha/\partial\sigma_{ij}$ are the components of the (fixed) outward normal vector to the plane $\phi_\alpha = 0$, whose original (prior to yielding) distance from the origin in stress space is denoted by r_α (positive, if the elastic domain for a virgin material contains the origin). The non-negative and non-decreasing quantities

$$\lambda_\alpha = \int \dot{\lambda}_\alpha\,dt \tag{13.24}$$

are internal variables, representing the total values of plastic multipliers associated to ϕ_α and developed up to the current stage. They make the yield plane $\phi_\alpha = 0$ translate because of yielding, a translation which is governed by the constant hardening coefficients $h_{\alpha\beta}$ (see fig.1).

Piecewise linear constitutive laws are conveniently represented by making use of a vector-matrix notation. Let the components σ_{ij} (ε_{ij}, p_{ij}) of stresses (total, plastic strains) be collected in vectors $\underline{\sigma}$ ($\underline{\varepsilon}$, $\underline{p}$) and let the Y-vectors $\underline{\phi}$ and $\underline{\lambda}$ collect the components ϕ_α and λ_α of yield functions and plastic multipliers, respectively. Moreover, let $\underline{D}$ be the square, symmetric and positive definite matrix of Hooke's law coefficients C_{ijhk} and let the following matrices be defined

$$\underline{N} = [\dots \underline{n}^\alpha \dots] \; ; \; \underline{H} = [h_{\alpha\beta}] \; ; \; \underline{R} = \{r_\alpha\} \qquad (13.25a\text{-}c)$$

Then Eq.(23), replacing Eq.(3), becomes

$$\underline{\phi} = \underline{N}^t \, \underline{\sigma} - \underline{H} \, \underline{\lambda} - \underline{R} \leq \underline{0} \qquad (13.26)$$

while Eqs.(1),(2) and (4-6) are now replaced by

$$\dot{\underline{\varepsilon}} = \dot{\underline{e}} + \dot{\underline{p}} \quad ; \quad \dot{\underline{\sigma}} = \underline{D} \, \dot{\underline{e}} \qquad (13.27a,b)$$

$$\dot{\underline{p}} = \underline{N} \, \dot{\underline{\lambda}} \quad ; \quad \dot{\underline{\phi}} = \underline{N}^t \, \dot{\underline{\sigma}} - \underline{H} \, \dot{\underline{\lambda}} \qquad (13.28a,b)$$

$$\dot{\underline{\lambda}} \geq \underline{0} \quad ; \quad \underline{\phi}^t \, \dot{\underline{\lambda}} = \underline{0} \quad ; \quad \dot{\underline{\phi}}^t \, \dot{\underline{\lambda}} = 0 \qquad (13.28c\text{-}e)$$

Eqs.(28) simply represent a particular form of associated flow rule of plasticity, the convexity of the elastic range being ensured by Eq.(26). Different hardening laws can be implemented by giving matrix $\underline{H}$ different expressions. When $\underline{H} = \underline{0}$, perfectly plastic behavior is reproduced. Kinematic and isotropic hardening are obtained with the following assumptions, respectively [2]

$$\underline{H} = c \, \underline{N}^t \, \underline{N} \quad \text{and} \quad \underline{H} = c \, \underline{R}^t \, \underline{R} \quad (c > 0) \qquad (13.29a,b)$$

When matrix $\underline{H}$ is symmetric and at least positive semidefinite, Drucker's postulate can be easily shown to be complied with. This condition replaces the requirement $h \geq 0$ in Sec.1.3. All previous results still hold, even if they must be (straightforwardly) extended to cover non-regular yield functions.

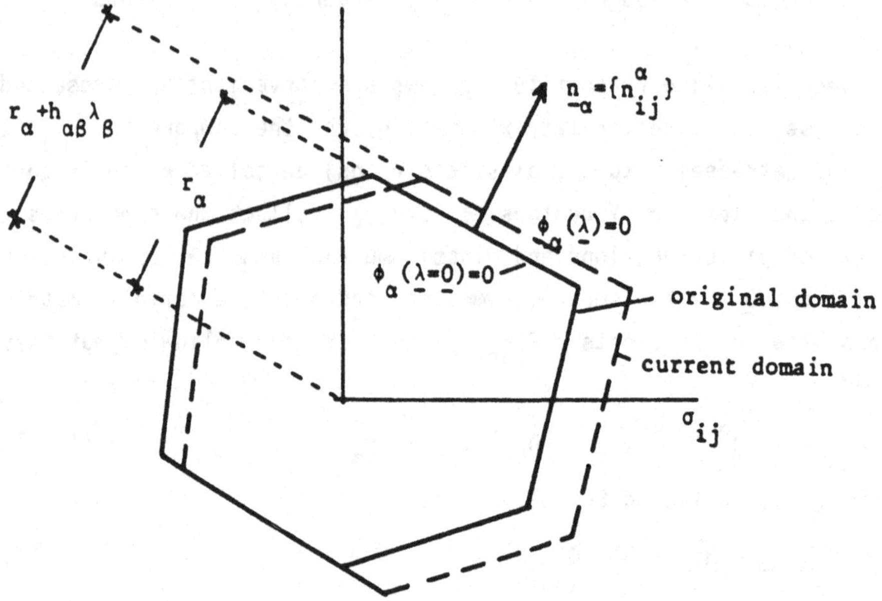

Fig.13.1. Piecewise linear yield function

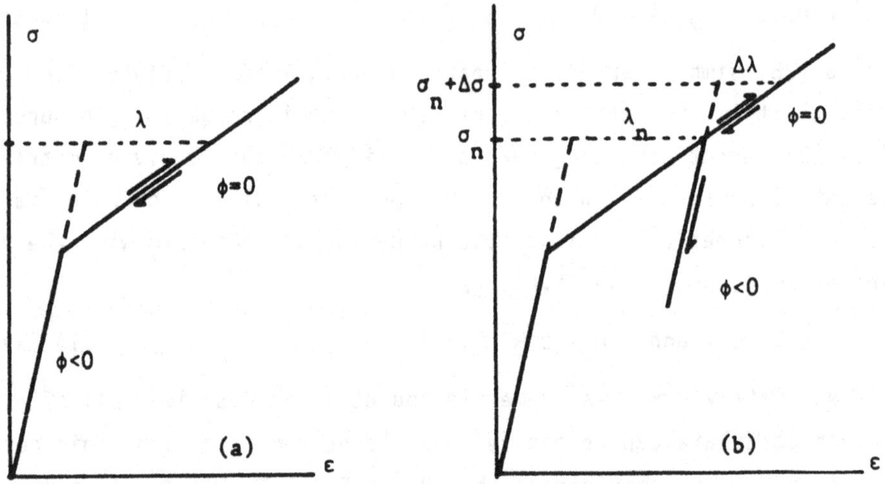

Fig.13.2. Holonomic and piecewise holonomic behavior

13.1.5. Holonomic and stepwise holonomic constitutive laws

One of the main advantages of a piecewise linear representation of the elastic-plastic laws rests on the possibility of establishing easily constitutive relationships in finite terms. In this way, however, an approximation is introduced, since the possibility of elastic unloading from a plastic state is either eliminated or limited to a finite number of instants during the elastic-plastic evolution.

The possibility of writing finite elastic-plastic laws is provided by Eq.(24), showing that a piecewise linear representation permits the definition of <u>finite</u> values of internal variables, each associated to a yield plane. In particular, Eq.(28) can be integrated and total plastic strains can be expressed as linear functions of $\underline{\lambda}$

$$\underline{p} = \underline{N}\,\underline{\lambda} \tag{13.30}$$

which permits the expression of stresses as

$$\underline{\sigma} = \underline{D}\,(\underline{\varepsilon} - \underline{N}\,\underline{\lambda}) \tag{13.31}$$

The simplest finite law is the fully reversible (or <u>holonomic</u>) relationship, which makes the elastic-plastic behavior essentially equal to a non-linear elastic one. Such law is enforced by the relations

$$\underline{\lambda} \geq \underline{0} \quad ; \quad \underline{\phi} \leq \underline{0} \quad ; \quad \underline{\phi}^t\,\underline{\lambda} = 0 \tag{13.32a-c}$$

with $\underline{\phi}$ given by Eq.(26). The completely reversible behavior is enforced by Eq.(32c), requiring that the <u>total</u> value of λ_α be equal to zero if $\phi_\alpha < 0$. For the unidimensional case, such a behavior is depicted in fig.2a.

Holonomy is a reasonable assumption when stresses increase nearly monothonically, but turns out to be grossly uncorrect when the stress history departs significantly from a radial path, as in cyclic cases. In these situations it is still possible to formulate elastic-plastic laws in finite terms, by subdividing the loading history into a finite number of steps. The material behavior is now holonomic within each step, but the loading-unloading alternative is allowed for at the beginning of each step.

To this purpose, suppose that at the end of the n-th step the material is in a state Σ_n , characterized by given values $\underline{\sigma}_n$ and $\underline{\lambda}_n$ of stresses and internal variables, which fulfill the inequality

$$\underline{\phi}_n = \underline{N}^t \underline{\sigma}_n - \underline{H}\,\underline{\lambda}_n - \underline{R} \leq \underline{0} \qquad (13.33)$$

Consider now a finite step $\Delta\Sigma$ from Σ_n , in which finite changes $\Delta\underline{\sigma}$ and $\Delta\underline{\lambda}$ are experienced. The final value of $\underline{\phi}$ can be expressed as

$$\underline{\phi} = \underline{N}^t(\underline{\sigma}_n + \Delta\underline{\sigma}) - \underline{H}\,(\underline{\lambda}_n + \Delta\underline{\lambda}) - \underline{R} =$$
$$= \underline{\phi}_n + \underline{N}^t\,\Delta\underline{\sigma} - \underline{H}\,\Delta\underline{\lambda} \leq \underline{0} \qquad (13.34)$$

or

$$\underline{\phi} = \underline{N}^t\,\Delta\underline{\sigma} - \underline{H}\,\Delta\underline{\lambda} - \underline{R}_n \leq \underline{0} \quad ; \quad \underline{R}_n = -\underline{\phi}_n \geq \underline{0} \qquad (13.35a,b)$$

The following inequalities

$$\Delta\underline{\lambda} \geq \underline{0} \quad ; \quad \underline{\phi} \leq \underline{0} \quad ; \quad \underline{\phi}^t\,\Delta\underline{\lambda} = 0 \qquad (13.36a\text{-}c)$$

now enforce the stepwise holonomic relationship for step $\Delta\Sigma$. As is illustrated in fig.2b, the loading-unloading alternative is allowed for at the beginning of the step, since $\phi_\alpha < 0$ now merely implies that $\Delta\lambda_\alpha = 0$, whithout affecting the value of $\lambda_{\alpha n}$ attained in previous history, which remains stored in the subsequent elastic-plastic evolution. Holonomy, however, is assumed within the step.

Stepwise holonomic laws allow for following any loading history with acceptable accuracy, provided that the step subdivision is such that a nearly monothonic increase of stresses can be reasonably foreseen within each step.

13.2. FORMULATION OF THE STRESS-ANALYSIS ELASTIC-PLASTIC PROBLEM

13.2.1. Basic relations and uniqueness of solution

Consider an elastic-plastic continuum made up of a material obeying the associated flow rule of classical plasticity. For simplicity, but without restricting the range of validity of the results, the yield function is first assumed to be regular. The continuum is in a known situation Σ , which was attained in the previous history at the price of the development of plastic strains in part of it. Its volume can be conceived as subdivided in two parts, namely: V_E, where $\phi < 0$ and no plastic strain rates can develop in the incremental process, and V_P, where $\phi = 0$ and non-vanishing $\dot{p}_{ij}$ are possible. Let rates of body forces $\dot{f}_i$ be applied on the volume V, of surface tractions $\dot{t}_i$ on the loaded portion S_F of the boundary and imposed velocities $\dot{u}_{oi}$ on the constrained part S_U of the boundary. The consequent velocity $(\dot{u}_i)$, stress rate $(\dot{\sigma}_{ij})$ and strain rate $(\dot{\varepsilon}_{ij})$ distributions are seeked.

In Cartesian coordinates and under the assumption of small strains, the equilibrium and compatibility equations read

$$\dot{\sigma}_{ij,i} + \dot{f}_j = 0 \quad \text{in } V \quad ; \quad \dot{\sigma}_{ij}\,\upsilon_i = \dot{t}_j \quad \text{on } S_F \tag{13.37a,b}$$

$$\dot{\varepsilon}_{ij} = \frac{1}{2}(\dot{u}_{i,j} + \dot{u}_{j,i}) \quad \text{in } V \quad ; \quad \dot{u}_i = \dot{u}_{oi} \quad \text{on } S_U \tag{13.38a,b}$$

with usual meaning of symbols. The above relations must be supplemented by the material constitutive law, Eqs.(13), which are rewritten as follows

$$\text{in } V_E : \quad \dot{\sigma}_{ij} = C_{ijhk}\,\dot{\varepsilon}_{hk} \tag{13.39}$$

$$\text{in } V_P : \quad \dot{\sigma}_{ij} = C_{ijhk}\,(\dot{\varepsilon}_{hk} - \dot{p}_{hk}) \tag{13.40a}$$

$$\dot{p}_{ij} = (\partial\phi/\partial\sigma_{ij})\,\dot{\lambda} \quad ; \quad \dot{\phi} = (\partial\phi/\partial\sigma_{ij})\,\dot{\sigma}_{ij} - h\,\dot{\lambda} \tag{13.40b,c}$$

$$\dot{\lambda} \geq 0 \quad ; \quad \dot{\phi} \leq 0 \quad ; \quad \dot{\phi}\,\dot{\lambda} = 0 \tag{13.40d-f}$$

A classical theorem of plasticity states that for materials obeying Drucker's postulate, the solution of the rate problem (if any) is unique in terms of stress rates. To prove the assertion, suppose that two solutions $(\dot{u}_i^1, \dot{\varepsilon}_{ij}^1, \dot{\sigma}_{ij}^1)$ and $(\dot{u}_i^2, \dot{\varepsilon}_{ij}^2, \dot{\sigma}_{ij}^2)$ exist complying with Eqs.(37-40). Then, from the principle of virtual velocities the following equations are readily obtained

$$\int_{S_U} (\dot{\sigma}_{ij}^1 - \dot{\sigma}_{ij}^2)\,\upsilon_i\,\dot{u}_{oj}\,dS = \int_V (\dot{\sigma}_{ij}^1 - \dot{\sigma}_{ij}^2)\,\dot{\varepsilon}_{ij}^1\,dV =$$

$$= \int_V (\dot{\sigma}_{ij}^1 - \dot{\sigma}_{ij}^2)\,\dot{\varepsilon}_{ij}^2\,dV$$

By subtracting term by term, one gets

$$\int_V (\dot{\sigma}^1_{ij} - \dot{\sigma}^2_{ij})(\dot{\varepsilon}^1_{ij} - \dot{\varepsilon}^2_{ij})\, dV = 0 \tag{13.41}$$

Because of Eqs.(1) and (2b) one can also write

$$\dot{\varepsilon}^{\alpha}_{ij} = A_{ijhk}\, \dot{\sigma}^{\alpha}_{hk} + \dot{p}^{\alpha}_{ij} \quad , \ \alpha = 1,2 \tag{13.42}$$

Hence, Eq.(41) becomes

$$\int_V A_{ijhk}(\dot{\sigma}^1_{ij} - \dot{\sigma}^2_{ij})(\dot{\sigma}^1_{hk} - \dot{\sigma}^2_{hk})\, dV + \int_V (\dot{\sigma}^1_{ij} - \dot{\sigma}^2_{ij})(\dot{p}^1_{ij} - \dot{p}^2_{ij})\, dV \tag{13.43}$$

Since A_{ijhk} is definite positive, the first addend is non-negative and vanishes only if $\dot{\sigma}^1_{ij} = \dot{\sigma}^2_{ij}$. The second term is non-negative because of Eq.(21a). Hence, the solution in terms of stress rates is unique.

If $h > 0$, the constitutive relation can be inverted and uniqueness is ensured in terms of strain rates as well. If $h = 0$, however (perfect plasticity), the strain rate solution is not necessarily unique. If multiple solutions exist, they must both comply with Eq.(42), in which $\dot{\sigma}_{ij}$ is unique; hence, they differ by a distribution of compatible, purely plastic strain rates only.

$$\dot{\varepsilon}^1_{ij} - \dot{\varepsilon}^2_{ij} = \Delta\dot{\varepsilon}_{ij} = \Delta\dot{p}_{ij} = \frac{1}{2}(\Delta\dot{u}_{i,j} + \Delta\dot{u}_{j,i}) \quad \text{in } V \tag{13.44a}$$

$$\Delta\dot{u}_i = 0 \quad \text{on } S_U \tag{13.44b}$$

Note also that Eq.(21b), when written for $\dot{\sigma}^1_{ij} = \dot{\sigma}^2_{ij} = \dot{\sigma}_{ij}$, becomes

$$\dot{\sigma}_{ij}\,(\dot{p}^1_{ij} - \dot{p}^2_{ij}) = \dot{\sigma}_{ij}\,\Delta\dot{\varepsilon}_{ij} \geq 0$$

Since superscripts can be interchanged, the above relation must hold as an equality. Therefore, if multiple strain rate solutions exist, they are such that

$$\dot{\sigma}_{ij}\,(\dot{\varepsilon}^1_{ij} - \dot{\varepsilon}^2_{ij}) = 0 \tag{13.44c}$$

13.2.2. Extremum theorems for Druckerian materials

If the material complies with Drucker's postulate requirements, the solution of the elastic-plastic rate problem exhibits a number of extremum properties. Two of them, of kinematic nature, are recalled.

THEOREM 1 (Greenberg [1]). The solution (if any) of the rate problem is characterized by the minimum of the functional

$$V(\dot{u}) = \frac{1}{2}\int_V \dot{\sigma}_{ij}\,\dot{\varepsilon}_{ij}\,dV - \int_V \dot{f}_i\,\dot{u}_i\,dV - \int_{S_F} \dot{t}_i\,\dot{u}_i\,dS \tag{13.45}$$

under the conditions

$$\dot{\varepsilon}_{ij} = \frac{1}{2}(\dot{u}_{i,j} + \dot{u}_{j,i}) \text{ in } V\ ;\quad \dot{u}_i = \dot{u}_{oi} \text{ on } S_U \tag{13.46a,b}$$

$$\dot{\sigma}_{ij} = \dot{\sigma}_{ij}(\dot{\varepsilon}_{ij}) \text{ through Eqs.(19)} \tag{13.47}$$

Eqs.(46) define a so-called kinematically admissible velocity field.

To prove the assertion, let superscript $(\ )^o$ indicate the values assumed by quantities at solution and $\Delta(\ )$ denote the difference between kinematic admissible and solution values. Then one can write

$$\Delta V = \frac{1}{2}\int_V(\dot{\sigma}_{ij}\dot{\varepsilon}_{ij} - \dot{\sigma}^o_{ij}\dot{\varepsilon}^o_{ij})dV - \int_V \dot{f}_i\Delta\dot{u}_i dV - \int_{S_F}\dot{t}_i\Delta\dot{u}_i dS \tag{13.48}$$

Account taken of the fact that $\Delta\dot{u}_i = 0$ on S_U, the principle of virtual velocities provides the following identity

$$\int_V \dot{f}_i\,\Delta\dot{u}_i\,dV + \int_{S_F}\dot{t}_i\,\Delta\dot{u}_i\,dS = \int_V \dot{\sigma}^o_{ij}\,\Delta\dot{\varepsilon}_{ij}\,dV \tag{13.49}$$

which, when introduced in Eq.(48), yields

$$\Delta V = \frac{1}{2}\int_V(\dot{\sigma}_{ij}\,\dot{\varepsilon}_{ij} + \dot{\sigma}^o_{ij}\,\dot{\varepsilon}^o_{ij} - 2\,\dot{\sigma}^o_{ij}\,\dot{\varepsilon}_{ij})\,dV \tag{13.50}$$

Eq.(42) is now used to eliminate both $\dot{\varepsilon}_{ij}$ and $\dot{\varepsilon}^o_{ij}$ from Eq.(50); note that $\dot{\sigma}^o_{ij}\,\dot{e}_{ij} = \dot{\sigma}_{ij}\,\dot{e}^o_{ij}$ because of the symmetry properties of tensor A_{ijhk}. Then one obtains

$$\Delta V = \frac{1}{2}\int_V A_{ijhk}\,\Delta\dot{\sigma}_{ij}\,\Delta\dot{\sigma}_{hk}\,dV + -$$

$$+\frac{1}{2}\int_V (\dot{\sigma}_{ij}\,\dot{p}_{ij} + \dot{\sigma}^o_{ij}\,\dot{p}^o_{ij} - 2\,\dot{\sigma}^o_{ij}\,\dot{p}_{ij})\,dV \qquad (13.51)$$

Due to the positive definiteness of A_{ijhk}, the first addend is positive and vanishes at solution only. the second addend is non-negative because of Eq.(21b). Hence, ΔV is non-negative for all kinematically admissible fields and vanishes at solution, which proves the assertion.

In the absence of plastic strain rates, Greenberg's theorem reduces to the classical principle of minimum potential energy and provides an extremum formulation of equilibrium, geometric compatibility and constitutive law being enforced by the constraints Eqs.(46) and (47) under which V is minimized. As an extension of this principle to rate plasticity, it deserves an outstanding conceptual interest, but its practical usefulness in numerical applications is jeopardized by the fact that V is a non-differentiable functional, due to the requirement that the <u>entire</u> constitutive relationship Eqs.(19) be present among the constraints. The following, alternative statement relaxes the above requirement.

THEOREM 2 (Capurso-Maier [3], Martin [4]). The solution (if any) of the rate problem is characterized by the minimum of the functional

$$\Psi(\dot{u}_i, \dot{\lambda}) = \frac{1}{2}\int_V C_{ijhk}\,\dot{e}_{ij}\,\dot{e}_{hk}\,dV + \frac{1}{2}\int_V h\,\dot{\lambda}^2\,dV -$$

$$-\int_V \dot{f}_i\,\dot{u}_i\,dV - \int_{S_F} \dot{t}_i\,\dot{u}_i\,dS \qquad (13.52)$$

subject to

$$\dot{e}_{ij} = \frac{1}{2}(\dot{u}_{i,j} + \dot{u}_{j,i}) - \frac{\partial\phi}{\partial\sigma_{ij}}\,\dot{\lambda} \text{ in } V\ ;\ \dot{u}_i = \dot{u}_{oi} \text{ on } S_U \qquad (13.53a,b)$$

$$\dot{\lambda} = 0 \text{ in } V_E \quad ;\quad \dot{\lambda} \geq 0 \text{ in } V_P \qquad (13.54)$$

Eqs.(53) still define a kinematically admissible field; Eq.(54) enforces only a part of the elastic-plastic constitutive law.

The proof of the theorem is again achieved by writing the expression of $\Delta\Psi$ and showing that it is non-negative and null at solution only. One has

$$\Delta\Psi = \int_V C_{ijhk}\, \dot{e}^0_{ij}\, \Delta\dot{e}_{hk}\, dV + \frac{1}{2}\int_V C_{ijhk}\, \Delta\dot{e}_{ij}\, \Delta\dot{e}_{hk}\, dV +$$

$$+ \frac{1}{2}\int_V h\,(2\,\dot{\lambda}^0\,\Delta\dot{\lambda} + \Delta\dot{\lambda}^2)\, dV - \int_V \dot{f}_i\, \Delta\dot{u}_i\, dV - \int_{S_F} \dot{t}_i\, \Delta\dot{u}_i\, dS$$

The following substitutions are legitimate in the first integral

$$\dot{\sigma}^0_{ij} = C_{ijhk}\, \dot{e}^0_{hk} \quad ; \qquad \Delta\dot{e}_{ij} = \Delta\dot{\varepsilon}_{ij} - (\partial\phi/\partial\sigma_{ij})\, \Delta\dot{\lambda}$$

then

$$\Delta\Psi = \int_V \dot{\sigma}^0_{ij}\, \Delta\dot{\varepsilon}_{ij} dV - \int_V \dot{f}_i\, \Delta\dot{u}_i\, dV - \int_{S_F} \dot{t}_i\, \Delta\dot{u}_i\, dS +$$

$$+ \frac{1}{2}\int_V (C_{ijhk}\, \Delta\dot{e}_{ij}\, \Delta\dot{e}_{hk} + h\, \Delta\dot{\lambda}^2)\, dV - \int_V (\frac{\partial\phi}{\partial\sigma_{ij}}\, \dot{\sigma}^0_{ij} - h\, \dot{\lambda}^0)\, \Delta\dot{\lambda}\, dV$$

Eq.(49) shows that the first line in the above equation vanishes identically. The last term can be simplified as follows

$$- \int_V (\frac{\partial\phi}{\partial\sigma_{ij}}\, \dot{\sigma}^0_{ij} - h\, \dot{\lambda}^0)\, \Delta\dot{\lambda}\, dV = - \int_V \dot{\phi}^0 (\dot{\lambda} - \dot{\lambda}^0)\, dV =$$

$$= - \int_{V_E} \dot{\phi}^0\, \dot{\lambda}\, dV - \int_{V_p} \dot{\phi}^0\, \dot{\lambda}\, dV + \int_V \dot{\phi}^0\, \dot{\lambda}^0\, dV = - \int_{V_p} \dot{\phi}^0\, \dot{\lambda}\, dV$$

In fact, $\dot{\lambda} = 0$ in V_E because of Eq.(54), $\dot{\phi}^0\, \dot{\lambda}^0 = 0$ because of Eq.(40f), which holds at solution. Hence

$$\Delta\Psi = \frac{1}{2}\int_V C_{ijhk} \Delta\dot{e}_{ij} \Delta\dot{e}_{hk}\, dV + \frac{1}{2}\int_V h\, \Delta\dot{\lambda}^2\, dV - \int_{V_p} \dot{\phi}^0\, \dot{\lambda}\, dV \qquad (13.55)$$

In Eq.(55) the first term is non-negative and vanishes at solution only; the second addend is also non-negative, since Drucker's postulate implies $h \geq 0$, and vanishes at solution, since two different solutions in terms of $\dot{\lambda}$ may exist only if $h = 0$; the last term is also non-negative ($\dot{\phi}^0 \leq 0$, $\dot{\lambda} \geq 0$ in V_p) and vanishes at solution ($\dot{\phi}^0 \dot{\lambda}^0 = 0$). Thus, the assertion is proved.

The above statement still represents an extension to rate plasticity of the minimum potential energy principle, in the sense that it reduces to the latter in the elastic case. However, it provides a variational formulation of both equilibrium and the remaining part of the constitutive relationship, namely $\dot{\phi} \leq 0$, $\dot{\phi} \dot{\lambda} = 0$ in V_p.

In spite of the fact that two independent variable fields (plastic multiplier rates in addition to velocities) are now present, from the point of view of numerical solutions this statememt is simpler than Greenberg's theorem, in that it requires the optimization of a quadratic functional, Eq.(52), under linear (equality and inequality) constraints. The application of this theorem spontaneously leads to discrete formulations of the rate problem as a Quadratic Programming problem [5].

For the sake of completeness, it must be mentioned that the static counterparts of the above kinematic statements are also available. That of theorem 1 is the classical Prager-Hodge statement [1], while that of theorem 2 was also proved by Capurso and Maier in ref.[3]. When plastic strain rates do not develop, they both reduce to the minimum complementary energy theorem of elasticity. Other variational statememts for the rate plasticity problem were also proposed (see, e.g., [6-8]), which are not discussed here. Recently [9] theorem 2 was generalized into a mixed-type variational statement through the addition of the mean stress as an independent variable field, to the purpose of treating appropriately the plastic incompressibility constraint which arises because of normality in conjunction with several yield conditions.

13.2.3. Piecewise linear, holonomic form of the two-field theorem

The statements discussed in the preceeding section hold essentially unaltered if a piecewise linear representation of the constitutive law is adopted. Proves follow conceptually the same path of reasoning, even if they are slightly more involved because of the necessity of accounting for non-regular yield functions.

Because of its particular interest in numerical applications, it is worth discussing briefly the form assumed by theorem 2 when the material behavior is considered as holonomic, i.e. is governed by Eqs.(26) and (30-32). By making use of the simbology introduced in Sects. 1.4, 1.5, the theorem can be stated as follows.

Under the assumption that $\underline{H}$ is symmetric and positive semidefinite, vectors $\underline{u}(\underline{x})$ and $\underline{\lambda}(\underline{x})$ solve the finite, holonomic, piecewise linear elastic-plastic problem if and only if they minimize the quadratic functional

$$\Psi(\underline{u}, \underline{\lambda}) = \frac{1}{2}\int_V \underline{e}^t\, \underline{D}\, \underline{e}\, dV + \frac{1}{2}\int_V \underline{\lambda}^t\, \underline{H}\, \underline{\lambda}\, dV + \int_V \underline{R}^t\, \underline{\lambda}\, dV - $$

$$- \int_V \underline{f}^t\, \underline{u}\, dV - \int_{S_F} \underline{t}^t\, \underline{u}\, dS \qquad (13.56)$$

subject to

$$\underline{e} = \underline{\varepsilon} - \underline{N}\, \underline{\lambda} \qquad (13.57)$$

$$\varepsilon_{ij} = \frac{1}{2}(u_{i,j} + u_{j,i}) \text{ in } V \; ; \quad u_i = u_{oi} \text{ on } S_U \qquad (13.58a)$$

$$\underline{\lambda} \geq \underline{0} \text{ in } V \qquad (13.58b)$$

It is worth underlining that the holonomy hypothesis rules out the necessity of handling separately the elastic and the plastic portion of the volume; The constraint Eq.(54) on $\dot{\lambda}$ is in fact replaced by Eq.(58b), imposed throughout the entire volume of the body. Note also that an additional, linear term has to be added to the functional.

Proof of the statement may be achieved either by following the same path of reasoning as in the previous section or (as it was originally done in [10]) by making use of Mathematical Programming concepts. The modifications required to handle stepwise holonomic relationships are straightforward. Often, Eq.(57) is used to eliminate $\underline{e}$ from the expression of Ψ . Thus, the functional assumes the form

$$\Psi(\underline{u}, \underline{\lambda}) = \frac{1}{2}\int_V (\underline{\varepsilon}^t \underline{D}\, \underline{\varepsilon} - 2\, \underline{\varepsilon}^t \underline{D}\, \underline{N}\, \underline{\lambda} + \underline{\lambda}^t(\underline{H} + \underline{N}^t \underline{D}\, \underline{N})\, \underline{\lambda}\,)\, dV + \int_V \underline{R}^t \underline{\lambda}\, dV - \int_V \underline{f}^t \underline{u}\, dV - \int_{S_F} \underline{t}^t \underline{u}\, dS \qquad (13.59)$$

to be minimized under the linear constraints Eqs.(58).

13.3. PIECEWISE LINEAR HOLONOMIC ANALYSIS OF DISCRETE SYSTEMS

13.3.1. Basic relations for a discrete system

Consider a structural system composed by the assemblage of a finite number of individual parts (bars, frame members, finite elements). The configuration of such system is governed by a finite number of free, displacement-type parameters (Lagrangean coordinates) collected in vector $\underline{U}$. For simplicity, imposed displacements are supposed to be zero and are from the beginning eliminated from the entries of $\underline{U}$. Let $\underline{F}$ denote the vector of generalized forces, defined in such a way that the work of external loads is given as

$$W_e = \underline{F}^t\, \underline{U} \qquad (13.60)$$

The behavior of each individual element is governed in terms of a finite number of generalized variables (stresses, strains, plastic multipliers, yield functions). The same symbols as in Sec.1.5 are used, but now they apply to generalized quantities. The element constitutive relation expressed in terms of generalized variables has the same formal aspect as in Section 1. When expressed in a piecewise linear, holonomic form, it reads

$$\underline{\sigma} = \underline{D}\, (\underline{\varepsilon} - \underline{N}\, \underline{\lambda}) \qquad (13.61)$$

$$\underline{\phi} = \underline{N}^t\, \underline{\sigma} - \underline{H}\, \underline{\lambda} - \underline{R} \qquad (13.62)$$

$$\underline{\lambda} \geq \underline{0} \quad ; \quad \underline{\phi} \leq \underline{0} \quad ; \quad \underline{\phi}^t\, \underline{\lambda} = 0 \qquad (13.63a\text{-}c)$$

In the above relations, quantities refer to the entire structure, i.e., vectors contain as non-intersecting subvectors the contributions of each element $n = 1,\ldots,N$ and matrices have a block-diagonal structure, each block referring to one element. For instance

$$\underline{\sigma}^t = \{ \cdots \underline{\sigma}_n^t \cdots \} \quad ; \quad \underline{\lambda}^t = \{ \cdots \underline{\lambda}_n^t \cdots \}$$

$$\underline{D} = \text{diag}\,[\underline{D}_n] \quad ; \quad \underline{N} = \text{diag}\,[\underline{N}_n]$$

For infinitesimal displacements, the geometric compatibility and equilibrium equations read

$$\underline{\varepsilon} = \underline{C}\,\underline{U} \quad ; \quad \underline{F} = \underline{C}^t\,\underline{\sigma} \qquad (13.64a,b)$$

The finite, holonomic, piecewise linear behavior of the discrete system is governed by Eqs.(61-64).

13.3.2. ($\underline{U}$ - $\underline{\lambda}$) formulation of the problem

Let Eqs.(61) and (64) be combined. One obtains

$$\underline{F} = \underline{C}^t\,\underline{D}\,\underline{C}\,\underline{U} - \underline{C}^t\,\underline{D}\,\underline{N}\,\underline{\lambda} \qquad (13.65)$$

Analogously, by introducing Eqs.(61) and (64a) into Eq.(62), the expression for generalized yield functions becomes

$$\underline{\phi} = \underline{N}^t\,\underline{D}\,\underline{C}\,\underline{U} - (\underline{H} + \underline{N}^t\,\underline{D}\,\underline{N})\,\underline{\lambda} - \underline{R} \qquad (13.66)$$

Let the following matrices be introduced

$$\underline{K}_{uu} = \underline{C}^t\,\underline{D}\,\underline{C} \quad ; \quad \underline{K}_{u\lambda} = [\underline{K}_{\lambda u}]^t = \underline{C}^t\,\underline{D}\,\underline{N} \qquad (13.67a,b)$$

$$\underline{K}_{\lambda\lambda} = \underline{H} + \underline{N}^t\,\underline{D}\,\underline{N} \qquad (13.67c)$$

Then, the finite, holonomic, piecewise linear elastic-plastic problem can be formulated as follows

$$\underline{F} = \underline{K}_{uu}\,\underline{U} - \underline{K}_{u\lambda}\,\underline{\lambda} \quad ; \quad \underline{\phi} = \underline{K}_{\lambda u}\,\underline{U} - \underline{K}_{\lambda\lambda}\,\underline{\lambda} - \underline{R} \qquad (13.68a,b)$$

$$\underline{\lambda} \geqq \underline{0} \quad ; \quad \underline{\phi} \leqq \underline{0} \quad ; \quad \underline{\phi}^t\,\underline{\lambda} = 0 \qquad (13.68c-e)$$

Note that matrix $\underline{K}_{uu}$, Eq.(67a), is the customary elastic stiffness matrix of a discrete system, symmetric and positive definite if rigid body motions are prevented. Matrix $\underline{K}_{\lambda\lambda}$, Eq.(67c), is symmetric if $\underline{H}$ is so and can be shown to be positive semidefinite even in the presence of softening, provided that is less than critical (see Eq.(18) for comparison). However, this circumstance does not ensure that the matrix

$$\underline{M} = \begin{bmatrix} \underline{K}_{uu} & -\underline{K}_{u\lambda} \\ -\underline{K}_{\lambda u} & \underline{K}_{\lambda\lambda} \end{bmatrix} \tag{13.69}$$

is positive semidefinite. This is the case only if $\underline{H}$ is so [11].

In Mathematical Programming theory, the problem governed by Eqs(68) is known as a Linear Complementarity Problem (L.C.P.), which must be solved in terms of $\underline{U}$ and $\underline{\lambda}$ in order to obtain the response of the structural system. When $\underline{H}$ is symmetric and positive semidefinite, its solution can be obtained by minimizing a convex quadratic function under linear sign constraints on some variables, i.e. it is provided by the optimal vectors of the following Quadratic Programming (Q.P.) problem

$$\min \{\tfrac{1}{2}\, \underline{U}^t\, \underline{K}_{uu}\, \underline{U} - \underline{U}^t\, \underline{K}_{u\lambda}\, \underline{\lambda} + \tfrac{1}{2}\, \underline{\lambda}^t\, \underline{K}_{\lambda\lambda}\, \underline{\lambda} - \underline{F}^t\, \underline{U} + \underline{R}^t\, \underline{\lambda}\} \tag{13.70a}$$

$$\text{subject to} \quad \underline{\lambda} \geqq \underline{0} \tag{13.70b}$$

The assertion can be proved by showing that Eqs.(68) are the Karush-Kuhn-Tucker (K.K.T.) conditions for the QP problem Eqs.(70), necessary and sufficient for optimality if the function to be minimized is convex. This is the case if matrix $\underline{M}$, Eq.(69), is positive semidefinite, a condition which is fulfilled if $\underline{H}$ is so.

The above theorem (due to Maier [12]) is the discrete counterpart of the statement Eqs.(59,58) or, equivalently. the discrete, finite, holonomic, piecewise linear counterpart of theorem 2 in Sec.2.2. As for these theorems, when $\underline{H}$ is not at least positive semidefinite, the K.K.T. conditions are necessary but no longer sufficient for optimality, i.e. solutions of Eqs.(68) may exist which do not correspond to a minimum of the QP problem Eqs.(70).

When convex, the QP problem can be formally dualized [13]. The statement so obtained can be interpreted as the static counterpart of the theorem [12].

13.3.3. $\underline{\lambda}$ formulation

Let Eq.(68a) be solved for $\underline{U}$. One obtains

$$\underline{U} = \underline{K}_{uu}^{-1}\,\underline{F} + \underline{K}_{uu}^{-1}\,\underline{K}_{u\lambda}\,\underline{\lambda} \tag{13.71}$$

If this expression is introduced in Eq.(68b), the following formulation of the problem is arrived at

$$\underline{\phi} = \underline{N}^t\,\underline{B}^t\,\underline{F} - \underline{A}\,\underline{\lambda} - \underline{R} \tag{13.72a}$$

$$\underline{\lambda} \geq \underline{0} \quad ; \quad \underline{\phi} \leq \underline{0} \quad ; \quad \underline{\phi}^t\,\underline{\lambda} = 0 \tag{13.72b-d}$$

where

$$\underline{A} = \underline{H} - \underline{N}^t\,\underline{Z}\,\underline{N} = \underline{K}_{\lambda\lambda} - \underline{K}_{\lambda u}\,\underline{K}_{uu}^{-1}\,\underline{K}_{u\lambda} \tag{13.73}$$

and

$$\underline{B} = \underline{K}_{uu}^{-1}\,\underline{C}^t\,\underline{D} \quad ; \quad \underline{Z} = \underline{D}\,\underline{C}\,\underline{K}_{uu}^{-1}\,\underline{C}^t\,\underline{D} - \underline{D} \tag{13.74a,b}$$

The mechanical meaning of the above matrices can be understood if Eq.(71) is introduced in Eq.(64a) and the result is used to evaluate the expression for $\underline{\sigma}$, Eq.(61). One obtains

$$\underline{\sigma} = \underline{B}^t\,\underline{F} + \underline{Z}\,\underline{N}\,\underline{\lambda} \tag{13.75}$$

Eq.(75) shows that the term $\underline{B}^t\,\underline{F}$ represents the elastic stress response of the discrete system to the given external loads, while matrix $\underline{Z}$ is the operator yielding the self-equilibrating stresses due to plastic strains $\underline{N}\,\underline{\lambda}$, conceived as known inelastic strains applied to an elastic body. Energy considerations show that matrix $\underline{Z}$ is symmetric and negative semidefinite [14]. Hence, matrix $\underline{A}$, Eq.(73), is symmetric and positive semidefinite if so is $\underline{H}$.

The LCP problem Eqs.(72) is a formulation of the discrete, finite, holonomic, piecewise linear elastic-plastic problem in terms of $\underline{\lambda}$ only. When $\underline{A}$ is positive semidefinite, its solution corresponds to the minimum of the following QP problem

$$\min\left\{\tfrac{1}{2}\,\underline{\lambda}^t\,\underline{A}\,\underline{\lambda} - (\underline{N}^t\,\underline{B}^t\,\underline{F} - \underline{R})\right\} \quad \text{subj.to} \quad \underline{\lambda} \geq \underline{0} \tag{13.76a,b}$$

Again, the proof of the statement can be achieved by showing that Eqs.(72) are the K.K.T. conditions of the QP problem Eqs.(76), both necessary and sufficient for optimality if $\underline{A}$ is positive semidefinite.

The above statement was proved by Maier [10] and can be regarded as the counterpart, valid in the present context, of Ceradini's theorem [6].

The QP problem Eqs.(76) can be formally dualized. The following statement (also proved in [10]) is arrived at

$$\min \{\frac{1}{2} \underline{\lambda}^t \underline{A} \underline{\lambda}\} \quad \text{subj.to} \quad \underline{N}^t \underline{B}^t \underline{F} - \underline{A} \underline{\lambda} - \underline{R} \le \underline{0} \qquad (13.77a,b)$$

As the above discussion shows, under the assumption that the material behavior is represented by a finite, holonomic, piecewise linear constitutive law with a hardening matrix $\underline{H}$ symmetric and at least positive semidefinite, the elastic-plastic analysis problem for a discrete system can be formulated as a QP problem, requiring the optimization of a convex quadratic function under linear inequality constraints. Attention was limited to two formulations, one involving both displacements $\underline{U}$ and internal variables $\underline{\lambda}$, the second $\underline{\lambda}$ only. The latter, however, requires a preliminary elastic solution, Eq.(71). From a computational point of view, this drawback is not significant, since the greatest computational burden is in both cases associated to the presence of $\underline{\lambda}$-variables.

For the sake of simplicity, only finite holonomic laws were explicitely considered. However, rate forms of the statements can easily be derived, representing the discrete counterparts of the continuous extremum theorems of Sec.2. It is worth mentioning that in this case the limitation to piecewise linear laws can be removed easily [5]. Also the stepwise holonomic laws introduced in Sec.1.5 can be incorporated at the price of a mere reinterpretation of symbols; in this form, the statement acquires a significant practical interest.

REFERENCES

1. Koiter,W.T.: General Theorems of Elastic Plastic Solids, in: Progress in Solid Mechanics, vol.I (ed. Sneddon and Hill), North-Holland, Amsterdam, 1960, 165-213.

2. Maier,G.: A Matrix Structural Theory of Piecewise Linear Elasto-Plasticity with Interacting Yield Planes, Meccanica, 5(1970) 54.

3. Capurso,M. and Maier,G.: Incremental Elasto-Plastic Analysis and Quadratic Optimization, Meccanica, 5(1970) 107.

4. Martin,J.: On the Kinematic Minimum Principle for the Rate Problem in Classical Plasticity, J.Mech.Phys.Solids, 23(1975) 123.

5. Maier,G.: Incremental Elastic-Plastic Analysis in the Presence of Large Displacements and Physical Instabilizing Effects, Int.J. of Solids and Structures, 7(1971) 345.

6. Ceradini,G.: Un principio di massimo per il calcolo dei sistemi elasto-plastici, Rend. Ist. Lombardo di Scienze e Lettere, A99 (1965), 125.

7. Nyssen,C. and Beckers,P.: A Unified Approach for Displacement, Equilibrium and Hybrid Finite Element Models in Elasto-Plasticity, Comp.Meth.Appl.Mech.Engn, 44(1984) 131.

8. Pereira,N.Z. and Fejiò,R.A.: On Kinematic Minimum Principles for Rates and Increments in Plasticity, Meccanica, 21(1986) 23.

9. Pinsky,P.M.: A Finite Element Formulation for Elasto-Plasticity Based on a Three-Field Variational Equation, Comp.Meth.Appl.Mech. Engn, 61(1987) 41.

10. Maier,G.: Teoremi di minimo in termini finiti per continui elasto-plastici con leggi costitutive linearizzate a tratti, Rend. Ist. Lombardo di Scienze e Lettere, A103(1969) 1066.

11. Corradi,L.: Stability of Discrete Elastic Plastic Structures with Associated Flow Laws, SM Archives, 3(1978) 201.

12. Maier,G.: Quadratic Programming and Theory of Elastic Plastic Structures, Meccanica, 3(1968) 265

13. Zangwill,W.S.: Nonlinear Programming, Prentice-Hall,New York 1969.

14. Maier,G.: A Quadratic Programming Approach for Certain Classes of Nonlinear Structural Problems, Meccanica, 3(1968) 121.

CHAPTER 14

FINITE ELEMENT MODELLING OF THE ELASTIC-PLASTIC PROBLEM

L. Corradi
Politecnico di Milano, Milan, Italy

ABSTRACT

A two-field finite element model for the elastic-plastic problem is presented, based on a kinematic minimum theorem due to Maier. The formulation rests on the independent discretization of the displacement and plastic multiplier fields and includes as particular cases existing finite element approaches, which can be conceived as based on implicit particular assumptions for the plastic multiplier model. The implications of such assumptions are discussed and an alternative formulation is presented, able to overcome some inconveniencies which might be experienced with traditional procedures. Mathematical Programming theory can be used to assess some features of the structural elastic-plastic behavior and can be exploited for numerical solution purposes. Some examples illustrate the essential aspects of the behavior of the model proposed and the flexibility of Mathematical Programming formulations.

14.1. INTRODUCTION

The discrete elastic-plastic problem formulated in Chapter 13 requires the definition of the constitutive relation for an element in terms of its generalized variables. Such a definition is straighforward for truss members or simple finite elements, such as constant strain plane triangles, the element behavior being an immediate consequence of the material law. For more complex finite elements, generalized stresses and strains can be defined on the basis of the natural mode concept first introduced by Argyris [1]; if a suitable definition for generalized yield functions and plastic multipliers is introduced, elastic-plastic laws for a finite element, in principle of any complexity, can be established and the discrete problem formulated.

In what follows, attention will be focused on the finite, holonomic, piecewise linear elastic-plastic problem. The finite element formulation proposed in [2] for this problem is first presented. Such formulation is based on the suitable modelling of the displacement field $\underline{u}(\underline{x})$ and of the plastic multiplier field $\underline{\lambda}(\underline{x})$.

14.2. A TWO-FIELD FINITE ELEMENT MODEL

14.2.1. Formulation

Let the structure be subdivided into N finite elements. Within each element n (n = 1,...,N) let the displacement field be expressed as function of the nodal displacements $\underline{q}_n$ of the element

$$\underline{u}(\underline{x}) = \underline{V}(\underline{x})\ \underline{q}_n \tag{14.1}$$

where $\underline{V}(\underline{x})$ is the customary shape function matrix used in displacement models, ensuring interelement continuity for $\underline{u}(\underline{x})$ once assemblage is performed. By making use of the strain-displacement relation one can write the classical equation

$$\underline{\varepsilon}(\underline{x}) = \underline{B}(\underline{x})\ \underline{q}_n \tag{14.2a}$$

expressing compatible strains within element n as functions of the nodal displacements. By making use of the principle of virtual work, the following relation is obtained

$$\underline{Q}_n = \int_{V_n} \underline{B}^t(\underline{x})\ \underline{\sigma}(\underline{x})\ dV - \underline{F}_n \tag{14.2b}$$

where $\underline{Q}_n$ is the element nodal force vector and

$$\underline{F}_n = \int_{V_n} \underline{V}^t(\underline{x})\ \underline{f}(\underline{x})\ dV + \int_{S_{Fn}} \underline{V}^t(\underline{x})\ \underline{t}(\underline{x})\ dS \tag{14.3}$$

represents the contribution to $\underline{Q}_n$ of the loads acting on the element.

Suppose now that the plastic multiplier field $\underline{\lambda}(\underline{x})$ is also interpolated as function of a finite number of parameters, collected in vector $\underline{\lambda}_n$

$$\underline{\lambda}(\underline{x}) = \underline{\Lambda}(\underline{x})\ \underline{\lambda}_n \tag{14.4a}$$

and let vector $\underline{\phi}_n$ be defined as follows

$$\underline{\phi}_n = \int_{V_n} \underline{\Lambda}^t(\underline{x})\ \underline{\phi}(\underline{x})\ dV \tag{14.4b}$$

Then, Eqs.(2-4) and the material law, Eqs.(13.26 and 30,31) permit the expression of the element behavior in the following form

$$\underline{Q}_n = \underline{k}^n_{uu}\ \underline{q}_n - \underline{k}^n_{u\lambda}\ \underline{\lambda}_n - \underline{F}_n \tag{14.5a}$$

$$\underline{\phi}_n = \underline{k}^n_{\lambda u}\ \underline{q}_n - \underline{k}^n_{\lambda\lambda}\ \underline{\lambda}_n - \underline{R}_n \tag{14.5b}$$

where

$$\underline{k}^n_{uu} = \int_{V_n} \underline{B}^t(\underline{x})\ \underline{D}\ \underline{B}(\underline{x})\ dV \tag{14.6a}$$

$$\underline{k}^n_{u\lambda} = [\underline{k}^n_{\lambda u}]^t = \int_{V_n} \underline{B}^t(\underline{x})\ \underline{D}\ \underline{N}\ \underline{\Lambda}(\underline{x})\ dV \tag{14.6b}$$

$$\underline{k}^n_{\lambda\lambda} = \int_{V_n} \underline{\Lambda}^t(\underline{x})\ (\underline{H} + \underline{N}^t\ \underline{D}\ \underline{N})\ \underline{\Lambda}(\underline{x})\ dV \tag{14.6c}$$

$$\underline{R}_n = \int_{V_n} \underline{\Lambda}^t(\underline{x})\ \underline{R}\ dV \tag{14.6d}$$

In addition to Eqs.(5), the following set of relations, formally analogous to Eqs.(13.32), is assumed to govern the element behavior

$$\underline{\lambda}_n \geq \underline{0} \quad ; \quad \underline{\phi}_n \leq \underline{0} \quad ; \quad \underline{\phi}_n^t \, \underline{\lambda}_n = 0 \qquad (14.7\text{a-c})$$

The equations governing the behavior of the entire system are derived by assembling the element contributions. The following, classical relations are used

$$\underline{q}_n = \underline{L}_n \, \underline{U} \quad ; \quad \sum_n \underline{L}_n^t \, \underline{Q}_n = \underline{0} \qquad (n = 1,\dots,N) \qquad (14.8\text{a,b})$$

In Eq.(8a) $\underline{L}_n$ represent the customary connectivity matrices, relating the element nodal displacements $\underline{q}_n$ to the nodal displacements $\underline{U}$ of the assembled system. Eq.(8b) enforces equilibrium, in the absence of external loads directly applied to the nodes.

Let now two vectors $\underline{\lambda}$ and $\underline{\phi}$ be introduced, collecting as non-intersecting subvectors the element contributions $\underline{\lambda}_n$ and $\underline{\phi}_n$, respectively

$$\underline{\lambda}^t = \{ \cdots \underline{\lambda}_n^t \cdots \} \quad ; \quad \underline{\phi}^t = \{ \cdots \underline{\phi}_n^t \cdots \} \qquad (14.9\text{a,b})$$

It is clearly possible to define matrices $\underline{M}_n$ so as to express Eqs.(9) in the form

$$\underline{\lambda}_n = \underline{M}_n \, \underline{\lambda} \quad ; \quad \underline{\phi} = \sum_n \underline{M}_n^t \, \underline{\phi}_n \qquad (n = 1,\dots,N) \qquad (14.10\text{a,b})$$

Eqs.(5),(7),(8) and (10) permit the derivation of the following set of relations governing the finite element formulation of the problem

$$\underline{F} = \underline{K}_{uu} \, \underline{U} - \underline{K}_{u\lambda} \, \underline{\lambda} \qquad (14.11\text{a})$$

$$\underline{\phi} = \underline{K}_{\lambda u} \, \underline{U} - \underline{K}_{\lambda\lambda} \, \underline{\lambda} - \underline{R} \qquad (14.11\text{b})$$

$$\underline{\lambda} \geq \underline{0} \quad ; \quad \underline{\phi} \leq \underline{0} \quad ; \quad \underline{\phi}^t \, \underline{\lambda} = 0 \qquad (14.11\text{c-e})$$

where

$$\underline{K}_{uu} = \sum_n \underline{L}_n^t \, \underline{k}_{uu}^n \, \underline{L}_n \quad ; \quad \underline{K}_{u\lambda} = [\underline{K}_{\lambda u}]^t = \sum_n \underline{L}_n^t \, \underline{k}_{u\lambda}^n \, \underline{M}_n \qquad (14.12\text{a,b})$$

$$\underline{K}_{\lambda\lambda} = \sum_n \underline{M}_n^t \underline{k}_{\lambda\lambda}^n \underline{M}_n \quad ; \underline{F} = \sum_n \underline{L}_n^t \underline{F}_n \quad ; \underline{R} = \sum_n \underline{M}_n^t \underline{R}_n \qquad (14.12c\text{-}e)$$

Except that for the definition of matrices, Eqs.(11) are identical to Eqs.(13.68), governing the discrete formulation of the problem. Hence, they are the K.K.T. conditions for the QP problem Eqs.(13.70) which, if $\underline{H}$ is positive semidefinite, is fully equivalent to them.

It is worth noting that the above QP problem could be formulated by introducing the assumed models for $\underline{u}(\underline{x})$ and $\underline{\lambda}(\underline{x})$ in the finite, piecewise linear, holonomic form, Eq.(13.59), of the functional Ψ, a procedure which would justify on variational ground the finite element proposed. In order to comply with the conditions, Eqs.(13.58), under which the functional is to be minimized, models should be able to ensure that displacements $\underline{u}(\underline{x})$ are continuous and plastic multipliers $\underline{\lambda}(\underline{x})$ are non-negative. The first of the above requirements is complied with if the element shape function matrices are correctly defined accordingly to displacement finite element formulation rules, so that interelement continuity for the displacement field is ensured by assemblage. In principle, the inequality constraint could be enforced by imposing that the entries of matrix $\underline{\Lambda}(\underline{x})$, Eq.(4a), are non-negative functions of $\underline{x}$

$$\underline{\Lambda}(\underline{x}) \geq \underline{0} \qquad (14.13)$$

so that it would result $\underline{\lambda}(\underline{x}) \geq \underline{0}$ everywhere within each element because of Eq.(7a). Clearly, such a definition is cumbersome and polynomial approximations turn out to be necessary in practice. Note that, in any case, interelement continuity for plastic multipliers is not required by the minimum principle.

14.2.2. Connections with traditional procedures

Interpolation functions for plastic multipliers are not explicitely introduced in current displacement formulations of the finite element elastic-plastic problem. The approach is usually based on Eqs.(2) written for each element n, supplemented by the material constitutive law (Eqs.(13.26 and 31,32) for the piecewise linear case) and by the assemblage relations Eqs.(8). A step-by-step analysis is usually performed and subsequent integrations of the current stiffness matrices are required. Among the numerical techniques available, the most popular one is based on Gaussian quadratures. A suitable number G of <u>Gauss points</u> $\underline{x}_g$ is defined within each element in such a way that an integral over the element volume can be evaluated (or approximated) by means of the following expression

$$\int_{V_n} f(\underline{x})\, dV = \sum_g W_g\, f(\underline{x}_g) \qquad (g = 1,\dots,G) \tag{14.14}$$

where W_g are suitably defined weighting factors [3].

Since it is difficult, if at all possible, to enforce the nonlinear constitutive law everywhere within elements, this is imposed at a finite number of points only. The most convenient locations prove to be the Gauss integration points, because the subsequent numerical integrations of the element matrices by means of Eq.(14) are facilitated by this choice.

This popular procedure, first proposed in [4], can be recovered as a particular case of the approach just presented. To prove the assertion, consider first the material constitutive law written for each Gauss point $\underline{x}_g$ in the element; by making use of Eq.(2a), Eq.(13.31) can be written as follows

$$\underline{\sigma}(\underline{x}_g) = \underline{D}\,(\underline{B}(\underline{x}_g)\,\underline{q}_n - \underline{N}\,\underline{\lambda}(\underline{x}_g)) \tag{14.15}$$

The element nodal forces $\underline{Q}_n$, Eq.(2b), can now be computed numerically by means of Eq.(14). One obtains

$$\underline{Q}_n = \sum_g W_g\, \underline{B}^t(\underline{x}_g)\,\underline{\sigma}(\underline{x}_g) - \underline{F}_n =$$

$$= \sum_g W_g\, \underline{B}^t(\underline{x}_g)\,\underline{D}\,\underline{B}(\underline{x}_g)\,\underline{q}_n - \sum_g W_g\,\underline{D}\,\underline{B}^t(\underline{x}_g)\,\underline{D}\,\underline{N}\,\underline{\lambda}(\underline{x}_g) - \underline{F}_n \tag{14.16a}$$

where Eq.(15) was used and vector $\underline{F}_n$ maintains the value given by Eq.(3), except that for possible approximations due to numerical integration. Analogously, Eq.(13.26), written for $\underline{x} = \underline{x}_g$, and Eq.(15) produce the following expression for $\underline{\phi}(\underline{x}_g)$

$$\underline{\phi}(\underline{x}_g) = \underline{N}^t\ \underline{D}\ \underline{B}(\underline{x}_g)\ \underline{q}_n - (\underline{H} + \underline{N}^t\ \underline{D}\ \underline{N})\ \underline{\lambda}(\underline{x}_g) - \underline{R} \tag{14.16b}$$

while Eqs.(13.32) at Gauss points become

$$\underline{\lambda}(\underline{x}_g) \geq \underline{0}\ ;\ \underline{\phi}(\underline{x}_g) \leq \underline{0}\ ;\ \underline{\phi}^t(\underline{x}_g)\ \underline{\lambda}(\underline{x}_g) = 0 \tag{14.16c-e}$$

When the procedure introduced in [4] is used in conjunction with the piecewise linear law, Eqs.(16) must be assumed to govern the element behavior. These relations become identical to Eqs.(5),(7) provided that matrix $\underline{\Lambda}(\underline{x})$ is defined as follows

$$\underline{\Lambda}(\underline{x}) = [\cdots\ \Lambda_g(\underline{x})\ \underline{I}_Y\ \cdots] \tag{14.17a}$$

$$\Lambda_i(\underline{x}_j) = \delta_{ij} = \begin{cases} 0 \text{ if } i \neq j \\ 1 \text{ if } i = j \end{cases} \quad (i,j = 1,\ldots,G) \tag{14.17b}$$

where $\underline{I}_Y$ is the $Y \times Y$ identity matrix and $\Lambda_g(\underline{x})$ are polynomial functions of the lowest possible order consistent with Eq.(17b). The following definitions are implied by Eqs.(17)

$$\underline{\lambda}_n^t = \{\cdots\ \underline{\lambda}^t(\underline{x}_g)\ \cdots\}\ ;\ \underline{\phi}_n^t = \{\cdots\ W_g\ \underline{\phi}^t(\underline{x}_g)\ \cdots\} \tag{14.18a,b}$$

Eq.(4a) immediatly shows that Eq.(18a) is implied by Eqs.(17). Eq.(18b) is obtained by numerically integrating Eq.(4b) by means of Eq.(14), account taken of the particular nature of matrix $\underline{\Lambda}(\underline{x})$, Eqs.(17).

When $W_g > 0$ (as in the large majority of cases), Eqs.(18) show that Eqs.(16c-e) and (7) coincide. Moreover, Eqs.(17) and (18) permit the replacement of Eqs.(16a,b) with the following relations

$$\underline{Q}_n = \left[\ \sum_g W_g\ \underline{B}^t(\underline{x}_g)\ \underline{D}\ \underline{B}(\underline{x}_g)\ \right] \underline{q}_n - \left[\ \sum_g W_g\ \underline{B}^t(\underline{x}_g)\ \underline{D}\ \underline{N}\ \underline{\Lambda}(\underline{x}_g)\right] \underline{\lambda}_n - \underline{F}_n \tag{14.19a}$$

$$\underline{\phi}_n = \left[\ \sum_g W_g\ \underline{\Lambda}^t(\underline{x}_g)\ \underline{N}^t\ \underline{D}\ \underline{B}(\underline{x}_g)\ \right] \underline{q}_n - \left[\ \sum_g W_g\ \underline{\Lambda}^t(\underline{x}_g)(\underline{H}+\underline{N}^t\underline{D}\underline{N})\ \underline{\Lambda}(\underline{x}_g)\ \right] \underline{\lambda}_n - \left[\ \sum_g W_g\ \underline{\Lambda}^t(\underline{x}_g)\ \underline{R}\ \right] \tag{14.19b}$$

which coincide with Eqs.(5), since the quantities in brackets are immediatly recognized to be those defined by Eqs.(6) when numerically integrated according to Eq.(14). It is worth noting that the order of the polynomial functions $\Lambda_g(\underline{x})$, Eq.(17b), is dictated by the number of Gauss points used. Because of this fact, the numerical integration of all matrices appearing in Eqs.(19) turns out to be exact for most elements when G is chosen so that the elastic stiffness matrix $\underline{k}^n_{uu}$, Eq.(6a), is integrated exactly by Eq.(14) [2].

At this stage, the assemblage relations Eqs.(8),(10) permit the establishment of Eqs.(11) and the formulation of the QP problem. Thus, it can be concluded that the widely used procedure described in ref.[4] can be interpreted as a particular finite element formulation of Maier's minimum principle, Eq.(13.59,58), based on the implicit assumption Eqs.(17) for modelling plastic multipliers. Note that, with the exception of the almost trivial case G = 1 , functions $\Lambda_g(\underline{x})$ are negative over part of the element volume, so that the sign constraint $\underline{\lambda}(\underline{x}) \geq \underline{0}$ is no longer enforced locally by Eq.(16c). Hence, one of the essential conditions under which the functional is to be minimized is removed and an additional approximation is introduced which, strictly speaking, relaxes the kinematic nature of the model.

14.2.3. Implicit assumptions for the element constitutive law

As it was already noted, Eqs.(11) are a L.C.P. identical to Eqs.(13.68), the only difference being the definition of matrices, now given by Eqs.(6),(12), replacing previous expressions Eqs.(13.67). In fact, the finite element model was established on the basis of constitutive laws for the <u>material</u> and no explicit use was made of relations expressing the behavior of the <u>finite element</u>, i.e., Eqs.(13.61-63) were not defined. Even if such relations do not appear to be required in order to establish the finite element formulation, their expressions permit gaining a deeper understanding on the behavior of the overall model and, for this reason, they are worth studying.

Let Eqs.(13.61,62) be rewritten with reference to the n-th finite element

$$\underline{\sigma}_n = \underline{D}_n(\underline{\varepsilon}_n - \underline{N}_n \underline{\lambda}_n) \quad ; \quad \underline{\phi}_n = \underline{N}_n^t \underline{\sigma}_n - \underline{H}_n \underline{\lambda}_n - \underline{R}_n \qquad (14.20a,b)$$

where $\underline{\varepsilon}_n$ and $\underline{\sigma}_n$ denote generalized strains and stresses for the element, $\underline{\lambda}_n$ and $\underline{\phi}_n$ generalized plastic multipliers and yield functions. Their definitions must comply with the following equivalence conditions [5,2]

$$\underline{\sigma}_n^t \, \underline{\varepsilon}_n = \int_{V_n} \underline{\sigma}^t(\underline{x}) \, \underline{\varepsilon}(\underline{x}) \, dV \; ; \; \underline{\phi}_n^t \, \underline{\lambda}_n = \int_{V_n} \underline{\phi}^t(\underline{x}) \, \underline{\lambda}(\underline{x}) \, dV \qquad (14.21a,b)$$

Note that vectors $\underline{q}_n$ and $\underline{Q}_n$, Eqs.(2), cannot be assumed as generalized strains and stresses, since rigid body motions and self-equilibrating nodal forces are included in them. From the element viewpoint, these vectors represent displacements and external forces, respectively, linearly related to generalized variables by means of the following equations, which play the role of Eqs.(13.64)

$$\underline{\varepsilon}_n = \underline{C}_n \, \underline{q}_n \quad ; \quad \underline{Q}_n = \underline{C}_n^t \, \underline{\sigma}_n - \underline{F}_n \qquad (14.22.a,b)$$

On this basis, Eqs.(5) can be derived, with the following expressions for matrices, alternative to Eqs.(6)

$$\underline{k}_{uu}^n = \underline{C}_n^t \, \underline{D}_n \, \underline{C}_n \quad ; \quad \underline{k}_{u\lambda}^n = [\underline{k}_{\lambda u}^n]^t = \underline{C}_n^t \, \underline{D}_n \, \underline{C}_n \qquad (14.23a,b)$$

$$\underline{k}_{\lambda\lambda}^n = \underline{H}_n + \underline{N}_n^t \, \underline{D}_n \, \underline{N}_n \qquad (14.23c)$$

Once the displacement model Eq.(1) is established, generalized strains can be defined by following the <u>natural</u> approach introduced by Argyris [1], who pointed out how the element nodal displacements $\underline{q}_n$ can be expressed as a linear combination of R rigid body motions $\underline{\rho}_n$ and of E natural or straining modes $\underline{\varepsilon}_n$ (generalized strains) through a non-singular matrix $\underline{\Gamma}$

$$\underline{q}_n = [\, \underline{\Gamma}_\rho \; \underline{\Gamma}_n] \begin{Bmatrix} \underline{\rho}_n \\ \underline{\varepsilon}_n \end{Bmatrix} \; ; \; \begin{Bmatrix} \underline{\rho}_n \\ \underline{\varepsilon}_n \end{Bmatrix} = \begin{bmatrix} \underline{C}_\rho \\ \underline{C}_n \end{bmatrix} \underline{q}_n \qquad (14.24a,b)$$

where $\underline{C} = \underline{\Gamma}^{-1}$. If Eq.(24a) is used to replace $\underline{q}_n$ in Eq.(2a), one gets

$$\underline{\varepsilon}(\underline{x}) = \underline{B}(\underline{x}) \, \underline{\Gamma}_\rho \, \underline{\rho}_n + \underline{B}(\underline{x}) \, \underline{\Gamma}_n \, \underline{\varepsilon}_n$$

The very definition of rigid body motion implies

$$\underline{B}(\underline{x})\ \underline{\Gamma}_\rho = 0$$

Hence, one can write

$$\underline{\varepsilon}(\underline{x}) = \underline{b}(\underline{x})\ \underline{\varepsilon}_n \tag{14.25}$$

with

$$\underline{b}(\underline{x}) = \underline{B}(\underline{x})\ \underline{\Gamma}_n \quad ; \quad \underline{B}(\underline{x}) = \underline{b}(\underline{x})\ \underline{C}_n \tag{14.26a,b}$$

Eq.(25) expresses local strains within the element as functions of the relevant generalized quantities. From Eq.(21a), the following definition for generalized stresses immediatly flows

$$\underline{\sigma}_n = \int_{V_n} \underline{b}^t(\underline{x})\ \underline{\sigma}(\underline{x})\ dV \tag{14.27}$$

For any assumptions for matrix $\underline{\Lambda}(\underline{x})$, Eqs.(4) introduce valid definitions of generalized yield functions and plastic multipliers for the element, consistent with each other in the sense of Eq.(21b).

It can be readily verified that Eqs.(23) become identical to Eqs.(6a-c) provided that the matrices governing the element behavior are defined as follows

$$\underline{D}_n = \int_{V_n} \underline{b}^t\ \underline{D}\ \underline{b}\ dV \quad ; \quad \underline{N}_n = \underline{D}_n^{-1} \int_{V_n} \underline{b}^t\ \underline{D}\ \underline{N}\ \underline{\Lambda}\ dV \tag{14.28a,b}$$

$$\underline{H}_n = \int_{V_n} \underline{\Lambda}^t\ \underline{H}\ \underline{\Lambda}\ dV + \int_{V_n} (\underline{b}^t\underline{N}_n - \underline{N}\ \underline{\Lambda})^t\ \underline{D}(\underline{b}^t\underline{N}_n - \underline{N}\ \underline{\Lambda})\ dV \tag{14.28c}$$

Eq.(28a) defines the so-called natural elastic stiffness matrix for the element [1]; matrices $\underline{N}_n$ and $\underline{H}_n$, together with vector $\underline{R}_n$ still expressed by Eq.(6d), define the piecewise linear yield function for the finite element, Eq.(20b). Of particular interest is the expression Eq.(28c) of the hardening matrix $\underline{H}_n$, which is composed of two addends. The first one accounts for the contribution of material hardening (if any), while the second is a positive semidefinite matrix introducing an additional hardening term due to stress redistribution within the element. Such a contribution is better evidentiated if the expression for local stresses is written as function of generilized variables. After some algebra, one obtains

$$\underline{\sigma}(\underline{x}) = \underline{D}\ \underline{b}(\underline{x})\ \underline{D}_n^{-1}\ \underline{\sigma}_n + \underline{D}\ (\underline{b}(\underline{x})\ \underline{N}_n - \underline{N}\ \underline{\Lambda}(\underline{x}))\ \underline{\lambda}_n \tag{14.29}$$

The first term in the r.h.side of Eq.(29) gives a local stress field governed by the generalized stress vector for the element; the second addend, corresponding to $\underline{\sigma}_n = \underline{0}$ (and, hence, to zero nodal forces), expresses a self-equilibrating or redundant stress distribution due to non-vanishing $\underline{\lambda}_n$.

The presence of this second term is not unexpected, since stress redistribution is naturally associated to the development and spreading of local plastic strains. However, the above stress modes, corresponding to vanishing nodal forces, are internal to each element and, as such, hardly can give a significant contribution to the overall stress representation. Actually, these modes are sometimes responsible for inconveniencies and errors in computed results.

A very simple example illustrates the effects of redundant modes. A cantilever beam with sandwhich cross section, elastic bending stiffness EI and limit moment M_o , is subject to a vertical load F at its free end. A single finite element is used with a cubic displacement field and, hence, a linear total curvature (strain) distribution, able to produce the exact solution in the elastic range.

Plastic multipliers are interpolated as functions of their values at J points. For J = 2 and 3 , the entries of $\underline{\Lambda}(\underline{x})$ are shown in figs.1a and 1b, for the assumptions Eqs.(13) (non-negative $\underline{\lambda}(\underline{x})$) and (17)(Gauss point interpolation, J = G), respectively. In both cases, for J = 3 redundant modes are introduced in the beam, as it appears from the force-displacement diagrams of figs.2a,b ; when yielding first occurs (for $F = 1.2(M_o/\ell)$ and $F = 1.127(M_o/\ell)$, respectively) the beam does not collapse but, because of the redundancies introduced, can carry additional load up to the values of $F = 1.5(M_o/\ell)$ and $F = 1.398(M_o/\ell)$ when a second yield function is activated. At this stage only, a compatible (linear) plastic strain rate distribution becomes possible and collapse occurs.

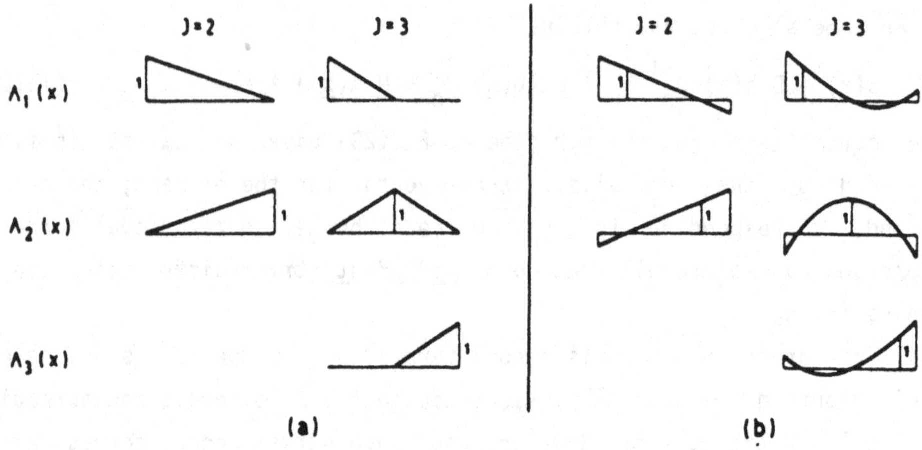

Fig.14.1. Plastic multiplier model (from [2])
(a) Non-negative $\underline{\Lambda}$, eq.(13); (b) Gauss point interpolation, eqs.(17)

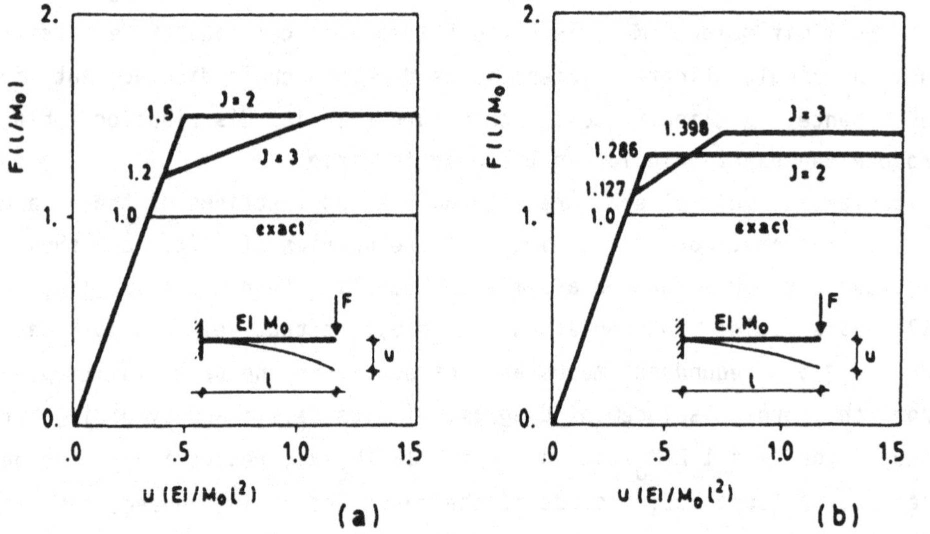

Fig.14.2. Cantilever beam example.
Interpolation functions as in figs.1a and 1b, respectively (from [2])

In the second case, redundant modes can be eliminated by using higher order displacement models. However, the phenomenon cannot be eliminated in this way when the interpolation functions of fig.1a are used, even if the flexibility of the beam in the plastic range is significantly increased. For instance, the same collapse load $F = 1.5(M_o/\ell)$ is predicted with a quadratic strain distribution, even if displacement at collapse increases by a factor of about 10 .

The example also shows that results improve when the interpolation functions of fig.1b are used; Gauss point interpolation provides not only computational advantages (due to the fact that matrix $\underline{\Lambda}(\underline{x})$, Eqs.(17), needs not to be explicitly introduced) but also better results. this feature is not unexpected and can be interpreted as an added flexibility due to the partial removal of the sign restriction $\underline{\lambda}(\underline{x}) \geq \underline{0}$. As is well known, similar advantages are often experienced also when the compatibility constraint is relaxed through the use of reduced integration techniques.

14.2.4. Construction of statically determinate elements

An alternative approach is now presented, able to eliminate redundant stress modes from the element behavior [2]. To this purpose, note first that such modes would not be present if it were possible to fulfill locally the following equality

$$(\underline{b}(\underline{x})\ \underline{N}_n - \underline{N}\ \underline{\Lambda}(\underline{x}))\ \underline{\lambda}_n = \underline{0} \tag{14.30}$$

However, the above condition happens to be fulfilled without approximations only in a limited number of simple elements [6]. In fact, Eq.(30) requires that the distribution of plastic strains

$$\underline{p}(\underline{x}) = \underline{N}\ \underline{\Lambda}(\underline{x})\ \underline{\lambda}_n \tag{14.31}$$

be <u>compatible</u> within the element. For plane or solid elements, this condition can be fulfilled if the total strain distribution is <u>at most linear</u>, which is the case only for linear or quadratic triangles or thetrahedra. Only in this case, in fact, Eqs.(17) provide compatible plastic strain fields.

A general procedure able to eliminate redundant stress modes at the price of an approximation was proposed in [2] and is briefly summarized here. It is based on the replacement of the expression Eq.(31) for local plastic strains with the following equation, enforcing compatibility with the assumed displacement model

$$\underline{p}(\underline{x}) = \underline{b}(\underline{x})\ \underline{N}_n\ \underline{\lambda}_n \tag{14.32}$$

and on the expression of local stresses as functions of the relevant generalized variables by means of a relation conceptually similar to Eq.(25), which reads

$$\underline{\sigma}(\underline{x}) = \underline{s}(\underline{x})\ \underline{\sigma}_n \tag{14.33}$$

where $\underline{s}(\underline{x})$ is a matrix consistent with the definition, Eq.(27), of generalized stresses, i.e., such as to comply with the following requirement

$$\int_{V_n} \underline{b}^t(\underline{x})\ \underline{s}(\underline{x})\ dV = \underline{I}_E \tag{14.34}$$

where $\underline{I}_E$ is the E×E identity matrix. Note that Eq.(34) does not uniquely define an expression for $\underline{s}(\underline{x})$. For instance, given any symmetric and positive definite matrix $\underline{\Omega}$, one can readily verify that the following definition

$$\underline{s}(\underline{x}) = \underline{\Omega}\ \underline{b}(\underline{x})\ \underline{\Omega}_n^{-1} \quad ; \quad \underline{\Omega}_n = \int_{V_n} \underline{b}^t(\underline{x})\ \underline{\Omega}\ \underline{b}(\underline{x})\ dV \tag{14.35a,b}$$

complies with the consistency requirement Eq.(34a). The choice $\underline{\Omega} = \underline{D}$ produces the expression

$$\underline{s}(\underline{x}) = \underline{D}\ \underline{b}(\underline{x})\ \underline{D}_n^{-1} \tag{14.36}$$

which yields, in the elastic range, the same stress distribution as traditional procedures. However, the use of Eq.(36) is not mandatory. In ref. [2] $\underline{\Omega} = \underline{I}$ was proposed, even if advantages with respect to Eq.(36) are questionable. As it was discussed in [6], legitimate expressions for $\underline{s}(\underline{x})$ can be derived for some elements, which cannot be cast in the form Eqs.(35); such expressions turn out to be convenient in that they are able to overcome some inconveniencies in stress computation, such as the presence of parasitic shears.

On the basis of Eqs.(32) and (33), the following expressions for matrices $\underline{N}_n$ and $\underline{H}_n$, alternative to Eqs.(28b,c), can be derived

$$\underline{N}_n = \int_{V_n} \underline{s}^t(\underline{x})\ \underline{N}\ \underline{\Lambda}(\underline{x})\ dV \quad ; \quad \underline{H}_n = \int_{V_n} \underline{\Lambda}(\underline{x})\ \underline{H}\ \underline{\Lambda}(\underline{x})\ dV \qquad (14.37a,b)$$

which, together with $\underline{D}_n$, Eq.(28a), and $\underline{R}_n$, Eq.(6b), permit the expression of the element behavior in the form of Eqs.(20) and, hence, the formulation of the L.C.P. or QP problem.

The expression Eq.(37a) for $\underline{N}_n$ becomes identical to Eq.(28b) if the particular asuumption Eq.(36) is made for $\underline{s}(\underline{x})$. On the other hand, matrix $\underline{H}_n$, Eq.(37b), coincides with the first addend only in the r.h.side of Eq.(28c), showing that fictitious hardening due to internal stress redistribution has been effectively eliminated. This fact is demonstrated by the solution of the beam example; while for $J = 2$ the same results as before are obtained, for $J = 3$ the present approach correctly predicts collapse when yielding first occurs, either for $F = 1.2(M_o/\ell)$ or $F = 1.127(M_o/\ell)$, depending on the assumed $\underline{\Lambda}(\underline{x})$.

The model produced by this procedure is built up by assembling <u>statically determinate</u> elements, in which the stress distribution is governed in terms of generalized stresses only through Eq.(33). The resulting picture is similar to that of a redundant truss, in which no self-equilibrating stress states are assumed to be present in the isolated bar and stress redistribution is allowed for only because of the finite number of redundancies introduced when assembling. The internal redundant stress modes implicitely introduced in the element behavior by traditional procedures play essentially the same role as possible self-equilibrating stress distributions in an isolated truss bar; they hardly contribute to the overall stress representation and often they are cause of inaccuracies at the local level.

The above result is achieved by replacing Eq.(31) with Eq.(32), which makes the plastic strain distribution within the element fully compatible with the assumed displacement model. The approximation so introduced amounts at relaxing the outward normality rule for plastic strains, which Eq.(31) enforces strictly and Eq.(32) only in a <u>weighted</u>

average sense over each element. In fact, by making use of Eqs.(34) and (37a) one can easily verify that the following identity holds

$$\int_{V_n} \underline{s}^t(\underline{x})(\underline{b}(\underline{x})\ \underline{N}_n - \underline{N}\ \underline{\Lambda}(\underline{x}))\ dV = \underline{0} \tag{14.38}$$

in each element, showing that a weighted integral of both expressions for $\underline{p}(\underline{x})$ have the same values.

Since the fulfillment of the local normality rule is an essential condition under which the functional Ψ, Eq.(13.59), is to be minimized, it may be hinted that the kinematic nature of the finite element formulation is destroyed by this procedure, which should more appropriately classified as a mixed method. A variational interpretation of the procedure supports in a sense this viewpoint [7]. Consider the functional

$$\Psi^*(\underline{u},\underline{\lambda},\underline{p},\underline{\sigma}) = \frac{1}{2}\int_V ((\underline{\varepsilon} - \underline{p})^t\ \underline{D}\ (\underline{\varepsilon} - \underline{p}) + \underline{\lambda}^t\ \underline{H}\ \underline{\lambda})\ dV + \int_V \underline{R}^t\ \underline{\lambda}\ dV$$

$$- \int_V \underline{f}^t\ \underline{u}\ dV - \int_{S_F} \underline{t}^t\ \underline{u}\ dS + \int_V \underline{\sigma}^t(\underline{p} - \underline{N}\ \underline{\lambda})\ dV \tag{14.39a}$$

subject to the constraints Eqs.(13.58). Then one could show that the solution of the finite, holonomic, piecewise linear elastic-plastic problem is characterized by the condition

$$\delta\Psi^* \geq 0 \tag{14.39b}$$

for all admissible variations of the independent fields. The proof is omitted for brevity; however, one can easily recognize in Eqs.(39) a variational inequality obtained from the original minimum principle by imposing the normality condition $\underline{p}(\underline{x}) = \underline{N}\ \underline{\lambda}(\underline{x})$ by means of Lagrange multipliers, which acquire the mechanical meaning of stresses.

The above inequality does not exhibit extremum properties, since the last term in Eq.(39a) destroies the convexity of the functional. Note, however, that when Eqs.(4a),(32) and (33) are used to model the distributions of $\underline{\lambda}(\underline{x})$, $\underline{p}(\underline{x})$ and $\underline{\sigma}(\underline{x})$, respectively, this term vanishes identically over each element by virtue of Eq.(38). Hence, the search for the solution still amounts to the minimization of the discrete form of the functional. Obviously, this circumstance does not imply the kinematic nature of the model, since the minimum so obtained is merely a feasible point for Ψ^*, the value of which is not necessarily higher than the optimal value of Ψ, Eq.(13.59).

However, as it was discussed in [7], it is possible to assess that the relaxation of the normality condition does not jeopardize in fundamental ways the essentially kinematic behavior of the model. The interpretation of the procedure as a mixed formulation is certainly legitimate, but does not appear to provide particular advantages for the interpretation of results. To this purpose, it must be noted that the use of Gauss point interpolation for $\underline{\lambda}(\underline{x})$, implicitely employed in traditional displacement approaches, is in itself an approximation relaxing an essential inequality constraint and, hence, strictly speaking the kinematic nature of the model.

Also with reference to the present formulation, the use of Eq.(17) provides significant operative advantages, in that matrices $\underline{N}_n$ and $\underline{H}_n$ need not to be defined explicitely. It could be shown easily [2] that Eqs.(20), (28a) and (37), governing the element behavior, reduce to the material law enforced at the Gauss points $\underline{x}_g$. Only, in contrast to traditional procedures, stresses at Gauss points are no longer treated as independent variables (governing possible redundant modes), but are now expressed as functions of the element generalized stresses $\underline{\sigma}_n$ by means of Eq.(33), which, when written at Gauss points, reads

$$\underline{\sigma}(\underline{x}_g) = \underline{s}(\underline{x}_g)\ \underline{\sigma}_n \qquad (14.40)$$

The use of Eq.(40) is the only difference with respect to conventional procedures and can easily be implemented into existing codes.

For simplicity, only the finite, holonomic, piecewise linear elastic-plastic problem was dealt with in this chapter. The extensions required to cover step-wise holonomic behavior are straightforward. Also the rate problem can be treated essentially in the same manner [8] and the formulation does not need piecewise linear constitutive laws. However, some care is required when dealing with the elastic-plastic boundary and additional approximations are introduced by numerical

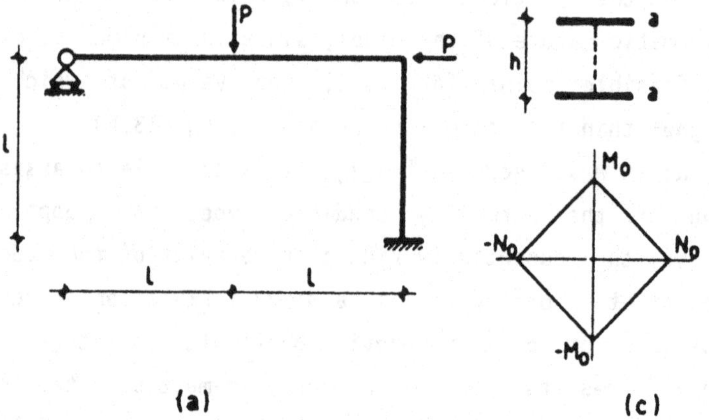

Fig.14.3. Frame example (from [2]). $\ell/h = 6$

TABLE 1 . Results for the frame problem

Number G of Gauss points	Elastic limit load $P_E(\ell/M_o)$	Collapse load $P_L(\ell/M_o)$ Conventional	Collapse load $P_L(\ell/M_o)$ Consistent
2	1.486	1.841	1.803
3	1.295	2.010	1.633
4	1.227	2.015	1.569
5	1.194	2.017	1.539
Exact values	1.131	1.480	1.480

integration, due to the fact that $\underline{N}$ and $\underline{H}$ are no longer constant if the constitutive law is not piecewise linear. Such approximations, however, do not alter the essential features of the formulation.

14.3. APPLICATIONS AND EXAMPLES

14.3.1. Differences between traditional and proposed procedures

The simply redundant frame of fig.3a (from [2]) will be considered firstly. All members have sandwhich cross section, as depicted in fig.3b. The structure is divided into three traditional beam elements, each with 6 degrees of freedom, including 3 rigid body motions. The $E = 3$ straining modes control the constant axial strain and the linearly varying curvature distributions. The corresponding stress components are the axial force $N(x)$ and the bending moment $M(x)$ along the element, whose distributions are also governed by generalized stress components. If Eqs.(33,36) are used, $N(x)$ and $M(x)$ have constant and linear distributions, respectively. As is well known, for perfectly plastic materials the elastic range in the N-M plane is given by the piecewise linear domain illustated in fig.3c.

Let the traditional procedure be used. When $G \geq 2$ Gauss points are selected within each element, $2G - 3$ redundant stress modes are introduced in the element behavior. The first two columns of Table 1 summarize the results obtained in this way. Increasing the number of Gauss points improves the estimate of the elastic limit load, due to the fact that plasticity is enforced closer to the element ends, where moment peaks are located. However, the computed collapse load turns out to increase with G , due to the artificially introduced possibility of stress redistribution.

The use of the alternative approach introduced in Sec.2 (called "consistent" in ref.[2], since it is based on the consistency condition Eq.(34)) removes this phenomenon, as shown by the last column in Table 1. The computed collapse load decreases with increasing G and eventually approaches the correct value of $P_L = 1.48(M_0/\ell)$. Clearly, for the same G both approaches provide the same elastic limit load, since no differences exist in the elastic range.

For $G = 5$, fig.4 shows the bending moment values at Gauss points when collapse is predicted by either approach. The consistent procedure forces moments to vary linearly within each element; on the contrary, traditional procedures allow for redundant stresses to be superimposed to the linear distribution, so as to push the stress state against the yield limit in a large number of points. The same effect is experienced for axial forces, which are no longer constant along the elements. This fictitious capability for stress redistribution is responsible for the increase of the predicted collapse load.

Note that the procedure enforces constant axial force and linear bending moment distribution, which are equilibrated with applied external loads. However, it is a mere coincidence, due to the particularly simple nature of the problem, that the stress distributions consistent through Eqs.(33,36) with the assumed displacement model comply with the equilibrium equations for the frame members. The consistent formulation mantains its kinematic nature and, in fact, it could be shown [7] that the same results would be obtained by using the traditional displacement procedure in conjunction with a displacement model rich enough to permit the fulfillment of Eq.(30), which turns out to be possible in this case.

The errors connected with internal redundant modes are dramatically evidentiated in the above example, because an isolated beam element must in fact be statically determinate in terms of its stress resultants N and M . In other circumstances, errors are not equally large or are not present in significant manner. Nevertheless, the consistent approach seems to provide advantages in any case, at least from the point of view of reducing the computational burden.

The assertion is demonstrated with reference to a few simple examples concerning square, sandwhich, elastic-perfectly plastic Kirchhoff plates with different boundary conditions, subjected to uniform load. Mises plasticity condition was considered. Melosh 12 d.o.f. rectangular element was employed; such element does not ensure C_1 continuity strictly and, hence, an approximation is introduced. However, it performs reasonably well as long the shape is not distorted

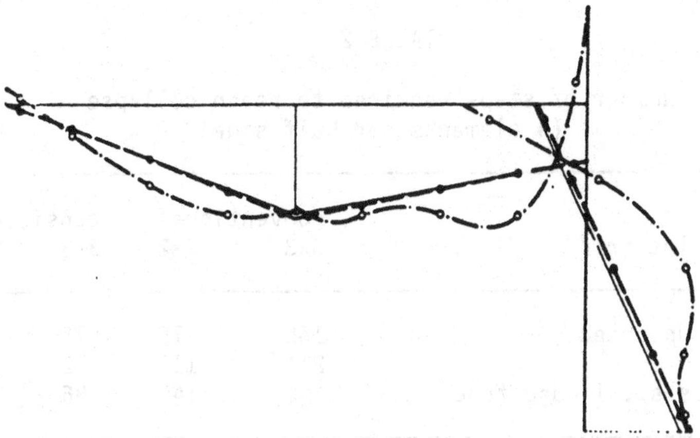

Fig.14.4. Bending moments at collapse for the frame of fig.3
Dashed line: consistent. Dash-dot line: conventional

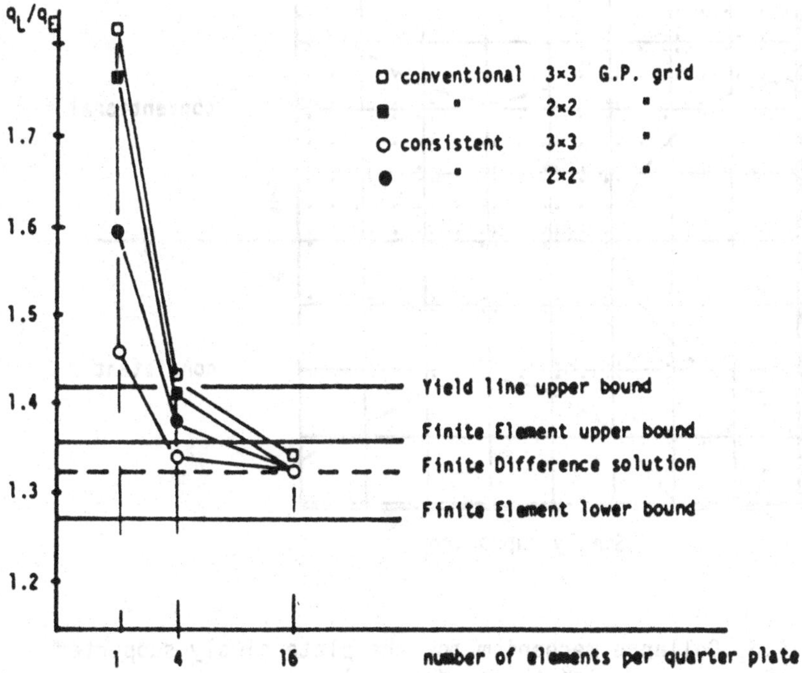

Fig.14.5. Computed collapse load for simply supported plate
Y.L. bound from [10], F.E. bounds from [11], F.D.solution from [12]
q_L = computed collapse load; q_E = closed form elastic limit

TABLE 2

Number of steps required to reach collapse
(4 elements per half side)

Gauss point grid	conventional 3×3	conventional 2×2	consistent 3×3	consistent 2×2
Simply supported	165	78	72	62
Clamped	211	113	72	60
3 edges s.s., 1 edge free	190	149	86	74

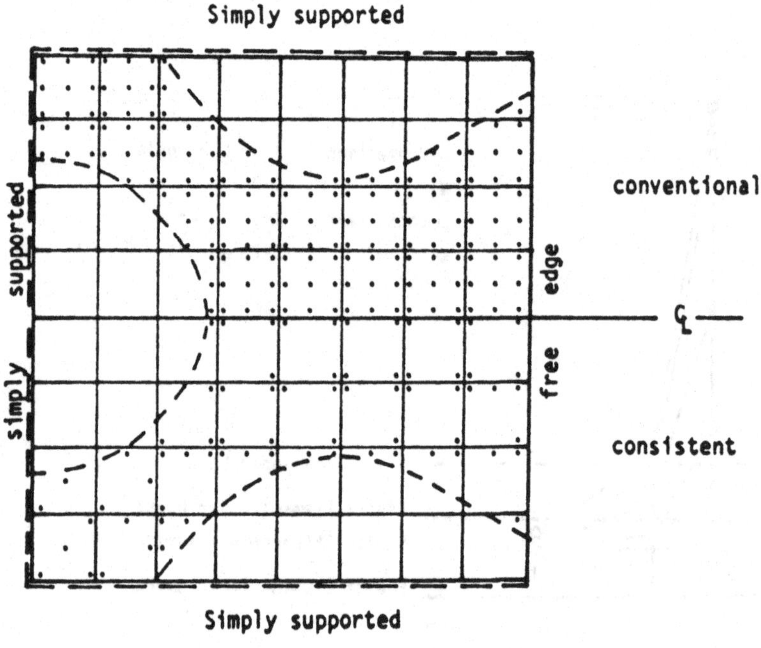

Fig.14.6. Collapse mechanism for the plate simply supported on three edges only

and it is felt to be adequate for comparison purposes. Both the traditional and the consistent procedure were used, each in connection with both a 2×2 and a 3×3 Gauss point grid in each element. Since the element has 9 generalized stress components, conventional procedures introduce up to 3 and 18 redundant stress modes, respectively. Analyses up to collapse were performed by using the code STRUPL2 [9], modified so as to enforce the consistency condition.

Fig.5 shows the computed collapse load factors for a simply supported plate and different meshes, compared with some classical results available in the literature. The following points are worth a comment.

The collapse load factors obtained with the consistent formulation are sistematically lower than those computed in conventional way.

When the procedure is used in connection with a 3×3 Gauss point grid, mesh convergence is remarkably faster. The same improvement is not experienced with traditional procedures; actually, for the example considered, as the number of Gauss points is increased stiffer results are obtained.

The proposed, consistent formulation also drastically decreases the computational burden, as it appears from Table 2 , where the number of steps required to reach collapse is indicated. The time required by each step solution is practically independent of the procedure used if the same Gauss point grid is adopted. Consistent figures not only look sistematically lower than conventional ones, but they are also nearly independent of the number of Gauss points. Thus, the improvement of results associated with a more refined grid does not entail a parallel increase in the computational effort.

The above remarks focus the central feature of the formulation. With conventional procedures an increase in the number of Gauss points produces a corresponding increase in the number of stress modes in the element; the possibility in stress redistribution is unnecessarily augmented and the number of steps required to reach a given load level is consequentially increased, with no significant improvement (if any) in computed solutions. On the other hand, with the present formulation the use of a larger number of Gauss points does not alter the internal stress distribution Eq.(33) and gives rise to a model which is actually more refined. The increase in computational efficiency that this model exhibits is due to the fact that only the essential stress modes are retained.

With consistent models, minor difficulties may arise when the extent of plastic zones is to be assessed. The collapse mechanism (i.e. the set of Gauss points which are plastic when collapse is predicted) obtained with either formulation for the plate simply supported on three edges only, are contrasted in fig.6. The rather awkward, "leopard skin" aspect of the consistent picture is explained by the fact that Eq.(33) does not allow local stresses to get in contact with the yield surface everywhere within the element (most of the additional steps required by conventional formulations are spent in performing this operation). When using the consistent approach, one ought to think in terms of "plastic elements" rather than of "plastic points"; on this ground the picture, even if rougher, appears identical to the conventional one. It is worth emphasizing that the accuracy in computed displacements and stresses is not jeopardized by the use of the consistent approach.

Even if the computational experience gained so far is limited, it seems possible to conclude that the use of this procedure provides significant advantages for the elastic-plastic analysis problem. The resulting formulation exhibits faster convergence, permits the refinement of the model by increasing the number of Gauss points without affecting the accuracy of results and reduces the computational burden associated to redundant stress modes.

14.3.2. A refined beam finite element model

As also previous examples have shown, independent assumptions for displacements and plastic multipliers (implicitly present in conventional approaches, even if usually not explicitly introduced) produce plastic strain distributions which are not compatible with total strains, giving rise to redundant, internal stress modes. The frame example of fig.3 illustrates clearly the unpleasant effects of this fictitiously introduced possibility of stress redistribution, which the present approach is able to remove.

However, stress redistribution over the cross section is an essential feature of elastic-plastic beam behavior, which a good finite element model must be able to account for. A beam element fulfilling this requirement was developed in [13] and it is based on the following idea; total strains are modelled on the ground of classical slender beam theory, while plastic multipliers are interpolated polynomially over the values assumed at a number of Gauss points located both over the cross section and along the element length. The consistency condition is enforced with reference to stress resultants $N(x)$ and $M(x)$ along the element axis but not over the cross section, where stress redistribution is permitted. The resulting model turns out to be rather refined and capable to accurately predict the response of elastic-plastic frames with an extremely limited number of elements, which often can be reduced down to the number of frame members.

Fig.7 shows the results obtained for a clamped beam of rectangular cross section under uniform transverse load. The displacement model assumes a fourth order transverse displacement field. In each element the plasticity law is enforced at $G \times H$ points, where G (= 7 in the example) denotes the number of Gauss points over the cross section and H (= 5) the number of Gauss points over the element length. Table 3 lists the percentage errors on the predicted collapse load value for a different number of elements and different Gauss point grids. A single element actually appears to be able to produce excellent results.

The four storey, one bay frame of fig.8 was also analized in [13]. 16 finite elements (one for each column and two for each beam) were used; for all of them $H = 5$ was assumed; in each cross section, plastic multipliers were interpolated over $G = 5$ points, a value which proves to be sufficiently accurate for I-beams. The resulting frame model was governed by 64 displacement and 800 λ-variables. The basic load conditions indicated in Table 4 were monothonically amplified up to collapse. For each of them, the computed elastic limit and collapse load factors are reported in Table 4 and compared with theoretical results.

TABLE 3
Number of variables involved by different models
for the clamped beam example (G = 7)

N	H	U	λ	ε(%)
9	3	33	378	+3.6
5	5	17	350	+0.8
3	7	9	294	+0.4
1	11	1	154	+1.3
1	13	1	182	+0.4

N = number of elements; H = number of Gauss points along the element axis; U = number of displacement variables; λ = number of plastic multiplier variables; ε = percentage error on collapse load value

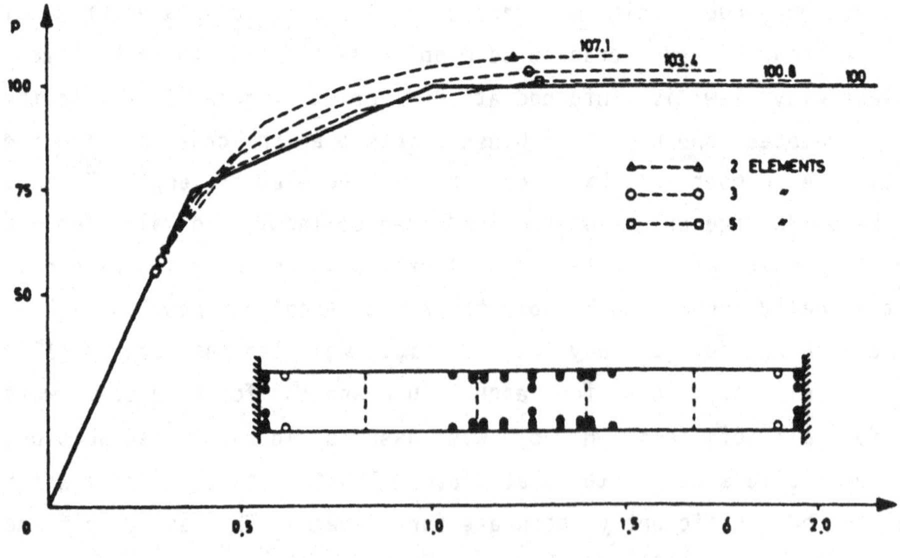

Fig.14.7. Clamped beam example (from [13])
(a) load-displacement curve; (b) collapse mechanism for 5 elements
black dots = plastic points, white dots = elastic unloading

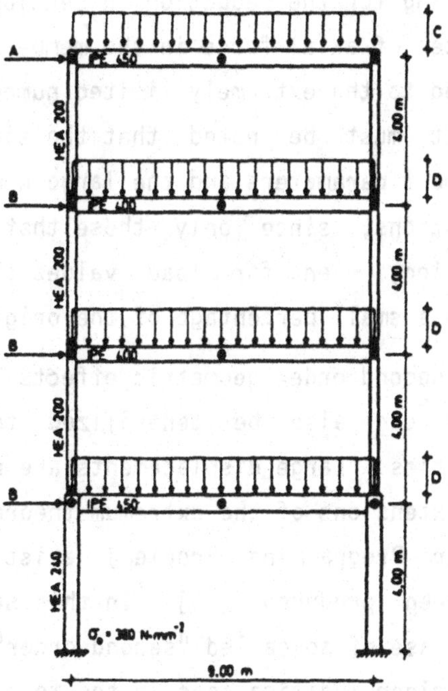

Fig.14.8. Frame example (from [13]). Black dots = element boundaries

TABLE 4
Load data and limit loads for the frame of fig.8

Load condition		1	2
A(KN)		1.50	1.50
B(KN)		2.87	2.87
C(KN/m)		3.46	6.90
D(KN/m)		3.01	3.01
Elastic limit	Computed	8.65	7.29
	Exact	8.33	7.00
Collapse load factor	Computed	15.85	10.34
	Exact	15.11	10.41

As theoretical collapse load an estimate was taken, based on a plastic hinge model, accounting for the reduction in sectional bending capacity due to the presence of axial force in the member. Results look good, above all if related to the extremely limited number of elements used. To this purpose, it must be noted that the size of the problem is governed by displacement parameters and the large number of λ-variables has limited implications, since only those that are actually active affect the computation; even for load values close to the collapse factor, they are only a small percentage of the original figure.

14.3.3. Extension to second order geometric effects

The formulation can also be generalized to cover the case of geometric nonlinearities. Large displacements are not considered here, even if both the extensions of the extremum theorem Eq.(13.70) (which becomes a Non-Linear Programming Problem) exist [14] and numerical applications have been produced [15]. In this section it is briefly summarized only the case of so-called "second order" geometric effects, which entails merely minor modifications in the formulation.

With "second order" geometry changes we refer to displacements from the undeformed configuration which are small enough for additive strain decomposition and for the linear compatibility relation Eq.(13.64a) to be used, but which affect significantly the equilibrium of the structural system. In this case [16] Eq.(13.64b) is to be replaced by the following relation

$$\underline{F} = \underline{C}^t \, \underline{\sigma} + \underline{K}_G \, \underline{U} \tag{14.41}$$

where $\underline{K}_G$ is the so-called geometric or stress stiffness matrix of the structure, depending on the current stress state.

Provided that Eq.(1) is used instead of Eq.(13.64b), the same path followed in Section 2 can be pursued and the elastic-plastic problem can be formulated in the L.C.P. forms Eqs.(13.68) or (13.72). Formally, the only change occurs in the expression Eq.(13.67a) of matrix $\underline{K}_{uu}$, which now includes the geometric effects contributions

$$\underline{K}_{uu} = \underline{C}^t \, \underline{D} \, \underline{C} + \underline{K}_G \tag{14.42}$$

In the finite element formulation, this contribution shows up at the element level and the expression for $\underline{k}_{uu}$ becomes

$$\underline{k}_{uu}^{n} = \underline{k}_{uu}^{on} + \underline{k}_{G}^{n} \qquad (14.43)$$

where $\underline{k}_{uu}^{on}$ is the same matrix as before, given by the fully equivalent definitions Eqs.(6a) or (23a) and $\underline{k}_{G}^{n}$ is the geometric stiffness for the element at hand, as it is defined for the study of buckling problems [16].

Even if formal changes are minor, the presence of geometric effects can alter the essential behavior of the resulting L.C. Problem. A comment on the following points is in order.

(a). Matrix $\underline{K}_{uu}$, Eq.(2), is still symmetric and positive definite if the structure, supposed to behave elastically, is stable under the given loads. However, matrices $\underline{M}$, Eq.(13.69), and $\underline{A}$, Eq.(13.73), might be indefinite in some circumstances. In this case, the relevant Q.P. formulations (Eqs.(13.70 and (13.76), respectively) cannot be established. As was proved in [17], when $\underline{K}_{uu}$ is positive definite, the above matrices are either both so or both do not fulfill this condition.

(b). When geometry effects are of destabilizing type (for instance, compression in a member), matrix $\underline{A}$ (and, hence, $\underline{M}$) is indefinite for perfectly plastic materials. This effect might be compensated by the presence of hardening. Analogously, possible softening behavior may be compensated by stabilizing second order effects.

(c). If an evolutive analysis under given loading hystories is performed, a unique solution to the elastic-plastic problem exists as long as a matrix $\underline{A}'$ is positive definite. $\underline{A}'$ is the principal sub-matrix of $\underline{A}$ made up by the rows and columns pertaining to the yield planes which are active at the current load level. When $\underline{A}'$ ceases to be positive definite (either because of geometry effects of increasing importance or due to the activation of additional yield planes) either bifurcation or (more frequently) collapse occurs [17].

When used in conjunction with the beam finite element model quoted in the preceeding section, second order formulations permit efficient numerical tools to be established, able to analyze beam-columns and

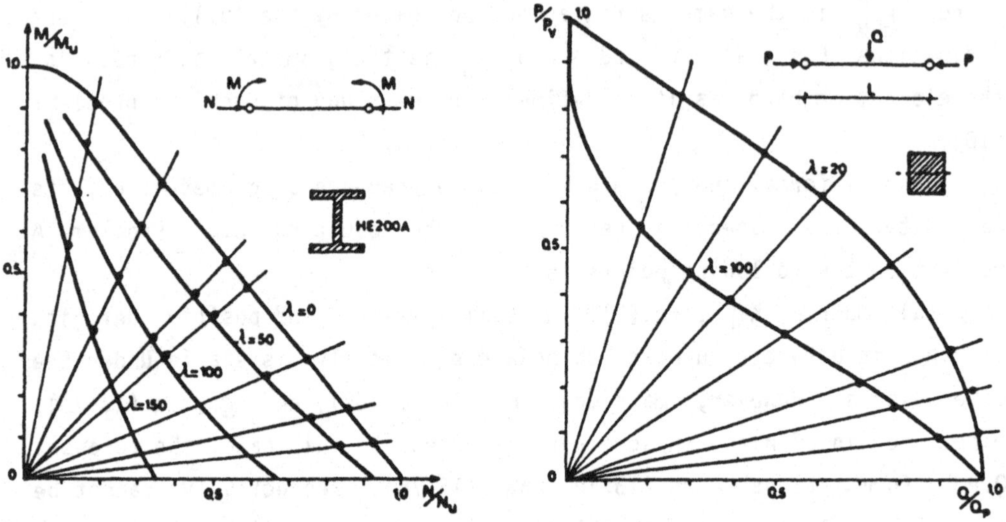

Fig.14.9. Beam-column interaction curves (from [18])
Dots = computed points. Reference curves from [19] and [20]

frames up to collapse, due to loss of stability of the structure deteriorated by the presence of plastic strains. Some results of this kind have been obtained and are illustrated in ref.[18]. Fig.9 shows the collapse load for two typical beam-column problems, which compare well with interaction curves available in the literature.

14.3.4. Structures under cyclic loads

The use of stepwise holonomic constitutive laws permits the analysis of elastic-plastic structures under cyclic load conditions. Holonomy can reasonably be assumed within each half cycle; obviously, elastic unloading from a plastic state must be allowed for when a change in the direction of the applied loads occurs in the loading history.

For perfectly plastic structures and in the absence of geometric effects, shakedown theory provides statements able to assess efficiently the structural safety under cyclic conditions. For piecewise linear plasticity, this theory was extended by Maier [21] to cover the case of hardening and of second order geometric effects, provided that matrix $\underline{A}$ is globally positive semidefinite or definite. However, results of limited significance are often obtained, since shakedown turns out to be nearly always ensured by the theory in this case; nevertheless, the amount of plastic strains required to attain shakedown might be large and, in some cases, such as to jeopardize the structural serviceability.

This fact is illustrated with reference to the two-bar example of fig.10. The structure is subjected to a constant mechanical load and to a cyclic temperature variation in the weakest bar. A bilinear hardening law is assumed for the material, governed by Eq.(13.1.28) in which matrix $\underline{H}$ is given the following expression

$$\underline{H} = h\,E \begin{bmatrix} 1 & -\beta \\ -\beta & 1 \end{bmatrix} \quad ; \quad -1 \leq \beta \leq 1 \qquad (14.44)$$

which can describe different behaviors with different assumptions for β. In particular, the extremes of the interval for β reproduce isotropic ($\beta = -1$) and kinematic ($\beta = 1$) hardening, respectively. Intermediate values are able to approximate cyclic hardening.

The behavior of the structure is illustrated by the Bree diagrams of fig.11. In the presence of hardening ($\beta \neq 1$) shakedown always occurs but ratchetting type deformations, slowly damped out by hardening, develop at each cycle in the regions labelled as $S_{1,3}$ and $S_{1,3,4}$ in fig.11b. Moreover, in the presence of hardening, shakedown is usually attained after a significant number of cycles, if not approached only asymptotically. Thus, divergence of the deformed configuration prior to shakedown may be experienced and checking against this occurrence by following the cyclic elastic-plastic evolution of the structure entails an extremely heavy computational effort for structures of meaningful size. On this ground, simplified procedures, such as those aiming at the computation of upper bounds to the amount of residual displacements, acquire a significant interest, even if the results they are able to provide are often comparatively coarse.

The discrete formulation described in this chapter permits the application of bounding theorems and simplified computations providing

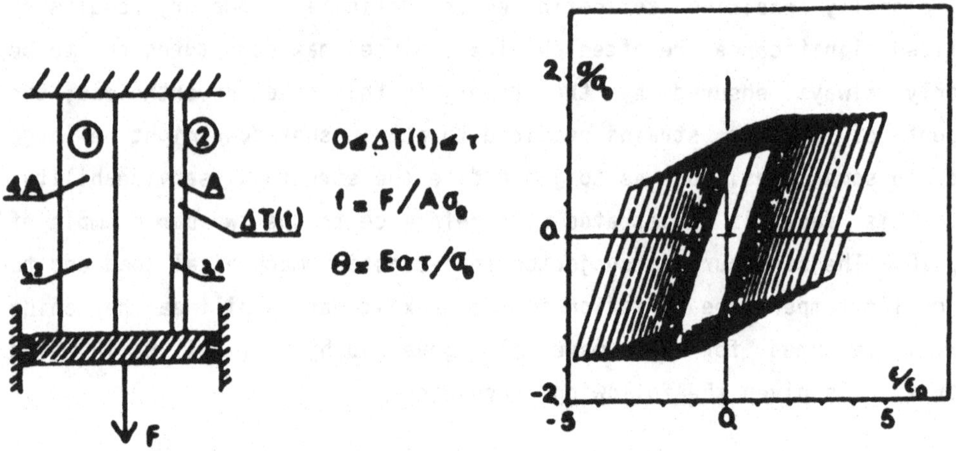

Fig.14.10. Two-bar example and bilinear cyclic hardening model for β = .95 (from [22])

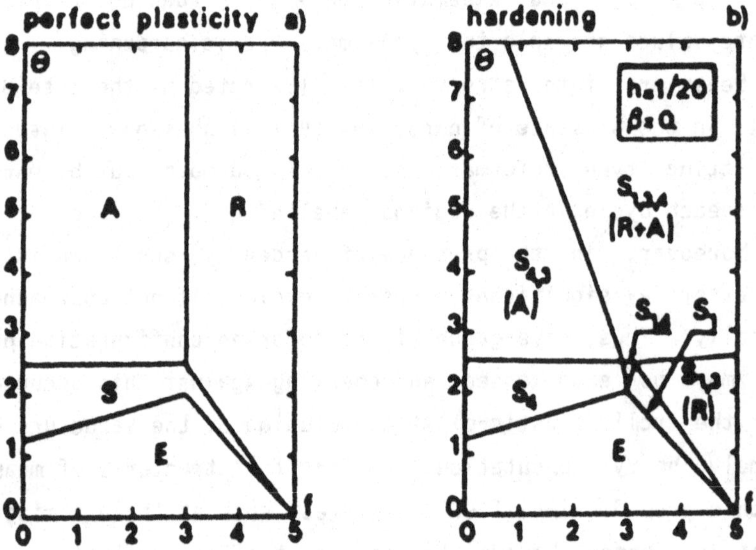

Fig.14.11. Bree diagrams for the two-bar example

essential information on the post-shakedown response at a reduced cost. A suitable procedure was recently proposed [22], which exploits some previous results in a particularly efficient manner. However, both shakedown and bounding theorems loose their validity when matrix $\underline{A}$ is no longer positive semidefinite, as in the case of perfectly plastic structures under geometric effects of destibilizing type. At present, the cyclic behavior of such structures can only be assessed by following the elastic-plastic evolution under cyclic loading histories.

The continuous beam of fig.12 was analyzed by making use of the beam finite element model described in Sec.3.2. Both first and second order analyses were performed. When geometric effects are considered, usually either shakedown occurs at first cycle or collapse is attained after a limited number of cycles, due to loss of stability associated to the gradual divergence of the deformed configuration. Some of the solutions computed are illustrated in fig.13, where plots refer to midspan deflections. In the first case, shakedown is predicted by first order theory, but collapse occurs during the third cycle if second order effects are accounted for. In the second case, a regime ratchetting situation is quickly attained if geometry effects are neglected, while collapse during the fourth cycle is predicted by a second order analysis.

Even if a regime ratchetting situation is no longer reached, the event of delayed collapse experienced in the presence of instabilizing effects can still be regarded as lack of shakedown, in that it is induced by the cyclic nature of loads. The structure, in fact, is able to withstand safely a single load application and only repeated actions are responsible for collapse. Computations of the shakedown boundary for the beam-column example of fig.12 were performed in [23] and results are depicted in fig.14. For $N \leq 0$ (tension) matrix $\underline{A}$ is positive semidefinite and the shakedown boundary can be obtained by exploting existing theorems. The discontinuity which shows up for $N = 0$ reflects the circumstance that tension prevents ratchetting and lack of shakedown can only occur because of reverse plasticity. For $N > 0$ matrix $\underline{A}$ becomes indefinite and the theory fails to apply. The boundary of the load domain: $\begin{cases} 0 \leq F_1 \leq 2\bar{F} \\ 0 \leq F_2 \leq \bar{F} \\ N \text{ fixed} \end{cases}$

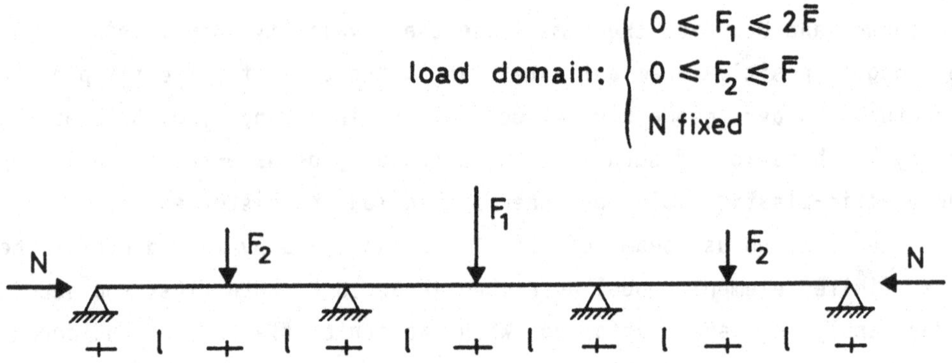

Fig.14.12. Continuous beam example

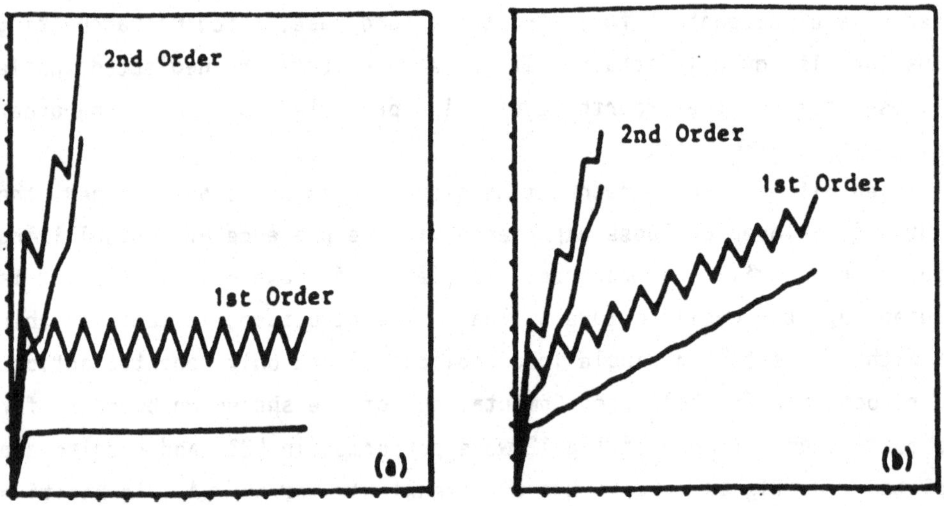

Fig.14.13. Total and permanent midspan deflection for the example of
Fig.12 as function of cycles. $F_1 = 2\bar{F}$ (fixed), $0 \leq F_2(t) \leq \bar{F}$
(a) $N/N_0 = .1$, $\bar{F}\ell/M_0 = 1.6$; (b) $N/N_0 = .05$, $\bar{F}\ell/M_0 = 1.8$

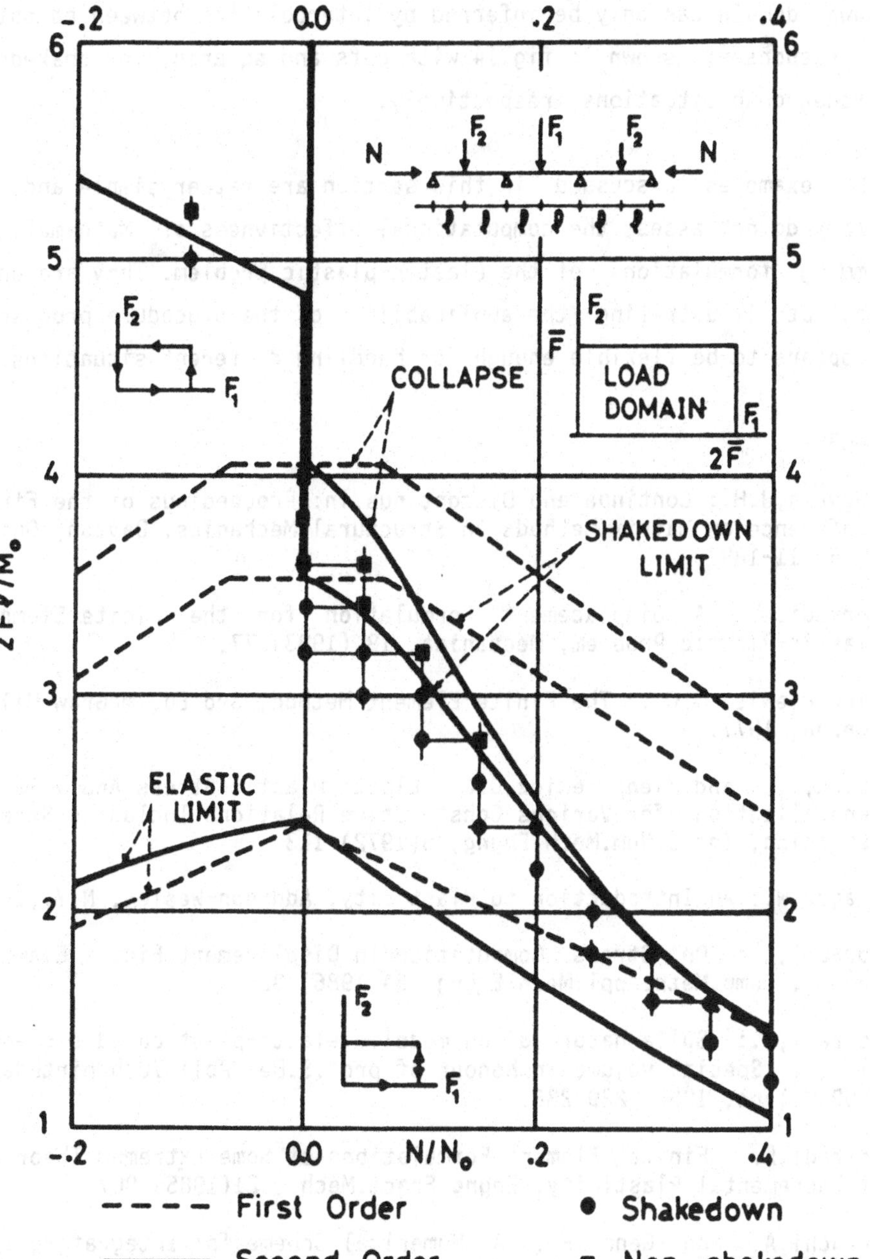

Fig.14.14. Shakedown boundary for the continuous beam example

shakedown domain can only be inferred by interpolation between computed cyclic responses, shown in fig.14 with dots and squares, for shakedown or non-shakedown situations, respectively.

The examples discussed in this section are rather simple and, as such, they do not assess the computational effectivness of Mathematical Programming formulations of the elastic-plastic problem. They are only intended at illustrating the applicability of the procedure proposed, which appears to be flexible enough for handling different situations.

REFERENCES

1. Argyris,J.H.: Continua and Discontinua,in: Proceedings of the First Conference on Matrix Methods in Structural Mechanics, Dayton, Ohio, 1976, 11-189.

2. Corradi,L.: A Displacement Formulation for the Finite Element Elastic-Plastic Problem, Meccanica, 18 (1983) 77.

3. Zienckiewicz,O.C.: The Finite Element Method, 3rd Ed. McGraw-Hill, London, 1977.

4. Nayak,G.C. and Zienckiewicz,O.C.: Elasto Plastic Stress Analysis. A Generalization for Various Constitutive Relations Including Strain Softening, Int.J.Num.Meth.Engng, 5(1972) 113

5. Prager,W.: An Introduction to Plasticity, Addison-Wesley, N.Y.,1959

6. Corradi,L.: On Stress Computation in Displacement Finite Element Models, Comp.Meth.Appl.Mech.Engng, 54(1986) 325

7. Corradi,L.: Sulla natura di un modello elasto-plastico ad elementi finiti, Special volume in honour of prof.S.Dei Poli 70th birthday, CLUP Milano, 1985, 229-236.

8. Corradi,L.: Finite Element Formulations of Some Extremum Theorems of Incremental Plasticity, Engng Fract.Mech., 21(1985) 807

9. Franchi,A. and Genna,F.: A Numerical Scheme for Integrating the Rate Plasticity Equations with an "A-Priori" Error Control, Comp. Meth.Appl.Mech.Engng, 60(1987) 317

10. Martin,J.B.: Plasticity, M.I.T. Press, Cambridge, Mass.,1975.

11. Hodge,P.G.Jr and Belytschko,T.:Numerical Methods for the Analysis of Plates, Trans.ASME,J.Appl.Mech., 35(1968) 796

12. Ang,A.H.S. and Lopez,L.A.:Discrete Model Analysis of Elastic-Plastic Plates, Proc.ASCE,J.Engng Mech.Div., 94(1968) 271

13. Corradi,L. and Poggi,C.: A Refined Finite Element Model for the Analysis of Elastic-Plastic Frames, Int.J.Num.Meth.Engng, 20(1984) 2155

14. Corradi,L. and Maier,G.: Extremum Theorems for Large Displacement Analysis of Discrete Elasto Plastic Structures with Piecewise Linear Yield Surfaces, Journal of Optimization Theory and Applications, 15(1975) 51

15. Contro,R. and Corradi,L., Non-Linear Programming Applications, in: Enginering Plasticity by Mathematical Programming (Eds. M.Z.Cohn and G.Maier), Pergamon Press, N.Y., 1979, 529-536

16. Maier,G.:Incremental Elastic Plastic Analysis in the Presence of Large Displacements and Physical Instabilizing Effects, Int.J.of Solids and Structures, 7(1971) 345

17. Corradi,L.: Stability of Discrete Elastic Plastic Structures with Associated Flow Laws, SM Archives, 3(1978) 201

18. Corradi,L. and Poggi,C.: An Analysis Procedure for Non-Linear Elastic Plastic Frames Accounting for the Spreading of Local Plasticity, Costruzioni Metalliche, n.1 (1985).

19. Ballio,G., Petrini,V. and Urbano,C.: The Effects of the Loading Process and Imperfections on the Load-Bearing Capacity of Beam Columns, Meccanica, 8(1973) 56

20. Chen,W.F. and Atsura,T.: Theory of Beam-Columns, vol.1, McGraw-Hill, :N.Y., 1976.

21. Maier,G.: A Shakedown Matrix Theory Allowing for Work-Hardening and Second Order Geometric Effects, Proc. Symp. on the Foundations of Plasticity, Warsaw, 1972

22. Corradi,L. and Poggi,C., Estimate of The Post-Shakedown Response of Hardening Structures by means of Simplified Computations, Meccanica 22(1987) 193

23. Cazzani,A. and Corradi,L.: Valutazione della risposta di telai elasto-plastici soggetti a carichi ciclici in presenza di effetti instabilizzanti, to be presented at the 9th AIMETA Conference, Bari, Italy, October 1988.

CHAPTER 15

RIGID PLASTIC DYNAMICS

D. Lloyd Smith and C. L. Sahlit (°)
Imperial College, London, U.K.

ABSTRACT

The problem of rigid-plastic framed structures subjected to load pulses of arbitrary form, and of intensity such that substantial plastic deformation takes place, is treated by approximate means. The investigation is restricted to the range of low intensity excitation which induces a global dynamic response.

Mesh and nodal descriptions of the kinetic and kinematic laws are given for a structure with discrete masses under the restriction of small displacements, while the structural material is assumed rigid, perfectly plastic and strain-rate insensitive.

The vectorial relations of the finite element representation of the structure are integrated numerically by means of Newmark's method. For each increment of time, there emerges a linear complementarity problem (LCP) which can be solved by a variant of Wolfe's algorithm.

It proves essential to incorporate a numerical procedure which can identify and properly account for the unstressing that may occur within any particular increment of time. Such unstressing is undoubtedly a major feature of rigid-plastic dynamics.

(°) On leave from Departamento de Engenharia Civil, Universidade de Brasilia, Brasil.

INTRODUCTION

Ductile structures impelled into motion by high-energy sources will tend to dissipate energy by inelastic deformation since their capacity for storing energy in elastic deformation is usually limited. With long-duration sources, in general, the interaction between elastic and plastic deformation is an important feature of the response: a process of adaptation occurs. For problems involving short-duration sources, which are becoming increasingly more important, much useful information can be gained by ignoring the elastic component of deformation and imposing rigid-plastic constitutive characteristics on the construction material.

The assumption of a rigid-plastic material has been taken up extensively in the field of dynamic plasticity, and particularly by Martin[1], Symonds[2], Jones[3] and Kaliszky[4]. Nevertheless, this idealisation seems to be a reasonable assumption only if the external dynamic energy is much larger than the maximum possible elastic strain energy which can be absorbed by the structure, and when the pulse duration is short compared with the fundamental natural period.

Other phenomena can also play an important rôle in the inelastic dynamic response. Large energy inputs may, in some cases, induce large displacements. Moreover, high velocities sensibly affect the value of the yield stress in most construction materials. These are refinements that will not be considered here.

As in the plastic analysis of structures under quasi-static loads, the application of mathematical programming to rigid-plastic dynamics seems to be playing its typical two-fold rôle of providing a tool for theoretical developments, Capurso[5], and for numerical solutions, Cannarozzi and Laudiero[6].

In this chapter, we shall introduce a method for establishing the diffuse (non-local) response of a rigid, perfectly plastic structure which is subjected to an arbitrary pressure pulse. It is therefore assumed that a local failure - by punching or tearing of the structural material - does not occur. The vectorial relations which govern the ensuing motion and which arise from the finite element modelling of the structure are integrated numerically by means of the well-known method of Newmark. For each increment of time, there emerges a linear complementarity problem which can be solved by a variation of the Wolfe algorithm.

KINETICS AND KINEMATICS

The simplest dynamic modelling of the structure is as a system of discrete masses. The gravity-centre of each mass can be assigned a number of permissible displacements, or degrees of freedom, both rectilinear and rotational, which can be collected into the column vector $\mathbf{u}_D$. Its components may be called *inertia coordinates*.

A system of discrete masses which is totally unconstrained - that is, where each mass is not connected to anything - would be governed by Newton's second law

$$\mathbf{F} - \mathbf{m}\,\ddot{\mathbf{u}}_D = \mathbf{0}. \tag{15.1}$$

In these fundamental relations, **F** is a column vector containing the component forces applied to the masses in directions which correspond with the inertia coordinates, and **m** is a diagonal *inertia matrix* containing the associated masses and mass moments of inertia. Newton's dot notation is used to denote differentiation with respect to time t. The components of the column vector μ defined by

$$\mu = -\mathbf{m}\,\ddot{\mathbf{u}}_D \tag{15.2}$$

are usually called the *inertia forces*. By considering these forces to be externally applied, we may interpret (15.1) as a system of equilibrium equations: $\mathbf{F} + \mu = \mathbf{0}$. The left side represents the total external forces on the masses, while the (null) right side gives the forces of constraint. These are the *kinetic equations* for an *unconstrained* system.

Now let the masses be affixed to, and therefore constrained by, a structure which is represented by a system of finite elements, the masses being sited at the terminals or node points of appropriately chosen elements. Then the usual equations of statics may be extended to encompass the kinetic case by the imposition of the inertia forces as an extra set of applied loads.

Let $\Delta t = t - t_n$ be a small increment of time, taken at time t_n. Then kinetic and kinematic descriptions can be written in terms of corresponding increments in the independent member forces ΔQ_I and deformations Δq_I, in a similar manner to that described in Chapter 4. The mesh description of kinetics and kinematics in incremental terms is therefore given by (15.3) and (15.4) respectively:

$$\Delta Q_I = \begin{bmatrix} B & B_D & B_o \end{bmatrix} \begin{bmatrix} \Delta p \\ \Delta \mu \\ \Delta F \end{bmatrix} \tag{15.3}$$

$$\begin{bmatrix} \Delta v = 0 \\ \Delta u_D \\ \Delta u_F \end{bmatrix} = \begin{bmatrix} B^T \\ B_D{}^T \\ B_o{}^T \end{bmatrix} \Delta q_I \tag{15.4}$$

In these relations, Δp are the current increments on the hyperstatic forces and Δv are the corresponding (or dual) discontinuity increments which must remain zero for the sake of compatibility. Δu_F are the increments in the displacements of the chords of the members which are dual to the current applied load increments ΔF.

The nodal description is then presented in a similar manner[7] through the equations (15.5) and (15.6).

$$\Delta U = 0 = \begin{bmatrix} A^T & A_D{}^T & A_o{}^T \end{bmatrix} \begin{bmatrix} \Delta Q_I \\ -\Delta \mu \\ -\Delta F \end{bmatrix} \tag{15.5}$$

$$\begin{bmatrix} \Delta q_I \\ \Delta u_D \\ \Delta u_F \end{bmatrix} = \begin{bmatrix} A \\ A_D \\ A_o \end{bmatrix} \Delta u \tag{15.6}$$

It should be noted that the hyperkinematic displacements u, or kinematic degrees of freedom, form the *generalised coordinates of Lagrange* for the kinematic description; their number (β) depends upon the finite element discretisation, and, in general, they need not coincide with the inertia coordinates of the masses, u_D. In other words, A_D need not be an identity matrix. Since the nodal constraint forces in the vector U will be zero when the structure is in dynamic equilibrium, the increment ΔU must also remain null.

CONSTITUTIVE RELATIONS

The appropriate relations for perfect plasticity appear in Chapter 5 as (5.21) to (5.25) inclusive. Briefly, they are

$$\{N^T Q + y = R, \quad N \dot{\lambda} = \dot{q}_P, \quad y \geq 0, \quad y^T \dot{\lambda} = 0, \quad \dot{\lambda} \geq 0\} \qquad (15.7\text{-}8\text{-}9\text{-}10\text{-}11)$$

Here, Q is the vector of stress resultants, measured at the critical sections; their interaction controls the plastic yielding which is assumed to be concentrated at those sections. Vector $\dot{q}_P$ contains the dual plastic strain resultant rates, y and $\dot{\lambda}$ are respectively vectors of plastic potentials and plastic multiplier rates, and R is the constant vector which defines the plastic capacity of all the critical sections.

To accommodate the incremental numerical process which is envisaged, it will be necessary to replace the rate of change of each kinematic variable by its increment. The plasticity relations are thereby approximated as follows:

$$\begin{bmatrix} 0 & N^T \\ N & 0 \end{bmatrix} \begin{bmatrix} \Delta\lambda \\ Q \end{bmatrix} + \begin{bmatrix} y \\ 0 \end{bmatrix} = \begin{bmatrix} R \\ \Delta q_P \end{bmatrix} \qquad \begin{matrix} (15.12) \\ (15.13) \end{matrix}$$

$$y \geq 0 \quad (15.14) \qquad y^T \Delta\lambda = 0 \quad (15.15) \qquad \Delta\lambda \geq 0 \quad (15.16)$$

It must be carefully noted that the plastic potentials y are total variables, not incremental. In this modified form, the relations will allow for unstressing at the commencement ($t = t_n$) of the current time increment, but not at any subsequent time within the increment. It is imperative that such an event be identified and its effects accounted for by the numerical process.

In general, the stress resultants Q at the critical sections need not coincide with the complete set of independent member forces Q_I. It may be necessary to provide the following transformations between the variables which govern plastic flow at the critical sections and the finite element system state variables, defined at the element termini.

$$Q = T^T Q_I, \qquad \Delta q_{IP} = T\, \Delta q_P \qquad (15.17\text{-}18)$$

Since the independent elastic deformations Δq_{IE} are suppressed in the rigid-plastic material model, it follows that

$$\Delta q_I = \Delta q_{IE} + \Delta q_{IP} = \Delta q_{IP} \qquad (15.19)$$

The deformation Δq_I in (15.6) is, for this class of problems, entirely plastic.

The numerical process which will be based upon the foregoing incremental relations is not self-starting; it will demand a special initiation or start-up routine through which relevant initial accelerations are calculated. It is therefore necessary to consider the time variation of the plasticity relations. If, for simplicity, we assume that N is a matrix of constants, such as would obtain if the yield surfaces were piecewise-linear, then equations (15.7), (15.8) and (15.10) are easily differentiated with respect to time.

$$N^T\dot{Q} + \dot{y} = 0, \qquad N\ddot{\lambda} = \ddot{q}_P, \qquad y^T\ddot{\lambda} + \dot{y}^T\dot{\lambda} = 0. \qquad (15.20\text{-}21\text{-}22)$$

Inequalities (15.9) and (15.11) must be satisfied at any time t; and they must clearly remain satisfied during any immediately subsequent small interval of time – that is,

$$y + h\dot{y} \geq 0, \qquad \dot{\lambda} + h\ddot{\lambda} \geq 0, \qquad (15.23\text{-}24)$$

where h is a small positive number. The implication of (15.9) and (15.23), considered jointly, and also of (15.11) and (15.24), likewise taken together, may be summarised simply in the following manner:

$$\left.\begin{array}{l} y_k = 0 \;\rightarrow\; \dot{y}_k \geq 0 \\ y_k > 0 \;\rightarrow\; \dot{y}_k \in U \end{array}\right\} (15.25) \qquad \left.\begin{array}{l} \dot{\lambda}_k = 0 \;\rightarrow\; \ddot{\lambda}_k \geq 0 \\ \dot{\lambda}_k > 0 \;\rightarrow\; \ddot{\lambda}_k \in U \end{array}\right\} (15.26)$$

where the symbol $\rightarrow$ should be read as "implies", and $\dot{y}_k \in U$ indicates that the plastic potential rate associated with the *kth* segment of some yield surface is a member of the set of unrestricted variables.

THE NODAL FORMULATION

The acceleration increments $\Delta\ddot{u}_D$ in the inertia coordinates and the inertia force increments $\Delta\mu$, by virtue of (15.6) and (15.2), may be written

$$\Delta\ddot{u}_D = A_D\,\Delta\ddot{u}, \qquad \Delta\mu = -m\,\Delta\ddot{u}_D. \qquad (15.27\text{-}28)$$

They enable the kinetic relations (15.5) to be set down as

$$A^T\,\Delta Q_I + M_u\,\Delta\ddot{u} = A_o^T\,\Delta F, \qquad (15.29)$$

where

$$M_u = A_D^T\,m\,A_D \qquad (15.30)$$

is a mass matrix referred to the acceleration increments $\Delta\ddot{u}$ of the generalised coordinates or kinematic degrees of freedom u. At time t, where $t > t_n$, plasticity relation (15.12) can be expressed in the form

$$N^T\ T^T\ \Delta Q_I + y = y_n, \tag{15.31}$$

where

$$y_n = R - N^T\ T^T\ Q_{In}, \tag{15.32}$$

and where y_n and Q_{In} are the values of the respective vectors at time t_n. By combining these relations with the compatibility relations (15.6) and with the flow rule of (15.13), we produce the system (15.33).

$$\begin{bmatrix} -M_u & 0 & 0 \\ 0 & 0 & 0 \\ 0 & 0 & 0 \end{bmatrix} \begin{bmatrix} \Delta\ddot{u} \\ \Delta\ddot{\lambda} \\ \Delta\ddot{Q}_I \end{bmatrix} + \begin{bmatrix} 0 & 0 & -A^T \\ 0 & 0 & N^T T^T \\ -A & TN & 0 \end{bmatrix} \begin{bmatrix} \Delta u \\ \Delta\lambda \\ \Delta Q_I \end{bmatrix} + \begin{bmatrix} 0 \\ y \\ 0 \end{bmatrix} = \begin{bmatrix} -A_o{}^T\Delta F \\ y_n \\ 0 \end{bmatrix} \tag{15.33}$$

$$y > 0 \qquad y^T \cdot \Delta\lambda = 0 \qquad \Delta\lambda > 0$$

This has been called a differential linear complementarity problem, Al-Samara[8]. No work on the mathematical basis of such problems is yet known to the writers. Therefore, an approximate solution procedure using numerical integration is suggested.

Numerical Solution by Newmark Integration

At time $t = t_{n+1}$, the acceleration and velocity components of the generalised coordinates u may be calculated approximately by means of the numerical integration scheme of *Newmark*:

$$\ddot{u}_{n+1} = a_0\Delta u - a_1\dot{u}_n - (a_2 - 1)\ddot{u}_n \tag{15.34}$$

$$\dot{u}_{n+1} = a_3\Delta u - a_4\dot{u}_n - a_5\ddot{u}_n \tag{15.35}$$

where the constants are

$$a_0 = \frac{1}{\beta\Delta t^2} \qquad a_1 = \frac{1}{\beta\Delta t} \qquad a_2 = \frac{1}{2\beta} \tag{15.36-37-38}$$

$$a_3 = \frac{\gamma}{\beta\Delta t} \qquad a_4 = \frac{\gamma}{\beta} - 1 \qquad a_5 = \left[\frac{\gamma}{2\beta} - 1\right]\Delta t \tag{15.39-40-41}$$

The values $\beta = 0.25$, $\gamma = 0.5$, which represent a constant average acceleration scheme, seem to give a stable and robust numerical procedure for rigid-plastic dynamics.

If we set $\Delta\ddot{u} = \ddot{u}_{n+1} - \ddot{u}_n$ and eliminate $\Delta\ddot{u}$ from (15.33) by means of (15.34), the approximating governing system (15.42) is obtained.

$$\left[\begin{array}{ccc} -a_0 M_u & 0 & -A^T \\ 0 & 0 & N^T T^T \\ \hline -A & TN & 0 \end{array}\right]\left[\begin{array}{c} \Delta u \\ \Delta\lambda \\ \hline \Delta Q_I \end{array}\right] + \left[\begin{array}{c} 0 \\ y_{n+1} \\ \hline 0 \end{array}\right] = \left[\begin{array}{c} -\Delta\bar{F}_n \\ y_n \\ \hline 0 \end{array}\right]$$

$$y_{n+1} \geq 0 \qquad y_{n+1}^T \Delta\lambda = 0 \qquad \Delta\lambda \geq 0 \qquad (15.42)$$

in which the right-hand side subvector $\Delta\bar{F}_n$ is given by

$$\Delta\bar{F}_n = A_o{}^T\Delta F + M_u[a_1\dot{u}_n + a_2\ddot{u}_n] \qquad (15.43)$$

The governing system is now a normal LCP. It must be solved for each time increment Δt, with consequent updating of the right hand side vector. In this way, the evolution of the continuous deformation process is traced through a sequence of discrete solutions of the governing LCP.

Let $[u_n, \dot{u}_n, \ddot{u}_n, Q_{In}, \lambda_n, q_{In}, y_n]$ be the known response of the structure at time $t = t_n$. With the applied loads F_{n+1} known at time $t = t_{n+1}$, the response can be advanced to $t = t_{n+1}$ through solution of the governing LCP by application of a variant of Wolfe's algorithm.

System (15.42) is partitioned so that its structure can be compared with that of the LCP(3.11). Variables Δu and ΔQ_I are unrestricted, however, and must be made permanent basic variables. This is usually achieved by the performing of a sequence of pivot operations to associate a canonical basis with the unrestricted variables. Unfortunately, the initial submatrix of (15.42) which relates to the unrestricted variables is generally only semidefinite, and simple pivot operations will not suffice for converting it into an identity matrix. Instead, whenever it becomes impossible to find a non-zero pivot, a double-pivot operation is rendered necessary.

In order to sustain accuracy, it is advisable to ensure satisfaction of the kinetic equations (15.29) at each increment of time. For example, in the special case where the generalised coordinates u coincide with the inertia coordinates u_D, then A_D is an identity matrix and we may simply calculate the accelerations by means of

$$\Delta\ddot{u} = m^{-1}[A_o{}^T\Delta F - A^T\Delta Q_I] \qquad (15.44)$$

rather than through the use of (15.34). In the general case, some further calculation is necessary for the obtaining of $\Delta\ddot{u}$.

Initiation

To initiate the sequence of solutions for LCP(15.42), the initial accelerations of the generalised coordinates – $\ddot{u}(t = 0) = \ddot{u}_0$ – are needed. They cannot be established through the kinetic equations because, in a rigid–plastic structure, the initial values of the independent member forces $Q_I(t = 0) = Q_{I0}$ are generally indeterminate. It therefore becomes necessary to re-establish the governing system in *total* variables at time $t = 0$.

For a structure which is coerced into motion by a pressure pulse, it is usual to consider that the generalised coordinates or kinematic degrees of freedom are subject to stationary initial conditions

$$u(t = 0) = u_0 = 0, \qquad \dot{u}(t = 0) = \dot{u}_0 = 0. \qquad (15.45\text{-}46)$$

The absence of initial velocity anywhere in the system implies that there is no initial plastic flow – that is, $\dot{\lambda}_0 = 0$. If this result is introduced into (15.22) and (15.26), it then follows that

$$y_0^T \ddot{\lambda}_0 = 0, \qquad \ddot{\lambda}_0 > 0. \qquad (15.47\text{-}48)$$

The appropriate kinetic relations for $t = 0$ are easily surmised from (15.29), the yield conditions from (15.7) and (15.9), while the combined conditions of compatibility and the flow rule are deducible from (15.6) and (15.21). These results will be found in the following statement of the initial governing system:

$$\left[\begin{array}{ccc} -M_u & 0 & -A^T \\ 0 & 0 & N^T T^T \\ \hline -A & TN & 0 \end{array}\right] \left[\begin{array}{c} \ddot{u}_0 \\ \ddot{\lambda}_0 \\ \hline Q_{I0} \end{array}\right] + \left[\begin{array}{c} 0 \\ y_0 \\ \hline 0 \end{array}\right] = \left[\begin{array}{c} -A_0^T F_0 \\ R \\ \hline 0 \end{array}\right]$$

$$y_0 > 0 \qquad y_0^T \ddot{\lambda}_0 = 0 \qquad \ddot{\lambda}_0 > 0 \qquad (15.49)$$

System (15.49) is an LCP for time $t = 0$. It must be solved for $\ddot{u}_0$, y_0 and Q_{I0}, and these data, introduced into (15.43), may be used to initiate the solution sequence of LCP(15.42).

THE MESH FORMULATION

Turning now to the mesh description, we obtain from (15.4), (15.13) and (15.18)

$$B^T TN\Delta\lambda = 0, \qquad \Delta\ddot{u}_D = B_D{}^T TN\Delta\ddot{\lambda}, \tag{15.50-51}$$

while, from (15.28) and (15.51), we have

$$\Delta\mu = -mB_D{}^T TN\Delta\ddot{\lambda}. \tag{15.52}$$

Relations (15.3), (15.12), (15.17) and (15.52) then give

$$N^T T^T [B\Delta p - B_D m B_D{}^T TN\Delta\ddot{\lambda} + B_0\Delta F] + y = y_n. \tag{15.53}$$

Let us introduce the notation

$$M_q = B_D m B_D{}^T \tag{15.54}$$

for the symmetric inertia matrix referred to the acceleration increments $\Delta\ddot{q}_I$ of the independent member deformations.

Now, by collecting together (15.50), (15.53) and (15.54) with the appropriate complementarity conditions (15.14), (15.15) and (15.16), we find the following differential LCP.

$$\begin{bmatrix} -N^T T^T M_q TN & 0 \\ 0 & 0 \end{bmatrix}\begin{bmatrix} \Delta\ddot{\lambda} \\ \Delta\ddot{p} \end{bmatrix} + \begin{bmatrix} 0 & N^T T^T B \\ B^T TN & 0 \end{bmatrix}\begin{bmatrix} \Delta\lambda \\ \Delta p \end{bmatrix} + \begin{bmatrix} y \\ 0 \end{bmatrix} = \begin{bmatrix} y_n - N^T T^T B_0\Delta F \\ 0 \end{bmatrix}$$

$$y \geq 0 \qquad y^T\Delta\lambda = 0 \qquad \Delta\lambda \geq 0 \tag{15.55}$$

Again, a numerical solution of this formulation must be contemplated.

Numerical Solution by Newmark Integration

The acceleration components of the inertia coordinates u_D can be expressed approximately by Newmark's scheme

$$\Delta\ddot{u}_D = a_0\Delta u_D - a_1\dot{u}_{Dn} - a_2\ddot{u}_{Dn}, \tag{15.56}$$

where the constants a_0, a_1 and a_2 are as given in (15.36), (15.37) and (15.38) respectively.

Consider the acceleration term in (15.55). From (15.54) and (15.51) we have

$$-N^T T^T M_q TN\Delta\ddot{\lambda} = -N^T T^T B_D m B_D{}^T TN\Delta\ddot{\lambda} = -N^T T^T B_D m\Delta\ddot{u}_D \tag{15.57}$$

Substituting (15.56) into (15.57) gives

$$-N^TT^TM_qTN\Delta\ddot{\lambda} = -N^TT^TB_Dm[a_0\Delta u_D - a_1\dot{u}_{Dn} - a_2\ddot{u}_{Dn}]$$

$$= -a_0N^TT^TB_DmB_D{}^TTN\Delta\lambda + N^TT^TB_Dm[a_1\dot{u}_{Dn} + a_2\ddot{u}_{Dn}] \qquad (15.58)$$

This latter expression may be introduced into (15.55) from which the following approximating system (15.59) is produced.

$$\begin{bmatrix} -a_0N^TT^TM_qTN & N^TT^TB \\ B^TTN & 0 \end{bmatrix} \begin{bmatrix} \Delta\lambda \\ \Delta p \end{bmatrix} + \begin{bmatrix} y_{n+1} \\ 0 \end{bmatrix} = \begin{bmatrix} \Delta\bar{y}_n \\ 0 \end{bmatrix}$$

$$y_{n+1} \geq 0 \qquad y_{n+1}{}^T \Delta\lambda = 0 \qquad \Delta\lambda \geq 0 \qquad (15.59)$$

In these relations, the right-hand side subvector $\Delta\bar{y}_n$ is given by

$$\Delta\bar{y}_n = y_n - N^TT^TB_o\Delta F - N^TT^TB_Dm[a_1\dot{u}_{Dn} + a_2\ddot{u}_{Dn}] \qquad (15.60)$$

System (15.59) is a normal LCP and, once again, it may be solved by the same variant of Wolfe's algorithm employed in solution of the nodal LCP(15.42).

The partitioning of the mesh LCP(15.59) identifies it with the LCP(3.11), studied in Chapter 3. There, the variables $\bar{\lambda}$ were non-negative, while the corresponding variables Δp in (15.59) are required to be unrestricted. Such variables must be made permanent basic variables, and, since the associated submatrix – E in (3.11) – is null in (15.59), a sequence of double pivot operations is once more rendered necessary.

Updating of the acceleration and velocity components of the inertia coordinates u_D may be performed through the Newmark scheme:

$$\ddot{u}_{Dn+1} = a_0\Delta u_D - a_1\dot{u}_{Dn} - (a_2 - 1)\ddot{u}_{Dn} \qquad (15.61)$$

$$\dot{u}_{Dn+1} = a_3\Delta u_D - a_4\dot{u}_{Dn} - a_5\ddot{u}_{Dn} \qquad (15.62)$$

where the constants are given by (15.36) to (15.41) and $\Delta u_D = B_D{}^TTN\Delta\lambda$.

Again, in the interest of maintaining an appropriate level of accuracy, it would be necessary to determine $\ddot{u}_{Dn+1}$ through the satisfaction of the kinetic equations

$$Q_{I\,n+1} = Bp_{n+1} + B_oF_{n+1} - B_Dm\ddot{u}_{Dn+1} \qquad (15.63)$$

at each time increment, rather than through use of (15.61).

Initiation

To initiate the solution sequence, the accelerations $\ddot{u}_D(t = 0) = \ddot{u}_{D0}$ are required. They cannot be established through the kinetic equations because the initial values of the independent member forces $Q_I(t = 0) = Q_{I0}$ are indeterminate in a rigid-plastic structure.

Once again, we must establish the governing system in total variables at time $t = 0$. This follows the same line of reasoning presented for the nodal formulation, and the outcome is:

$$\left[\begin{array}{c|c} -N^TT^TM_qTN & N^TT^TB \\ \hline B^TTN & 0 \end{array}\right]\left[\begin{array}{c} \ddot{\lambda}_0 \\ \hline p_0 \end{array}\right] + \left[\begin{array}{c} y_0 \\ \hline 0 \end{array}\right] = \left[\begin{array}{c} R \\ \hline 0 \end{array}\right]$$

$$y_0 \geq 0 \qquad y_0{}^T\ddot{\lambda}_0 = 0 \qquad \ddot{\lambda}_0 \geq 0 \qquad (15.63)$$

System (15.63) is an LCP for time $t = 0$ which should be solved for $\ddot{\lambda}_0$, y_0 and p_0. The solution sequence we should like to obtain from (15.59) can then be inititated with

$$\ddot{u}_{D0} = B_D{}^TTN\ddot{\lambda}_0 \qquad (15.64)$$

$$Q_{I0} = Bp_0 + B_0F_0 - B_Dm\ddot{u}_{D0} \qquad (15.65)$$

PLASTIC UNSTRESSING OR DE-ACTIVATION

Suppose that the yield condition associated with the *kth* segment of some yield surface is currently active at time t – that is, $y_k(t) = 0$. If, at time t, it also happens that $\dot{y}_k(t) > 0$, then the system exhibits an event that we may call *plastic unstressing* or *yield-mode de-activation*. It turns out that the plasticity relations (15.9),(15.10) and (15.11), together with (15.25) and (15.26), imply that each of the terms in the scalar product (15.22) must vanish component-wise. Thus, $y_k\ddot{\lambda}_k = 0$ and $\dot{y}_k\dot{\lambda}_k = 0$. If plastic unstressing occurs, then $\dot{y}_k > 0$ implies that $\dot{\lambda}_k = 0$, and plastic flow must instantly cease.

The approximating plasticity relations (15.12) to (15.16) allow such unstressing to occur at the time t_n of commencement of each time increment $\Delta t = t_{n+1} - t_n$, but not at any subsequent time within the increment. Suppose that plastic unstressing is known to occur at some time $t_{n+\varepsilon}$ interior to the interval $[t_n, t_{n+1}]$. The solution process described above will only permit unstressing to happen at the time of commencement t_{n+1} of the next time increment. Plastic flow, as calculated in terms of the components $\Delta\lambda_k$, may therefore have been allowed to progress throughout the sub-interval $[t_{n+\varepsilon}, t_{n+1}]$ in contravention of the requirement that such flow should instantly cease at an unstressing event.

Rigid–plastic structural systems are exceedingly sensitive to such contraventions of the constitutive laws, and they manifest this sensitivity by wild, spurious oscillations in the calculated stress resultants. It is consequently imperative to incorporate a procedure within the numerical process which will correctly identify the instant $t_{n+\varepsilon}$ at which plastic unstressing occurs. The process should then be terminated temporarily at time $t_{n+\varepsilon}$ and the increments for all the relevant system variables should be determined for the sub–interval $[t_n, t_{n+\varepsilon}]$. Then, with $t_{n+\varepsilon}$ as the start–time t_0 for the next phase of the dynamic response, the numerical process may be continued with a return to the initiation routine.

EXAMPLES

Simply Supported Beam

This first example serves to test the proposed formulations and associated numerical procedures through a comparison with the exact solution previously obtained by Capurso[5]. The structure to be analysed is a simply supported beam with three equal and equispaced point masses, symmetrically loaded by two concentrated and rectangular pulses, as shown in Figure 15.1. The uniform and inextensible beam has plastic moment of resistance M_p, and the bending moment caused by the dead weight of the masses will be ignored.

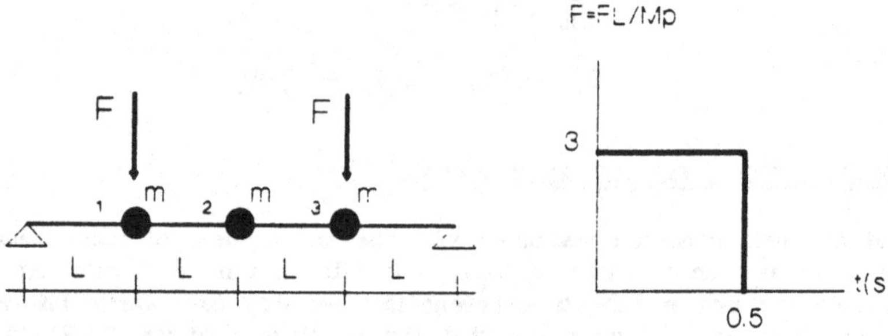

Figure 15.1

For the special case where the load is of magnitude $3M_p/L$ and duration 0.5 s, the ensuing motion – initiated from stationary conditions – occupies three distinct phases. The corresponding sequence of mechanisms – or velocity fields – is that depicted in Figure 15.2, where t1, t2 and t3 indicate the times of termination of the respective phases. The displaced configurations of the beam, taken at equal intervals of time throughout the motion, are shown in Figure 15.3.

Relevant components of the complete response are displayed in Figure 15.4. A notable feature of the transverse accelerations $\ddot{u}(1)$, $\ddot{u}(2)$, $\ddot{u}(3)$, corresponding to the similarly numbered masses and critical sections, is the instantaneous "jump" or change in value which occurs whenever a plastic hinge is activated or de–activated.

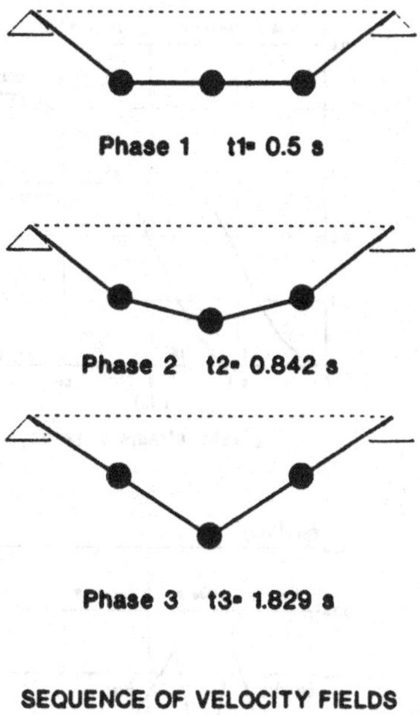

Figure 15.2

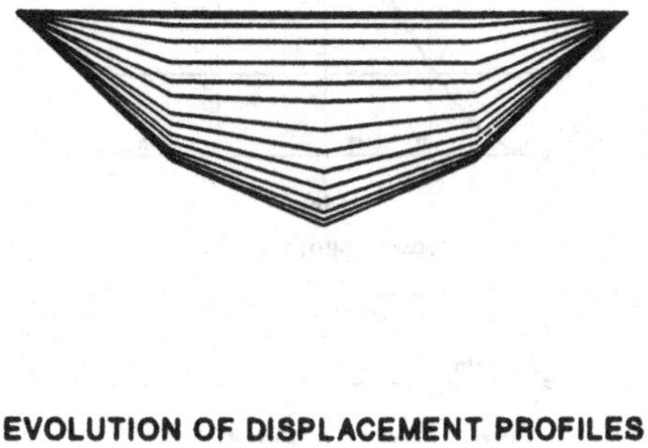

Figure 15.3

When the motion of the beam subsides, the transverse displacements of the masses are those given in Table 15.1. In this same table, a comparison with the results of Capurso[5], for the particular magnitude and duration of the loading pulse adopted herein, is also presented.

	Capurso	This approach	Difference (%)
$u(1)mL/M_p$	0.5000	0.4997	-0.060
$u(2)mL/M_p$	0.7222	0.7165	-0.789
Final time (s)	1.8333	1.8290	-0.235

Table 15.1 - Final Results

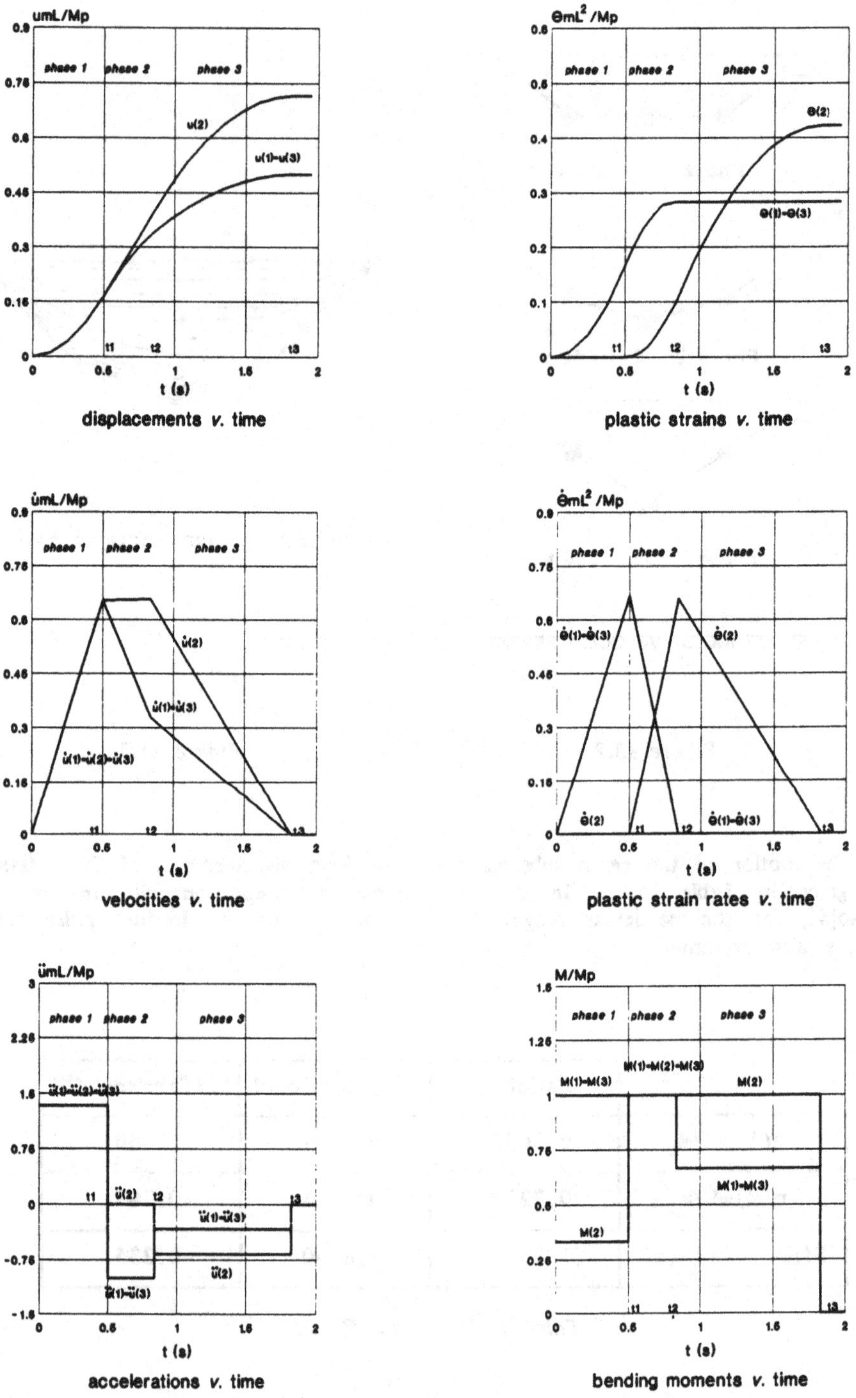
umL/Mp
phase 1
phase 2
phase 3
u(2)
u(1)=u(3)
t1
t2
t3
t (s)
displacements v. time
ΘmL²/Mp
Θ(2)
Θ(1)=Θ(3)
plastic strains v. time
u̇mL/Mp
u̇(2)
u̇(1)=u̇(3)
u̇(1)=u̇(2)=u̇(3)
velocities v. time
Θ̇mL²/Mp
Θ̇(1)=Θ̇(3)
Θ̇(2)
plastic strain rates v. time
üMl/Mp
ü(1)=ü(2)=ü(3)
ü(2)
ü(1)=ü(3)
accelerations v. time
M/Mp
M(1)=M(2)=M(3)
M(1)=M(3)
M(2)
bending moments v. time

Figure 15.4

Portal Frame with Hinged Bases

Capurso's elegant variational formulation describes the rigid-plastic dynamic response of structures in which motion is induced by *constant* pulse loads. The mesh and nodal formulations presented in this chapter, although not exact, are more straightforward, more automatic and can deal effectively with pulse loads of general form.

As a second example, the dynamic response of a single storey portal frame with hinged bases has been calculated for applied pulse loads of triangular form, as shown in Figure 15.5. The frame, which is fabricated from uniform and inextensible members having identical plastic moments of resistance M_p, supports a discrete mass m for which the rotational inertia is considered negligible. There are thus two inertia coordinates u_1 and u_2, identified in Figure 15.6, and the same figure shows the location of the three critical sections where discrete plastic hinges may develop.

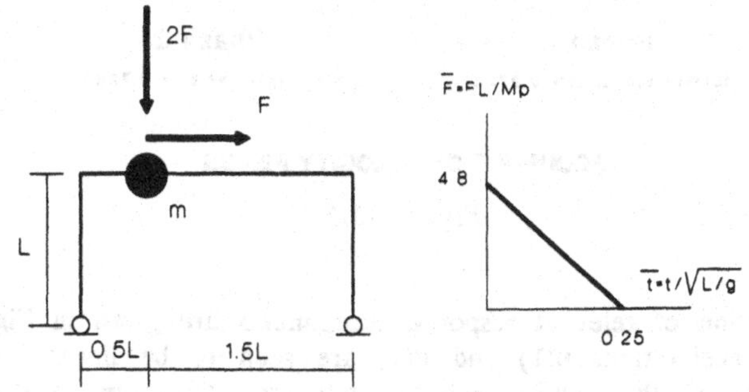

Figure 15.5

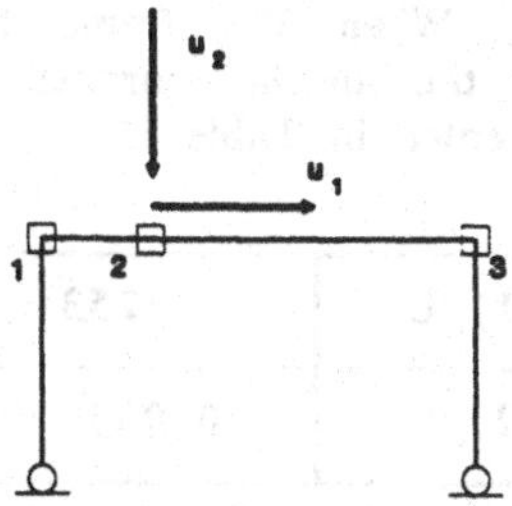

Figure 15.6

Stationary initial conditions are assumed, and the static bending moments

$$M_s(1) = M_s(3) = -0.281M_p, \quad M_s(2) = 0.469M_p,$$

caused by the dead weight mg of the mass (on frame members assumed to have the

same constant flexural rigidity, whatever its magnitude) are introduced into (15.7) to calculate the initial values of y_0. The rigid-plastic dynamic response is governed by the non-dimensional parameter $\beta_1 = mgL/M_p$ which is set at the specific value $\beta_1 = 2$ for this example. Furthermore, to preserve non-dimensionality throughout the calculation, the time parameter $\tau = t / \surd(L/g)$ is employed.

The impressed motion of the frame comprises two distinct phases for which the respective velocity fields are those shown in Figure 15.7.

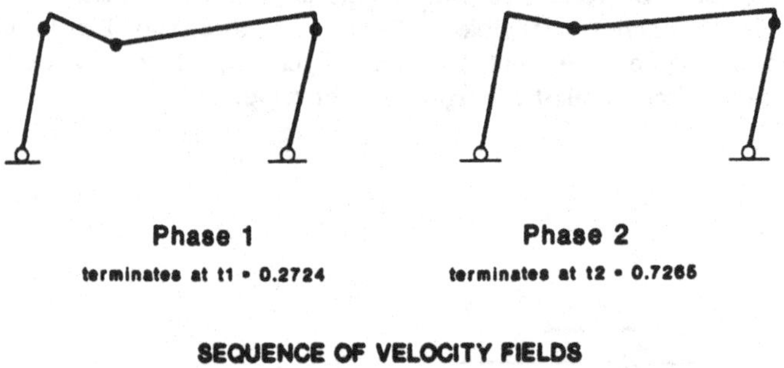

Figure 15.7

Details of the evolution of relevant response components are given in Figure 15.8. For this example, the accelerations $\ddot{u}(1)$ and $\ddot{u}(2)$ are seen to be linearly varying for the duration $\tau = 0.25$ of the pulse loads – with corresponding quadratic and cubic variations in the velocities and displacements respectively – and then they remain constant until the termination of Phase 1. Characteristic acceleration "jumps" occur as the structure switches mechanisms of motion from that of Phase 1 to that of Phase 2. The total bending moments at the three critical sections are represented by the sum of the dynamic bending moments shown in Figure 15.8 and the corresponding static bending moments given above. When the frame has come to rest, the final displacements in the directions of the inertia coordinates, as given by the mesh and nodal formulations, are those presented in Table 15.2.

u(1)/L	0.1253
u(2)/L	0.0857

Table 15.2 – Final Displacements

It should be noted that the formulations and associated numerical procedures can take full account of the effects of dead weight and of any pulse shape. Furthermore, the procedure also furnishes the complete evolution of all response parameters. This makes it possible to accommodate easily the effects of work hardening and strain rate sensitivity on the material constitutive relations.

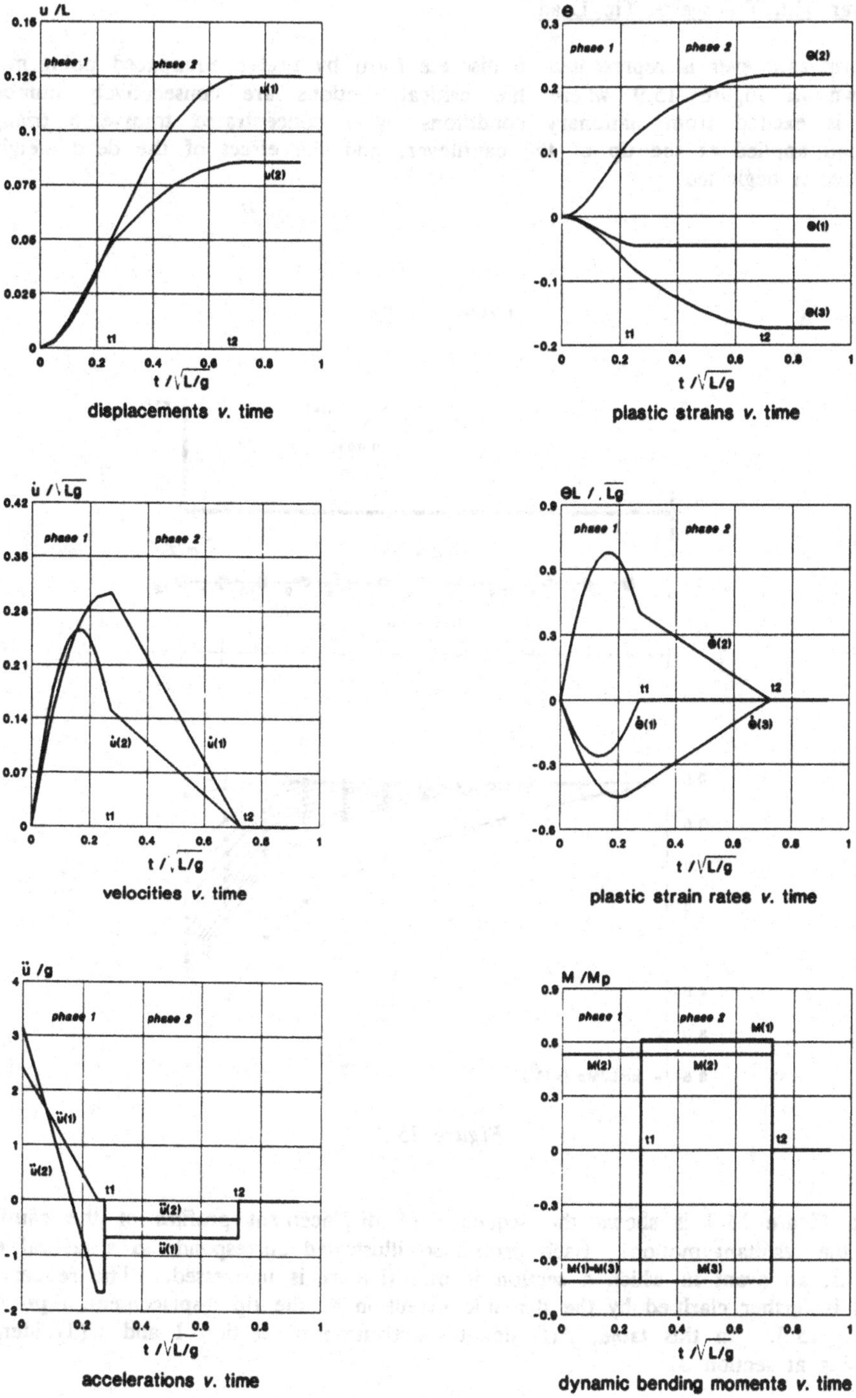
u /L
phase 1
phase 2
u(1)
u(2)
t1
t2
t /√L/g
displacements v. time
Θ
phase 1
phase 2
Θ(2)
Θ(1)
Θ(3)
t1
t2
t /√L/g
plastic strains v. time
u̇ /√Lg
phase 1
phase 2
u̇(2)
u̇(1)
t1
t2
t /√L/g
velocities v. time
Θ̇L /√Lg
phase 1
phase 2
Θ̇(2)
Θ̇(1)
Θ̇(3)
t1
t2
t /√L/g
plastic strain rates v. time
ü /g
phase 1
phase 2
ü(1)
ü(2)
t1
t2
t /√L/g
accelerations v. time
M /Mp
phase 1
phase 2
M(1)
M(2)
M(2)
M(1)-M(3)
M(3)
t1
t2
t /√L/g
dynamic bending moments v. time

Figure 15.8

Cantilever with Transverse Tip Load

A uniform cantilever is represented in discrete form by twelve equispaced point masses, as shown in Figure 15.9 where the critical sections are consecutively numbered. Motion is excited from stationary conditions by a concentrated transverse triangular pulse load applied at the tip of the cantilever, and the effect of the dead weight of the masses is neglected.

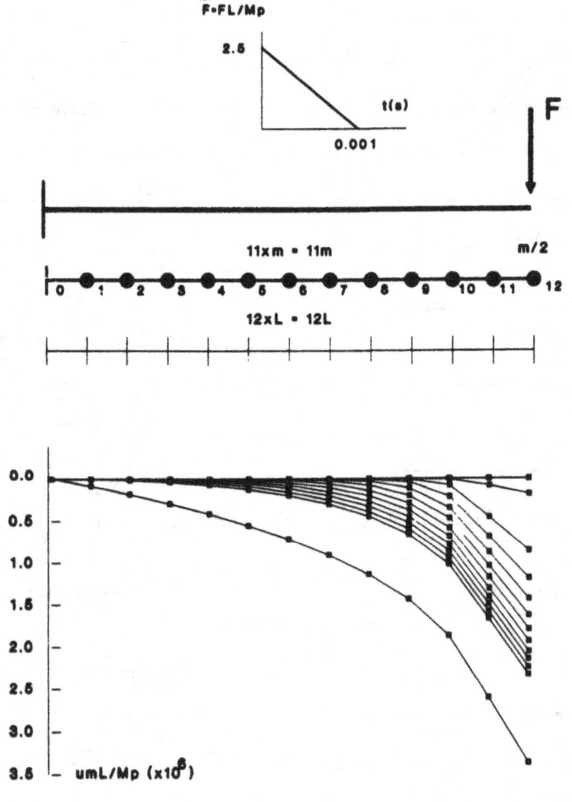

Figure 15.9

Also in Figure 15.9 is shown the sequence of displacement profiles of the cantilever during the resultant motion. Each profile so illustrated corresponds to a critical event – that is, an event in which a section is plastified or is unstressed. The sequence of profiles is further clarified by the dynamic evolution of the tip displacement u presented in Table 15.3. In this table, A(I) denotes activation of section I and U(J) identifies unstressing at section J.

Event	time (ms)	$umL/M_p \times 10^6$
A(11), A(10)	0.000	0.000
U(11), A(9)	0.400	0.185
U(10), A(8)	1.111	0.864
U(9), A(7)	1.562	1.193
U(8), A(6)	2.000	1.439
U(7), A(5)	2.431	1.636
U(6), A(4)	2.857	1.802
U(5), A(3)	3.281	1.944
U(4), A(2)	3.704	2.068
U(3), A(1)	4.125	2.179
U(2), A(0)	4.545	2.279
U(1)	4.965	2.371
U(0) - standstill	14.99	3.415

Table 15.3 - Tip displacement

The evolution of plastic deformation is conveniently illustrated in Figure 15.10. Initially, plastic hinges form simultaneously at sections 10 and 11. Thereafter, as each succeeding section becomes plastified, the active section nearest to the tip unstresses. In this way, at each instant, two adjacent sections will exhibit active plastic hinges, and this pair appears to "walk" towards the fixed end of the cantilever.

If the number of masses and critical sections were to be increased indefinitely, this modelling would, in the limit, identify a single travelling plastic hinge. It would commence at some internal point of the cantilever's span, dependent upon the initial magnitude of the load pulse, and would move smoothly to the fixed end. With the vanishing of the load upon the uniform cantilever, the velocity of propagation of the travelling hinge would assume a constant value. This latter proposition is confirmed,

for instance, by the theoretical solution of Hodge[9] for the classical problem of a uniform cantilever subjected to an impulse at its tip.

For the present example, the initial value $\bar{F}$ = 2.5 of the load has been chosen so that plastic deformation would commence at the section nearest to the tip, i.e section 11. After the elapsing of 1 ms, when the load is no longer present, it will be observed in Figure 15.10 that the active pair of plastic hinges advance towards the fixed end section 0 with a constant velocity of propagation.

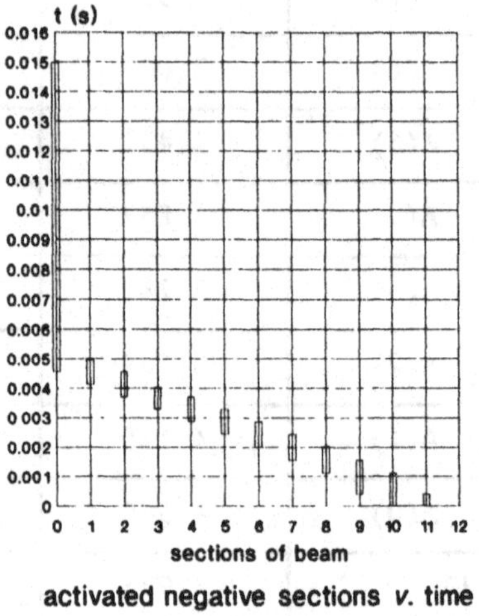

Figure 15.10

It is seen that the rigid–plastic response for this discrete modelling of a one–dimensional continuum comprises two distinct phases. At the commencement, there is a transient phase, of 4.96 ms duration, in which the plastic hinges are "walking" towards the root of the cantilever. There follows a concluding phase in which a single plastic hinge is active at the root of the cantilever; the resulting motion exhibits a *mode form* – that is, it develops a velocity profile whose characteristic shape is sustained througout the phase. For this example, the shape of the velocity profile associated with the mode form coincides with that of the mechanism of static plastic collapse. As Table 15.3 indicates, the corresponding mode form motion endures from 4.96 ms to 14.99 ms.

For interest, let us consider the dynamic response of a 254 × 102 British Universal Beam section in Grade 43 steel for which the unit mass is 22 kg/m and M_p = 65375 Nm. For $\bar{F}$ = 2.5, and assuming that the cantilever is 4 m long, the maximum value of the applied load F will be 490.31 kN. Then the final or permanent tip displacement and the maximum tip velocity will be 9.13 cm and 27.52 m/s respectively. In addition, the velocity of longitudinal propagation of the "walking" hinges during the no–load interval of the transient phase will be 780 m/s approximately.

CLOSING REMARKS

For the two formulations – mesh and nodal – presented herein, that associated with the mesh description has the smaller LCP tableau. Consequently, the mesh formulation attracts a more rapid numerical solution since most of the computational effort is expended in solving the LCP at each time step.

In such calculations of rigid–plastic dynamic response, the size of the time increment Δt needs to satisfy only two criteria: that the load pulse is effectively digitised and that plastic unstressing may be accurately detected. Failure to re–start the solution process from the instant of each unstressing can lead to wild, spurious oscillations in some of the response parameters with a consequential accumulation of errors in others.

In the authors' experience, the Newmark parameters $\beta = 0.25$, $\gamma = 0.5$, which represent a constant average acceleration scheme for numerical integration, appear to give a stable and robust procedure for rigid–plastic dynamic response calculations. This particular choice of the parameters is strongly reinforced by a simple observation: for most of the motion impressed by pulse loading, the accelerations are piecewise constant and Newmark's constant average acceleration scheme gives exact results.

REFERENCES

1. Martin, J. B., Extremum principles for a class of dynamic rigid–plastic problems, Int. J. Solids Structures, 8 (1972) 1185–1204.

2. Symonds, P. S., Survey of methods of analysis for plastic deformation of structures under dynamic loading, Brown University Report BU/NSRDC/1–67, 1967.

3. Jones, N., Response of structures to dynamic loading, Inst. Phys. Conf. Ser., No. 47 Chapter 3 (1979) 254–276.

4. Kaliszky, S., Dynamic plastic response of structures, in: Plasticity Today: Modelling, Methods and Applications, (Eds. A. Sawczuk and G. Bianchi) Elsevier Applied Science Publishers, London 1986, 787–820.

5. Capurso, M., A quadratic programming approach to the impulsive loading analysis of rigid–plastic structures, Meccanica, 7 (1972) 45–57.

6. Cannarozzi, A. A. and Laudiero, F., On plastic dynamics of discrete structural models, Meccanica, 11 (1976) 23–35.

7. Lloyd Smith, D., Plastic limit analysis and synthesis of structures by linear programming, PhD Thesis, University of London 1974.

8. Al–Samara, M. A., Elastoplastic dynamics of skeletal structures by mathematical programming, PhD Thesis, University of London 1986.

9. Hodge, P. G., Jr., Plastic Analysis of Structures, McGraw–Hill 1959.

CHAPTER 16

BOUNDING TECHNIQUES AND THEIR APPLICATION TO SIMPLIFIED PLASTIC ANALYSIS OF STRUCTURES

T. Panzeca, C. Polizzotto and Ș. Rizzo
University of Palermo, Palermo, Italy

ABSTRACT

In the framework of the simplified analysis methods for elastoplastic analysis problems, the bounding techniques possess an important role. A class of these techniques, based on the so-called perturbation method, are here presented with reference to finite element discretized structures. A general bounding principle is presented and its applications are illustrated by means of numerical examples.

1. INTRODUCTION

In modern technology and industrial activities, structures are quite often required to undergo complex programmes of loadings, as repeated cycles of loadings or random sequences of loadings within a fixed load domain. Then, the complete analysis of the structures, which likely operate beyond their own elastic limit, turns out to be a very hard computational task even for rather simple structures. For this reason, the existence and availability of simplified methods of analysis play a crucial role in structural design, at least in the preliminary phase of element or component proportioning and the study and development of such methods at different levels of applicability and approximation deserve much interest in structural engineering practice.

One category of simplified methods of analysis is constituted by the so-called "bounding techniques". These are appropriate analytical-numerical procedures leading to the evaluation of a scalar quantity, B, which bounds from above some specific scalar measure, η, of the inelastic deformation experienced by the structure, in such way that the following inequality holds:

$$\eta \leqq B. \qquad (1.1)$$

η can have a variety of meanings in practice, such as components of inelastic strain and residual displacements, inelastic dissipation work and the like; also, η can have "local" or "global" character, according to whether it pertains to a specific point of the body, or to a region which is a finite portion of, or even coincides with, the entire volume of the body.

Since η is related to the actual behaviour of the loaded structure and is therefore to be considered as a time function $\eta(t)$, it can in principle be exactly evaluated only at the cost of a complete evolutive analysis to be pursued from the initial time $t = 0$ up to some $t > 0$, following the actual or the most effective loading path. Without this analysis, a piece of information over the unknown deformation measure η can be provided, in armony with eq. (1.1), through the knowledge of the appropriate upper bound B. Obviously, the closer B to η, the more useful the piece of information it provides.

A bounding technique can be distinguished for the following two features:

a) the simplicity of the related analytical-numerical procedure, and

b) the stringentness of the upper bounds B provided with respect to the actual deformation measures η.

But these features are obviously in conflict with each other, for it is reasonable to conjecture that the simpler the procedure applied, the less stringent the upper bound one can obtain. To give an idea, one can say that an upper bound to be useful must possess an order of magnitude not greater than that of the related actual deformation measure and, also, that an upper bound a few times greater than the related actual deformation measure may pratically be regarded as a not bad bound.

Within a class of bounding techniques, (see, e.g. [1-8]) holding in the commonest case of repeated loadings, the upper bound B does not depend on the loading path producing the actual deformation, nor on the time t at which η is referred to. Therefore, B turns out to be an upper bound upon the peak value η_{mx} which η takes on within the interval (0,t). For this class of bounding techniques there is perhaps a good compromise between simplicity of the related analytical-numerical procedures and stringentness of the bounds provided. However, there are some drawbacks with this type of bounding techniques, namely the bounds provided rapidly diverge as the load intensity approaches the relevant shakedown limit and this type of bounding technique is in fact applicable only for load intensities smaller than the above limit.

In the following, bounding techniques for repeated loadings will be treated. This topic will be addressed through the Perturbation Method introduced by Polizzotto [7-11] and constituting a generalization of the Dummy Load Method due to Ponter [3,5] and studied also by others [1,6,2].

Although the Perturbation Method, as pointed out in [7,11], applies also to the case of elastic-plastic-creeping materials, for the sake of brevity the discussion will here be confined to elastic, perfectly plastic materials. Moreover, in order to emphasize the computational aspects of the method, discrete structural models will be treated. Quasi-static loandings will be considered first, then an extension to dynamic loadings will be shown.

2. THE STRUCTURAL MODEL

The typical structure considered in the following is an aggregate of n finite elements connected at their nodes with one another and with fixed supports to prevent any rigid motion. Every element of the system is assumed elastic, perfectly plastic. With reference to the i-th element, let $\underset{\sim}{q}_i$ be the vector (of m_i components) which describes the intrinsic deformation of the element, and let $\underset{\sim}{Q}_i$ be the vector (of m_i components, too) which describes the forces applied at the element "free" nodes (the element is supposed to be restrained against rigid motion in a statically determinate manner) and called (element) "generalized stresses" or simply "stresses" in the following. In other words, $\underset{\sim}{q}_i$ and $\underset{\sim}{Q}_i$ describe natural deformation and stress modes of the element [12-14].

As usual within the framework of small deformations, the (total) deformation q_i can be divided into three parts, i.e.

$$\underset{\sim}{q}_i = \underset{\sim}{e}_i + \underset{\sim}{p}_i + \underset{\sim}{\theta}_i \tag{2.1}$$

where $\underset{\sim}{e}_i$ is the elastic part, $\underset{\sim}{p}_i$ the plastic part, while $\underset{\sim}{\theta}_i$ denotes the imposed deformation, arising for instance from temperature variations of the element. The elastic deformation $\underset{\sim}{e}_i$ is related to $\underset{\sim}{Q}_i$ by

$$\underset{\sim}{e}_i = \underset{\sim}{A}_i \underset{\sim}{Q}_i, \quad \text{or} \quad \underset{\sim}{Q}_i = \underset{\sim}{D}_i \underset{\sim}{e}_i \tag{2.2}$$

in which $\underset{\sim}{A}_i$ is the deformability matrix of the element and $\underset{\sim}{D}_i = \underset{\sim}{A}_i^{-1}$ is its stiffness matrix. Both $\underset{\sim}{A}_i$ and $\underset{\sim}{D}_i$ are symmetric, positive definite and of order m_i.

The elastic domain of the i-th element, i.e. the region of the $\underset{\sim}{Q}_i$-space in which the element deforms elastically, is assumed to be a hyperpolyhedron of ν_i faces each of which is specified by its unit external normal vector $\underset{\sim}{N}_i^h$ and by its distance $\underset{\sim}{R}_i^h$ from the stress point $\underset{\sim}{Q}_i = \underset{\sim}{0}$ (Fig. 1).

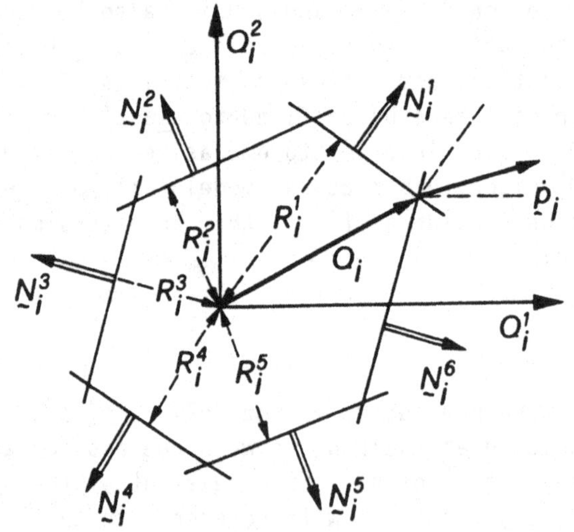

Fig. 1 - Typical element elastic domain ($m_i = 2$, $\nu_i = 6$).

The condition that the point $\underset{\sim}{Q}_i$ be inside the elastic domain is therefore expressed by the following ν_i inequalities

$$\underset{\sim}{N}_i^{hT} \underset{\sim}{Q}_i - \underset{\sim}{R}_i^h \leq 0, \quad (h = 1,2,\ldots,\nu_i) \tag{2.3}$$

where the superscript T is the matrix transposition operator. In matrix form, eq. (2.3) can be written

$$\begin{bmatrix} \underset{\sim}{N}_i^{1T} \\ \underset{\sim}{N}_i^{2T} \\ \vdots \\ \underset{\sim}{N}_i^{\nu_i T} \end{bmatrix} \underset{\sim}{Q}_i - \begin{bmatrix} R_i^1 \\ R_i^2 \\ \vdots \\ R_i^{\nu_i} \end{bmatrix} \leq \begin{bmatrix} 0 \\ 0 \\ \vdots \\ 0 \end{bmatrix} \tag{2.4}$$

i.e.

$$\underset{\sim}{N}_i^T \underset{\sim}{Q}_i - \underset{\sim}{R}_i \leq \underset{\sim}{0}, \tag{2.5}$$

where $\underset{\sim}{N}_i$ is the $m_i \times \nu_i$ matrix whose ν_i columns identify with the ν_i unit external normals $\underset{\sim}{N}_i^h$, respectively, and $\underset{\sim}{R}_i$ is the vector of the ν_i "plastic resistances" R_i^h.

When eqs. (2.5) are all satisfied as strict inequalities, the stress point $\underset{\sim}{Q}_i$ is internal to the elastic domain and the i-th element can deform only elastically. The element can also deform plastically when the stress point $\underset{\sim}{Q}_i$ is on the yield surface (i.e. the boundary surface of the hyperpolyhedron), that is when at least one of the ν_i inequalities (2.5) is satisfied as an equality. Whenever this is the case, the element plastic deformation rate vector is given by

$$\dot{\underset{\sim}{p}}_i = \underset{\sim}{N}_i^1 \dot{\lambda}_i^1 + \underset{\sim}{N}_i^2 \dot{\lambda}_i^2 + \ldots + \underset{\sim}{N}_i^{\nu_i} \dot{\lambda}_i^{\nu_i}, \quad \dot{\lambda}_i^h \geq \underset{\sim}{0} \quad \text{all } h \tag{2.6}$$

namely it is expressed as a linear combination of the unit external normal vectors (the $\underset{\sim}{N}_i^h$) with nonnegative coefficients (the $\dot{\lambda}_i^h$); in compact form, eq. (2.6) reads

$$\dot{\underset{\sim}{p}}_i = \underset{\sim}{N}_i \dot{\underset{\sim}{\lambda}}_i, \quad \dot{\underset{\sim}{\lambda}}_i \geq \underset{\sim}{0} \tag{2.7}$$

where $\dot{\underset{\sim}{\lambda}}_i$ is, by definition, the vector of the "plastic activation coefficient". In order that the definition (2.7) be consistent, the vector $\dot{\underset{\sim}{\lambda}}_i$ must also satisfy the following complementarity condition

$$\dot{\underset{\sim}{\lambda}}_i^T (\underset{\sim}{N}_i^T \underset{\sim}{Q}_i - \underset{\sim}{R}_i) = 0, \tag{2.8}$$

which, in virtue of eq. (2.5) and of the nonnegativeness of $\dot{\lambda}_i$, implies that a plastic activation coefficient $\dot{\lambda}_i^h$ can be different from zero only if the corresponding yield face is "activated", i.e. it contains the stress point Q_i.

The above rules for plastic flow of the element are in perfect agreement with the well known laws of plasticity theory [15,16]. In fact, introducing the vector of the "plastic potentials" ϕ_i given by

$$\phi_i = N_i^T Q_i - R_i \tag{2.9}$$

and then applying the usual plastic flow rules gives

$$\dot{p}_i = \frac{\partial \phi_i}{\partial Q_i} \dot{\lambda}_i = N_i \dot{\lambda}_i \tag{2.10}$$

which must hold together with the conditions

$$\phi_i \leq 0, \quad \dot{\lambda}_i \geq 0, \quad \phi_i^T \dot{\lambda}_i = 0, \quad \dot{\phi}_i^T \dot{\lambda}_i = 0. \tag{2.11}$$

The first three of these respectively identify with eq. (2.5), the second of eqs. (2.7) and eq. (2.8), while the last of eqs. (2.11) is a new one which takes into account the "unloading" or "elastic recovery" of the element [15,16].

In conclusion, the plastic behaviour of the i-th element is described by eqs. (2.9)-(2.11). Since the element is assumed to behave as elastic-perfectly plastic, both the matrix N_i and the vector R_i are constant so that the element elastic domain does not change its size and shape during plastic deformation. It is worth noting that not always the piecewise representation of the element elastic domain as in eq. (2.5) is exact; in most cases it is the result of a piecewise linearization in which the real smooth yield surface of the element is substituted by a polyhedral surface with more or less faces: obviously, the more the polyhedron faces, the better the resulting approximation. For more details on this topic, see [11,14,17,18].

The constitutive behaviour of the n structural elements considered altogether, but not yet assembled, can be described in a useful compact matrix form if appropriate supervectors and supermatrices are introduced. So we can write for the entire set of elements:

$$p = e + p + \theta \tag{2.12}$$

$$e = A\,Q, \quad \text{or} \quad Q = D\,e \tag{2.13}$$

$$\underset{\sim}{\phi} = \underset{\sim}{N}^T \underset{\sim}{Q} - \underset{\sim}{R} \tag{2.14}$$

$$\dot{\underset{\sim}{p}} = \underset{\sim}{N} \dot{\underset{\sim}{\lambda}} \tag{2.15}$$

$$\underset{\sim}{\phi} \leq \underset{\sim}{0}, \quad \dot{\underset{\sim}{\lambda}} \geq 0 \tag{2.16}$$

$$\underset{\sim}{\phi}^T \dot{\underset{\sim}{\lambda}} = 0, \quad \dot{\underset{\sim}{\phi}}^T \dot{\underset{\sim}{\lambda}} = 0. \tag{2.17}$$

Here $\underset{\sim}{A}$ and $\underset{\sim}{N}$ are block-diagonal matrices, i.e.

$$\underset{\sim}{A} = \lceil \underset{\sim}{A}_1 \; \underset{\sim}{A}_2 \ldots \underset{\sim}{A}_n \rfloor, \quad \underset{\sim}{N} = \lceil \underset{\sim}{N}_1 \; \underset{\sim}{N}_2 \ldots \underset{\sim}{N}_n \rfloor, \tag{2.18}$$

which have orders $m \times m$ and $m \times \nu$, with

$$m = m_1 + m_2 + \ldots + m_n, \quad \text{(total number of stress components)} \tag{2.19}$$

$$\nu = \nu_1 + \nu_2 + \ldots + \nu_n, \quad \text{(total number of yield faces)} \tag{2.20}$$

and $\underset{\sim}{D} = \underset{\sim}{A}^{-1}$; analogously, $\underset{\sim}{q}$, $\underset{\sim}{Q}$, $\dot{\underset{\sim}{\lambda}}$, $\underset{\sim}{R}$, etc. are supervectors collecting, one after the other and in the same order, the analogous vectors of all the elements, for instance

$$\underset{\sim}{q} = \begin{bmatrix} \underset{\sim}{q}_1 \\ \underset{\sim}{q}_2 \\ \vdots \\ \underset{\sim}{q}_n \end{bmatrix}, \quad \underset{\sim}{Q} = \begin{bmatrix} \underset{\sim}{Q}_1 \\ \underset{\sim}{Q}_2 \\ \vdots \\ \underset{\sim}{Q}_n \end{bmatrix}, \quad \dot{\underset{\sim}{\lambda}} = \begin{bmatrix} \dot{\underset{\sim}{\lambda}}_1 \\ \dot{\underset{\sim}{\lambda}}_2 \\ \vdots \\ \dot{\underset{\sim}{\lambda}}_n \end{bmatrix}, \quad \underset{\sim}{R} = \begin{bmatrix} \underset{\sim}{R}_1 \\ \underset{\sim}{R}_2 \\ \vdots \\ \underset{\sim}{R}_n \end{bmatrix} \tag{2.21}$$

When the n structural elements are assembled to generate the structure, two sets of linear equation must be satisfied by the deformation vector $\underset{\sim}{q}$ and the stress vector $\underset{\sim}{Q}$, respectively. Precisely, $\underset{\sim}{q}$ is expressed in terms of the vector $\underset{\sim}{u}$ of the n_f degrees of freedom of the system (i.e. system node displacements) by means of the compatibility equations

$$\underset{\sim}{q} = \underset{\sim}{C}\, \underset{\sim}{u}, \quad \underline{\text{(compatibility)}} \tag{2.22}$$

while $\underset{\sim}{Q}$ is required to be in static equilibrium with the applied loads, i.e.

$$\underset{\sim}{C}^T \underset{\sim}{Q} = \underset{\sim}{F}, \quad \underline{\text{(static equilibrium)}} \tag{2.23}$$

where $\underset{\sim}{F}$ is the load vector. Note that eq. (2.23) is equivalent to n_f scalar equilibrium equations, that is to as many as the number of the system degrees of freedom. The matrix $\underset{\sim}{C}$, of order $m \times n_f$, depends only on the overall geometric properties of the undeformed system, not on the geometric or physical properties of the elements. In eq. (2.23) inertia and damping forces are assumed to be so small as to be negligible. If this is not the case, the static equilibrium equations (2.23) transform into the dynamic ones

$$\underset{\sim}{C}^T \underset{\sim}{Q} + \underset{\sim}{V}\dot{\underset{\sim}{u}} + \underset{\sim}{M}\ddot{\underset{\sim}{u}} = \underset{\sim}{F} \quad \text{(\underline{dynamic equilibrium})} \tag{2.24}$$

where $\dot{\underset{\sim}{u}}$ and $\ddot{\underset{\sim}{u}}$ are the velocity and acceleration vectors, While $\underset{\sim}{V}$ and $\underset{\sim}{M}$ are the appropriate damping and mass matrices, both symmetric and positive definite (and of order n_f).

Once the load history is specified through the vector-valued time functions $\underset{\sim}{F} = \underset{\sim}{F}(t)$ and $\underset{\sim}{\theta} = \underset{\sim}{\theta}(t)$, and the appropriate initial conditions at $t = 0$ assigned, the response of the above elastoplastic system can be obtained solving the governing equations (2.12)-(2.17), and (2.22) together with (2.23) or (2.24) according to whether the problem is a quasi-static problem or a dynamic one. In any case, the solving procedure must be evolutive in nature, that is the actual response of the structure must be constructed step by step through a sequence of small increments of the loads along the given path. Solution methods of this problem are not the object of the present notes and the interested reader can profitably consult the valuable book of J.B. Martin [15].

Before proceeding further, we remark that the structural model outlined hereabove applies to a wide class of structures, both of discrete and continuous type, but an appropriate discretization is required in the latter case. Just to illustrate the concepts, let us consider the truss structure of Fig. 2. Every bar element is in uniform states of stress and strain which can be characterized by only one stress component (the axial force, Q_i) and only one deformation component (the total elongation, q_i), so that $m_i = 1$ for all elements (Fig. 2b). The deformability matrix $\underset{\sim}{A}_i$ is now a scalar, i.e.

$$A_i = \frac{\ell_i}{E\Omega_i}$$

where ℓ_i is the bar length and Ω_i its cross section area, while E is the Young modulus. Denoting the material yield limit by σ_0, the bar plastic resistance (or yield axial force) is $R_i = \sigma_0 \Omega_i$ and therefore the bar elastic domain is defined by the inequalities

$$\begin{bmatrix} 1 \\ -1 \end{bmatrix} Q_i - \begin{bmatrix} R_i \\ R_i \end{bmatrix} \leq \begin{bmatrix} 0 \\ 0 \end{bmatrix}$$

Then we have

$$\underset{\sim}{N} = [\,1 \quad -1\,], \qquad \underset{\sim}{R}_i = \sigma_0\, \Omega_i \begin{bmatrix} 1 \\ 1 \end{bmatrix}$$

The instantaneous plastic deformation is expressed as

$$\dot{\underset{\sim}{p}}_i = \underset{\sim}{N}_i\, \dot{\underset{\sim}{\lambda}}_i = \dot{\lambda}_i^1 - \dot{\lambda}_i^2$$

and there are two plastic activation coefficients for every bar ($\nu_i = 2$). The truss of Fig. 2a possesses $n_f = 4$ degrees of freedom (the node displacements u_k, with $k = 1,2,3,4$).

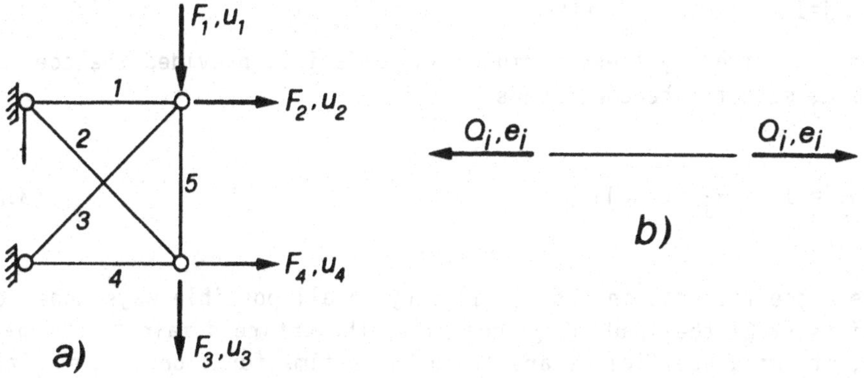

Fig. 2 - Truss structure: a) Assembled elements; b) Single element.

The compatibility matrix $\underset{\sim}{C}$, of order 5×4, is as follows:

$$\underset{\sim}{C} = \begin{bmatrix} 0 & 1 & 0 & 0 \\ 0 & 0 & 1/\sqrt{2} & 1/\sqrt{2} \\ -1/\sqrt{2} & 1/\sqrt{2} & 0 & 0 \\ 0 & 0 & 0 & 1 \\ -1 & 0 & 1 & 0 \end{bmatrix}$$

Note that the k-th column of $\underset{\sim}{C}$ represents the vector $\underset{\sim}{q}$ corresponding to $u_k = 1$, all other components of $\underset{\sim}{u}$ being assumed zero, remark which may be useful in constructing $\underset{\sim}{C}$.

3. A BOUNDING PRINCIPLE FOR REPEATED LOADINGS

In the following, we assume that the loads vary in a quasi-static manner within a specified multidimensional domain Π. The latter, by hypothesis, is shaped as a hyperpolyhedron of the load space and determined by the "basic loads" $\underset{\sim}{F}_j$, $\underset{\sim}{\theta}_j$ ($j = 1,2,\ldots, b$), which represent its b vertices. Any point inside Π or on its boundary surface represents an "admissible" load, i.e. a load which is potentially active, and it can obviously be represented as a convex linear interpolation of the basic loads, namely

$$\underset{\sim}{F} = \sum_{j=1}^{b} \underset{\sim}{F}_j \alpha_j \qquad \underset{\sim}{\theta} = \sum_{j=1}^{b} \underset{\sim}{\theta}_j \alpha_j . \tag{3.1}$$

The load $\underset{\sim}{F}$, $\underset{\sim}{\theta}$ given by these formulas is admissible provided the coefficients α_j satisfy the conditions

$$\sum_{j=1}^{b} \alpha_j = 1, \quad \alpha_j \geq 0 \quad \text{all } j. \tag{3.2}$$

When the α coefficients in eqs. (3.1) vary in all possible ways under the constraints (3.2) the load ($\underset{\sim}{F}$, $\underset{\sim}{\theta}$) describes the entire domain Π. In particular, if the α coefficient are assigned as time functions, $\alpha_j = \alpha_j(t)$, such as to satisfy (3.2) for every t, an "admissible loading history" is generated.

Now we consider an elastic-perfectly plastic structure modelled as said in the previous Section 2. When this structure is subjected to an admissible loading history, that is a sequence of loads inside Π, plastic deformation is likely produced and we want information about it without resorting to any step-by-step analysis. This is possible through the following procedure.

Besides the given structure, which is governed by eqs. (2.12)-(2.17), (2.22) and (2.23), let us consider a fictitious (or modified) structure differing from the real one only for its plastic resistance vector $\underset{\sim}{R}^*$. These are given by

$$\underset{\sim}{R}^* = \underset{\sim}{R} - \omega \underset{\sim}{d} \tag{3.3}$$

in which $\underset{\sim}{d}$ is an arbitrary vector (of ν components like $\underset{\sim}{R}$), hereafter called "perturbation vector", and ω is a nonnegative scalar, hereafter called "perturbation multiplier". The fictitious structure may therefore be thought of as being derived from the real structure through the application of a "perturbation" which modify the plastic resistance vector, but not the matrix $\underset{\sim}{N}$ of the outward normals. The elastic domain of the fictitious structure is therefore:

$$\underset{\sim}{N}^T \underset{\sim}{Q} - \underset{\sim}{R}^* \leqq \underset{\sim}{0}. \tag{3.4}$$

Let $\underset{\sim}{p}^*$ be the state of initial plastic deformation in the fictitious structure and let $\underset{\sim}{S}$ be the self-stresses uniquely corresponding to $\underset{\sim}{p}^*$, such that $\underset{\sim}{C}^T \underset{\sim}{S} = \underset{\sim}{0}$. The basic loads, separately applied on this structure, and in the hypothesis that <u>no further plastic deformation is produced</u>, give rise to the stresses, respectively;

$$\underset{\sim j}{Q}^* = \underset{\sim j}{Q}^E + \underset{\sim}{S}, \quad (j = 1,2,\ldots, b) \tag{3.5}$$

where $\underset{\sim j}{Q}^E$ is the purely elastic response both of the real and fictitious structures to the j-th basic load. Obviously, $\underset{\sim j}{Q}^*$ is inside the elastic domain (3.4), that is:

$$\underset{\sim}{N}^T \underset{\sim j}{Q}^* - \underset{\sim}{R} + \omega \underset{\sim}{d} \leqq \underset{\sim}{0}, \quad (j = 1,2,\ldots, b). \tag{3.6}$$

Let us now consider an admissible load history specified in the time interval $0 \leqq t \leqq t_1$, given by the time functions $\alpha_j = \alpha_j(t)$ satisfying the constraints (3.2) for every t. Multiplying eq. (3.6) by $\alpha_j(t)$ and summing with respect to j gives

$$\underset{\sim}{N}^T \left[\sum_{j=1}^{b} \underset{\sim j}{Q}^E \alpha_j(t) + \underset{\sim}{S} \right] - \underset{\sim}{R} + \omega \underset{\sim}{d} \leqq \underset{\sim}{0}. \tag{3.7}$$

Since the vector

$$\underset{\sim}{Q}^E(t) = \sum_{j=1}^{b} \underset{\sim j}{Q}^E \alpha_j(t) \tag{3.8}$$

is the elastic stress response of both the real and fictitious structures to the admissible load (3.1) acting at time t, if in analogy with eqs. (3.5) we set

$$\underset{\sim}{Q}^* = \underset{\sim}{Q}^*(t) \equiv \underset{\sim}{Q}^E(t) + \underset{\sim}{S}, \tag{3.9}$$

then eq. (3.7) becomes

$$\underset{\sim}{N}^T \underset{\sim}{Q}^* - \underset{\sim}{R} + \omega \underset{\sim}{d} \leq \underset{\sim}{0} \tag{3.10}$$

which holds for $0 \leq t \leq t_1$. Due to the arbitrariness of the admissible load history, we can state:

If for a given $\underset{\sim}{d}$ there exist $\underset{\sim}{S}$ and ω which satisfy eq. (3.6) at the b vertices of a convex hyperpolyhedral load domain, then they satisfy also eq. (3.10) at any point of this domain.

When the same load history is applied on the real structure, eqs. (2.12)-(2.17), (2.22) and (2.23) hold. In particular we have, at any t inside $(0, t_1)$:

$$\underset{\sim}{\phi}^T \dot{\underset{\sim}{\lambda}} = \underset{\sim}{Q}^T \dot{\underset{\sim}{p}} - \underset{\sim}{R}^T \dot{\underset{\sim}{\lambda}} = 0. \tag{3.11}$$

Subtracting the latter equality from the inequality which is easily derived by multiplying eq. (3.10) by $\dot{\underset{\sim}{\lambda}} \geq \underset{\sim}{0}$, i.e.

$$\underset{\sim}{Q}^{*T} \dot{\underset{\sim}{p}} - \underset{\sim}{R}^T \dot{\underset{\sim}{\lambda}} + \omega \underset{\sim}{d}^T \dot{\underset{\sim}{\lambda}} \leq 0, \tag{3.12}$$

we obtain

$$(\underset{\sim}{Q}^* - \underset{\sim}{Q})^T \dot{\underset{\sim}{p}} + \omega \underset{\sim}{d}^T \dot{\underset{\sim}{\lambda}} \leq 0. \tag{3.13}$$

On the other hand, the total deformations relative to the two structures considered are

$$\underset{\sim}{q}^* = \underset{\sim}{A} \underset{\sim}{Q}^* + \underset{\sim}{p}^* + \underset{\sim}{\theta}, \tag{3.14}$$

$$\underset{\sim}{q} = \underset{\sim}{A} \underset{\sim}{Q} + \underset{\sim}{p} + \underset{\sim}{\theta}, \tag{3.15}$$

so that subtracting one from the other gives

$$\underset{\sim}{q} - \underset{\sim}{q}^* = -\underset{\sim}{A}(\underset{\sim}{Q}^* - \underset{\sim}{Q}) + \underset{\sim}{p} - \underset{\sim}{p}^*. \tag{3.16}$$

By a differentiation with respect to t, since $\dot{\underset{\sim}{p}}^* = 0$, we obtain from eq. (3.16)

$$\dot{\underset{\sim}{p}} = \dot{\underset{\sim}{q}} - \dot{\underset{\sim}{q}}^* + \underset{\sim}{A}(\dot{\underset{\sim}{Q}}^* - \dot{\underset{\sim}{Q}}) \tag{3.17}$$

which can be substituted into eq. (3.13) to give

$$(\underset{\sim}{Q}^* - \underset{\sim}{Q})^T (\dot{\underset{\sim}{q}} - \dot{\underset{\sim}{q}}^*) + (\underset{\sim}{Q}^* - \underset{\sim}{Q})^T \underset{\sim}{A}(\dot{\underset{\sim}{Q}}^* - \dot{\underset{\sim}{Q}}) + \omega \underset{\sim}{d}^T \dot{\lambda} \leq 0. \tag{3.18}$$

Observing that $\underset{\sim}{Q}^* - \underset{\sim}{Q}$ is in equilibrium with zero external loads and that $\dot{\underset{\sim}{q}} - \dot{\underset{\sim}{q}}^*$ is a compatible instantaneous deformation, so that their scalar product vanishes, eq. (3.18) can be integrated over the time interval $(0, t_1)$ to give

$$\omega \underset{\sim}{d}^T \underset{\sim}{\lambda}(t_1) \le L(0) - L(t_1) \tag{3.19}$$

where $L(t)$ is, by definition, the positive definite quadratic form

$$L(t) = \frac{1}{2} (\underset{\sim}{Q}^* - \underset{\sim}{Q})^T \underset{\sim}{A} (\underset{\sim}{Q}^* - \underset{\sim}{Q}). \tag{3.20}$$

Finally, supposing that $\omega > 0$ and dropping the nonnegative subtractive term in the r.h. side of eq. (3.19) -- so enforcing the inequality -- we can write

$$\underset{\sim}{d}^T \underset{\sim}{\lambda}(t_1) \le \frac{1}{\omega} L(0) \quad \text{with} \quad L(0) = \frac{1}{2} \underset{\sim}{S}^T \underset{\sim}{A} \underset{\sim}{S}. \tag{3.21}$$

This is a bound of the form (1.1), provided that we set

$$\eta = \underset{\sim}{d}^T \underset{\sim}{\lambda}(t_1), \qquad B = \frac{1}{\omega} L(0) = \frac{1}{\omega} \frac{1}{2} \underset{\sim}{S}^T \underset{\sim}{A} \underset{\sim}{S}. \tag{3.22}$$

Inequality (3.21) constitutes a quite general Bounding Principle since it holds for any perturbation vector $\underset{\sim}{d}$, provided the conditions (3.5), and (3.6) are satisfied with $\omega > 0$.

4. A FEW IMPLICATIONS OF THE BOUNDING PRINCIPLE

The r.h. member of eq. (3.21) actually depends only on the basic loads which define the load domain Π, while the ℓ.h. member of the same eq. (3.21) --which represents some measure of the actual deformation-- depends on the load history considered inside Π and which causes this deformation. We shall see later on that a deformation measure in the form $\eta = \underset{\sim}{d}^T \lambda(t_1)$ is actually quite effective as it can take different practical meanings according to the way $\underset{\sim}{d}$ is chosen. Here we want remove or clarify a couple of limitations introduced in the derivation of the Bounding Principle of Section 3, that is the polyhedral form of the load domain Π and its convexity.

In effect, the actual load domain is often in the shape of a hyper-polyhedron, but if this is not the case, we can resort to the typical hyperpolyhedral domain by means of appropriate piecewise linearizations of the actual domain and at the cost of acceptable errors. To stay on the safe side, it is sufficient that the hyperpolyhedral surface be circumscribed to the actual boundary surface.

As for the convexity requirement of Π, this is actually not a limitation. To clarify this point, let us consider the nonconvex load domain of Fig. 3a, specified by the basic loads 1 to 7. To demonstrate the bounding inequality (3.21), it is strictly required that the inequality (3.10) be satisfied at <u>all</u> points of Π and hence also at the basic loads, 1,2,3 and 4, so that it is satisfied too at <u>all</u> points of the <u>convex</u> domain 1-2-3-4. As an immediate consequence, we can state:

a) The basic loads to consider as associated with a given load domain Π (convex or not) are by rule all those load points strictly required to define the convex hull Π_c of Π (i.e. the smallest convex domain circumscribed to Π);

b) The bounding inequality (3.21), which is founded on the satisfaction of eqs. (3.5) and (3.6) for all the basic loads of Π, is concerned with the actual plastic deformation produced as a result of any load history (even cyclic, with any number of cycles), provided it belongs to the convex hull Π_c of Π.

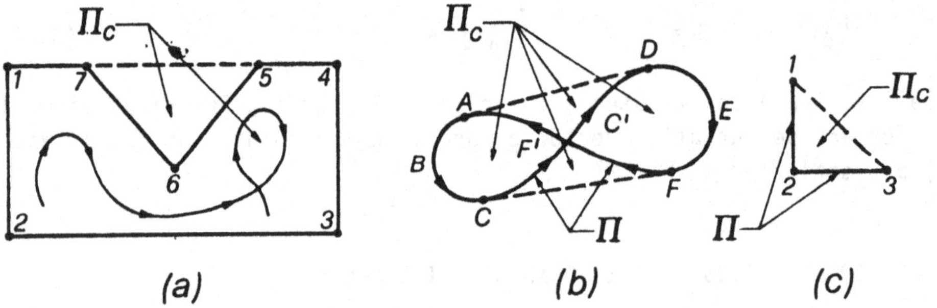

Fig. 3 - Typical load domains.

For example, for the load domain of Fig. 3b, shaped as the closed line ABCC'DEFF'A, the basic loads are the two arch segments ABC and DEF, while the convex hull is the region with boundary line ABCFEDA; similarly, for the bilateral domain Π of Fig. 3c, constituted by the two straight segments 1-2 and 2-3, the basic loads are the three isolated points 1,2 and 3, while the convex hull Π_c is the triangle 1-2-3.

One third consequence can be derived in relation to dynamic loads. A dynamic load history $\{F(t), \theta(t)\}$, not necessarily inside the load domain Π, is to be considered as "admissible" when the "equivalent load history" $\{\underset{\sim}{F}_{eq}(t), \underset{\sim}{\theta}(t)\}$ with $\underset{\sim}{F}_{eq}(t) = \underset{\sim}{F}(t) - \underset{\sim}{M}\,\ddot{\underset{\sim}{u}}^E(t) - \underset{\sim}{V}\,\dot{\underset{\sim}{u}}^E(t)$, is inside Π. For instance, the load history represented by the curve C of Fig. 4, would be not admissible as a quasi-static load history, but it is admissible as a dynamic load history provided its equivalent load history (curve C_{eq} in Fig. 3) is inside Π. It can be easily shown that the bounding inequality (3.21) holds true also for any dynamic load hystory corresponding to equivalent dynamic load histories inside Π_c.

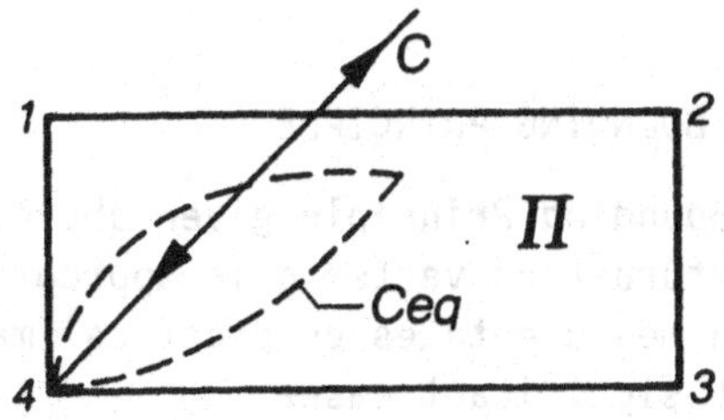

Fig. 4 - Typical admissible dynamic load history.

Since the proof of this third consequence is not so obvious, it may be useful to report it in short. The procedure followed in Section 3 remains unaltered up to the inequality (3.18), with the only difference that now the stress vector $\underset{\sim}{Q}^E(t)$ is to be meant as the dynamic elastic stress response to the given load history (and initial conditions), or, what is the same, as the quasi-static elastic response to the equivalent load history. The product $(\underset{\sim}{Q}^* - \underset{\sim}{Q})^T(\dot{\underset{\sim}{q}} - \dot{\underset{\sim}{q}}^*)$ is no longer zero now, because $\underset{\sim}{Q}^* - \underset{\sim}{Q}$ is in equilibrium with the inertia and damping forces, i.e.

$$\underset{\sim}{C}^T(\underset{\sim}{Q}^* - \underset{\sim}{Q}) = -\underset{\sim}{M}(\ddot{\underset{\sim}{u}}^* - \ddot{\underset{\sim}{u}}) - \underset{\sim}{V}(\dot{\underset{\sim}{u}}^* - \dot{\underset{\sim}{u}}), \tag{4.1}$$

such that, being $\dot{\underset{\sim}{q}} - \dot{\underset{\sim}{q}}^* = \underset{\sim}{C}(\dot{\underset{\sim}{u}} - \dot{\underset{\sim}{u}}^*)$, we have:

$$(\underset{\sim}{Q}^* - \underset{\sim}{Q})^T(\dot{\underset{\sim}{q}} - \dot{\underset{\sim}{q}}^*) = (\ddot{\underset{\sim}{u}}^* - \ddot{\underset{\sim}{u}})^T\underset{\sim}{M}(\dot{\underset{\sim}{u}}^* - \dot{\underset{\sim}{u}}) + (\dot{\underset{\sim}{u}}^* - \dot{\underset{\sim}{u}})^T\underset{\sim}{V}(\dot{\underset{\sim}{u}}^* - \dot{\underset{\sim}{u}}). \tag{4.2}$$

Introducing the positive definite quadratic form

$$G(t) = \frac{1}{2}(\dot{\underset{\sim}{u}}^* - \dot{\underset{\sim}{u}})^T\underset{\sim}{M}(\dot{\underset{\sim}{u}}^* - \dot{\underset{\sim}{u}}) + \int_0^t (\dot{\underset{\sim}{u}}^* - \dot{\underset{\sim}{u}})^T\underset{\sim}{V}(\dot{\underset{\sim}{u}}^* - \dot{\underset{\sim}{u}})\,d\tau, \tag{4.3}$$

eq. (4.2) can be rewritten as

$$(\underset{\sim}{Q}^* - \underset{\sim}{Q})^T(\dot{\underset{\sim}{q}} - \dot{\underset{\sim}{q}}^*) = \dot{G}(t) \tag{4.4}$$

such that the inequality (3.18), through an integration over the time interval $(0, t_1)$, and remembering (3.20), becomes

$$\omega \underset{\sim}{d}^T \underset{\sim}{\lambda}(t_1) \leq L(0) + G(0) - L(t_1) - G(t_1) \leq L(0) + G(0) \tag{4.5}$$

Since $G(0) = 0$ because $\dot{\underset{\sim}{u}} = \dot{\underset{\sim}{u}}^*$ at $t = 0$, eq. (4.5) coincides with eq. (3.21) for $\omega > 0$ and the statement given before is proved.

5. SPECIALIZATIONS OF THE BOUNDING PRINCIPLE

The interest of the Bounding Principle given above lies on the possibility of choosing the perturbation vector $\underset{\sim}{d}$ in appropriate ways such as the associated deformation measure takes on practical meanings. Hereafter we consider the most significant cases.

5.1 Bounds on the plastic dissipation work

Choosing $\underset{\sim}{d} = \underset{\sim}{R}$, remembering eq. (3.11) we can write:

$$\underset{\sim}{d}^T \underset{\sim}{\lambda}(t_1) = \underset{\sim}{R}^T \underset{\sim}{\lambda}(t_1) = \int_0^{t_1} \underset{\sim}{Q}^T \dot{\underset{\sim}{p}}\, d\tau. \tag{5.1}$$

Therefore the deformation measure $\eta = \underset{\sim}{d}^T \underset{\sim}{\lambda}(t_1)$ identifies with the plastic dissipation work produced in the entire actual structure during any admissible loading history.

If $\underset{\sim}{d}$ is chosen in such a way that its subvector $\underset{\sim}{d}_i$, relative to the i-th structural element, is equal to the relevant plastic resistance vector, i.e. $\underset{\sim}{d}_i = \underset{\sim}{R}_i$, while all the other subvectors of $\underset{\sim}{d}$ are vanishing, i.e. $\underset{\sim}{d}_r = \underset{\sim}{0}$ for all $r \neq i$, then we have:

$$\underset{\sim}{d}^T \underset{\sim}{\lambda}(t_1) = \underset{\sim}{R}_i^T \underset{\sim}{\lambda}_i(t_1) = \int_0^{t_1} \underset{\sim}{Q}_i^T \dot{\underset{\sim}{p}}_i\, d\tau. \tag{5.2}$$

In this way the deformation measure η turns out to coincide with the plastic dissipation work actually produced within the i-th structural element alone.

5.2 Bounds on the plastic deformation

Let $\hat{\underset{\sim}{Q}}$ be an arbitrary stress vector and let $\underset{\sim}{d}$ be chosen in the form $\underset{\sim}{d} = \underset{\sim}{N}^T \hat{\underset{\sim}{Q}}$. Thus, in virtue of eq. (2.15), we have

$$\underset{\sim}{d}^T \underset{\sim}{\lambda}(t_1) = \hat{\underset{\sim}{Q}}^T \underset{\sim}{N} \underset{\sim}{\lambda}(t_1) = \hat{\underset{\sim}{Q}}^T \underset{\sim}{p}(t_1) \tag{5.3}$$

which shows that the deformation measure η now takes on the form of a linear combination of the plastic deformation components accumulated at time t_1 in the actual structure, the combination coefficient being arbitrary. In particular, if $\hat{Q}$ possesses a single component different from zero, $\hat{Q}_i^h = 1$ say, eq. (5.3) becomes

$$d^T \lambda(t_1) = p_i^h(t_1) \tag{5.4}$$

and η coincides with the h-th component of the deformation vector $p_i(t_1)$ relative to the i-th element.

5.3 Bounds on the residual displacements

Let the stress vector $\hat{Q}$ of the previous case be the elastic stress response to some arbitrary static load $\hat{F}$ and let $d = N^T \hat{Q}$ again. Eq. (5.3) still holds. Since

$$C u^R = q^R = A Q^R + p, \qquad \text{all } t \geq 0, \tag{5.5}$$

where u^R, Q^R are elastic responses to p -- applied statically as imposed strain -- we can eliminate $p(t_1)$ from eq. (5.3), so obtaining

$$d^T \lambda(t_1) = \hat{Q}^T C u^R(t_1) - \hat{Q}^T A Q^R(t_1) = \hat{F}^T u^R(t_1) \tag{5.6}$$

since $\hat{Q}^T A Q^R = 0$ (Q^R is self-equilibrated, $A\hat{Q}$ is compatible). Eq. (5.6) indicates that in this case the deformation measure η identifies with a linear combination of the residual displacement components, with arbitrary combination coefficients. In particular, if we take $\hat{F}$ having a single component different from zero, $\hat{F}_k = 1$ say, eq. (5.6) gives

$$d^T \lambda(t_1) = u_k^R(t_1) \tag{5.7}$$

that is η coincides with the k-th component of the actual residual displacement vector evaluated at time t_1.

The important result of this Section is that the perturbation vector d must be specified in an appropriate manner taking into account the particular deformation measure of interest [9].

6. DIFFERENT USES OF THE BOUNDING PRINCIPLE

The Bounding Principle -- in virtue of its generalized formulation -- allows one to use it for approaching a number of different problems of relevant interest in the study of inelastic structures. Among them the following are worth being outlined:

(i) Assessment of plastic deformations developed along an assigned loading history, or during repeated admissible loading paths inside a given domain Π.

(ii) Evaluation of load domain multipliers s^* for bounded deformation, i.e. such that for any $s < s^*$ the actual plastic deformation is bounded.

(iii) Determination of the shakedown safety factor s_B, in presence of assigned limits on plastic deformations.

(vi) Optimal shakedown design of structures with constraints on deformations.

While the latter topic does not concern with this paper and the Seminar purposes and it will not herein treated, the first three problems deserve attention and will be therefore studied. Let us, to this purpose, make a few preliminary mathematical semplifications.

The vectors $\underset{\sim}{Q}_j^E$, $(j = 1,2,\dots,b)$, being known, let

$$\underset{\sim}{P} = \max_j \underset{\sim}{N}^T \underset{\sim}{Q}_j^E \tag{6.1}$$

where the maximum operation is to be applied component-wise. In addition we can express the self-stresses $\underset{\sim}{S}$ in terms of the redundant forces, i.e.

$$\underset{\sim}{S} = \underset{\sim}{L}\underset{\sim}{X}, \quad \underset{\sim}{N}^T\underset{\sim}{S} = \underset{\sim}{N}^T\underset{\sim}{L}\underset{\sim}{X} = \underset{\sim}{W}^T\underset{\sim}{X}, \quad \underset{\sim}{W} = \underset{\sim}{L}^T\underset{\sim}{N} \tag{6.2}$$

where $\underset{\sim}{L}$ is the relevant self-stress matrix. Then the modified or perturbed yield inequalities

$$\left.\begin{array}{ll} \underset{\sim}{N}^T(\underset{\sim}{Q}_j^E + \underset{\sim}{S}) + \omega\underset{\sim}{d} - \underset{\sim}{R} \le \underset{\sim}{0}, & (j = 1,2,\dots,b) \\ \underset{\sim}{C}^T\underset{\sim}{S} = \underset{\sim}{0}, \quad \omega > 0 & \end{array}\right\} \tag{6.3}$$

simplify as follows:

$$\underset{\sim}{P} + \underset{\sim}{W}^T\underset{\sim}{X} + \omega\underset{\sim}{d} - \underset{\sim}{R} \le \underset{\sim}{0}, \qquad \omega > 0. \tag{6.4}$$

6.1 <u>Assessment of plastic deformation via bound computation</u>

Once the vector $\underset{\sim}{d}$ has been suitably chosen, i.e. accordingly to the particular deformation measure η of interest, if a solution $(\hat{\underset{\sim}{X}}, \hat{\omega})$ exists such as to satisfy the linear inequalities (6.4), we can compute an upper bound $\hat{B}$ to η as follows:

$$\eta(t) \le \hat{B} \equiv \frac{1}{\hat{\omega}}\frac{1}{2}\,\hat{\underset{\sim}{S}}^T\underset{\sim}{A}\hat{\underset{\sim}{S}} = \frac{1}{\hat{\omega}}\,\frac{1}{2}\,\hat{\underset{\sim}{X}}^T\underset{\sim}{a}\hat{\underset{\sim}{X}} \tag{6.5}$$

in which we have set the matrix

$$\underset{\sim}{a} = \underset{\sim}{L}^T\underset{\sim}{A}\underset{\sim}{L} \tag{6.6}$$

as the flexibility matrix of the assembled primary structure. Such an approach gives answer to the problem (i) before postulated, no matter what the loading history, but belonging to the convex loading domain Π. Existance of the solution set $(\hat{X}, \hat{\omega})$ is ensured when the structure is able to adapt to variable repeated loading histories in Π, since -- accordingly to the well-known Bleich-Melan adaptation theorem -- at least one solution set $(X_0, \omega_0 = 0)$ exists, such as the inequalities (6.4)

$$P + W^T X_0 - R \leq 0 \tag{6.7}$$

particularized as adaptation conditions. But in this event the bound (6.5) diverges, unluckily. However, in order to obtain a solution $(\hat{X}, \hat{\omega})$ the following approach can be followed. Let us solve the two LP problems

$$\min_{(X,\omega)} \left\{ \omega \,\|\, P + W^T X + \omega d - R \leq 0, \quad \omega \geq 0 \right\} \tag{6.8}$$

$$\max_{(X,\omega)} \left\{ \omega \,\|\, P + W^T X + \omega d - R \leq 0, \quad \omega \geq 0 \right\} \tag{6.9}$$

Here, as usual, the symbol $\|$ separates the objective function on the left from the constraint equations on the right. The shakedown or adaptation solution $(X_0, \omega_0 = 0)$ is an admissible one for problem (6.8), while both (6.8) and (6.9) have not admissible solutions in the inadaptation case, which is therefore prevented from further discussion. In the former case, LP problem (6.9) can produce an unbounded solution, difficulty which is suitably removed by appending to the constraints in (6.9) the further constraint $\omega \leq \bar{\omega}$, where $\bar{\omega}$ is given a very large appropriate value. Once the sets $(X_0, \omega_0 = 0)$, $(X_1, \omega_1 \leq \bar{\omega})$ have been found, solutions to LP problems (6.8) and (6.9) respectively, a continuous set $(X_\alpha, \omega_\alpha)$ of admissible solutions to inequalities (6.4) can be obtained as

$$X_\alpha = \alpha X_1 + (1-\alpha) X_0, \quad \omega_\alpha = \alpha \omega_1, \quad (0 < \alpha \leq 1) \tag{6.10}$$

and therefore the inequality

$$\eta(t_1) = d^T \lambda(t_1) \leq \beta(\alpha) \tag{6.11}$$

holds with

$$\beta(\alpha) = \frac{B_\alpha}{\omega_\alpha} = \frac{1}{\alpha \omega_1} \left[\alpha^2 B_0 + (1-\alpha)^2 B_1 + \alpha(1-\alpha) X_0^T a X_1 \right] \tag{6.12}$$

and

$$B_0 = \frac{1}{2} \underset{\sim}{X}_0^T \underset{\approx}{a} \underset{\sim}{X}_0 \qquad B_1 = \frac{1}{2} \underset{\sim}{X}_1^T \underset{\approx}{a} \underset{\sim}{X}_1 . \tag{6.13}$$

The minimum β value is shown to be obtained for

$$\alpha = \alpha_{opt} = \sqrt{B_0/\Delta B} \tag{6.14}$$

with

$$\Delta B = \frac{1}{2} (\underset{\sim}{X}_1 - \underset{\sim}{X}_0)^T \underset{\approx}{a} (\underset{\sim}{X}_1 - \underset{\sim}{X}_0) . \tag{6.15}$$

6.2 The Shakedown Analysis Problem

6.2.0 The Shakedown Safety Factor s_{AD}

Once we have assumed some domain Π of loadings which act on a structure of given geometry and physical properties, we may be interested in knowing if, for arbitrary repeated quasi-static admissible loading programmes, the structure is able to adapt to such actions or not. The adaptation or shakedown phenomenon constitutes a vital event since: i) plastic deformations of limited amount may develop during the initial stages of the loading programmes and then they stop being developed -- if shakedown occurs -- the subsequent structural response being purely elastic; ii) such occurrence prevents collapse of structure eventually, since neither incremental collapse nor alternating plasticity collapse modes may take place. The well-known shakedown theorem [11,16], in this discrete framework, is expressed as:

The structure adapts if, and only if, there exists a time-independent force vector $\underset{\sim}{X}$, such as the inequality

$$\underset{\sim}{P} + \underset{\approx}{W}^T \underset{\sim}{X} - \underset{\sim}{R} \leq \underset{\sim}{0} \tag{6.16}$$

is satisfied.

Let us now introduce a load domain multiplier $s \geq 0$ by which proportional expansion or contraction of the load domain is produced, but its shape being not altered.

Let us define as <u>adaptation safety</u> factor s_{AD} the maximum value of the load domain multipliers satisfying the inequality

$$s\underset{\sim}{P} + \underset{\approx}{W}^T \underset{\sim}{X} - \underset{\sim}{R} \leq \underset{\sim}{0} . \tag{6.17}$$

The s_{AD} is obtained by solving the Linear Programming (LP) problem following:

$$\max_{(s,\underset{\sim}{X})} \left\{ s \;\|\; s\underset{\sim}{P} + \underset{\sim}{W}^T\underset{\sim}{X} - \underset{\sim}{R} \le \underset{\sim}{0} \right\}. \qquad (6.18)$$

Since the optimal solution (s_{AD}, X_{AD}) to the LP (6.18) exists, the structure is able to adapt to any loading programme inside Π_{AD}, i.e. the original domain Π amplified by s_{AD}. The LP problem (6.4) is a discrete formulation of the so-called shakedown analysis problem.

6.2.1 Safety factor s^* for bounded deformation

Let us now perturbate the yield surface by introduction of a suitable vector $\underset{\sim}{d}$, associated with a plastic deformation parameter of interest and amplified by a given fixed scalar $\omega > 0$.

The LP problem in (6.18) is therefore rewritten as

$$\max_{(s,\underset{\sim}{X})} \left\{ s \;\|\; s\underset{\sim}{P} + \underset{\sim}{W}^T\underset{\sim}{X} - (\underset{\sim}{R} - \omega\underset{\sim}{d}) \le \underset{\sim}{0} \right\}. \qquad (6.19)$$

From the solution set (s^*, $\underset{\sim}{X}^*$) to the LP problem (6.19) we can obtain an upper bound B to the relevant plastic deformation, i.e.

$$\eta = \underset{\sim}{d}^T \underset{\sim}{\lambda}(t_1) \le B \equiv \frac{1}{\omega}\frac{1}{2}\, \underset{\sim}{X}^{*T} \underset{\sim}{a}\, \underset{\sim}{X}^*. \qquad (6.20)$$

This use of the Bounding Principle is not a trivial one since it can provide, with very small further computational effort, information of interest -- namely boundedness of selected plastic deformation parameters -- for all loading histories inside Π^*, i.e. amplified by s^*.

6.2.2 Accounting for Limited Ductility Requirements

The assessment of the plastic deformations developed during complex loading programme constitutes a crucial point of the shakedown occurrence, from which stems the importance of the bounding techniques. Plastic deformations although finite may result in fact excessive to cope with appropriate serviceability or ductility requirements. In order to account for such requirements the analysis problem (6.18) needs to be suitably reformulated.

Let $\underset{\sim}{d}_i$, ($i = 1,\dots,m$), be a set of m perturbation vectors, suitably chosen in relation to a given set of deformation measures $\eta_i = \underset{\sim}{d}_i^T \underset{\sim}{\lambda}$. The latter are required to satisfy the inequalities

$$\eta_i = \underset{\sim}{d}_i^T \underset{\sim}{\lambda}(t_1) \le U_i, \quad (i = 1,\dots,m) \qquad (6.21)$$

where the U_i are some given upper limits to be imposed on the deformation measures η_i, respectively. Let s_ℓ be the maximum load amplifier for which the constraint in eq. (6.21), as well as the shakedown requirement, are contemporarily satisfied. The above s_ℓ is by definition the shakedown safety factor (SF) for limited ductility behaviour. When $m = 1$, $\underset{\sim}{d}_1 = \underset{\sim}{0}$ and U_1 is given a very large value, the simple shakedown SF will be obtained, i.e. $s_\ell = s_{AD}$, while in general the inequality $s_\ell \leq s_{AD}$ holds. The determination of s_ℓ requires knowledge of the actual elastoplastic response of the structure, for which a step-by-step analysis should be carried out. This analysis can be however avoided by using the Bounding Principle. Let a vector $\underset{\sim}{X}_i$ and a scalar $\omega_i > 0$ exist -- for any $i = 1,\ldots,m$ -- such as to satisfy

$$s\underset{\sim}{P} + \underset{\sim}{W}^T \underset{\sim}{X}_i - \underset{\sim}{R} + \omega_i \underset{\sim}{d}_i \leq \underset{\sim}{0}, \tag{6.22}$$

where s is a positive load amplifier. Hence, in virtue of the Bounding Principle (eq. 3.21), the following is also made to hold

$$\eta_i = \underset{\sim}{d}_i^T \underset{\sim}{\lambda}(t) \leq \frac{1}{\omega_i} \frac{1}{2} \underset{\sim}{X}_i^T \underset{\approx}{a} \underset{\sim}{X}_i \equiv B_i \tag{6.23}$$

The history-dependent constraints in eq. (6.21) can be therefore replaced -- in virtue of eq. (6.23) -- by

$$B_i \equiv \frac{1}{\omega_i} \frac{1}{2} \underset{\sim}{X}_i^T \underset{\approx}{a} \underset{\sim}{X}_i \leq U_i, \quad (i = 1,\ldots,m) \tag{6.24}$$

which are history-independent although more restrictive than eq. (6.21). Denoting with s_B the maximum load multiplier for which all the constraints in eq. (6.24) are satisfied, the inequality $s_B \leq s_\ell$ holds, in general as a strict inequality, provided that the simple shakedown requirement is satisfied, for instance by considering it as a particular specification of the perturbation vector, e.g. $\underset{\sim}{d}_1 = \underset{\sim}{0}$, $U_1 \gg 0$. Thus,

$$s_B = \max_{(s, \underset{\sim}{X}_i, \omega_i)} \left\{ s \;\middle\|\; \begin{array}{l} s\underset{\sim}{P} + \underset{\sim}{W}^T \underset{\sim}{X}_i - \underset{\sim}{R} + \omega_i \underset{\sim}{d}_i \leq \underset{\sim}{0} \\ \underset{\sim}{X}_i^T \underset{\approx}{a} \underset{\sim}{X}_i - 2\,\omega_i U_i \leq 0, \quad \omega_i \geq 0, \\ (i = 1,\ldots,m) \end{array} \right\} \tag{6.25}$$

which is nonlinear through the presence of m convex quadratic constraints derived from eq. (6.24). The nonlinear problem in eq. (6.25) can be attacked by the use of a NLP algorithm (e.g. [19,20]), all the nonlinear constraints being acccounted for. The following alternative computational strategies can be outlined.

(i) A very useful decomposition procedure of the NLP in eq. (6.25) was envisaged and implemented for the numerical applications herein included. This decomposition method consists in solving m separate similar NLP subproblems, each of them incorporating a single quadratic constraint. Let s_{Bi} denote the safety factor obtained for the i-th decomposed NLP problem; the set of s_{Bi}, $(i = 1,\dots,m)$, will be such as to hold the following

$$s_B = \min_{(i)} s_{Bi}, \quad (i = 1,\dots,m) \tag{6.26}$$

(see e.g., Polizzotto [9], Appendix C). The solution of m decomposed NLP problem shows to give computational advantages, because of the reduction of nonlinear constraints, with respect to the complete NLP in eq. (6.25), but still NLP algorithms need to be used as numerical tool. However, the above decomposed NLP problem can be transformed through the following step. Let us solve the following linear subproblem

$$\max_{(s,\underset{\sim}{X}_i)} \{s \parallel s\underset{\sim}{P} + \underset{\sim}{W}^T\underset{\sim}{X}_i + \omega_i \underset{\sim}{d}_i \leq \underset{\sim}{R}\} \tag{6.27}$$

for each $i = 1,2,\dots,m$, where the scalar ω_i is taken constant during the maximum operation and plays the role of parameter. Solution of the above parametric linear programming (PLP) problem [21] will give the piece-wise linear functions $s_i = s(\omega_i)$, $\underset{\sim}{X}_i = \underset{\sim}{X}(\omega_i)$. To each value of s a pair $(\underset{\sim}{X}_i, \omega_i)$ corresponds in the PLP solution, and from these the function $\bar{B}_i(s)$ is computed:

$$\bar{B}_i(s) = \frac{1}{2\omega_i(s)} \underset{\sim}{X}_i^T(s)\, \underset{\sim}{a}\, \underset{\sim}{X}_i(s). \tag{6.28}$$

Let $\bar{s}_{Bi}$ denote the value of s such as the following holds

$$\bar{B}_i(\bar{s}_{Bi}) = U_i; \tag{6.29}$$

since in general in the solution to the NLP subproblem the nonlinear constraint is satisfied as an equality, it results that $\bar{s}_{Bi} = s_{Bi}$. In any case the vector $\underset{\sim}{X}_i(\bar{s}_{Bi})$ and the scalar $\omega_i(\bar{s}_{Bi})$ together with $\bar{s}_{Bi}$ are a feasible solution to the NLP subproblem (6.25): hence, $\bar{s}_{Bi} \leq s_{Bi}$, and as a results $\bar{s}_B \leq s_B$ where $s_B = \min_{(i)} s_{Bi}$ and $\bar{s}_{Bi} = \min_{(i)} \bar{s}_{Bi}$. Therefore,

$$\bar{s}_B \leq s_B \leq s_\ell, \tag{6.30}$$

but in general $\bar{s}_B = s_B$. The decomposed PLP problems have the advantages of being linear.

(ii) To compute the load multiplier s_B in presence of m quadratic convex constraints in the problem (6.25), a different procedure consists in transforming the last mentioned constraints so as to have "separate variables" and subsequently in making a piecewise linearization of the obtained quadratic functions. This change of the nonlinear problem makes it possible to find the safety factor s_B in presence of limited ductility requirements by step-by-step procedures, the latter factor being not greater than the shakedown safety factor s_{AD}.

Let us introduce into the problem (6.25) but written for a single nonlinear constraint, new variables by using the following linear transformation

$$\underset{\sim}{X} = \underset{\sim}{T}\,\underset{\sim}{Y} \tag{6.31}$$

where $\underset{\sim}{T}$ is the eingenvector matrix of $\underset{\sim}{a}$. Thus the quadratic form of the nonlinear problem can be defined in this way:

$$\underset{\sim}{X}^T \underset{\sim}{a}\,\underset{\sim}{X} = \underset{\sim}{Y}^T \underset{\sim}{T}^T \underset{\sim}{a}\,\underset{\sim}{T}\,\underset{\sim}{Y} = \underset{\sim}{Y}^T \underset{\sim}{\Lambda}\,\underset{\sim}{Y} \tag{6.32}$$

with $\underset{\sim}{\Lambda} = \underset{\sim}{T}^T \underset{\sim}{a}\,\underset{\sim}{T}$ as the eigenvalue matrix.

The nonlinear problem therefore takes on the following form:

$$s_B = \max_{(s,\underset{\sim}{Y},\omega)} \left\{ s \;\middle\|\; \begin{array}{l} s\underset{\sim}{P} + \underset{\sim}{\bar{N}}^T \underset{\sim}{Y} - \underset{\sim}{R} + \omega \underset{\sim}{d} \leq \underset{\sim}{0} \\ \underset{\sim}{Y}^T \underset{\sim}{\Lambda}\,\underset{\sim}{Y} - 2\omega U \leq 0, \quad \omega \geq 0 \end{array} \right\} \tag{6.33}$$

where $\underset{\sim}{\bar{N}}^T = \underset{\sim}{N}^T \underset{\sim}{L}\,\underset{\sim}{T} = \underset{\sim}{W}^T \underset{\sim}{T}$.

The quadratic function, present in the problem (6.33), is a linear combination of square single variables, namely

$$\underset{\sim}{Y}^T \underset{\sim}{\Lambda}\,\underset{\sim}{Y} = \sum_{k=1}^{n} \lambda_k Y_k^2 = \sum_{k=1}^{n} f_k \tag{6.34}$$

with $f_k = \lambda_k Y_k^2$, where n is the redundant force number of the structure. Each f_k function may be submitted to appropriate piecewise linearization, like that shown in Fig. 5. It consists in choosing for each function the number and size of the intervals, so obtaining the following equations:

$$Y_k = b_{ok} + \sum_{r=1}^{\nu} b_{rk} Z_{rk}, \qquad f_k = h_{ok} + \sum_{r=1}^{\nu} h_{rk} Z_{rk} \tag{6.35}$$

where the nonnegative variables Z_{rk} are required to satisfy some constraints depending on the particular interval (working interval) where f_k is computed. Namely, if j is the working interval, the Z_{rk} variables may be classified and constrained as follows:

-- "Exhausted" variables Z_{rk}: $Z_{rk} = 1$ for all $r < j$;

-- "Working" variables Z_{jk}: $0 \le Z_{jk} \le 1$; (6.36)

-- "Nonworking" variables Z_{rk}: $Z_{rk} = 0$ for all $r > j$.

With the aid of these piecewise linearizations of the quadratic forms in the problem (6.33), the latter can be reduced to the simpler form:

$$s_B = \max_{(s,\underset{\sim}{Y},\underset{\sim}{Z},\omega)} \left\{ s \;\middle\|\; \begin{array}{l} s\underset{\sim}{P} + \underset{\sim}{\tilde{N}}^T \underset{\sim}{Y} - \underset{\sim}{R} + \omega \underset{\sim}{d} \le \underset{\sim}{0} \\ \sum_{k=1}^{\nu} f_k - 2\omega U \le 0, \quad \omega \ge 0 \end{array} \right\} \qquad (6.37)$$

in which the unknowns $\underset{\sim}{Y}$ and $\underset{\sim}{Z}$ must comply with the constraints (6.35) and (6.36). The solution procedure is iterative in nature, each iteration being characterized by the choice of the working intervals relative to the f_k functions, and the best choice is that which maximizes s_B. For every nonlinear constraint in (6.25), a problem like (6.37) must be solved and then eq. (6.26) is to be applied; but all the m nonlinear constraints may by accounted for simultaneously.

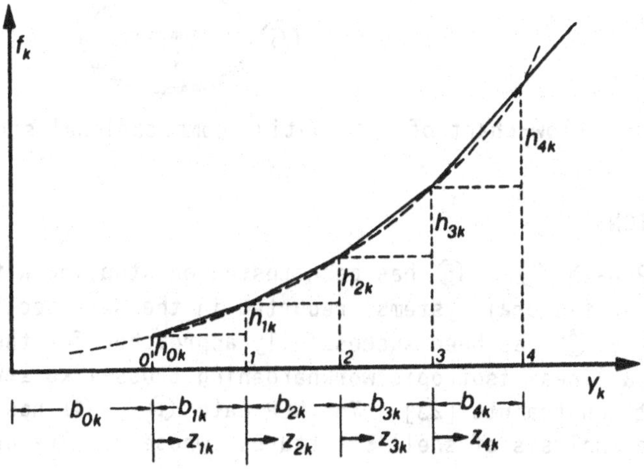

Fig. 5 - Linearization of quadratic function $f_k = \lambda_k y_k^2$.

The last procedure, which allows the piecewise linearization of quadratic functions as in (6.34), is implemented in an IBM program and is called "Delta Method" [26].

The flow-chart in Fig. 6 summarizes the three alternative computational strategies of the shakedown analysis problem in presence of ductility constraints.

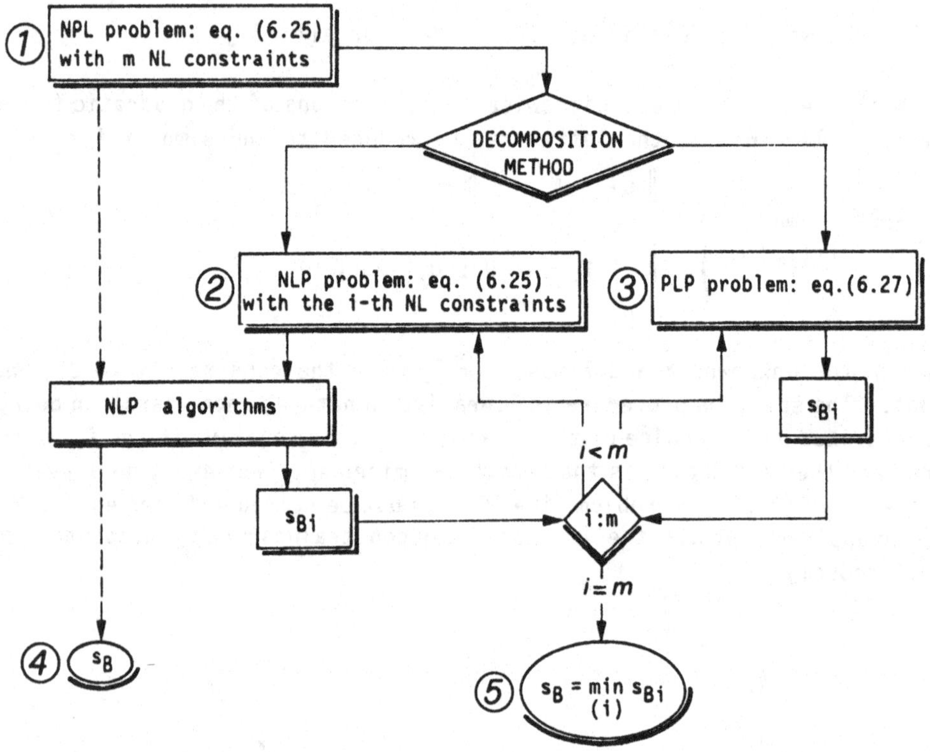

Fig. 6 - Flow-chart of alternative computational strategies.

7. APPLICATIONS

The PLP path ① - ⑤ has been tested on studying a few examples of truss-like and flexural systems, reported in the next section [23]. The NLP path ① - ⑤ has been successfully approached for the shakedown analysis of a linear isotropic workhardening truss-like structure with displacement constraints [23]. The NLP path ① - ④ has been also used [24] for the analysis of shell structures, whose results are herein reported.

7.1 PLP - Approach

The decomposed PLP approach was therefore used to study a flexural system, suitably discretized into a finite number of elements (FE), and truss-like structures.

The first example regards a two-bays one-storey rigid-joints steel frame, Fig. 7a), subjected to vertical loads and to a horizontal load. The cross-section of columns and beams are given in Fig. 7a), the height $\ell = 300$ cm. being assumed. The frame was discretized into 7 beam-column bending elements, the three columns and the beams being subdivided into half-span elements. The 15 DOF of the FE scheme were reduced to only 8 by disregarding the deformations produced by the axial forces. The yield bending moments were computed by $M_0 = 2W_0\sigma_0$, with M_0 the first area moment of half cross-section with respect to the neutral bending axis, and with the steel yield stress $\sigma_0 = 40$ kN/cm². The intensity of the applied loads were $F_1 = 25$ kN and $F_2 = 40$ kN. The problem was written in terms of the bending moments, the tableau of the linear constraints having a matrix of 24 rows by 8 columns. A primary structure was obtained by introducing hinges at the joints B,C,D,E,F in Fig. 7a); upon application of self equilibrated unit couples at these joints, the self-stress matrix $\underset{\sim}{L}_{6\times 12}$ was easily computed. Fig. 7c) reports the plots of $U_1(s)$ and $U_2(s)$, i.e. of the upper limits imposed on the residual vertcal displacement u_H and the absolute value of the relative rotation Φ_H at the plastic hing H, mid of the larger span beam CE. The structure exhibits an elastic limit $s_{EL} = 1.9601$ and a shakedown safety factor $s_{AD} = 2.1307$. These plots cannot be used however for $U_1 < 2.3$ and $U_2 < .13$, since otherwise a reduced SF smaller than s_{EL} would be obtained. As example, for assigned $U_1 = 3$ cm. the curve $U_1(s)$ gives a SF $s_B = 2.01$, such that for any $s < 2.01$ it results $u_H \leq 3$ cm.

The second example regards the truss-like structure, the simple seven-bars truss in Fig. 8a). The bars have the same cross-section, for simplicity, and the load can vary according to the load domain in b), same Fig. 8, given in dimensionless form by the scalar variable ratios $\tau_i = F_i/F_0$ $(i = 1,2)$, of the applied forces F_i with respect to a given reference force $F_0 = \sigma_0 A_0$. A yield stress $\sigma_0 = 40$ kN/cm², reference cross-section $A_0 = 2$ cm², Young Modulus $E = 20\cdot 10^6$ N/cm², were assumed. Loads are made acting at the nodes. In Fig. 8c) are given the plots of the upper limits upon the residual displacements of the nodes, both vertical and horizontal displacement components. The upper limits are given in dimensionless form (U/ℓ). The structure exhibits an elastic limit $s_{EL} = 1.3691$ and a shakedown safety factor $s_{AD} = 1.4204$. If $U_1/\ell = U_2/\ell = U_3/\ell = U_4/\ell = 0.002$, i.e. assuming a common value of upper limits to all the residual displacements, it results that the curve 4 gives the most severe safety factor, namely $s_B = s_{B4} = 1.4035$, with a 33% reduction magnitude on the elastic-shakedown SF interval $[(s_{AD} - s_{B4})/(s_{AD} - s_{EL})]$. Available FORTRAN-IV Codes for the PLP problem were used to compute the bound functions before described: the QLP Code [25] for the two truss-like examples, and the MPSX Code [26], for the rigid joints frame. Both programs were run a 3033-IBM computer, the latter however showing a faster computational speed.

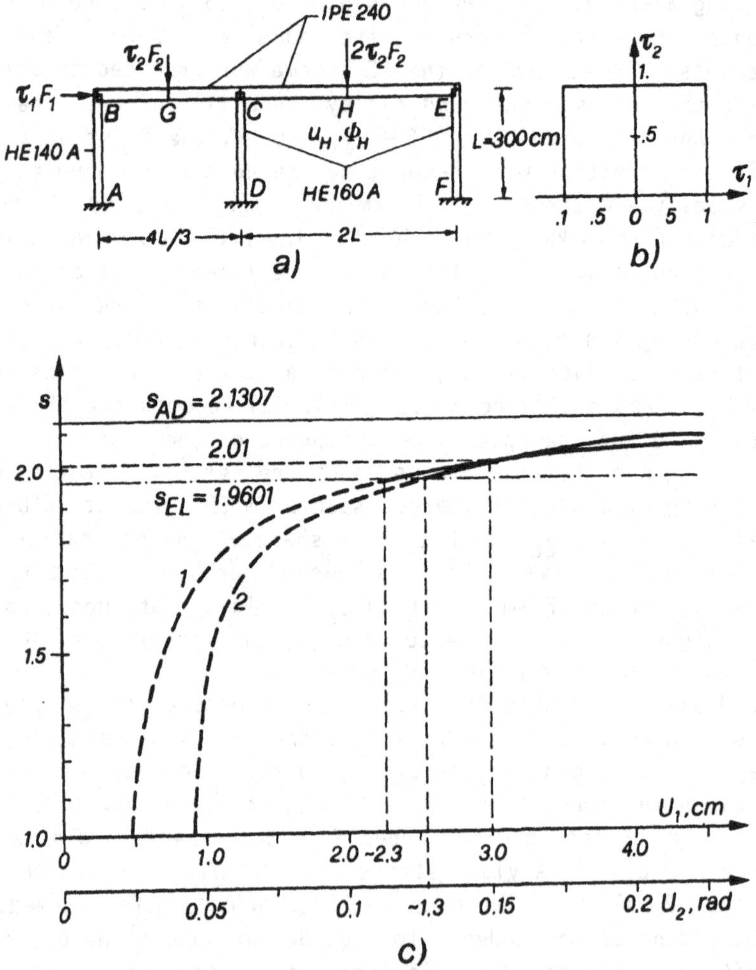

Fig. 7 - Rigid-joints steel frame subjected to quasi-static vertical and horizontal loads. Shakedown analysis, limits being imposed on vertical residual displacement u_H and relative rotation Φ_H of cross-section H. a) geometry and load scheme; b) dimensionless loading domain; c) plots of U_1 and U_2, upper limits to u_H and Φ_H, as functions of s.

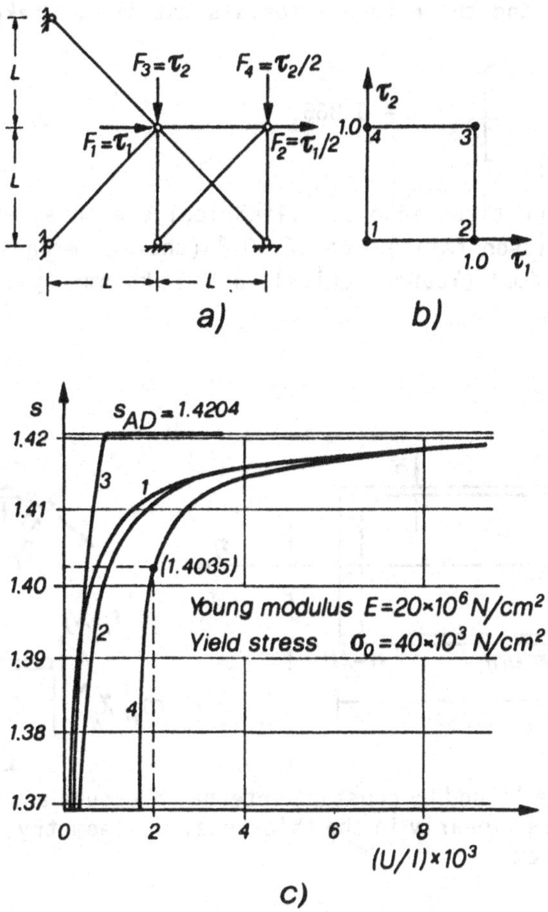

Fig. 8 - Simple truss-like structure (Ex. n. 2) subjected to vertical and horizontal nodal loads, all the nodal displacements being restricted. a) geometry and load scheme; b) dimensionless loading domain; c) plots of U_1, U_2, U_3 and U_4 as functions of s.

7.2 NLP - Approach

The non-linear approach has been utilized in the study of rotationally symmetric shells [24-27], discretized by the substitution of the meridian curve by a broken line, thus obtaining n conical frusta with interposed ring nodes. In particular, the cantilever pipe of Fig. 9a) was studied, with internal pressure sq and temperature variation sT linearly variable in the thickness h. The loading domain was assumed in the shape depicted in Fig. 10a), with the relevant basic loads identified with the vertices O A B C.

The smooth yield surface of the pipe [28] was made piecewise linear as shown in Fig. 10b) and the external normals matrix was taken as:

$$N = \begin{bmatrix} 1 & -1 & -1 & 1 \\ \gamma & \gamma & -\gamma & -\gamma \end{bmatrix}, \qquad \gamma = 0.866.$$

The pipe was discretized into 10 cylindrical elements, with $r = 0.3\,\ell$, $h = 0.03\,\ell$, $\mu = 0.3$ (Poisson ratio), $\alpha = 1.2\cdot 10^{-5}$ (thermal expansion coefficient), E= 21.000 kN/cm² (Young modulus), $\sigma_0 = 42$ kN/cm² (yield stress).

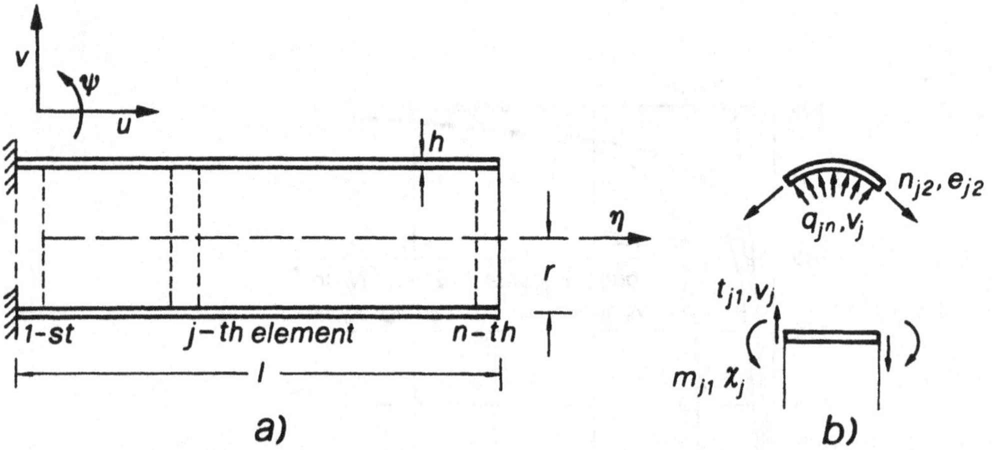

Fig. 9 - Cantilever pipe loaded by constant internal pressure and by temperature T variable linearly in the thickness: a) geometry; b) element reference system.

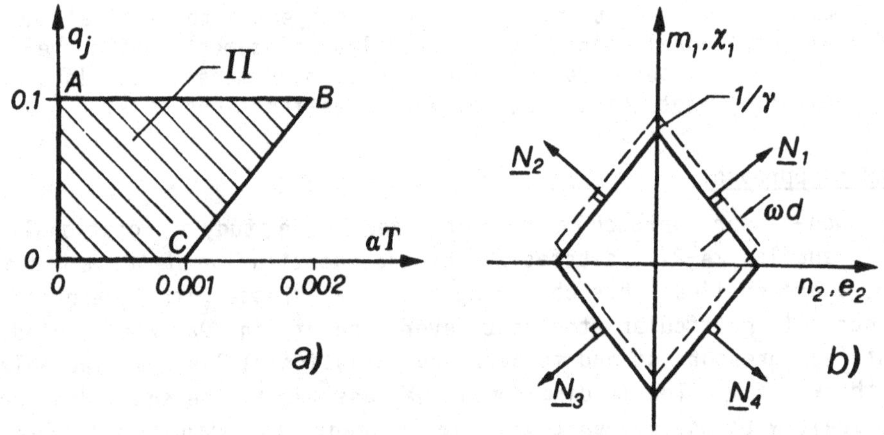

Fig. 10 - a) Loading domain; b) Piecewise linear yield surface.

The shakedown safety factor $s_{B1} = s_{AD}$ was obtained first ($\underset{\sim}{d}_1 = 0$, $U_1 \gg 0$). Then the plastic curvature χ_1 at theclamped end section of the pipe was constrained and the relevant safety factor s_B computed. Thus, as shown in [24], the perturbation mode to consider is in the form $\underset{\sim}{d}_2 = [\gamma\ \gamma\ -\gamma\ -\gamma]^T$, that is a rigid traslation of the yield surface in the direction of χ_1, as shown in Fig. 10b. Once a value to the upper limit U is assigned, the safety factor s_B can be obtained by solving the linear programming problem (6.37) through a step-by-step procedure shown in Sec. 6.2.2. (ii). In Fig. 11 the shakedown safety factors s_{AD} and s_B are reported, the latter being plotted as a function of U. It is easy to see that s_B is not greater than the shakedown safety factor s_{AD}, and that the more stringent U, the smaller s_B.

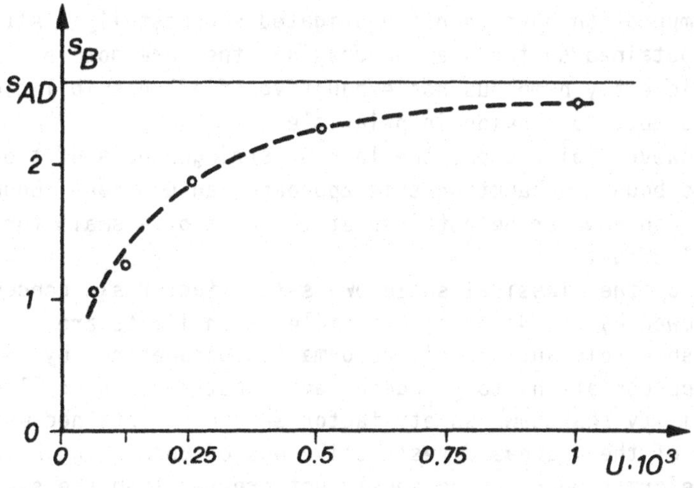

Fig. 11 - Load multiplier for the pipe structure of Fig. 9a: Comparison between the shakedown safety factor s_{AD} and the safety factor in presence of limited ductility requirements (upper limit on the plastic curvature in the clamped section).

8. CONCLUSION

The inelastic analysis of structures always requires -- as a necessary corollary -- the assessment of inelastic deformations which may develop during complex and severe programmes of actions which they undergo more and more frequently, nowadays. To this purpose, a Bounding Principle has been herein illustrated which -- in virtue of its generalized character -- is shown to be well suited to give answer to a few problems of relevant interest within such a framework. Let us observe then, as conclusions:

(i) The perturbation method possesses an advantageous unified formulation which allows to provide upper bounds to a number of physical parameters of technical interest apt to describe the inelastic behaviour of the structures. The advantages of such an approach stem from both theoretical and computational points of view.

(ii) This method shows its wide range of applicability, but for variable repeated load programmes which are safely inside the adaptation domain Π_{AD}.

(iii) Suitably discretized elasto-plastic structures -- of various geometry, typology and subjected to different actions -- have been numerically analized. The discrete formulation of the Bounding Principle leads to Nonlinear Mathematical Programming problems, having a discrete number of variables. Various alternative strategies of numerical computation have been investigated successfully. Altough the results obtained so far are encouraging, they are however neither sufficiently numerous nor exhaustive of all possibilities the method is able to provide in principle.

(vi) It must be however pointed out the lack of stringentness of the values of the bounding function this approach can provide. Bound optimization can however be obtained at the cost of a small further computational effort.

(v) Reassessment of the classical shakedown safety factor s_{AD} concept has been allowed by the Bounding Principle, when limits are assigned on some relevant plastic deformation parameter. By introduction of appropriate history-independent constraints so-called limited ductility shakedown safety factor s_B can be obtained which -- in virtue of the increased restrictiveness of constraint being imposed on deformation -- is obviously not greater than the s_{AD}.

(vi) Applications for elastic-isotropic linear workhardening structures as well as for optimal shakedown design have been also provided.

9. AKNOWLEDGEMENT.

This paper is part of a research programme financially supported by the Italian CNR (National Research Council) and by the MPI (Public Education Ministry), whose contributions are herein acknowledged.

REFERENCES

1. Capurso, M., Corradi, L., and Maier, G.: Bounds on deformations and displacements in shakedown theory, in: Proc. of the Seminar on Materials and Structures under cyclic loads, Laboratoire de Mécanique des Solides, Ecole Polytechnique, Palaiseau, (France), Sept. 28 and 29, 1978, 231-244.
2. Leckie, F.A.: Review of bounding techniques in shakedown and ratchetting at elevated temperature, Welding Research Council Bullettin, 195, (1974), p. 1.
3. Ponter, A.R.S.: An upper bound on the small displacements of elastic, perfectly plastic structures, J. Appl. Mech., 39, (1972), 959-963.
4. König, J.A., and Maier, G.: Shakedown analysis of elastoplastic structures: a review of recent developments, Nucl. Engrg. Des., 66, (1981), 81-95.
5. Ponter, A.R.S.: General displacement and work bounds for dynamically loaded bodies, J. Mech. Phy. Solids, 23, (1975), 157-163.
6. König, J.A.: On some recent developments in the shakedown theory, Advances in Mechanics, 5, (1982), 237-258.
7. Polizzotto, C.: Bounding principles for elastic-plastic-creeping solids loaded below their shakedown limits, Meccanica, 17, (1982), 143-148.
8. Polizzotto, C.: On shakedown of structures under dynamic agencies, in: Inelastic structures under variable loads, (eds. C. Polizzotto and A. Sawczuk), Cogras, Palermo, (1984), 5-29.
9. Polizzotto, C.: A unified treatment of shakedown theory and related bounding techniques, SM Archives, 7, (1982), 19-75.
10. Polizzotto, C.: A bounding technique for dynamic plastic deformations of damped structures, Nucl. Engrg. Des., 79, (1984), 363-376.
11. Polizzotto, C.: Deformation bounds for elastic-plastic solids within and out of the creep range, Nucl. Engrg. Des., 83, (1984), 293-301.
12. Argyris, J.H.: Continua and Discontinua, Proc. 1st. Conf. on Matrix Methods in Structural Mechanics, Wright Patterson Air Force Base, Dayton, Ohio, (1965).
13. Corradi, L.: On compatible finite element models for elastic plastic analysis, Meccanica, 13, (1978), 133-150.
14. Panzeca, T., and Polizzotto, C.: A finite element model for dynamic elastoplastic analysis, in: Computational Plasticity, (ed. by D.R.J. Owen, E. Hinton, E. Onate), Pineridge Press, Swansea, U.K. (1987), Vol. II, 1247-1262.
15. Martin, J.B.: Plasticity: Fundamentals and General Results, The MIT Press, Cambridge Mass., (1975).

16. Koiter, W.T.: General theorems of elastic-plastic solids, in: Progress in Solid Mechanics, (eds. I.N. Sneddon and R. Hill), North-Holland, Amsterdam, (1964), Vol. 1, 167-221.
17. Maier, G.: Piecewise linearization of yield criteria in structural plasticity, SM Archives, 1, (1976), 239-281.
18. Cohn, M.Z, and Maier, G., (eds.): Engineering Plasticity by Mathematical Programming, Pergamon Press, (1978).
19. Fiacco, A.V., and McCormik, G.P.: Nonlinear programming sequential unconstrained minimization techniques, J. Wiley, New York, (1968).
20. Best, M.J., and Bowler, A.T.: ACDPAC: A Fortran - IV subroutine to solve differentiable mathematical programmes, University of Waterloo, Canada, (1978).
21. Hadley, G.: Linear Programming, Addison -Wesley, Reading, Mass. (USA), (1962).
22. Rizzo, S., and Giambanco, F.: Shakedown analysis of limited-ductility structures, Meccanica, 19, (1984), 151-157.
23. Rizzo, S.: Shakedown analysis of discrete elastic-workhardening structures with displacement constraints, Proc. of SMIRT-7, Struct. Mech. in Reactor Techn., Chicago (USA), 22-26 Aug. 1983, Paper L 9-2.
24. Mazzarella, C., and Panzeca, T.: The safety factor of shell structures under variable loads with constraints on deformations, Proc. of the Euromech Coll. on Inelastic Structures under Variable Loads, Palermo, Italy, 10-14 Oct., 1983, 435-450.
25. Land, A., and Powell, S.: Fortran codes for mathematical programming, J. Wiley, New York, (1973).
26. IBM-Progr., Rod SH 20-0968-1, Mathematical Programming System Extended (MPSX), and Generalized Upper Bounding (GUB), Program Description, (1972).
27. Nguyen Dang Hang, Aspects of analysis and optimization of structures under proportional and variable loadings, Eng. Opt., 7, (1983), 35-57.
28. Mazzarella, C.: I domini di plasticizzazione per le lastre formate da materiale con limiti plastici a trazione ed a compressione diversi in condizione di simmetria radiale, Giornale del Genio Civile, Fasc. 8, (1967), 511-523.

CHAPTER 17

MATHEMATICAL PROGRAMMING METHODS FOR THE EVALUATION OF DYNAMIC PLASTIC DEFORMATIONS

G. Borino, S. Caddemi and C. Polizzotto
University of Palermo, Palermo, Italy

ABSTRACT

Dynamic plastic deformation can be evaluated with two accuracy levels, nemely either by a full analysis making use of a step-by-step procedure, or by a simplified analysis making use of a bounding technique. Both procedures can be achieved by means a unified mathematical programming approach here presented. It is shown that for a full analysis both the direct and indirect methods of linear dynamics coupled with mathematical programming methods can be successfully applied, whereas for a simplified analysis a convergent bounding principle, holding both below and above the shakedown limit, can be utilized to produce an efficient linear programming-based algorithm.

1. INTRODUCTION

In a variety of problems of engineering practice, structures operate within the time-independent plasticity range under the action of dynamic loadings. Plastic deformations then produced may cause demages of various nature, such as excessive permanent deflections (producing unserviceability), excessive plastic strain (producing exhaustion of the material ductily capacity with consequent localized fracture), excessive dissipation of plastic work (producing material fatigue), etc. Therefore, within this context, correct structural design requires knowledge of plastic deformations produced during the application of a loading programme, with more or less accuracy according to the specific problem treated.

Like for quasi-static problems, dynamic elastoplastic analysis can be achieved at two different levels, namely:

a) Evolutive, step-by-step analysis following the given load history, for instance by means of mathematical programming (MP), and mostly quadratic programming (QP), methods;

b) Application of appropriate bounding techniques to derive upper bounds on deformation quantities of interest.

Methods of type a) have received a great deal of attention in recent years, basing on the concepts of "tangent stiffness", "initial strain" and "initial stress" (see e.g. [1 - 4]). More recently, MP methods have been shown to be most suitable to solve either types of problems (see e.g. [5-13]. In such a way, well-known MP methods for quasi-static analysis (see e.g. [14-19]) were extended to the dynamic elastoplastic field, with a unified view that turns out to be both conceptually and practically valuable. The key concept for this unification is the well-known Colonnetti principle [20,21]. The latter, phrased in dynamical terms, states that the actual elastoplastic response to some dynamic loads may be expressed -- within the infinitesimal displacement range -- as the sum of two elastic responses, namely the (elastic) response to the given loads and the (elastic) response to the unknown plastic strains treated as initial strains dynamically applied upon the (elastic) structure. Colonnetti's principle enables one to treat the structure as purely elastic, with the basic nonlinear problem being shifted to an ensuing moment when the relevant plasticity laws, suitably discretized in space and time, are employed to determine the plastic strains, giving rice to a set of QP problems to be solved sequentially.

As for the methods of type b), two categories can be envisaged, namely:

b1) Bounding techniques for repeated loads;

b2) Bounding techniques so-called "convergent".

The techniques of type b1) show the drawback of becoming too big with the loads approaching the shakedown limit [22-25]; they are quite similar to those used in statics, which are the object of another paper in this book. The techniques of type b2) do not show the above drawback and in fact may be applied both below and above the shakedown limit [22-25]. A linear programming (LP) procedure has been shown to be useful for the latter techniques [30].

The purpose of the present paper is to review methods of type a) and b2).

2. THE GOVERNING EQUATIONS

The structural model is assumed as an assemblage of finite elements (FE), each of which possesses an elastic perfectly plastic behaviour described in terms of generalized variables only. Consistent formulations of elastic plastic FE's were proposed first by Corradi [31-33] and then by Panzeca et al. [34]. Here the same FE constitutive equations reported in the paper by Panzeca et al. in this book and devoted to the bounding techniques are adopted, for simplicity sake. These equations are as follows:

$$\underset{\sim}{q} = \underset{\sim}{A}\underset{\sim}{Q} + \underset{\sim}{p} + \underset{\sim}{\theta} \tag{2.1}$$

$$\underset{\sim}{\Psi} = \underset{\sim}{N}^T \underset{\sim}{Q} - \underset{\sim}{R}, \quad \dot{\underset{\sim}{p}} = \underset{\sim}{N}\dot{\underset{\sim}{\lambda}}, \tag{2.2}$$

$$\underset{\sim}{\Psi} \leq \underset{\sim}{0}, \quad \dot{\underset{\sim}{\lambda}} \geq 0, \quad \underset{\sim}{\Psi}^T \dot{\underset{\sim}{\lambda}} = 0 \tag{2.3}$$

$$\dot{\underset{\sim}{\Psi}}^T \dot{\underset{\sim}{\lambda}} = 0 \tag{2.4}$$

which may be interpreted either at the element level, or at the assembled system level. Eq. (2.1) represents the total (generalized) strain $\underset{\sim}{q}$ as the sum of the elastic part, $\underset{\sim}{A}\underset{\sim}{Q}$, the plastic part, $\underset{\sim}{p}$, and the initial or imposed strain, $\underset{\sim}{\theta}$. The element yield surfaces have the shape of hyperpolyhedra, characterized by the matrix $\underset{\sim}{N}$ of the unit external normals to the yield hyperplanes, as well as the vector $\underset{\sim}{R}$ of the plastic resistances, identifying with the distances of these planes from the origin of the (generalized) stress space, $\underset{\sim}{Q}$. The vector $\dot{\underset{\sim}{\lambda}}$ collects internal variables, called plastic activation coefficients. Eqs. (2-2)-(2.4) describe the plastic flow-law rules for piecewise associated plasticity [14-16].

The motion equations of the assembled system can be written in the form

$$\underset{\sim}{M}\ddot{\underset{\sim}{u}} + \underset{\sim}{V}\dot{\underset{\sim}{u}} + \underset{\sim}{K}\underset{\sim}{u} - \underset{\sim}{B}\underset{\sim}{p} = \underset{\sim}{f} + \underset{\sim}{B}\underset{\sim}{\theta} \tag{2.5}$$

$$Q = B^T u - D p - D \theta \tag{2.6}$$

where M, V, are the mass and damping matrices, K is the system stiffness matrix, D is the element stiffness matrix, B is the pseudo-force matrix and f is the load vector. Once the initial conditions, namely

$$u = u_0, \quad \dot{u} = \dot{u}_0, \quad p = p_0, \quad \text{at } t = 0 \tag{2.7}$$

are assigned, the system motion can be found, at least in principle, by solving the set of governing equations, i.e. Eqs. (2.1)-(2.6), with $f = f(t)$ and $\theta = \theta(t)$ being vector-valued functions specified in some time interval $0 \leq t \leq t_f$.

In practice, only a numerical solution can be pursued by time integration procedures which require a discretization of the time axis into small steps. Within the framework of the MP approach to dynamic plasticity, two methods can be recognized [7,9], namely:

a) The indirect method, which operates by superposition of the natural vibration modes of the (elastic) system. It thus requires a preliminary modal analysis in order to provide the dynamic characteristics of the system considered as elastic, i.e. displacement modes, frequencies, etc. and leads to exact integration operators, provided that all the natural modes are taken into account, with the only source of errors arising from the time modelling of the plastic strain rates within every step.

b) The direct method, which utilizes the known time integration procedures of linear dynamics, such as Wilson's, Newmark's, etc. [3,4,35-39]. It leads to approximate integration operators.

3. THE STEP-BY-STEP ANALYSIS

If $t_0 = 0, t_1, t_2, .., t_n, .., t_{nf} = t_f$ denote the subdivision times, the integration formulas can be written in the form (see Appendices A and B):

$$X_{(n)} = \Gamma X_{(n-1)} + X^L_{(n)} + W_1 p_{(n-1)} + W_2 \Delta p_{(n)} \tag{3.1}$$

$$Q_{(n)} = \Pi X_{(n-1)} + Q^L_{(n)} + Z_1 p_{(n-1)} + Z_2 \Delta p_{(n)} \tag{3.2}$$

in which X is the state vector, namely

$$X = [u^T \dot{u}^T \ddot{u}^T \ldots \overset{(m-1)}{u}{}^T]^T \tag{3.3}$$

with the integer m suitably chosen, and the subscript (n) signifies that the relevant quantity is evaluated at t_n. The matrices Γ, W_1, W_2, Π, Z_1, Z_2 are operators which depend on the specific method used, and $X^L_{(n)}$, $Q^L_{(n)}$ denote the effects produced by the loadings. It worths noting that the above matrix operators remain constant all through the integration procedure if the step length h is taken constant.

In Eqs. (3.1) and (3.2), which are a direct consequence of the Eqs. (2.5) and (2.6) respectively, the vector

$$\Delta p_{(n)} = N \Delta\lambda_{(n)} = N \int_{t_{n-1}}^{t_n} \dot{\lambda}\, dt \tag{3.4}$$

represents the unknown plastic strain increment of the n-th step and for its evaluation the plasticity laws must be taken into account. The time discretization used implies that the unknown time function $\dot{\lambda}(t)$ be in some way modelled within the step, for instance by assuming that it is constant, i.e.

$$\dot{\lambda}(t) = \frac{1}{h} y_{(n)}, \quad \text{for} \quad t_{n-1} < t \leq t_n, \tag{3.5}$$

and that the plasticity laws be given the appropriate forms consistent with the adopted time modelling. If Eq. (3.5) is used, this consistency requirement is supposed to be accomplished by simply enforcing the yielding laws at the step end and ignoring Eq. (2.4), such that we can write:

$$\Psi_{(n)} = N^T Q_{(n)} - R, \quad \Delta p_{(n)} = N y_{(n)}, \tag{3.6}$$

$$\Psi_{(n)} \leq 0, \quad y_{(n)} \geq 0, \quad \Psi^T_{(n)} y_{(n)} = 0. \tag{3.7}$$

If $X_{(n-1)}$ is known, Eqs. (3.1), (3.2), (3.6) and (3.7) enable one to solve the step problem. This in practice can be achieved by two different ways, according to whether the plasticity laws (3.6) and (3.7) are satisfied contemporaneously for all the FE's (system optimization technique), or they are satisfied separately, element-by-element, (element optimization technique).

With the system optimization technique (SOT), we substitute from Eq. (3.2) into the first of Eq. (3.6) and set [5,9]:

$$S = -N^T Z_2 N, \tag{3.8}$$

$$b_{(n)} = R - N^T(\Pi X_{(n-1)} + Q^L_{(n)} + Z_1 p_{(n-1)}) \tag{3.9}$$

such that the yielding laws (3.6) and (3.7) take on the form of a linear complementarity problem (LCP), namely

$$\underset{\sim}{S}\,\underset{\sim}{y}_{(n)} + \underset{\sim}{b}_{(n)} \geq \underset{\sim}{0}, \quad \underset{\sim}{y}_{(n)} \geq \underset{\sim}{0}, \quad \underset{\sim}{y}^{T}_{(n)}(\underset{\sim}{S}\,\underset{\sim}{y}_{(n)} + \underset{\sim}{b}_{(n)}) = 0 \tag{3.10}$$

which yields the unknown vector $\underset{\sim}{y}_{(k)} = \Delta\underset{\sim}{\lambda}_{(k)}$, hence $\Delta\underset{\sim}{p}_{(n)} = \underset{\sim}{N}\,\underset{\sim}{y}_{(n)}$. Since $\underset{\sim}{S}$ is symmetric and positive semidefinite, the above LCP (3.10) is equivalent to two QP problems [16,40], i.e.

$$\left.\begin{array}{l} \min \Omega_1 = \frac{1}{2}\,\underset{\sim}{y}^{T}_{(n)}\,\underset{\sim}{S}\,\underset{\sim}{y}_{(n)} + \underset{\sim}{b}^{T}_{(n)}\,\underset{\sim}{y}_{(n)} \\ \quad \text{s.t.}\ \underset{\sim}{y}_{(n)} \geq \underset{\sim}{0} \end{array}\right\} \text{(QPP1)} \tag{3.11}$$

$$\left.\begin{array}{l} \min \Omega_2 = \frac{1}{2}\,\underset{\sim}{y}^{T}_{(n)}\,\underset{\sim}{S}\,\underset{\sim}{y}_{(n)} \\ \quad \text{s.t.}\ \underset{\sim}{S}\,\underset{\sim}{y}_{(n)} + \underset{\sim}{b}_{(n)} \geq \underset{\sim}{0} \end{array}\right\} \text{(QPP2)} \tag{3.12}$$

which are dual of each other and the Kuhn-Tucker conditions of which are easily recognized to coincide with Eqs. (3.10). Basing on the fact that the matrix $\underset{\sim}{Z}_2$ is negative definite, it can be proved that the above MP problems, either (3.10) or (3.11) and (3.12), admit a unique solution in terms of strains $\Delta\underset{\sim}{p}_{(n)}$ and stresses $\underset{\sim}{Q}_{(n)}$, but not in terms of $\underset{\sim}{y}_{(n)}$ [9].

With the SOT the elastoplastic analysis will consist in a sequence of n_f = number of step QP problems, starting from the first step for which the initial conditions are known (or easily provided). At every step, the plastic strain increments $\Delta\underset{\sim}{p}_{(n)}$ are computed from $\underset{\sim}{y}_{(n)}$ and then, through Eqs. (3.1) and (3.2), $\underset{\sim}{X}_{(n)}$ and $\underset{\sim}{Q}_{(n)}$ are deduced to shift to the next step. A limitation to this analysis method may arise from the likely excessive number of $\underset{\sim}{y}$ variables in every step MP problem, although procedures were already envisaged in order to cope with a relatively large number of variables (as few hundreds) [41].

The <u>element optimization technique</u> (EOT), proposed first by Feijòo and Zouain [42,43], enables one to notably reduce the problem dimension, but at the moderate cost of a few iterations at every step. With this technique, the yielding laws (3.6) and (3.7) are enforced separately element-by-element. It is therefore more convenient to rewrite Eqs.(3.6) and (3.7) for every individual element, i.e.

$$\underset{\sim}{\Psi}_{e(n)} = \underset{\sim}{N}^{eT}\,\underset{\sim}{Q}_{e(n)} - \underset{\sim}{R}_{e}, \quad \Delta\underset{\sim}{p}_{e(n)} = \underset{\sim}{N}^{e}\,\underset{\sim}{y}_{e(n)} \tag{3.13}$$

$$\underset{\sim}{\Psi}_{e(n)} \leq \underset{\sim}{0}, \quad \underset{\sim}{y}_{e(n)} \geq \underset{\sim}{0}, \quad \underset{\sim}{\Psi}^{T}_{e(n)}\,\underset{\sim}{y}_{e(n)} = 0 \tag{3.14}$$

where $\underset{\sim}{Q}_{e(n)}$ is now provided from Eq. (2.6) written for $t = t_n$, i.e.

$$\underset{\sim}{Q}_{e(n)} = \underset{\sim}{B}^{eT}\underset{\sim}{u}_{(n)} - \underset{\sim}{D}^{e}\underset{\sim}{p}_{e(n)} - \underset{\sim}{D}^{e}\underset{\sim}{\theta}_{e(n)}$$

$$= \underset{\sim}{B}^{eT}\underset{\sim}{u}_{(n)} - \underset{\sim}{D}^{e}\underset{\sim}{p}_{e(n-1)} - \underset{\sim}{D}^{e}\underset{\sim}{\theta}_{e(n)} - \underset{\sim}{D}^{e}\Delta\underset{\sim}{p}_{e(n)}. \tag{3.15}$$

On substitution from the latter into the first of Eqs. (3.13) and with the aid of the positions

$$\underset{\sim}{S}^{e} = -\underset{\sim}{N}^{eT}\underset{\sim}{D}^{e}\underset{\sim}{N}^{e}, \tag{3.16}$$

$$\underset{\sim}{b}_{e(n)} = \underset{\sim}{R}_{e} + \underset{\sim}{N}^{eT}\underset{\sim}{D}^{e}(\underset{\sim}{p}_{e(n-1)} + \underset{\sim}{\theta}_{e(n)}) - \underset{\sim}{N}^{eT}\underset{\sim}{B}^{eT}\underset{\sim}{u}_{(n)}, \tag{3.17}$$

we easily obtain:

$$\underset{\sim}{S}^{e}\underset{\sim}{y}_{e(n)} + \underset{\sim}{b}_{e(n)} \geq \underset{\sim}{0}, \quad \underset{\sim}{y}_{e(n)} \geq \underset{\sim}{0}, \quad \underset{\sim}{y}^{T}_{e(n)}(\underset{\sim}{S}^{e}_{(n)}\underset{\sim}{y}_{(n)} + \underset{\sim}{b}_{e(n)}) = 0. \tag{3.18}$$

which are equivalent to Eqs. (3.13) and (3.14). Eqs. (3.18) look like Eqs. (3.10), but they do not constitute a LCP owing to the circumstance that the "load" vector $\underset{\sim}{b}_{e(n)}$ is actually dependent on $\underset{\sim}{y}_{e(n)}$ through the displacement $\underset{\sim}{u}_{(n)}$. However, Eqs. (3.18) can be effectively utilized in an iterative procedure in which, at the k-th iteration:

i) An estimated value of $\Delta\underset{\sim}{p}_{(n)}$, say $\Delta\underset{\sim}{p}^{k-1}_{(n)}$, is introduced into Eq. (3.1), to obtain

$$\underset{\sim}{X}^{k-1}_{(n)} = \underset{\sim}{X}^{o}_{(n)} + \underset{\sim}{W}_{2}\Delta\underset{\sim}{p}^{k-1}_{(n)}, \quad \text{where} \tag{3.19a}$$

$$\underset{\sim}{X}^{o}_{(n)} = \underset{\sim}{\Gamma}\underset{\sim}{X}_{(n-1)} + \underset{\sim}{X}^{L}_{(n)} + \underset{\sim}{W}_{1}\underset{\sim}{p}_{(n-1)}, \tag{3.19b}$$

and thus the approximate displacement $\underset{\sim}{u}^{k-1}_{(n)}$ from $\underset{\sim}{X}^{k-1}_{(n)}$, which is used in Eq. (3.17) to give

$$\underset{\sim}{b}^{k-1}_{e(n)} = \underset{\sim}{\beta}_{e(n)} + \underset{\sim}{\chi}^{k-1}_{e(n)}, \quad \text{where} \tag{3.20a}$$

$$\underset{\sim}{\beta}_{e(n)} = \underset{\sim}{R}_{e} + \underset{\sim}{N}^{eT}\underset{\sim}{D}^{e}(\underset{\sim}{p}_{e(n-1)} + \underset{\sim}{\theta}_{e(n)}), \tag{3.20b}$$

$$\underset{\sim}{\chi}^{k-1}_{e(n)} = -\underset{\sim}{N}^{eT}\underset{\sim}{B}^{eT}\underset{\sim}{u}^{k-1}_{(n)}. \tag{3.20c}$$

ii) The LCP's (3.18), with the $\underset{\sim}{b}^{k-1}_{e(n)}$ in Eq. (3.20a) as the appropriate load vectors, namely

$$\underset{\sim}{S}^{e}\underset{\sim}{y}^{k}_{e(n)} + \underset{\sim}{b}^{k-1}_{e(n)} \geq \underset{\sim}{0}, \quad \underset{\sim}{y}^{k}_{e(n)} \geq \underset{\sim}{0}, \quad \underset{\sim}{y}^{kT}_{e(n)}(\underset{\sim}{S}^{e}\underset{\sim}{y}^{k}_{e(n)} + \underset{\sim}{b}^{k-1}_{e(n)}) = 0, \tag{3.21}$$

are solved to obtain $\underset{\sim}{y}^{k}_{(n)}$, separately for every FE, such that new, likely better, values of the plastic strain increments, $\Delta\underset{\sim}{p}^{(k)}_{e(n)} = \underset{\sim}{N}^{e}\underset{\sim}{y}^{(k)}_{e(n)}$, can be computed and, if necessary, a new iteration can be started. In the first iteration, we set $\Delta p^{0}_{(n)} = 0$. An appropriate error measure together with some tolerance will enable the iteration procedure to be stopped.

The main advantage of the EOT is in the small number of dimensions of every MP problem and the numerical applications so far accomplished confirmed the more computational convenience of the EOT with respect to the SOT [44].

4. THE CONVERGENT BOUNDING PRINCIPLE

Often in engineering practice, mainly in the early stage of the design process, only a rough evaluation of a few plastic deformation quantities is required. When this is the case, the application of a bounding technique may be more convenient than an expensive full analysis.

We assume that the plastic deformation quantities which may be of interest may be represented in the form

$$d^T \lambda(t_1) = d^T \int_0^{t_1} \dot{\lambda}(t)\, dt\,, \tag{4.1}$$

where d is the "perturbation vector", which is arbitrary, but can in practice be chosen in such a way as the product $d^T\lambda$ have some desired physical meaning (such as residual displacement, plastic strain, plastic work, etc.) [25]. With reference to the time discretization of Sec. 3, and taking $t_1 = n_1 h$, Eq. (4.1) becomes:

$$d^T \lambda(t_1) = \sum_{j=1}^{n_1} d^T y_{(j)}. \tag{4.2}$$

An upper bound to the latter quantity can be constructed in the form (see Appendix C):

$$d^T \lambda_{(n_1)} \le d^T \sum_{j=1}^{n_1} y^*_{(j)} + \frac{1}{\omega} \sum_{j=1}^{n_1} z^{*T}_{(j)} y^*_{(j)} \tag{4.3}$$

where $z^*_{(j)}$, $y^*_{(j)}$ are vector variables satisfying the conditions:

$$z^*_{(j)} = S y^*_{(j)} + \hat{b}^*_{(j)}, \quad z^*_{(j)} \ge 0, \quad y^*_{(j)} \ge 0 \tag{4.4}$$

where

$$\hat{b}^*_{(j)} = b^*_{(j)} - d\,\omega, \tag{4.5}$$

and ω is some positive scalar ("perturbation multiplier"). Further, the vector $b^*_{(j)}$ in Eq. (4.5) is defined as in Eq. (3.9), but $p^*_{(n-1)}$ in place of $p_{(n-1)}$, and $X^*_{(n-1)}$ in place of $X_{(n-1)}$.

A solution $(z^*_{(j)}, y^*_{(j)})$ to Eqs. (4.4) is here provided as a feasible solution to the following linear programming (LP) problem:

$$\left.\begin{aligned} &\min \Omega_3 = a^T y^*_{(j)}, \quad \text{subject to} \\ &S\, y^*_{(j)} + \hat{b}^*_{(j)} \geq 0, \quad y^*_{(j)} \geq 0 \end{aligned}\right\} \text{(LPP)} \tag{4.6}$$

where a is a vector suitably chosen, usually as

$$a = [1 \;\; 1 \;\; \dots \;\; 1]^T. \tag{4.7}$$

Once $y^*_{(j)}$ is obtained from the above LPP relative to the j-th step, we compute $\Delta p^*_{(j)} = N\, y^*_{(j)}$ and thus:

$$X^*_{(j)} = \Gamma X^*_{(j-1)} + X^L_{(j)} + W_1 p^*_{(j-1)} + W_2 \Delta p^*_{(j)} \tag{4.8}$$

$$b^*_{(j+1)} = R - N^T(\Pi X^*_{(j)} + Q^L_{(j)} + Z_1 p^*_{(j)}) \tag{4.9}$$

such that a new LPP, relative to the next step, can be solved. This sequential procedure is quite similar to that used for a full analysis. At the first step, we set $p^*_{(0)} = 0$. If d is changed into $-d$, the upper bound (4.3) trasforms into a lower bound.

The procedure described here above appeals to the SOT of the previous Section, and analogous computational difficulties may arise due to the likely excessive problem dimensions, but with comparatively reduced stringentness due to the more efficiency of LP computer packages than the QP ones. The EOT can also be utilized for providing bounds to plastic deformations, with a sequential procedure similar to the one employed for the full analysis. We disregard the details for lack of space.

5. NUMERICAL APPLICATIONS

The cantilever structure of Fig. 1(a) was considered for numerical applications. The plate has constant thickness and is loaded by cyclically variable vertical load $f = \mu(t)\, q(z)$, where $\mu = \sin \alpha t$ and

$q = (3/2)(Q_0/L)(1 - 4z^2/L^2)$, with $\alpha = 4$ rad/sec, $Q_0 = 260$ kN. The material data are: $\sigma_0 = 20$ kN/cm^2 (yield stress), E = 1,000 kN/cm^2 (Young modulus), $\mu = 0.25$ (Poisson ratio), $\rho = 8{,}011 \cdot 10^{-5}$ kg/cm^3 (density).

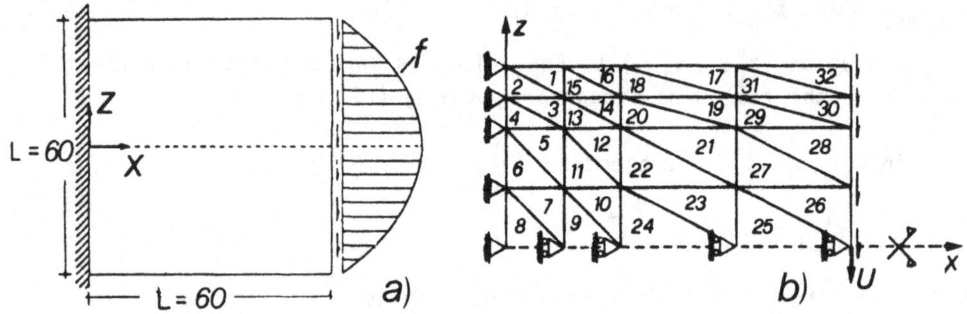

Fig. 1. Elastic perfectly plastic cantilever plate under cyclically variable loads: a) Geometrical and loading scheme; b) FE model.

Due to symmetry, half plate was considered and discretized by 32 uniform stress triangular FE's, as shown in Fig. 1(b), in which the continuously distributed load is replaced by the corresponding nodal loads. Every FE possesses an elastic perfectly plastic behaviour with a Von Mises yield surface piecewise linearized by 14 planes of the three-dimensional stress space, as depicted in Fig. 2.

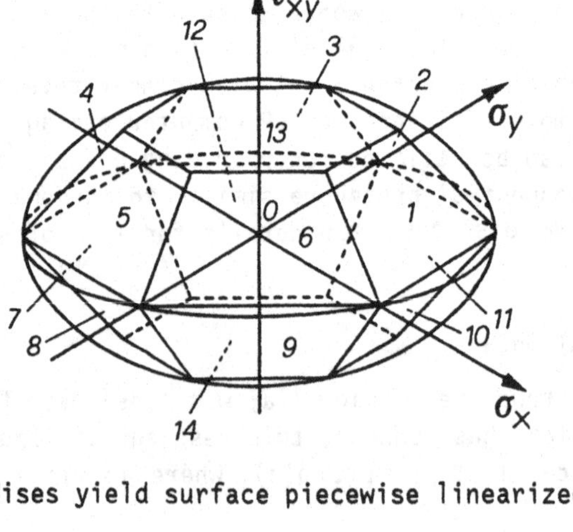

Fig. 2. Von Mises yield surface piecewise linearized by 14 planes.

The full step-by-step analysis was first performed by the indirect method and the element optimization technique. Then, upper and lower bounds to the free edge mid-point residual deflection and to the plastic strain components relative to the FE number 1 (see Fig. 1(b)) were computed. These bounds are reported in Figs. 3, 4 and 5 as functions of time t (dotted lines) besides the actual response values (solid lines). The bounds were obtained by means of the LP-approach. Each bound curve required a computer time about 25% of that required by the full analysis.

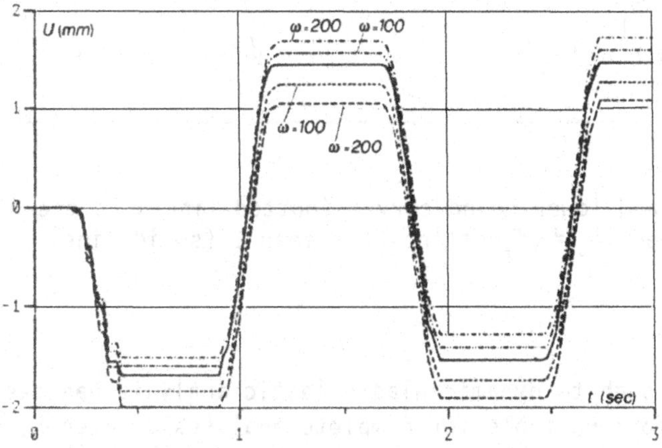

Fig. 3. Upper and lower bound curves (dotted lines) to the residual deflection (solid line) of the free-edge mid point.

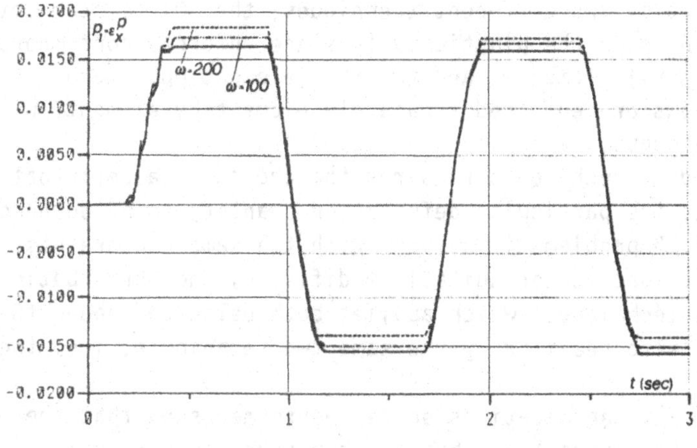

Fig. 4. Upper bound curves (dotted lines) to the plastic strain component $p_1 = \varepsilon_x^p$ of the FE number 1 (solid line).

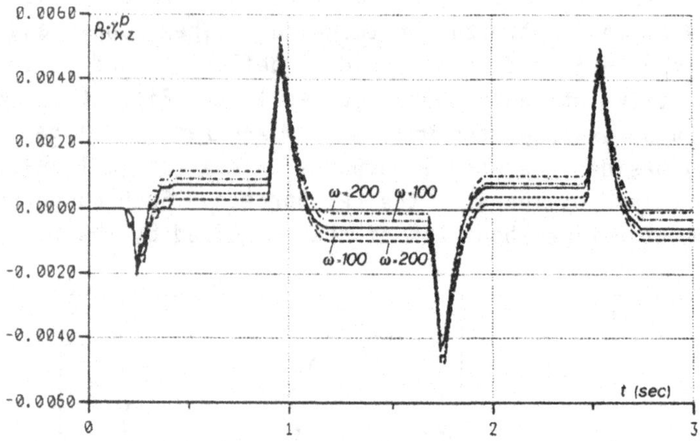

Fig. 5. Upper and lower bound curves (dotted lines) to the plastic strain component $p_3 = \gamma^p_{xz}$ of the FE number 1 (solid line).

6. CONCLUSIONS

A MP approach to dynamic elastoplastic analysis has been presented, in a unified form suitable for complete analyses by step-by-step procedures, and for simplified analyses by bounding techniques.

The complete analysis consists in solving a sequence of linear complementarity problems (equivalent QP problems), each of which gives the step plastic strain increment. It has been shown that this procedure can be realized by two different techniques, the SOT (system optimization technique), in which the plasticity laws are enforced contemporaneously for all structural elements, and the EOT (element opt. technique), in which these laws are enforced separately element-by-element, with computational advantages.

The bounding techniques requires the choice of a perturbation vector $\underset{\sim}{d}$ according to the particular deformation quantity to be bounded. Then, a sequence of LP problems is solved, with the same constraints as the LCP's, but the load vector suitably modified by the perturbation vector. This bounding techniques, which applies both below and above the shakedown limit, may be also realized by the same two techniques, i.e. the SOT and the EOT.

The numerical applications so far performed show that the above methods are rather attractive, but further study is hoped for.

7. AKNOWLEDGEMENT

This paper is part of a research programme financially supported by the Italian CNR (National Research Council) and by the MPI (State Education Ministry), whose contributions are herein acknowledged.

REFERENCES

1. Zienkiewicz, O.C.: The finite element method in engineering science, McGraw-Hill, 1971.
2. Owen, D.R.J., Hinton, E.: Finite elements in plasticity, Pineridge Press Ltd., Swansea, U.K., 1980.
3. Owen, D.R.J.: Implicit finite element methods for the dynamic transient analysis of solids with particular reference to nonlinear situations, Advanced Structural Analysis, ed. by J. Donea, Applied Science Publs., London, 1980, pp. 123-152.
4. Hughes, T.J.R.: Analysis of transient algorithms with particular reference to stability behaviour, Computational Methods for Transient Analysis, ed. by T. Belytschko and T.J.R. Hughes, North-Holland, Amsterdam, 1983, chap. 2.
5. Polizzotto, C.: Elastoplastic analysis method for dynamic agencies, Jour. Eng. Mech., Div. ASCE, March 1986, pp. 293-310.
6. Polizzotto, C.: A solution method for elastoplastic structures under dynamic agencies, Numerical Methods in Engineering: Theory and applications, ed. by J. Middleton and G.N. Pande, A.A. Balkema, Rotterdam, 1985, pp. 243-251.
7. Polizzotto, C., Borino, G.: Mathematical programming formulation of dynamic elastoplasticity analysis problems, Proc. Int. Conf. on Computational Mechanics, Tokyo, May 25-29, 1986, pp. VI/105-110.
8. Borino, G., Polizzotto, C.: Time integration algorithms and quadratic programming in the dynamic analysis of elastoplastic structures, Proc. 8-th Congress of the Associazione Italiana di Meccanica Teorica ed Applicata, AIMETA, Torino, Sept. 29 - Oct. 3, 1986, Vol. 2, pp. 731-736.
9. Borino, G., Polizzotto, C.: A mathematical programming approach to dynamic elastoplastic structural analysis, Meccanica (to appear).
10. Polizzotto, C.: A bounding technique for dynamic plastic deformations of damped structures, Nuclear Engineering and Design, June 1984, Vol. 79, No. 3, pp. 363-376.
11. Polizzotto, C.: A quadratic programming approach to dynamic elastoplasticity, Trans. Int. Conf. on Structural Mechanics in Reactor Technology, SMIRT-8, Brussels, August 19-23, 1985, paper B4/5.

12. Muscolino, G., Polizzotto, C.: Un approccio alla dinamica elastoplastica basato sulla programmazione matematica, (in italian), VII Nat. Congress of the Associazione Italiana di Meccanica Teorica ed Applicata, AIMETA, Trieste, Oct. 1984, Vol. 5, pp. 181-192.
13. Di Paola, M., Polizzotto, C.: Metodo per la determinazione delle deformazioni plastiche in una struttura soggetta ad azioni sismiche, (in italian), Proc. II Nat. Congress "L'Ingegneria Sismica in Italia", Rapallo, June 6-9, 1984, pp. 7/143-161.
14. Capurso, M., Maier, G.: Incremental elastoplastic analysis and quadratic optimization, Meccanica, Vol. 5, No. 2, June 1970, pp. 107-116.
15. Maier, G.: Mathematical programming methods in structural analysis, Proc. Int. Conf. on Variational Methods in Engineering, Vol. II, 8, Southampton Univ. press, August 1978.
16. Cohn, M.Z., Maier, G., (eds.): Engineering plasticity and mathematical programming, Pergamon Press, New York, 1979.
17. De Donato, O., Maier, G.: Historical deformations analysis of elastoplastic structures as a parametric linear complementarity, Meccanica, Vol. 11, No. 3, 1976, pp. 166-171.
18. Hodge, P.G., Belytschko, T., Merakovich, C.T.: Quadratic programming and plasticity, ASME, Computational Approaches in Applied Mechanics, ed. by E. Sevin, New York, 1969, p. 73.
19. Sayegh, A.F., Rubistein, N.F.: Elastic-plastic analysis by quadratic programming, Journal of the Engineering Mechanics Division, ASCE, No. EM6, Dec. 1972, pp. 1547-1572.
20. Colonnetti, G.: L'equilibre des corps deformables, Dunod, Paris, 1955.
21. Polizzotto, C.: Dynamic shakedown by modal analysis, Meccanica, Vol. 17, No. 2, 1984, pp. 133-144.
22. Capurso, M., Corradi, L., Maier, G.: Bounds on deformations and displacements in shakedown theory, in Proc. of the Seminary on Materials and Structures Under Cyclic Loads, Laboratoire de Mécanique des Solides, Ecole Polytechnique, Palaiseau, France, Sept. 28-29, 1978, pp. 231-144.
23. König, J.A., Maier, G.: Shakedown analysis of elastoplastic structures: A review of recent developments, Nucl. Engng. Des., 66, 1981, pp. 81-95.
24. Ponter, A.R.S.: General displacements and work bounds for dynamically loaded bodies, J. Mech. Phys. Solids, 23, 1975, pp. 157-163.
25. Polizzotto, C.: A unified treatment of shakedown theory and related bounding techniques, S.M. Archives, 7, 1982, pp. 19-75.
26. Polizzotto, C.: Bounding principles for elastic-plastic-creeping solids loaded below and above the shakedown limits, Meccanica, 17, 1982, pp. 143-148.

27. Polizzotto, C.: Deformation bounds for elastic-plastic solids within and out of the creep range, Nuclear Engng. Design, Vol. 83, 1984, pp. 293-301.
28. Polizzotto, C.: On shakedown of structures under dynamic agencies, in Polizzotto C. and Sawczuk A., (eds.), Inelastic Structures under Variable Loads, Cogras, Palermo, 1984, Proc. Euromech Colloquium, Palermo, 1983.
29. Polizzotto, C.: A convergent bounding principle for a class of elasto-plastic strain-hardening solids, J. of Plasticity, Vol. 2, 1986, pp. 359-370.
30. Borino, G., Caddemi, S., Polizzotto, C.: A linear programming method for bounding plastic deformations, International Conference on Computational Engineering Science, ICES '88, Atlanta, Georgia, U.S.A., April 10-14, 1988.
31. Corradi, L.: On compatible finite element methods for elastic-plastic analysis, Meccanica, Vol. 13, 1978, p. 133.
32. Corradi, L.: A displacement formulation for finite element elasto-plastic problems, Meccanica, Vol. 18, 1983, pp. 77-91.
33. Corradi, L., Maier, G.: Finite element elasto-plastic and limit analysis, in Nonlinear F.E. Analysis in Structural Mechanics, ed. by Winderlich, E. Stein, K.J. Bathe, Springer-Verlag, 1981, p. 290.
34. Panzeca, T., Polizzotto, C.: A finite element model for dynamic elastoplastic structural analysis, in Computational Plasticity, ed. by D.R.J. Owen, E. Hinton, E. Onate, Swansea, Pineridge Press, 1987, pp. 1247-1262.
35. Bathe, K.J.: Computational methods in structural dynamics, in R.F. Hartung (ed.), Computing in Applied Mechanics, The Am. Society of Mechanical Engineers, AMD-18, 1976, pp. 163-176.
36. Bathe, K.J., Wilson, E.L.: Numerical methods in finite element analysis, Englewood Cliffs, N.J., Prentice-Hall, 1976.
37. Belytschko, T.: Explicit time integration of structure - Mechanical Systems, in J. Donea (ed.), Advanced Structural Dynamics, Appl. Science Pubs., London, 1980, pp. 97-122.
38. Zienkiewicz, O.C., Wood, W.L., Hine, N.W., Taylor, R.L.: A unified set of single step algorithms, Part 1, Int. J. Num. Meth. Engng., Vol. 20, 1984, pp. 1529-1552.
39. Wood, W.L.: A unified set of single step algorithms, Part 2, Int. J. Num. Meth. Engng., Vol. 20, 1984, pp. 2303-2309.
40. Kuhn, H.W., Tucker, A.W.: Linear inequalities and related systems, Ann. Math. Stat., 1956.
41. De Donato, O., Franchi, A.: Elastic-plastic analysis by finite elements, Engineering Plasticity and Mathematical Programming, ed. by Cohn M.Z. and Maier G., Pergamon Press, New York, 1979, pp. 413-432.

42. Feijòo, R.A.: Variational methods in the theory of plasticity, Seminar held at the Dept. of Structural and Geotechnical Engineering, University of Palermo, Oct. 1986.
43. Feijòo, R.A., Zouain, N.: Variational formulation for rates and increments in plasticity, Proc. Int. Conf. on Computational Plasticity, Barcellona, 1987, pp. 33-57.
44. Borino, G., Caddemi, S., Polizzotto, C.: Mathematical programming methods for evaluating dynamic deformations of elasto-plastic structures, Proc. Int. Conf. on Computational Plasticity, Barcellona, 1987, pp. 1231-1245.

APPENDIX A. INTEGRATION FORMULAS--INDIRECT METHOD

With the indirect method [5-7,9,12,13], a preliminary modal analysis is performed to obtain the displacement modes, $\underset{\sim}{\Phi}_i$, the natural frequencies, ω_i, damping ratios ξ_i and damped frequencies, $\omega_i^D = \omega_i(1-\xi_i^2)^{\frac{1}{2}}$, for $i = 1,2,..,\nu$, with ν = number of degrees of freedom. The elastic response of the system to mechanical and/or imposed-strain-like load histories can then represented by making use of the dynamic influence matrices:

$$\underset{\sim}{A}(t) = \underset{\sim}{K}^{-1} - \underset{\sim}{G}(t), \tag{A.1}$$

$$\underset{\sim}{L}(t) = \underset{\sim}{A}(t)\,\underset{\sim}{B}, \tag{A.2}$$

$$\underset{\sim}{Z}(t) = \underset{\sim}{B}^T \underset{\sim}{A}(t)\,B - \underset{\sim}{D}, \tag{A.3}$$

where the matrix $\underset{\sim}{G}(t)$ is defined as

$$\underset{\sim}{G}(t) = \sum_{i=1}^{\nu} \underset{\sim}{\Phi}_i\, g_i(t)\, \underset{\sim}{\Phi}_i^T, \tag{A.4a}$$

$$g_i(t) = \frac{1}{\omega_i^2} \exp(-\xi_i\,\omega_i\,t)[\cos\omega_i^D t + \frac{\xi_i\omega_i}{\omega_i^D} \sin\omega_i^D t]. \tag{A.4b}$$

In fact, we can write:

$$\underset{\sim}{u}(t) = \underset{\sim}{G}(t)\,\underset{\sim}{K}\,\underset{\sim}{u}_0 - \dot{\underset{\sim}{G}}(t)\,\underset{\sim}{M}\,\dot{\underset{\sim}{u}}_0 + \underset{\sim}{A}(t)\,\underset{\sim}{f}(0) + \underset{\sim}{L}(t)\,\underset{\sim}{\theta}(0) + \underset{\sim}{L}(t)\,\underset{\sim}{p}_0$$
$$+ \int_0^t [\underset{\sim}{A}(t-\bar{t})\,\dot{\underset{\sim}{f}}(\bar{t}) + \underset{\sim}{L}(t-\bar{t})\,\dot{\underset{\sim}{\theta}}(\bar{t}) + \underset{\sim}{L}(t-\bar{t})\,\dot{\underset{\sim}{p}}(\bar{t})]\,d\bar{t} \tag{A.5}$$

$$Q(t) = B^T[G(t)\,K\,u_0 - \dot{G}(t)\,M\,\dot{u}_0] + L^T(t)\,f(0) + Z(t)\,\theta(0) + Z(t)\,p_0$$
$$+ \int_0^t [L^T(t-\bar{t})\,\dot{f}(\bar{t}) + Z(t-\bar{t})\,\dot{\theta}(\bar{t}) + Z(t-\bar{t})\,\dot{p}(\bar{t})]\,d\bar{t}. \tag{A.6}$$

The state vector is chosen for $m = 2$, i.e. $X = [u^T, \dot{u}^T]^T$, such that from Eqs. (A.5) and (A.6) we easily obtain

$$X(t) = \Gamma(t)\,X_0 + \Omega(t)\,F_0 + W(t)\,p_0 + \int_0^t \Omega(t-\bar{t})\,\dot{F}(\bar{t})\,dt$$
$$+ \int_0^t W(t-\bar{t})\,\dot{p}(\bar{t})\,dt \tag{A.7}$$

$$Q(t) = \Pi(t)\,X_0 + \Lambda(t)\,F_0 + Z(t)\,p_0 + \int_0^T \Lambda(t-\bar{t})\,\dot{F}(\bar{t})\,d\bar{t}$$
$$+ \int_0^t Z(t-\bar{t})\,\dot{p}(\bar{t})\,d\bar{t} \tag{A.8}$$

where

$$\Gamma(t) = \begin{bmatrix} G(t)\,K & -\dot{G}(t)\,M \\ \dot{G}(t)\,K & -G(t)\,M \end{bmatrix}, \qquad W(t) = \begin{bmatrix} L(t) \\ \dot{L}(t) \end{bmatrix}$$

$$\Omega(t) = \begin{bmatrix} A(t) & L(t) \\ \dot{A}(t) & \dot{L}(t) \end{bmatrix}, \qquad F(t) = \begin{bmatrix} f(t) \\ \theta(t) \end{bmatrix} \tag{A.9}$$

$$\Pi(t) = B^T[G(t)\,K \mid -\dot{G}(t)\,M], \qquad \Lambda(t) = [L^T(t) \mid Z(t)].$$

Applying Eqs. (A.7) and (A.8) to the n-th step with the initial conditions specified at $t = t_{n-1}$ yields

$$X_n = \Gamma(h)\,X_{n-1} + X_n^L + W(h)\,p_{n-1} + \bar{W}\,\Delta p_n \tag{A.10}$$

$$Q_n = \Pi(h)\,X_{n-1} + Q_n^L + Z(h)\,p_{n-1} + \bar{Z}\,\Delta p_n \tag{A.11}$$

where

$$X_n^L = \Omega(h)\,F_{n-1} + \bar{\Omega}\,\Delta F_n, \qquad Q_n^L = \Lambda(h)\,F_{n-1} + \bar{\Lambda}\,\Delta F_n \tag{A.12}$$

and the upper bar means mean value within the step, for instance

$$\bar{\underset{\sim}{W}} = \frac{1}{h} \int_0^h \underset{\sim}{W}(\bar{t})\, d\bar{t}. \tag{A.13}$$

Eqs. (A.10) and (A.11) are the relevant integration formulas and they have the form of Eqs. (3.1) and (3.2).

APPENDIX B. INTEGRATION FORMULAS--DIRECT METHOD

Following a work by Zienkiewicz et al. [38-39], the displacement response of an elastic structural system can be set in the form of a Taylor expansion of order m, i.e.

$$\underset{\sim}{u}(t_{n-1}+\tau) = \sum_{j=1}^{m-1} \overset{(j)}{\underset{\sim}{u}_n} \gamma_j(\tau) + \underset{\sim}{v}_n \gamma_m(\tau), \qquad (0 \le \tau \le h) \tag{B.1a}$$

$$\gamma_j(\tau) \equiv \tau^j/j!, \qquad (j = 0,1,2,..,m;\ \gamma_0(\tau) \equiv 1) \tag{B.1b}$$

where $\overset{(j)}{\underset{\sim}{u}_n} = d^j \underset{\sim}{u}(t_{n-1}+\tau)/d\tau^j|_{\tau=0}$ and $\underset{\sim}{v}_n$ is an unknown vector. The substitution of (B.1a) into Eq. (2.5) and using a weighted residual procedure gives:

$$\int_0^h w(\tau)\{\underset{\sim}{K}\gamma_m(\tau) + \underset{\sim}{V}\gamma_{m-1}(\tau) + \underset{\sim}{M}\gamma_{m-2}(\tau)\}\, d\tau\, \underset{\sim}{v}_n$$

$$+ \sum_{j=0}^{m-1} \int_0^h w(\tau)\{\underset{\sim}{K}\gamma_j(\tau) + \underset{\sim}{V}\gamma_{j-1}(\tau) + \underset{\sim}{M}\gamma_{j-2}(\tau)\}\, d\tau\, \overset{(j)}{\underset{\sim}{u}_n}$$

$$= \int_{t_{n-1}}^{t_n} w(t)\{\underset{\sim}{f} + \underset{\sim}{B}\underset{\sim}{p}\}\, dt. \tag{B.2}$$

The imposed strains $\underset{\sim}{\theta}$ have been disregarded for simplicity. Setting

$$\int_0^h w(t)\, dt = 1 \quad \text{(unit total weight)} \tag{B.3a}$$

$$\beta_j = \int_0^h \left(\frac{\tau}{h}\right)^j w(\tau)\, d\tau, \qquad (j = 0,1,..,m;\ \beta_0 \equiv 1) \tag{B.3b}$$

and assuming $\underset{\sim}{f}(t)$ and $\underset{\sim}{p}(t)$ to be linear within the step, i.e.

$$\mathbf{f} = \mathbf{f}_{n-1} + \Delta\mathbf{f}_n \tau/h, \qquad \mathbf{p} = \mathbf{p}_{n-1} + \Delta\mathbf{p}_n \tau/h,$$

Eq. (B.2) can be rewritten as

$$\hat{\mathbf{K}}_m \mathbf{v}_n = -\sum_{j=0}^{m-1} \hat{\mathbf{K}}_j \overset{(j)}{\mathbf{u}_n} + (\mathbf{f}_{n-1} + \beta_1 \Delta\mathbf{f}_n) + \mathbf{B}(\mathbf{p}_{n-1} + \beta_1 \Delta\mathbf{p}_n) \tag{B.4}$$

where the $\hat{\mathbf{K}}_j$ are modified stiffness matrices given by

$$\hat{\mathbf{K}}_j = \mathbf{K}\beta_j \gamma_j(h) + \mathbf{V}\beta_{j-1}\gamma_{j-1}(h) + \mathbf{M}\beta_{j-2}\gamma_{j-2}(h) \tag{B.5}$$

where, by definition, $\beta_{-1} = \beta_{-2} = 0$, $\gamma_{-1}(h) = \gamma_{-2}(h) = 0$, such that

$$\hat{\mathbf{K}}_0 = \mathbf{K}, \quad \hat{\mathbf{K}}_1 = \mathbf{K}\beta_1 h + \mathbf{V}. \tag{B.6}$$

Solving Eq. (B.4) for $\mathbf{v}_n$ and substituting into Eq. (B.1a) gives

$$\mathbf{u}(t_{n-1} + \tau) = \sum_{j=0}^{m-1} \{\mathbf{I}\gamma_j(\tau) - \hat{\mathbf{K}}_m^{-1} \hat{\mathbf{K}}_j \gamma_m(\tau)\} \overset{(j)}{\mathbf{u}_n}$$
$$+ \hat{\mathbf{K}}_m^{-1}\{(\mathbf{f}_{n-1} + \beta_1 \Delta\mathbf{f}_n) + \mathbf{B}(\mathbf{p}_{n-1} + \beta_1 \Delta\mathbf{p}_n)\}\gamma_m(\tau). \tag{B.7}$$

The latter can be differentiated with respect to τ, up to the order m-1, to construct the state vector $\mathbf{X}_n = \mathbf{X}(t_n)$. The formula is so obtained:

$$\mathbf{X}_n = \mathbf{\Gamma}\mathbf{X}_{n-1} + \mathbf{P}(\mathbf{f}_{n-1} + \beta_1 \Delta\mathbf{f}_n) + \bar{\mathbf{W}}\mathbf{p}_{n-1} + \hat{\mathbf{W}}\Delta\mathbf{p}_n, \tag{B.8}$$

where the matrices $\mathbf{\Gamma}$, $\mathbf{P}$, $\bar{\mathbf{W}}$, $\hat{\mathbf{W}}$, partitioned in agreement to the partition of $\mathbf{X}$, have the following typical blocks:

$$\mathbf{\Gamma}_{ij} = \mathbf{I}\gamma_{j-1}(h) - \hat{\mathbf{K}}_m^{-1}\hat{\mathbf{K}}_{j-1}\gamma_{m-i+1}(h), \quad (i,j = 1,2,..,m) \tag{B.9a}$$

$$\mathbf{P}_i = \hat{\mathbf{K}}_m^{-1}\gamma_{m-i+1}(h), \quad (i = 1,2,..,m) \tag{B.9b}$$

$$\bar{\mathbf{W}}_i = \mathbf{P}_i \mathbf{B}, \quad \hat{\mathbf{W}}_i = \beta_1 \bar{\mathbf{W}}_i . \tag{B.9c}$$

The stress $\underset{\sim}{Q}_n$ associated to $\underset{\sim}{X}_n$ are

$$\underset{\sim}{Q}_n = \underset{\sim}{B}^T \underset{\sim}{u}_n - \underset{\sim}{D}(\underset{\sim}{p}_{n-1} + \Delta\underset{\sim}{p}_n) \tag{B.10}$$

and, in virtue of Eq. (B.7) evaluated for $\tau = h$,

$$\underset{\sim}{Q}_n = \underset{\sim}{\Pi}\underset{\sim}{X}_{n-1} + \underset{\sim}{\bar{W}}_1^T(\underset{\sim}{f}_{n-1} + \beta_1 \Delta\underset{\sim}{f}_n) + \underset{\sim}{\bar{Z}}\, \underset{\sim}{p}_{n-1} + \underset{\sim}{\hat{Z}}\, \Delta\underset{\sim}{p}_n \tag{B.11}$$

where

$$\underset{\sim}{\Pi} = \underset{\sim}{B}^T[\underset{\sim}{\Gamma}_{11}\ \underset{\sim}{\Gamma}_{12}\ \cdots\ \underset{\sim}{\Gamma}_{1m}], \tag{B.12a}$$

$$\underset{\sim}{\bar{Z}} = \underset{\sim}{B}^T \underset{\sim}{\hat{K}}_m^{-1} \underset{\sim}{B}\, \gamma_m(h) - \underset{\sim}{D}, \tag{B.12b}$$

$$\underset{\sim}{\hat{Z}} = \underset{\sim}{B}^T \underset{\sim}{\hat{K}}_m^{-1} \underset{\sim}{B}\, \beta_1 \gamma_m(h) - \underset{\sim}{D}. \tag{B.12c}$$

Eqs. (B.8) and (B.11), with

$$\underset{\sim}{X}_n^L = \underset{\sim}{P}(\underset{\sim}{f}_{n-1} + \beta_1 \Delta\underset{\sim}{f}_n), \quad \underset{\sim}{Q}_n^L = \underset{\sim}{\bar{W}}_1^T(\underset{\sim}{f}_{n-1} + \beta_1 \Delta\underset{\sim}{f}_n), \tag{B.13}$$

show the same forms as Eqs. (3.1) and (3.2). Choosing appropriately m and the β coefficients, almost all known integration procedures are obtained, as Newmark's, Wilson's, etc.

APPENDIX C. PROOF OF THE BOUNDING PRINCIPLE

Let us consider two deformation processes, one of which is the actual one described by Eqs. (2.1) to (2.7), the other is a fictitious one characterized by starred quantities. The two processes are coincident with each other in all, except in the following:

i) the elastic domain in the fictitious process has the form:

$$\underset{\sim}{\Psi}^* := \underset{\sim}{N}^T \underset{\sim}{Q}^* - \underset{\sim}{R} + \underset{\sim}{d}\,\omega \le \underset{\sim}{0} \tag{C.1}$$

where $\underset{\sim}{d}$ is an arbitrary time-independent perturbation vector, and $\omega > 0$;

ii) the complementarity conditions are not obeyed in the fictitious process, such that in general the inequality holds:

$$\underset{\sim}{\Psi}^{*T}\dot{\underset{\sim}{\lambda}}^{*} \leq 0. \tag{C.2}$$

At any instant t during the two processes, we can write:

$$\dot{\underset{\sim}{\lambda}}^{T}(\underset{\sim}{N}\underset{\sim}{Q} - \underset{\sim}{R}) = 0, \quad \dot{\underset{\sim}{\lambda}}^{T}(\underset{\sim}{N}^{T}\underset{\sim}{Q}^{*} - \underset{\sim}{R} + \underset{\sim}{d}\omega) = \underset{\sim}{\Psi}^{*T}\dot{\underset{\sim}{\lambda}}, \tag{C.3}$$

which, by subtracting one from the other, give

$$\dot{\underset{\sim}{\lambda}}^{T}\underset{\sim}{N}^{T}(\underset{\sim}{Q}^{*} - \underset{\sim}{Q}) + \omega\underset{\sim}{d}^{T}\dot{\underset{\sim}{\lambda}} = \underset{\sim}{\Psi}^{*T}\dot{\underset{\sim}{\lambda}}; \tag{C.4}$$

further, at the same time t, it is:

$$\dot{\underset{\sim}{\lambda}}^{*T}(\underset{\sim}{N}^{T}\underset{\sim}{Q} - \underset{\sim}{R}) \leq 0, \quad \dot{\underset{\sim}{\lambda}}^{*T}(\underset{\sim}{N}^{T}\underset{\sim}{Q}^{*} - \underset{\sim}{R} + \underset{\sim}{d}\omega) = \underset{\sim}{\Psi}^{*T}\dot{\underset{\sim}{\lambda}}^{*} \tag{C.5}$$

which, subtracting the latter from the former, give

$$-\dot{\underset{\sim}{\lambda}}^{*T}\underset{\sim}{N}^{T}(\underset{\sim}{Q}^{*} - \underset{\sim}{Q}) - \omega\underset{\sim}{d}^{T}\dot{\underset{\sim}{\lambda}}^{*} \leq -\underset{\sim}{\Psi}^{*T}\dot{\underset{\sim}{\lambda}}^{*}. \tag{C.6}$$

Summing Eqs. (C.4) and (C.6) with each other yields

$$\omega\underset{\sim}{d}^{T}\dot{\underset{\sim}{\lambda}} \leq \omega\underset{\sim}{d}^{T}\dot{\underset{\sim}{\lambda}}^{*} + (\underset{\sim}{Q}^{*} - \underset{\sim}{Q})^{T}\underset{\sim}{N}(\dot{\underset{\sim}{\lambda}}^{*} - \dot{\underset{\sim}{\lambda}}) - \underset{\sim}{\Psi}^{*T}\dot{\underset{\sim}{\lambda}}^{*} + \underset{\sim}{\Psi}^{*T}\dot{\underset{\sim}{\lambda}} \tag{C.7}$$

which is a relationship linking the two deformation processes at the same instant $t \geq 0$.

Since

$$\underset{\sim}{N}\dot{\underset{\sim}{\lambda}} = \dot{\underset{\sim}{\varepsilon}} - \underset{\sim}{A}\dot{\underset{\sim}{Q}}, \quad \underset{\sim}{N}\dot{\underset{\sim}{\lambda}}^{*} = \dot{\underset{\sim}{\varepsilon}}^{*} - \underset{\sim}{A}\dot{\underset{\sim}{Q}}^{*}, \tag{C.8}$$

where $\dot{\underset{\sim}{\varepsilon}}$, $\dot{\underset{\sim}{\varepsilon}}^{*}$ are the relevant total strain rate vectors, and thus

$$\underset{\sim}{N}(\dot{\underset{\sim}{\lambda}}^{*} - \dot{\underset{\sim}{\lambda}}) = \dot{\underset{\sim}{\varepsilon}}^{*} - \dot{\underset{\sim}{\varepsilon}} - \underset{\sim}{A}(\dot{\underset{\sim}{Q}}^{*} - \dot{\underset{\sim}{Q}}), \tag{C.9}$$

substituting from the latter into Eq. (C.7) gives

$$\omega\underset{\sim}{d}^{T}\dot{\underset{\sim}{\lambda}} \leq \omega\underset{\sim}{d}^{T}\dot{\underset{\sim}{\lambda}}^{*} + (\underset{\sim}{Q}^{*} - \underset{\sim}{Q})^{T}(\dot{\underset{\sim}{\varepsilon}}^{*} - \dot{\underset{\sim}{\varepsilon}}) - (\underset{\sim}{Q}^{*} - \underset{\sim}{Q})^{T}\underset{\sim}{A}(\dot{\underset{\sim}{Q}}^{*} - \dot{\underset{\sim}{Q}}) - \underset{\sim}{\Psi}^{*T}\dot{\underset{\sim}{\lambda}}^{*} + \underset{\sim}{\Psi}^{*T}\dot{\underset{\sim}{\lambda}}. \tag{C.10}$$

By the virtual work principle, we write

$$(\underset{\sim}{Q}^* - \underset{\sim}{Q})(\dot{\underset{\sim}{\varepsilon}}^* - \dot{\underset{\sim}{\varepsilon}}) = -(\ddot{\underset{\sim}{u}}^* - \ddot{\underset{\sim}{u}})^T \underset{\sim}{M} (\dot{\underset{\sim}{u}}^* - \dot{\underset{\sim}{u}}) - (\dot{\underset{\sim}{u}}^* - \dot{\underset{\sim}{u}})^T \underset{\sim}{V} (\dot{\underset{\sim}{u}}^* - \dot{\underset{\sim}{u}}); \tag{C.11}$$

therefore, setting

$$J(t) = \frac{1}{2} (\underset{\sim}{Q}^* - \underset{\sim}{Q})^T \underset{\sim}{A} (\underset{\sim}{Q}^* - \underset{\sim}{Q}) + \frac{1}{2} (\dot{\underset{\sim}{u}}^* - \dot{\underset{\sim}{u}})^T \underset{\sim}{M} (\dot{\underset{\sim}{u}}^* - \dot{\underset{\sim}{u}}) + \int_0^t (\dot{\underset{\sim}{u}}^* - \dot{\underset{\sim}{u}})^T \underset{\sim}{V} (\dot{\underset{\sim}{u}}^* - \dot{\underset{\sim}{u}})\, d\bar{t}, \tag{C.12}$$

Eq. (C.10) becomes after an integration over the interval $(0, t_1)$:

$$\omega \underset{\sim}{d}^T \int_0^{t_1} \dot{\underset{\sim}{\lambda}}\, d\bar{t} \le \omega \underset{\sim}{d} \int_0^{t_1} \dot{\underset{\sim}{\lambda}}^* d\bar{t} + J(0) - J(t_1) - \int_0^{t_1} \underset{\sim}{\psi}^{*T} \dot{\underset{\sim}{\lambda}}^* d\bar{t} + \int_0^{t_1} \underset{\sim}{\psi}^{*T} \dot{\underset{\sim}{\lambda}}\, d\bar{t}. \tag{C.13}$$

Since $J(t_1) \ge 0$ and further, in virtue of the inequality in Eq. (C.1),

$$\int_0^{t_1} \underset{\sim}{\psi}^{*T} \dot{\underset{\sim}{\lambda}}\, d\bar{t} \le 0, \tag{C.14}$$

these two terms can be dropped in Eq. (C.13) so enforcing the inequality, which so reads:

$$\omega \underset{\sim}{d}^T \int_0^{t_1} \dot{\underset{\sim}{\lambda}}\, d\bar{t} \le \omega \underset{\sim}{d}^T \int_0^{t_1} \dot{\underset{\sim}{\lambda}}^* d\bar{t} + J(0) - \int_0^{t_1} \underset{\sim}{\psi}^{*T} \dot{\underset{\sim}{\lambda}}^* d\bar{t}, \tag{C.15}$$

where

$$J(0) = \frac{1}{2} (\underset{\sim}{Q}^* - \underset{\sim}{Q})^T \underset{\sim}{A} (\underset{\sim}{Q}^* - \underset{\sim}{Q})\Big|_{t=0} + \frac{1}{2} (\dot{\underset{\sim}{u}}^* - \dot{\underset{\sim}{u}})^T \underset{\sim}{M} (\dot{\underset{\sim}{u}}^* - \dot{\underset{\sim}{u}})\Big|_{t=0}, \tag{C.16}$$

that is $J(0)$ depends on the initial values of the stresses and the velocities in the two processes. Assuming the same initial conditions for all of them, with $\underset{\sim}{\lambda} = \underset{\sim}{\lambda}^* = \underset{\sim}{0}$ at $t = 0$, we finally get

$$\underset{\sim}{d}^T \underset{\sim}{\lambda}(t_1) \le \underset{\sim}{d}^T \underset{\sim}{\lambda}^*(t_1) - \frac{1}{\omega} \int_0^{t_1} \underset{\sim}{\psi}^{*T} \dot{\underset{\sim}{\lambda}}^* d\bar{t} \tag{C.17}$$

which is the desired bounding principle.

If now the same time discretization as in Sec. 3 is used and we assume $t_1 = n_1 h$, we can write:

$$\underset{\sim}{d}^T \underset{\sim}{\lambda}(t_1) = \underset{\sim}{d}^T \sum_{j=1}^{n_1} \underset{\sim}{y}_{(j)}, \quad \underset{\sim}{y}_{(j)} \geq \underset{\sim}{0}, \tag{C.18a}$$

$$\underset{\sim}{d}^T \underset{\sim}{\lambda}^*(t_1) = \underset{\sim}{d}^T \sum_{j=1}^{n_1} \underset{\sim}{y}^*_{(j)}, \quad \underset{\sim}{y}^*_{(j)} \geq \underset{\sim}{0}, \tag{C.18b}$$

$$- \int_0^{t_1} \underset{\sim}{\Psi}^{*T} \dot{\underset{\sim}{\lambda}}^* d\bar{t} = - \sum_{j=1}^{n_1} \left(\frac{1}{h} \int_{t_{j-1}}^{t_j} \underset{\sim}{\Psi}^* d\bar{t} \right)^T \underset{\sim}{y}^*_{(j)} \cong \sum_{j=1}^{n_1} \underset{\sim}{z}^{*T}_{(j)} \underset{\sim}{y}^*_{(j)}, \tag{C.18c}$$

where

$$\underset{\sim}{z}^*_{(j)} = -\underset{\sim}{N}^T \underset{\sim}{Q}^*_{(j)} + \underset{\sim}{R} - \underset{\sim}{d}\,\omega \geq \underset{\sim}{0}, \tag{C.19}$$

such that the bound relationship (C.15) takes on the form

$$\underset{\sim}{d}^T \sum_{j=1}^{n_1} \underset{\sim}{y}_{(j)} \leq \underset{\sim}{d}^T \sum_{j=1}^{n_1} \underset{\sim}{y}^*_{(j)} + \frac{1}{\omega} \sum_{j=1}^{n_1} \underset{\sim}{z}^*_{(j)} \underset{\sim}{y}^*_{(j)} \tag{C.20}$$

which coincides with Eq. (4.3). Since, by Eq. (3.2),

$$\underset{\sim}{Q}^*_{(j)} = \underset{\sim}{\Pi}\, \underset{\sim}{X}^*_{(j-1)} + \underset{\sim}{Q}^L + \underset{\sim}{Z}_1 \underset{\sim}{p}^*_{(j-1)} + \underset{\sim}{Z}_2 \Delta \underset{\sim}{p}^*_{(j)}, \tag{C.21}$$

substituting from the latter into Eq. (C.19) gives

$$\underset{\sim}{z}^*_{(j)} = \underset{\sim}{S}\, \underset{\sim}{y}^*_{(j)} + \underset{\sim}{b}^*_{(j)} - \underset{\sim}{d}\,\omega \geq \underset{\sim}{0}, \quad \underset{\sim}{y}^*_{(j)} \geq \underset{\sim}{0} \tag{C.22}$$

where

$$\underset{\sim}{b}^*_{(j)} = \underset{\sim}{R} - \underset{\sim}{N}^T (\underset{\sim}{\Pi}\, \underset{\sim}{X}^*_{(j-1)} + \underset{\sim}{Q}^L + \underset{\sim}{Z}_1 \underset{\sim}{p}^*_{(j-1)} \tag{C.23}$$

and thus Eqs. (C.22) coincide with Eqs. (4.4).

The approximation in the last of Eqs. (3.18c) implies that the left-hand member of Eq. (C.14) transforms as

$$\int_0^{t_1} \underset{\sim}{\psi}^{*T} \dot{\underset{\sim}{\lambda}} \, d\bar{t} = \sum_{j=1}^{n_1} \left(\frac{1}{h} \int_{t_{j-1}}^{t_j} \underset{\sim}{\psi}^{*} d\bar{t} \right)^T \underset{\sim}{y}_{(j)} \cong - \sum_{j=1}^{n_1} \underset{\sim}{z}^{*T}_{(j)} \underset{\sim}{y}_{(j)} \leq 0$$

in virtue of Eqs. (C.19).

CHAPTER 18

STRUCTURAL ANALYSIS FOR NONLINEAR MATERIAL BEHAVIOUR

J. A. Teixeira de Freitas
Istituto Superior Tecnico, Lisbon, Portugal

Abstract: A method for the analysis of structures in the presence of non-linear physical effects is presented. The technique consists in replacing by asymptotic approximations the nonlinear functions intervening in the governing system. Procedures for detecting and solving situations of plastic straining and unstressing and of elasto-plastic instabilization of the structure are incorporated in the algorithm suggested to perform the analysis of the response of the structure.

Introduction

Presented next is a method for the analysis of structures in the presence of nonlinear physical effects.

It is assumed that the displacements and the deformations developing in the structure are very small. Linear descriptions are therefore used to represent the equilibrium and compatibility conditions of the structure, which may however be sujected to a piecewise-nonlinear loading programme.

The constitutive relations are nevertheless the essential source of nonlinearity. Nonlinear elasticity and plasticity conditions relating stress- and strain-resultants through nonlinear hardening, yield and flow laws are herein accepted in the representation of the behaviour of the materials constituting the structure building elements.

The system thus found to govern the response of the structure is highly nonlinear. To overcome the inherent numerical implementation difficulties, two linearization techniques are often adopted.

One approach consists in replacing the nonlinear functions describing the constitutive relations by piecewise-linear approximations [1]. As the degree of accuracy is increasingly refined, this technique tends to generate a rapid growth in the number of variables and constraints required to represent the response of the structure.

The second technique consists in creating an iterative sequence based on estimates for the entries of the nonlinear structural operators [2,3]. The dimensions of the governing system remain unchanged but very sensitive procedures are required to guarantee an acceptable rate of convergence.

The alternative technique described herein consists in replacing the intervening nonlinear functions by asymptotic approximations [4]. The governing system is then replaced by a recursive, convergent sequence of linear complementarity problems involving the original number of variables and constraints.

This technique, which has already been applied in the implementation of geometrically nonlinear structural analysis problems [5], is herein illustrated using a displacement method based formulation. The procedures described next for detecting and solving situations of plastic straining and unstressing and elastoplastic instabilization of the structure are however readily applicable to any other formulation for structural analysis problems [6].

Equilibrium and Compatibility Conditions

Let the structure independent stress-{strain-}resultants be collected in array $\mathbf{X}$ $\{\mathbf{u}\}$. Listing in vector λ $\{\mathbf{q}\}$ the applied forces {indeterminate displacements} and denoting by δ $\{\mathbf{Q}\}$ the associated displacements {forces}, the following description is found the equilibrium and compatibility conditions of the structure:

$$\mathbf{Q} = \mathbf{0} = \begin{bmatrix} \tilde{\mathbf{A}} & \tilde{\mathbf{A}}_0 \end{bmatrix} \begin{Bmatrix} -\mathbf{X} \\ \lambda \end{Bmatrix}, \tag{1}$$

$$\begin{Bmatrix} \mathbf{u} \\ \delta \end{Bmatrix} = \begin{bmatrix} \mathbf{A} \\ \mathbf{A}_0 \end{bmatrix} \mathbf{q}, \qquad \begin{matrix} (a) \\ (b) \end{matrix} \tag{2}$$

As the displacements and the deformations are assumed to be infinitesimal , the entries of operators $\mathbf{A}$ and $\mathbf{A}_0$ depend solely on the initial topography of the structure. The compatibility conditions (2) are therefore linear and the equilibrium conditions (1) will not be so only if the loading programme λ is assumed to be nonlinear. Whatever the circumstance, the loading can always be expressed in the form

$$\lambda = \lambda^0 + \Lambda \ \Delta\lambda + \mathbf{R}_\lambda, \tag{3}$$

wherein $\Delta\lambda$ denotes the parameter selected to describe a finite load increment. In definition (3), vector

$$\Lambda = \Lambda\ (\lambda^0),$$

represents the loading gradient at the instant increment $\Delta\lambda$ takes place, and array

$$\mathbf{R}_\lambda = \mathbf{R}_\lambda\ \left(\lambda^0,\ \Delta\lambda\right),$$

collects all the intervening terms nonlinear in the load parameter $\Delta\lambda$.

Elastoplastic Constitutive Relations

Assume that the materials constituting the structural building elements follow elastic-plastic laws described by conditions (4) to (7) as suggested in [6]:

$$\mathbf{u} = \mathbf{u}_e + \mathbf{u}_p, \tag{4}$$

$$\mathbf{X} = \mathbf{K}\ \mathbf{u}_e, \tag{5}$$

$$\begin{Bmatrix} \phi_* \\ \mathbf{u}_p \end{Bmatrix} = \begin{bmatrix} -\mathbf{H}' & \tilde{\mathbf{N}}' \\ \mathbf{N}' & \mathbf{0} \end{bmatrix} \begin{Bmatrix} \mathbf{u}_* \\ \mathbf{X} \end{Bmatrix} - \begin{Bmatrix} \mathbf{X}_* \\ \mathbf{0} \end{Bmatrix} \tag{6}$$

$$\phi_* \leq \mathbf{0}, \tag{7}$$

$$\mathbf{u}_* \geq \mathbf{0}, \tag{8}$$

$$\tilde{\phi}_*\ \mathbf{u}_* = 0. \tag{9}$$

Definition (4) decomposes the total deformation into elastic and plastic addends, and arrays ϕ_* and $\mathbf{u}_*$ collect the yield functions and the plastic parameters required to describe the plastic phase of the response.

To model materials with nonlinear behaviour, the entries of the elastic stiffness and plastic hardening matrices of the finite elements into which the structure is descretized are now allowed to depend on the current state of stress and strain:

$$\mathbf{K}' = \mathbf{K}'\,(\mathbf{X},\,\mathbf{u})\ \ ,\ \ \mathbf{H}' = \mathbf{H}'\,(\mathbf{X},\,\mathbf{u}).$$

In definition (6), the columns of matrix $\mathbf{N}'$ still represent vectors orthogonal to the associated yield loci, the initial distances of which to the origin of the stress-resultant space are collected in array $\mathbf{X}_*$; their entries are now functions of the current state of stress:

$$\mathbf{N}' = \mathbf{N}'\,(\mathbf{X})\ \ ,\ \ \mathbf{X}_* = \mathbf{X}_*\,(\mathbf{X}).$$

Incremental Formulation

Let $\mathbf{v}$ denote a generic variable and assume that its finite increment,

$$\mathbf{v} = \mathbf{v}^0 + \Delta\,\mathbf{v}, \tag{10}$$

is expressed in a power series on a non-negative parameter ϵ:

$$\Delta\,\mathbf{v} = \mathbf{v}^{(n)}\,\frac{\epsilon^n}{n!}. \tag{11}$$

Taking finite increments (10,11) in (1.2) and equating terms of the same order in parameter ϵ, the following asymptotic description is found for the structure equilibrium and compatibility conditions,

$$\mathbf{0} = \begin{bmatrix} \tilde{\mathbf{A}} & \mathbf{a} \end{bmatrix} \begin{Bmatrix} -\mathbf{X} \\ \lambda \end{Bmatrix}^{(n)} + \mathbf{R}_Q^{(n)}, \tag{12}$$

$$\begin{Bmatrix} \mathbf{u} \\ W \end{Bmatrix}^{(n)} = \begin{bmatrix} \mathbf{A} \\ \tilde{\mathbf{a}} \end{bmatrix} \mathbf{q}^{(n)}, \qquad \begin{matrix} (a) \\ (b) \end{matrix} \tag{13}$$

wherein

$$\mathbf{a} = \tilde{\mathbf{A}}_0\,\Lambda,$$

and

$$\mathbf{R}_Q^{(n)} = \tilde{\mathbf{A}}_0\,\mathbf{R}_\lambda^{(n)},$$

according to definition (3). In definition (13b), $W^{(n)}$ represents the n-th order term in the series expansion (11) of the external work rate:

$$\Delta W = \Delta\,\tilde{\lambda}\ \ \Delta\,\tilde{\delta}/\Delta\lambda.$$

Processing definitions (4) to (6) in a similar manner, the folowing expressions are found,

$$\mathbf{u}^{(n)} = \mathbf{u}_e^{(n)} + \mathbf{u}_p^{(n)}, \tag{14}$$

$$\mathbf{X}^{(n)} = \mathbf{K}\,\mathbf{u}_e^{(n)} + \mathbf{P}_x^{(n)}, \tag{15}$$

$$\begin{Bmatrix} \phi_* \\ \mathbf{u}_p \end{Bmatrix}^{(n)} = \begin{bmatrix} -\mathbf{H} & \tilde{\mathbf{N}} \\ \mathbf{N} & \mathbf{0} \end{bmatrix} \begin{Bmatrix} \mathbf{u}_* \\ \mathbf{X} \end{Bmatrix}^{(n)} - \begin{Bmatrix} \mathbf{R}_\phi \\ \mathbf{R}_* \end{Bmatrix}^{(n)}. \tag{16}$$

The infinite sequence of systems (12-17) is linear as the intervening operators depend on the state of stress, strain and loading prior to implementation of the finite increments $\Delta\mathbf{v}$:

$$\mathbf{a} = \mathbf{a}\,(\lambda^0), \quad \mathbf{K} = \mathbf{K}(\mathbf{X}^0,\, \mathbf{u}^0),$$
$$\mathbf{N} = \mathbf{N}\,(\mathbf{X}^0), \quad \mathbf{H} = \mathbf{H}(\mathbf{X}^0,\, \mathbf{u}^0).$$

Moreover, the sequence is recursive since the n-th order residuals $\mathbf{R}^{(n)}$ are functions of variables appearing in the $(n-1)$-th order and prior systems in the sequence, and are thus known at the outset in the n-th order system:

$$\mathbf{R}^{(n)} = \mathbf{R}^{(n)}\left(\dot{\mathbf{v}}\,,\, \ddot{\mathbf{v}}\,,\, \mathbf{v}^{(n-1)}\right).$$

The yield and flow rules (7,8) of plasticity are now expressed in form:

$$\phi_* = \phi_*^0 + \phi_*^{(n)}\,\frac{\epsilon^n}{n!} \le 0, \qquad (a)$$
$$\phi_*^0 \le 0, \qquad (b) \tag{17}$$

$$\mathbf{u}_* = \mathbf{u}_*^0 + \mathbf{u}_*^{(n)}\,\frac{\epsilon^{(n)}}{n!} \ge 0, \qquad (a) \tag{18}$$

with

$$\dot{\mathbf{u}}_* \ge \mathbf{0}, \qquad (b) \tag{19}$$

to ensure the irreversibility of plastic straining.

The holonomy hypothesis implied by the plastic association condition (9) is abandoned in incremental procedures in favour of the following complementarity conditions:

$$\tilde{\phi}_*^0\,\dot{\mathbf{u}}_* = 0\ ,\quad \tilde{\dot{\phi}}_*\,\dot{\mathbf{u}}_* = 0. \tag{20}$$

System (21) is obtained manipulating definitions (12-16) as to eliminate the stress-resultants and the elastic component of the strain-resultants:

$$\begin{bmatrix} \mathbf{S} & -\mathbf{C} & -\mathbf{a} \\ -\tilde{\mathbf{C}} & \mathbf{D} & \mathbf{0} \\ -\tilde{\mathbf{a}} & 0 & 0 \end{bmatrix} \begin{Bmatrix} \mathbf{q} \\ \mathbf{u}_* \\ \lambda \end{Bmatrix}^{(n)} + \begin{Bmatrix} \mathbf{0} \\ \phi_* \\ W \end{Bmatrix}^{(n)} = \begin{Bmatrix} \mathbf{R}_1 \\ \mathbf{R}_2 \\ 0 \end{Bmatrix}^{(n)} \cdot \quad \begin{matrix} (a) \\ (b) \\ (c) \end{matrix} \qquad (21)$$

The following notation is used for the structural matrices and stipulation vectors:

$$\begin{aligned} \mathbf{S} &= \tilde{\mathbf{A}}\,\mathbf{K}\,\mathbf{A}, & \mathbf{R}_1^{(n)} &= \mathbf{R}_Q^{(n)} - \tilde{\mathbf{A}}\,\mathbf{R}_0^{(n)}, \\ \mathbf{C} &= \tilde{\mathbf{A}}\,\mathbf{K}\,\mathbf{N}, & \mathbf{R}_2^{(n)} &= -\mathbf{R}_\phi^{(n)} - \tilde{\mathbf{N}}\,\mathbf{R}_0^{(n)}; \\ \mathbf{D} &= \tilde{\mathbf{N}}\,\mathbf{K}\,\mathbf{N} + \mathbf{H}, & \mathbf{R}_0^{(n)} &= \mathbf{R}_x^{(n)} - \mathbf{K}\,\mathbf{R}_*^{(n)}. \end{aligned} \qquad (22)$$

The Governing System

The incremental procedures, by their very nature, allow for a direct control of those yield modes which are currently active.

The variables describing the plastic phase may therefore be partitioned into two groups associated respectivelly with the currently active and inactive yield modes. Identifying these sets by subscripts 1 and 2,

$$\begin{aligned} \phi_{*1}^{0} &= \mathbf{0} \quad , \quad \mathbf{u}_{*1}^{0} \geq \mathbf{0}, \\ \phi_{*2}^{0} &< \mathbf{0} \quad , \quad \mathbf{u}_{*2}^{0} > \mathbf{0}, \end{aligned}$$

the following conditions are found for the higher order terms,

$$\begin{aligned} \phi_{*1}^{(n)} &= \mathbf{0} \quad , \quad \dot{\mathbf{u}}_{*1} \geq \mathbf{0} \quad , \quad \mathbf{u}_{*1}^{(n)} \;\; \mathbf{0} \quad , \quad n \geq 2, \\ \phi_{*2}^{(n)} & \;\; \mathbf{0} \quad , \quad \dot{\mathbf{u}}_{*2} = \mathbf{0} \quad , \quad \mathbf{u}_{*2}^{(n)} = \mathbf{0} \quad , \quad n \geq 2, \end{aligned}$$

thus simplifying constraints (17 - 20) to the following:

$$\phi_{*2} = \phi_{*2}^{0} + \phi_{*2}^{(n)} \frac{\epsilon^{(n)}}{n!} \leq \mathbf{0}, \qquad (23)$$

$$\mathbf{u}_{*1} = \mathbf{u}_{*1}^{0} + \mathbf{u}_{*1}^{(n)} \frac{\epsilon^{(n)}}{n!} \geq \mathbf{0}, \qquad (24)$$

$$\dot{\mathbf{u}}_{*1} \geq \mathbf{0}. \qquad (25)$$

The system governing the response of the structure to incremental action described by parameter ϵ is obtained combining conditions (23-25) with the sequence of recursive linear equations (21) in which only the variables and constraints associated with the currently active mode need be considered:

$$\begin{bmatrix} \mathbf{S} & -\mathbf{C}_1 & -\mathbf{a} \\ -\tilde{\mathbf{C}}_1 & \mathbf{D}_{11} & \mathbf{0} \\ -\tilde{\mathbf{a}} & \mathbf{0} & 0 \end{bmatrix} \begin{Bmatrix} \mathbf{q} \\ \mathbf{u}_{*1} \\ \lambda \end{Bmatrix}^{(n)} = \begin{Bmatrix} \mathbf{R}_1 \\ \mathbf{R}_2 \\ -W \end{Bmatrix}^{(n)} . \quad \begin{matrix} (a) \\ (b) \\ (c) \end{matrix} \tag{26}$$

Traditional Solution Procedures

Most formulation in nonlinear structural analysis are based in the first order approximation $\left(\dot{\mathbf{R}}_1, \dot{\mathbf{R}}_2 = \mathbf{0}\right)$ of system (26a,b):

$$\begin{bmatrix} \mathbf{S} & -\mathbf{C}_1 \\ -\tilde{\mathbf{C}}_1 & \mathbf{D}_{11} \end{bmatrix} \begin{Bmatrix} \dot{\mathbf{q}} \\ \dot{\mathbf{u}}_{*1} \end{Bmatrix} = \begin{Bmatrix} \mathbf{a}\,\dot{\lambda} \\ \mathbf{0} \end{Bmatrix} . \quad \begin{matrix} (a) \\ (b) \end{matrix} \tag{27}$$

The common pratice is to modify the elastic stiffness matrix of the structure to take into account the effects of plasticity,

$$\mathbf{S}'\,\dot{\mathbf{q}} = \mathbf{a}\,\dot{\lambda}, \tag{28}$$

with

$$\mathbf{S}' = \mathbf{S} - \mathbf{C}_1\,\mathbf{D}_{11}^{-1}\,\tilde{\mathbf{C}}_1 .$$

The triple product present in the above definition is avoided has it is first implemented at element level and then allocated to the systems using the standard assembly procedures of finite element structural analysis.

Systems (27) and (28) are usually solved for fixed increments of the load parameter; displacements and plastic parameters may however be used as control variables.

As these increments are treated as finite quantities, the algorithm generates a sequence of increasingly diverging solution points unless numerically expensive acceleration techniques are continuously enforced. Also expensive is the implementation of predictor and corrector techniques [2,3,7] required in fixed-step programs to detect the activation of new yield modes. Moreover, the analysis of the response of the structure in the vicinity of limit and branching points becomes particularly difficult

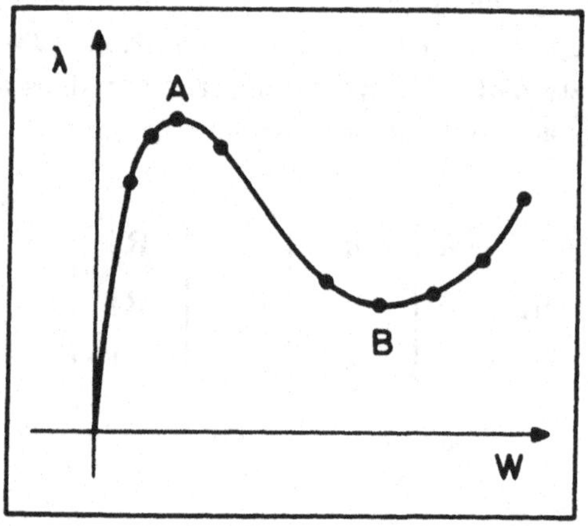

Figure 1

to handle as it requires checks on the monotonicity of variables described by increments which by then can be as significant as the numerical noise accumulated in the process.

In the procedure suggested herein, the user is allowed to control directly the degree of acuracy of the output by selecting the higher-order terms of the sequence of systems (26) he wishes to implement. The identification of the external work rate with the perturbation parameter in the series expansion (11), to yield,

$$\Delta W \equiv \epsilon \ , \quad W^{(n)} = \delta_{1n}, \tag{29}$$

wherein δ_{mn} ($m = 1$) denotes the Kronecker symbol, circumvents the need for monotonicity checks on the control variable and simplifies the detection of yielding and unstressing points, as well as of limit and branching points the structure equilibrium path may exhibit.

Truncation of the Series Expansion

The algorithm assumes that the power series expansion (11) is truncated to a finite number k of terms:

$$\Delta v_m \simeq v_m^{(n)} \frac{\epsilon^n}{n!} \ , \quad n = 1, 2 \ldots . k. \tag{30}$$

If the series expansion is convergent, the contribution of the truncated terms can be rendered negligible by requiring that the highest order term in (30) does not exceed a prescribed tolerance τ,

$$v_m^{(k)} \frac{\epsilon^k}{k!} \leq \tau,$$

thus providing the following bound for the step increment:

$$\epsilon \leq \min_m \left[\frac{k! \ \tau}{| \ v_m^{(k)} \ |} \right]^{1/k} . \qquad (31)$$

As the step increment is inversely proportional to the k-th order derivatives of the problem variables, the above expression adapts automatically the size of the increment to the degree of nonlinearity manifested by the response of the structure. This is illustrated in figure 1; quasi-linear phases in the response are solved with large step lenghts which are then progressively reduced as the nonlinearity of the response becomes increasingly pronounced.

Plastic Straining

Consider one of the yield functions contained in the non-active set of plastic modes (23):

$$\phi_{*2} = \phi_{*2}^{0} + \phi_{*2}^{(n)} \frac{\epsilon^n}{n!} \leq 0.$$

Assume that the step increment produced by definition (31) violates the above yield condition, as illustrated by transition from state A to C in figure 2. The step

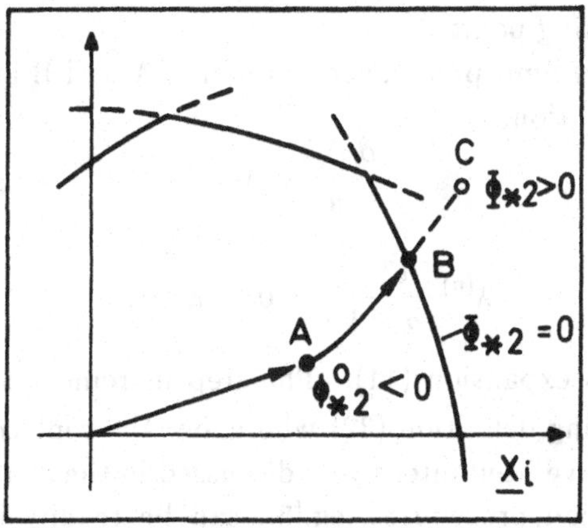

Figure 2

lenght that exposes the activation of the yield mode at state B is defined by condition

$$\phi_{*2} + \phi_{*2}^{(n)} \frac{\epsilon^n}{n!} = 0,$$

which can be solved to give the following expression for the step increment,

$$\epsilon = \phi_n \, \gamma_0^n, \tag{32}$$

with

$$\gamma_k = -\phi_{*2}^{(k)} / \dot{\phi}_{*2},$$

and

$$\phi_1 = 1 \; , \; \phi_2 = 1/2 \, \gamma_2 \; , \; \phi_3 = 2\phi_2^2 - 1/6 \, \gamma_3 \; , \; \ldots$$

The same type of control has to be implemented on all piecewise defined functions which intervene in the problem, namely those describing the loading and the elastic and plastic hardening modes.

Assume, for instance, that λ_{max} represents the bound on a current loading function (3). The step increment that exposes this limit is obtained through definition (32), wherenow λ_{max}^0 and $\lambda^{(n)}$ replace variables ϕ_{*2}^0 and $\phi_{*2}^{(2)}$, respectively.

Branching of the Equilibrium Path

Whenever the structural matrix in the governing system (26) becomes singular, the values taken by the intervening variables are rendered indefinite, thus revealing the occurence of a critical stress/strain distribution. If the analysis is to be extended beyond this critical point it becomes necessary to determine its nature - that of a limit point or a branching point.

The identification of limit points, such as points A and B in figure 1 is controlled by the sationarity condition,

$$\frac{\partial \, \Delta\lambda}{\partial \epsilon} = 0,$$

or

$$\lambda^{(n)} \frac{\epsilon^{n-1}}{(n-1)!} = 0 \; , \; n \geq 1, \tag{33}$$

according to the series expansion (11). The step increment that exposes the limit point is determined using definition (32) wherenow $\lambda^{(n)}$ replaces $\phi_{*2}^{(n-1)}$.

Branching points have been intensively discussed in the context of elastic systems and the existing solution procedures, eg.[8], can be readily incorporated into the formulation being presented [9]. In fact system (26) may well be interpreted as the governing system of an elastic structure with additional kinematic indeterminacies $\mathbf{u}_{*2}$.

Plastic Unstressing

Plastic unstressing is a common occurence in elastic-plastic systems subject to discontinuous loading programs but may also arise in structures under proportional loading.

A unified treatment of the different modes of plastic unstressing emerged with the application of mathematical programming theory to structural plasticity [5,10-13].

Plastic unstressing is typically associated witn non-monotonic plastic straining modes, as illustrated by graph OAA' in figure 3. When maximum straining is attained at configuration A, the associated yield mode has to be deactivated, so that the plastic deformation is allowed to remain constant in the ensuing incremental action. Configuration A is characterized by the following stationarity conditions:

$$\frac{\partial \Delta\, u_{*2}}{\partial \epsilon} = u_{*2}^{(n)} \frac{\epsilon^{n-1}}{(n-1)!} = 0 \ , \quad n \geq 1. \tag{34}$$

The step increment that exposes this form of plastic unstressing is determined through definition (32) wherenow $\varphi_{*2}^{(k)}$ reads $\mathbf{u}_{*1}^{(k+1)}$.

Plastic unstressing is also exposed when the first-order solution of systems (26) generates a set of plastic multipliers which contravene the non-negativity condition (25), as illustrated by graph OC in figure 3. This is a rather more complex situation as it may simulate any of the four configurations represented in figure 1: besides unstressing, plastic locking and branching modes may be simultaneously involved.

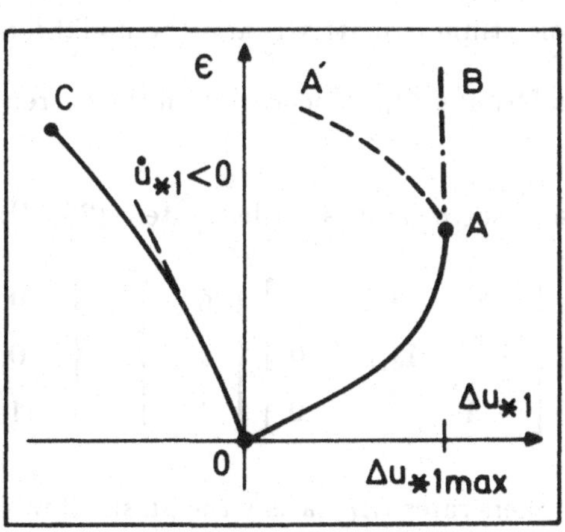

Figure 3

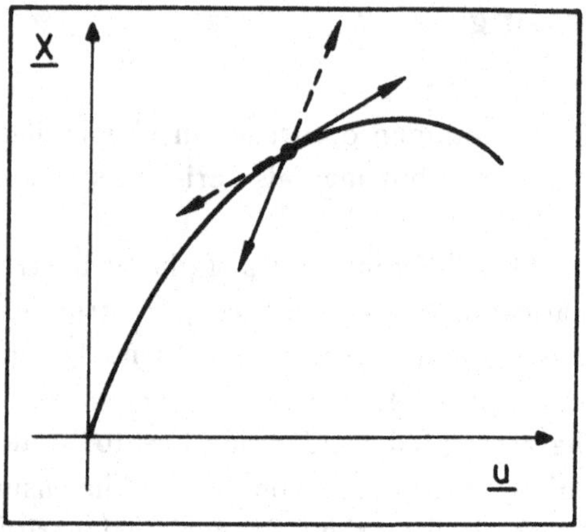

Figure 4

This problem is analysed in detail in [5] and the simplest of the procedures suggested therein is included in the algorithm described below.

Numerical Implementation Procedure

The asymptotic method for the analysis of structures in the presence of physical instabilizing effects can be summarized as follows:

1. Assemble the elastic stiffness matrix $\mathbf{S}$ and the nodal forces vector $\mathbf{a}$.

2. Assemble matrices $\mathbf{D}_{11}$ and $\mathbf{C}_1$ associated with the currently active plastic yield modes.

3. For $n = 1$ set-up and solve the first-order system (26,27):

$$\begin{bmatrix} \mathbf{S} & -\mathbf{C}_1 & -\mathbf{a} \\ -\tilde{\mathbf{C}}_1 & \mathbf{D}_{11} & \mathbf{0} \\ -\tilde{\mathbf{a}} & \tilde{\mathbf{0}} & 0 \end{bmatrix} \begin{Bmatrix} \dot{\mathbf{q}} \\ \dot{\mathbf{u}}_{*1} \\ \dot{\lambda} \end{Bmatrix} = \begin{Bmatrix} \mathbf{0} \\ \mathbf{0} \\ -1 \end{Bmatrix}. \qquad (35)$$

4. If the plastic multipliers rates $\dot{\mathbf{u}}_{*1}$, satisfy the plastic flow condition (25) proceed to step 6.

5. Remove from the list of the currently active plastic yield modes the one associated with the most negative plastic multiplier rate and return to step 3.

6. To obtain the higher-order solutions, set $n = n + 1$ and assemble residuals $\mathbf{R}_1^{(n)}$ and $\mathbf{R}_2^2$.

7. Solve the n-th order system (26,27):

$$\begin{bmatrix} \mathbf{S} & -\mathbf{C}_1 & -\mathbf{a} \\ -\tilde{\mathbf{C}}_1 & \mathbf{D}_{11} & \mathbf{0} \\ -\tilde{\mathbf{q}} & \tilde{\mathbf{0}} & 0 \end{bmatrix} \begin{Bmatrix} \mathbf{q} \\ \mathbf{u}_{*1} \\ \lambda \end{Bmatrix}^{(n)} = \begin{Bmatrix} \mathbf{R}_1 \\ \mathbf{R}_2 \\ 0 \end{Bmatrix}^{(n)} . \tag{36}$$

8. Return to step 7 if $n \leq k$ $(k = 3$, in general$)$.

9. Compute the plastic potentials $\varphi_{*2}^{(n)}$ $(n = 1, 2, \ldots, k)$ associated with the non-active yield modes.

10. Evaluate the largest step increment that complies with controls on the series truncation error (31), the activation of new yield modes (32), the identification of a limit point (33) and the occurence of plastic unstressing (34).

11. Update the list of currently active yield modes, if appropriate.

12. Update the system variables using definitions (10,11) and return to step 1 after checking the stoping rule.

The operations described in steps 1, 2 and 6 are implemented using the standard finite element techniques for direct allocation of the individual contributions of the structural building elements.

The solution of the symmetric, linear systems of equations (35) and (36) can be performed using direct and/or iterative methods adapted to the particular structure presented by the plastic-elastic stiffness matrices; matrix $\mathbf{S}_{11}$ is highly sparse and matrix $\mathbf{D}_{11}$ is block-diagonal.

The additional computational effort required by the solution of the higher-order systems (36) is rewarded by the degree of accuracy that can thus be achieved and compensated by the possibility of enforcing larger step increments.

The procedure described above can be readily adapted to perform the analysis of structures modeled by piecewise-linear elastic-plastic laws. Steps 6, 7 and 8 are now omited as the second- higher-order variables are null $\left(\mathbf{v}^{(n)} \quad 0 \, , \, n \geq 2\right)$ and, in step 10, the only relevant criterion for determining the step increment ϵ is the control on the activation of new yield modes. Piecewise-linear elastic modes can be described using holonomic plasticity models in which strain reversal is allowed for; step 1 in the algorithm is implemented only once and the "plastic" parameters describing the deformations associated with the subsquent elastic modes are released from the plastic unstressing control implemented in steps 4 and 5.

REFERENCES

1. Cohn, M.Z. and G. Maier (Eds): Engineering plasticity by mathematical programming, Pergamon, 1979

2. Zienkiewicz, O.C., R. Valliapans and I.P. King: Elastoplastic solutions of engineering problems. Int. J. Num. Meth. Engrg.. 1, 1969

3. Argyris, J.H.: Methods of elastoplstic analysis, J. Appl. Math. Phys., 23, 1972

4. Freitas, J.A.T.: Análise estrutural com materiais não-lineares

5. Freitas, J.A.T. and D.L. Smith: Plastic straining, unstressing and branching in large displacement perturbation analysis, Int. J. Num. Meth. Engrg., 20, 1984

6. Freitas, J.A.T.: Elastoplastic analysis of skeletal structures, Mathematical Programming Methods in Structural Plasticity, CISM, Udine. 1986

7. Jennings. A. and K. Majid: An elastic-plastic analysis by computer for framed structures loaded up to collapse. Struct. Eng., 43, 1965

8. Thompson. J.M.T. and G.W. Hunt: A General Theory of Elastic Instability , Wiley, 1973.

9. Freitas, J.A.T. and J.P.B.M. Almeida: A nonlinear projection method for constrained optimization, Civ. Engrg. Syst., 1, 1984

10. De Donato, O. and G. Maier: Finite element elastoplastic analysis by quadratic programming: the multistage method, 2nd SMIRT Conf. Berlin. 1973

11. Maier, G., S. Giacomini and F. Paterlini: Combined elastoplastic and limit analysis via restricted basis linear programming, Comp. Meths. Appl. Mech. Eng., 17-18, 1975

12. De Donato, O. and G. Maier: Local unloading in piecewise linear plasticity. Proc. ASCE, J. Eng. Mech. Div., 102, 1976

13. Smith, D.L.: The Wolfe-Markowitz algorithm for nonholonomic elastoplastic analysis, Eng. Struct., 1, 1978

CHAPTER 19

LARGE DISPLACEMENT ELASTOPLASTIC ANALYSIS OF STRUCTURES

J. A. Teixeira de Freitas
Istituto Superior Tecnico, Lisbon, Portugal

Abstract: A unified approach for treating nonlinear structural analysis problems is presented. It combines four fundamental ingredients, namely structural discretization, substitution of structures by graphs, static-kinematic duality and mathematical programming. Graph theory is called upon to exhaust the alternative processes through which the finite elements can be assembled to implement a governing system featuring symmetry; reciprocity in the constitutive relations and duality in the descriptions of equilibrium and compatibility are artificially preserved. Use is made of mathemetical programming theory and algorithms to complement the resulting discrete representation with a variational interpretation and to develop numerically efficient solution procedures.

Introduction

When the displacements of an engineering structure become sufficiently large to merit a consideration of a nonlinear description of its equilibrium and compatibility, it is almost certain that the deformation of its elements will then be large enough to have induced considerable inelasticity. Inelasticity is concomitant with large displacements, and its effect is usually represented mathematically by the theory of plasticity.

With nonlinearity appearing in every aspect of the problem, it is necessary to impose some regularizing feature. In linear structural analysis the relations describing equilibrium and compatibility exhibit duality - the static-kinematic duality. In the presence of large displacements and deformations this duality can be preserved by the incorporation of fictitious forces and deformations. This, together with the further device of separating the general constitutive law describing the elastoplastic material into forms which exhibit reciprocity and nonlinear residuals which do not, confers upon the problem an unique mathematical formalism: the structural problem is represented by a symmetric, nonlinear complementarity problem. Such a formalism offers a mathematical framework within which not only pratical computations can be performed, but also structural theorems obtained.

This approach is summarized in the flowchart presented next. In this flowchart it is stressed the central objective of extending into the nonlinerar domain [1] the methodology described in [2,3] for formulating and solving problems in linear structural mechanics.

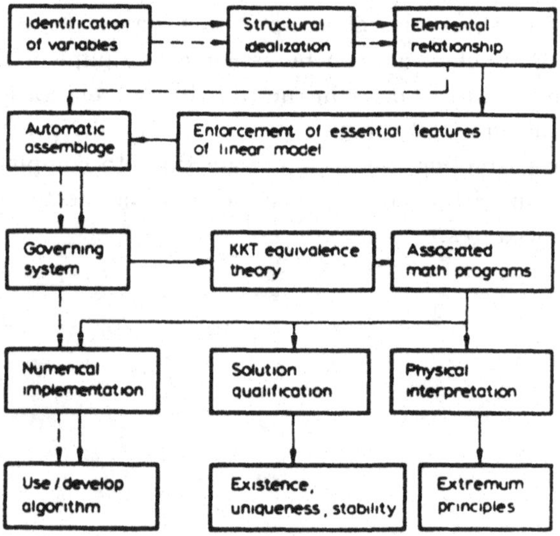

Lagrangian Description of Statics and Kinematics

Let the structure under analysis be discretized into finite elements. Assume that a typical element e is dissected from the displaced configuration of the structure.

Let $\mathbf{q}^e$ be the array that collects the generalized nodal displacements selected to describe the element displacement field. Similarly, list in array $\mathbf{u}^e$ the generalized strain-resultants used to quantify the strain field.

These two quantities are related through compatibility conditions,

$$\mathbf{u}^e = \mathbf{C}^e\, \mathbf{q}^e, \tag{1}$$

which can be derived from geometrical considerations. The entries of the compatibility operator, $\mathbf{C}^e$, can be expressed as nonlinear functions of topographic parameters, $\mathbf{t}$, that define the initial configuration of the element, and kinematic quantities, $\mathbf{k}$, associated with the element motion:

$$\mathbf{C}^e = \mathbf{C}^e\,(\mathbf{t},\, \mathbf{k}). \tag{2}$$

Let $\mathbf{Q}^e$ and $\mathbf{X}^e$ be the static duals of variables $\mathbf{q}^e$ and $\mathbf{u}^e$, respectively; $\mathbf{Q}^e$ is the array that collects the generalized nodal forces selected to describe the element force field and arrray $\mathbf{X}^e$ lists the generalized stress-resultants used to quantify the stress field.

As the equilibrium conditions have to be enforced on the displaced and deformed configuration of the element,

$$\mathbf{Q}^e = \mathbf{E}^e\, \mathbf{X}^e, \tag{3}$$

the entries of the equilibrium operator, $\mathbf{E}^e$, are also found to be nonlinear functions of the topographic and kinematic parameters describing the modified geometry of the element:

$$\mathbf{E}^e = \mathbf{E}^e\,(\mathbf{t},\, \mathbf{k}). \tag{4}$$

If the displacements and deformations are assumed infinitesimal, the compatibility and equilibrium operators (2) and (4) become linear,

$$\lim_{k\to 0} \mathbf{C}^e(\mathbf{t},\, \mathbf{k}) = \mathbf{C}^e_0(t), \tag{5}$$

$$\lim_{k\to 0} \mathbf{E}^e(\mathbf{t},\, \mathbf{k}) = \mathbf{E}^e_0(t), \tag{6}$$

since in analysis problems the topography of the structure is known a priori. Furthermore, the linear equilibrium and compatibility operators become the transpose of each other:

$$\mathbf{E}^e_0 = \tilde{\mathbf{C}}^e_0. \tag{7}$$

Let the linear equilibrium operator (6) be forcibly introduced in condition (3):

$$\mathbf{Q}^e + \mathbf{Q}^e_\pi = \mathbf{E}^e_0\, \mathbf{X}^e. \tag{8}$$

The condition above remains exact if the following definition is used for the additional nodal forces:

$$\mathbf{Q}_\pi^e = (\mathbf{E}_0^e - \mathbf{E}^e)\ \mathbf{X}^e. \tag{9}$$

The exact equilibrium conditions (8) can be written in the alternative form,

$$\mathbf{Q}^e = \left[\ \mathbf{E}_0^e \ \ \mathbf{E}_\pi^e\ \right] \left\{ \begin{array}{c} \mathbf{X}^e \\ -\pi^e \end{array} \right\}, \tag{10}$$

wherein

$$\pi^e = \mathbf{Z}^e\ \mathbf{X}^e, \tag{11}$$

is the subset of linearly independent entries of array $\mathbf{Q}_\pi^e$ defined by (9). The associated equilibrium operator is linear,

$$\mathbf{E}_\pi^e = \mathbf{E}_\pi^e(\mathbf{t}), \tag{12}$$

since all linear entries present in condition (10) are now considered in definition (11):

$$\mathbf{Z}^e = \mathbf{Z}^e\ (\mathbf{t}, \mathbf{k}).$$

It may therefore be stated that the role of the fictitious forces π^e is that of reporting the stress-resultants $\mathbf{X}^e$ from the initial (undeformed) configuration of he element to its actual one (deformed and displaced).

Consider now the dual transformation of the exact equilibrium condition (10),

$$\left\{ \begin{array}{c} \mathbf{u}^e + \mathbf{u}_\pi^e \\ \delta_\pi^e \end{array} \right\} = \left[\begin{array}{c} \tilde{\mathbf{E}}_0^e \\ \tilde{\mathbf{E}}_\pi^e \end{array} \right] \mathbf{q}^e, \tag{13}$$

wherein δ_p^e denotes the fictitious forces displacements.

If use is made of (5) and (7), it is easily found that condition (13) identifies with the element exact compatibility conditions (1) if the fictitious deformations are defined as follows:

$$\mathbf{u}_\pi^e = (\mathbf{C}_0^e - \mathbf{C}^e)\, \mathbf{q}^e.$$

Removing the linearly dependent columns present in the above definitions, the alternative expression is found,

$$\mathbf{u}_\pi^e = \mathbf{T}^e\ \delta_\pi^e, \tag{14}$$

wherein $\mathbf{T}^e$ is the operator that embodies the nonlinear effects associated with the large displacements and deformations the element has endured during its motion:

$$\mathbf{T}^e = \mathbf{T}^e(\mathbf{t}, \mathbf{k}). \tag{15}$$

Hence, the role of the fictitious deformations $\mathbf{u}_\pi^e$ can be interpreted as that of reporting the strain-resultants $\mathbf{u}^e$ from the quasi-initial (infinitesimally deformed) configuration of the element to its actual one (grossly deformed and displaced).

The equilibrium and compatibility conditions (16) of the structure under analysis can be obtained processing the elementary relationships (10) and (13) through an assemblage procedure identical to the one adopted in linear structural analysis [3].

Lagrangian description		
Statics	$\begin{bmatrix} \mathbf{0} & \mathbf{E}_0 \\ \mathbf{C}_0 & \mathbf{0} \end{bmatrix} \begin{Bmatrix} \mathbf{k}_1 \\ \mathbf{s}_2 \end{Bmatrix} = \begin{Bmatrix} \mathbf{s}_1 \\ \mathbf{k}_2 \end{Bmatrix}$	(a)
Kinematics		(b)

(16)

Summarized in table 1 are the variables **s** and **k** selected to describe the static and kinematic fields. Also given in the table is the identification of these variables when the structure, idealized as a connected graph, is interpreted on a mesh basis [3].

Table 1	Operators intervening in system (16)				
Description	$\mathbf{s}_1$	$\mathbf{k}_1$	$\mathbf{s}_2$	$\mathbf{k}_2$	$\mathbf{E}_0 = \tilde{\mathbf{C}}_0$
Nodal	$\mathbf{Q} = \mathbf{0}$	$\mathbf{q}$	$\mathbf{X}, \ \pi, -\lambda$	$\mathbf{u} - \mathbf{u}_\pi, \delta_\pi, \delta$	$\tilde{\mathbf{A}}\ \tilde{\mathbf{A}}_\pi \tilde{\mathbf{A}}_0$
Mesh	$\mathbf{X}$	$\mathbf{u} - \mathbf{u}_\pi$	$\mathbf{p}, \pi, \lambda$	$\mathbf{v} = \mathbf{0}, \delta_\pi, \delta$	$\mathbf{B}\ \mathbf{B}_\pi \mathbf{B}_0$

Exact definitions for the fictitious forces and deformations associated with the large displacement behaviour of skeletal structures can be found in [4-6]. Sustitution in system (16) of the first-order terms in the series expansion of the expressions given therein for definitions (11) and (14) generates the approximate formulations presented in [7-12]. The linear description of Statics and Kinematics [3] is recovered by setting π and $\mathbf{u}_\pi$ to zero and removing from system (16b) the definition for δ_π thus rendered irrelevant.

Incremental Description of Statics and Kinematics

The definitions (11) and (14) for the fictitious forces and deformations can be shown [4-6] to give the finite increments,

$$\Delta\pi = \mathbf{P}\Delta\delta_\pi - \mathbf{Z}\Delta\mathbf{X} - \Delta\mathbf{R}_\pi, \tag{17}$$

$$\Delta\mathbf{u}_\pi = \tilde{\mathbf{Z}}\Delta\delta_\pi + \Delta\mathbf{R}_{u\pi}, \tag{18}$$

wherein $\Delta\mathbf{R}_\pi$ and $\Delta\mathbf{R}_{u\pi}$ collect the nonlinear incremental terms.

The implicity nonlinear system (16) can be replaced by the equivalent infinite sequence of systems (20) by substituting definitions (17) and (18) into the incremental version of system (16), and equating next the same order terms in the power series expansion of the intervening variables, say **v**, on a generic perturbation parameter, ϵ:

$$\Delta\mathbf{v} \quad \mathbf{v}^{(n)}\frac{\epsilon^{n}}{n!}. \tag{19}$$

Incremental description		
Statics Kinematics	$\begin{bmatrix} \mathbf{K}_G & \mathbf{E}_* \\ \mathbf{C}_* & \mathbf{F}_G \end{bmatrix} \begin{Bmatrix} \mathbf{k}_1 \\ \mathbf{s}_2 \end{Bmatrix}^{(n)} = \begin{Bmatrix} \mathbf{s}_1 \\ \mathbf{k}_2 \end{Bmatrix}^{(n)} + \begin{Bmatrix} \mathbf{R}_s \\ \mathbf{R}_k \end{Bmatrix}^{(n)}$	(a) (b)

(20)

System (20) is linear and recursive as the entries of the equilibrium {compatibility} matrices $\mathbf{E}^*$ and $\mathbf{K}_G$ $\{$ $\mathbf{C}^*$ and $\mathbf{F}_G$ $\}$ are functions of the state of stress and strain prior to the implementation of the finite increments $\Delta\mathbf{s}$ and $\Delta\mathbf{k}$; moreover, the n-th order residuals $\mathbf{R}_s^{(n)}$ and $\mathbf{R}_s^{(k)}$ are functions of variables appearing in the (n-1)-th order and prior systems in the sequence, and are thus known at the outset in the n-th order system (20):

$$\mathbf{R}^{(n)} = \mathbf{R}_n \left(\dot{\mathbf{v}}, \ \ddot{\mathbf{v}}, \ \ldots, \ \mathbf{v}^{(n-1)} \right). \tag{21}$$

The expressions for the operators intervening in system (20) can be derived [4,13,14] using the identifications given in table 2 for the variables intervening in the description of the nodal and mesh structural models. If is found that Static-Kinematic Duality is preserved in system (20),

$$\mathbf{K}_G = \tilde{\mathbf{K}}_G \ , \ \mathbf{E}_* = \tilde{\mathbf{C}}_* \ , \ \mathbf{F}_G = \tilde{\mathbf{F}}_G,$$

in consequence of the symmetry manifested by matrix $\mathbf{P}$ present in definition (17).

In system (20) Statics and Kinematics now appear to be directly linked through the geometric stiffness matrix $\mathbf{K}_G$ in the nodal description, or through the geometric flexibility matrix $\mathbf{F}_G$ in the mesh description.

Table 2	n-th order variables intervening in system (20)			
Description	$\mathbf{s}_1^{(n)}$	$\mathbf{k}_1^{(n)}$	$\mathbf{s}_2^{(n)}$	$\mathbf{k}_2^{(n)}$
Nodal	$\mathbf{Q} = \mathbf{0}$	$\mathbf{q}$	$\mathbf{X}$	$\mathbf{u}$
Mesh	$\mathbf{X}$	$\mathbf{u}$	$\{p, \pi\}$	$\{\mathbf{v} = \mathbf{0}, \delta_\pi\}$

Constitutive Relations

It is assumed that the material constituting the finite elements into which the structure is discretized follows a nonlinear elastic-plastic law relating stress- and strain-resultants; the identification of the operators intervening in systems (22) and

(23) is the same as that given in [2,3].

Lagrangian description of elasticity	
$\mathbf{u}_e = \mathbf{F}_0\mathbf{X} + \mathbf{R}_{ue}$	$\mathbf{X} = \mathbf{K}_0 u_e + \mathbf{R}_{xe}$
Flexibility	Stiffness

(22)

Lagrangian description of plasticity		
Statics / Kinematics	$\begin{bmatrix} -\mathbf{H}_0 & \tilde{\mathbf{N}}_0 \\ \mathbf{V}_0 & \mathbf{0} \end{bmatrix} \begin{Bmatrix} \mathbf{u}_* \\ \mathbf{X} \end{Bmatrix} = \begin{Bmatrix} \phi_* \\ \mathbf{u}_p \end{Bmatrix} + \begin{Bmatrix} \mathbf{X}_* - \mathbf{R}'_s \\ -\mathbf{R}'_n \end{Bmatrix}$ (a) (b)	
$\phi_* \leq \mathbf{0}$ (c)	$\phi_*\mathbf{u}_* = 0$ (d)	$\mathbf{u}_* \geq \mathbf{0}$ (e)
Yield rule	Complementarity	Flow rule

(23)

In the lagrangian description of elasticity in equations (22) matrices $\mathbf{F}_0$ and $\mathbf{K}_0$ are block diagonal, collecting the flexibility and stiffness matrices of the unassembled finite elements. They are obtained from a solution of the nonlinear equation of the Elastica, which also accounts for the effects of axial and shear deformability [15].

Collected in the nonlinear residuals $\mathbf{R}'_x$ and $\mathbf{R}'_{ue}$ are those terms that would otherwise destroy the reciprocity of the elastic causality operators that one desires to maintain, i.e.,

$$\mathbf{F}_0 = \tilde{\mathbf{F}}_0 \ , \ \mathbf{K}_0 = \tilde{\mathbf{K}}_0 .$$

These operators are not necessarily positive definite since their entries depend on the current values taken by the element stability functions, the familiar definitions for which [16-21] can be recovered by specialization of the more general relations given in [15].

The description of the plastic phase in form (23) based upon Maier's piecewise-linear discretization of the yielding and hardening functions is not necessarily assumed in the description adopted herein [3]. Residuals $\mathbf{R}'_s$ and $\mathbf{R}'_k$ include those terms that otherwise would tend to destroy the assumed relationships of reciprocity and duality, respectively,

$$\mathbf{H}_0 = \tilde{\mathbf{H}}_0 \ , \ \mathbf{N} = \mathbf{V} .$$

between the static (23a) and kinematic (23b) phases of plasticity; reciprocity and associated flow laws are thus not necessarily implied in description (23).

The incremental description of elastoplasticity [3] is summarized in systems (24) and (25). As before, the nonlinear residual terms are so defined as to preserve symmetry of the elastoplastic operators:

$$\mathbf{F} = \tilde{\mathbf{F}}_{*}, \quad \mathbf{K} = \tilde{\mathbf{K}}, \quad \mathbf{H} = \tilde{\mathbf{H}}, \quad \mathbf{N} = \mathbf{V}.$$

Except for the complementarity conditions (25-d), the perturbed finite-incremental descriptions of the elastic (24) and plastic (25) phases are linear and recursive. This follows because the intervening constitutive operators depend on the state of stress and strain immediately prior to the occurence of the finite variation, and residual terms present therein comply with definition (21).

Incremental description of elasticity	
$\Delta\mathbf{u}_e = \mathbf{F}\mathbf{X} + \Delta\mathbf{R}_{ue}$ (a)	$\Delta\mathbf{X} = \mathbf{K}\,\Delta\mathbf{u}_e + \Delta\mathbf{R}_{xe}$ (b)
Flexibility	Stiffness

(24)

Incremental description of plasticity		
$\begin{bmatrix} -\mathbf{H} & \tilde{\mathbf{N}} \\ \mathbf{V} & \mathbf{0} \end{bmatrix} \begin{Bmatrix} \Delta\mathbf{u}_* \\ \Delta\mathbf{X} \end{Bmatrix} = \begin{Bmatrix} \Delta\varphi_* \\ \Delta\mathbf{u}_p \end{Bmatrix} - \begin{Bmatrix} \Delta\mathbf{R}'_s \\ \Delta\mathbf{P}'_k \end{Bmatrix}$ (a) (b)		
$\phi_* + \Delta\phi_* \leq \mathbf{0}$ (c)	$\tilde{\phi}_*\Delta\mathbf{u}_* = 0 - \Delta\dot{\phi}_*\Delta u_*$ (d)	$\Delta\mathbf{u}_* \geq \mathbf{0}$ (e)
Yield rule	Complementarity	Flow rule

(25)

The Governing System

After combining the equilibrium and compatibility conditions (16) with the constitutive relations (22,23), the nonlinear complementarity problem (26) is obtained as the system governing the response of the structure.

Lagrangian governing system

$$\begin{bmatrix} \mathbf{D}_{11} & \mathbf{D}_{12} & \mathbf{M}_1 \\ \tilde{\mathbf{D}}_{12} & \mathbf{D}_{22} & \mathbf{M}_2 \\ \tilde{\mathbf{M}}_1 & \tilde{\mathbf{M}}_2 & \mathbf{C} \end{bmatrix} \begin{Bmatrix} \mathbf{y}_1 \\ \mathbf{y}_2 \\ \mathbf{x} \end{Bmatrix} \begin{matrix} = \\ \geq \\ = \end{matrix} \begin{Bmatrix} \mathbf{d}_1 \\ \mathbf{d}_2 \\ \mathbf{c} \end{Bmatrix} \quad \begin{matrix} (a) \\ (b) \\ (c) \end{matrix} \qquad (26)$$

$$\mathbf{y}_2 \geq \mathbf{0} \quad (d)$$

$$\mathbf{y}_2(\tilde{\mathbf{D}}_{12}\mathbf{y}_1 + \tilde{\mathbf{D}}_{22}\mathbf{y}_2 + \mathbf{M}_2\mathbf{x} - \mathbf{d}_2) = 0 \quad (e)$$

Two terms can be distinguished in the stipulation vector,

$$\left\{ \begin{array}{c} \mathbf{d} \\ \mathbf{c} \end{array} \right\} = \left\{ \begin{array}{c} \mathbf{d}_0 \\ \mathbf{c}_0 \end{array} \right\} \lambda + \left\{ \begin{array}{c} \mathbf{R}_y \\ \mathbf{R}_y \end{array} \right\}, \tag{27}$$

one dependent on the applied load vector, which is assumed to be parametric in a common load factor λ, the other involving the nonlinear residuals.

In system (26), the entries of the stipulation vector and of the structural matrix, which is symmetric as a consequence of SKD and material reciprocity, can be derived [13,14] from the information contained in table 3; four alternative formulations are possible depending on the descriptions adopted for the state conditions.

Table 3	Variables intervening in system (26)			
Formulation	Nodal–Stiffness	Nodal–Flexibility	Mesh–Stiffness	Mesh–Flexibility
$\mathbf{y}_1$	$\mathbf{q}$	$\mathbf{q}$	$\mathbf{u}_E$	-
$\mathbf{y}_2$	$\mathbf{u}_*$	$\mathbf{u}_*$	$\mathbf{u}_*$	$\mathbf{u}_*$
$\mathbf{x}$	-	$\mathbf{X}$	$\mathbf{p}$	$\mathbf{p}$

The system governing the response of the displaced structure to a finite load or displacement increment, which is then identified with the perturbation parameter ϵ in expansion (19), is obtained by combining equations (20) with the incremental version of the elastoplasticity conditions (24,25). The nonlinear complementarity problem (26) is thus replaced by an infinite, recursive sequence of linear complementarity problems.

The n-th order incremental governing system presents the general format (26). As before, a specific identification results for the intervening operators and variables, depending on the adopted description for the state conditions [13,14]. Those summarized in table 3 are still applicable, except for the mesh descriptions wherein $\mathbf{p}_* = \{\mathbf{p}, \pi\}$ replaces $\mathbf{p}$.

Structural Theorems

The theorems of virtual displacements and forces can be preserved in the nonlinear domain if they are interpreted as an energetic representation of SKD [14].

The theorem of virtual displacements {forces} equation (28) {(29)}

$$\tilde{\mathbf{X}}\, \Delta \mathbf{u} + \tilde{\mathbf{X}}\, \mathbf{R}_{u\pi} = \tilde{\lambda}\, \Delta \delta, \tag{28}$$

$$\tilde{\mathbf{u}}\, \Delta \mathbf{X} + \tilde{\mathbf{u}}_\pi \Delta \mathbf{X} = \tilde{\delta}\, \Delta \lambda + \tilde{\delta}_\pi\, \Delta \pi, \tag{29}$$

is obtained by performing the inner product of the equilibrium {compatibility} condition (16a) {(16b)} with the finite increment version of the compatibility {equilibrium} condition (16b) {(16a)}.

Even if $\mathbf{u}_\pi$ is eliminated through (14) and $\Delta\pi$ through (17), it is impossible to express (29) solely in terms of the virtual static field $\{\Delta\mathbf{X},\ \Delta\lambda\}$. For this reason the theorem of virtual forces is seldom applied in nonlinear mechanics.

The theorems of virtual displacements and forces in linear mechanics are recovered by setting to zero in equations (28) and (29) the fictitious forces and deformations, as well as its residual $\mathbf{R}_{u\pi}$.

The Haar-Karman and Kachanov-Hodge theorems on the minimum potential co-energy and minimum potential energy can be extended to nonlinear mechanics by identification of the mathematical programs associated with the Karush-Kuhn-Tucker problem (26):

$$\text{Min z} = 1/2\tilde{\mathbf{y}}\,\mathbf{D}\,\mathbf{y} + 1/2\tilde{\mathbf{x}}\,\mathbf{C}\,\mathbf{x} + \tilde{\mathbf{x}}\,\mathbf{c}\ :\ \mathbf{D}\,\mathbf{y} + \mathbf{M}\,\mathbf{x}\left\{\begin{matrix} = \\ \geq \end{matrix}\right\}\mathbf{d}, \tag{30}$$

$$\text{Min w} = 1/2\tilde{\mathbf{y}}\,\mathbf{D}\,\mathbf{y} + 1/2\tilde{\mathbf{x}}\,\mathbf{C}\,\mathbf{x} - \tilde{\mathbf{y}}\,\mathbf{d}\ :\ \tilde{\mathbf{M}}\,\mathbf{y} - \mathbf{C}\,\mathbf{x} = \mathbf{c},\ \mathbf{y} \geq \mathbf{0}. \tag{31}$$

The physical interpretation of program (31) {(30)} shows [14] that the objective function represents the incremental potential energy {co-energy} of the structure and that its constraints implement kinematic {static} admissibility conditions.

The role of the complementarity energy principles (30) in kinematically nonlinear mechanics has been a polemic issue, as is apparent in the works by Zubov [23], de Veubeke [24] and Koiter [25] devoted to this subject. It is worth stressing that it is the preservation of SKD and reciprocity in the constitutive relations, and the processing through mathematical programming theory of the resulting (exact) symmetric governing system, that enable the extension into the linear domain of the dual role of energy and complementarity energy demonstrated by Westergaard [26] and Argyris [27] for linear behaviour.

The minimum potential theorems proposed in [28-31] can be obtained if in the nodal stiffness description of statement (31) the finite increment on the generic (kinematic) variable $\Delta\mathbf{v}$ is replaced by its first-order variation, which according to expansion (19) is defined by:

$$\dot{\mathbf{v}} = \lim_{\epsilon\to 0}\frac{\Delta\mathbf{v}}{\epsilon}.$$

Sufficient conditions for a kinematically nonlinear solution to exist and to be unique are obtained by applying to statements (30) and (31) the mathematical programming theorems on duality and uniqueness. Multiple solutions are characterized by applying the theorems on multiplicity to the composite form of these programs in order to guarantee that such solutions are simultaneously statically and kinematically admissible. Stability criteria are established by processing the formulation through the stability postulates of Drucker.

The solution qualification statements thus obtained [13] involve all state variables, namely generalized displacements and their associate forces, as well as strain- and stress-resultants, since the procedure is applied to each of the forces equivalent formulations of elastoplastic structural analysis. Such statements contain, therefore, those proposed by Corradi [32] and Maier [28], who first suggested this method of approach, working on tangent stiffness formulations.

Numerical Implementation

The determination of the failure load is the major objective in performing the analysis of the nonlinear response of a structure.

The simplest method for obtaining this information consists in solving system (26) through an iterative procedure. Let $\mathbf{x}$ collect from the variables intervening in system (26) those which are unrestricted in sign and reintroduce the plastic potentials ϕ_* in the yield condition (26b). Then the problem of tracing the sequence of values taken by the variables $\mathbf{x}$, $\mathbf{u}_*$ and ϕ_* as the load parameter is increased from zero can be represented as the following iterative linear program:

$$\text{Max}\ \lambda^{\bullet}: \begin{bmatrix} \mathbf{S}_{11} & \mathbf{S}_{12} \\ \tilde{\mathbf{S}}_{12} & \mathbf{S}_{22} \end{bmatrix} \begin{Bmatrix} \mathbf{x} \\ \mathbf{u}_* \end{Bmatrix} + \begin{bmatrix} \mathbf{0} \\ \mathbf{I} \end{bmatrix} \phi_* - \begin{bmatrix} \mathbf{S}_{0_1} \\ \mathbf{S}_{0_2} \end{bmatrix} \lambda = \begin{Bmatrix} \mathbf{R}_1 \\ \mathbf{R}_2 \end{Bmatrix},$$

$$\mathbf{u}_* \geq \mathbf{0}\ ,\quad \phi_* \leq \mathbf{0}.$$

The complementarity condition (26e) is simply enforced by preventing the simultaneous presence of $\mathbf{u}_{*i}$ and ϕ_{*i} in any pivoting basis. The elastoplastic failure load can be obtained by repetitive application of an adaptation of the Wolfe-Markowitz algorithm described by Smith [33], with recurrent reevaluation of operators $\mathbf{S}$ and $\mathbf{R}$. The structure, at each iteration, is subjected to the (constant) instabilizing effects found when attaining $\lambda_{\max}$ in the immediately preceding iteration. Eventually, in general 3 to 4 iterations prove to be sufficient, the solutions are found to converge. A similar procedure has been discussed by Corradi [34].

The adapted Wolfe- Markowitz algorithm can still be applied if the static and kinematic configurations corresponding to the correct sequence of activation of the yield modes are to be obtained. Instead of iterating on the elastoplastic failure load, convergence is now required upon each basic solution, which defines the activation of a yield mode. The post-collapse phase is investigated, if required, by switching from the maximization of λ to is minimization. By recalculating the load parameter λ at each iteration using the virtual work equation (29), a significant improvement in the rate of convergence of the algorithm can be achieved.

Solution of the problem of the yield mode activation sequence establishes a skeleton for the equilibrium path of the structure. Detail may be added within each stage applying the incremental procedure described in [35] to the perturbation analysis formulation of the governing system (26).

The essential advantage in using algorithms based on the pertubation technique [36] consists in their ability to provide monotonically improving approximations to the solution, and to easily incorporate procedures that detect and accommodate situations of plastic straining and unstressing, overall instability, and branching of the equilibrium path. This technique should be used whenever highly accurate results are needed, as is often the case in research activities.

Most engineering structures do however attain the maximum load-carrying capacity for relatively small displacements, and the post-failure behaviour is in most cases irrelevant for pratical purposes. The first-order nonlinear iterative analysis programs should perform successfully in such situations. The solutions they provide are fairly accurate and the rate of convergence is good. Serious difficulties have, however, been experienced in calculating the post-collapse phase by programs of iterative analysis. Yet another limitation of the simplicial algorithm is the inability to identify the occurence of critical (branching and/or limit) points between the activation of two consecutive yield modes. On the other hand, it is capable of dealing efficiently with the more frequent situations of multiple plastic straining and unstressing.

REFERENCES

1. Freitas, J.A.T. and D.L. Smith: A general methodology for nonlinear structural analysis by mathematical programming, Eng. Struct., 6, 1984

2. Freitas, J.A.T.: Elastic-plastic analysis of structural cross-sections, Mathematical Programming Methods in Structural Plasticity, CISM, Udine 1986

3. Freitas, J.A.T.: Elastic-plastic analysis of skeletal structures, Mathematical Programming Methods in Structural Plasticity, CISM, Udine 1986

4. Freitas, J.A.T. and D.L. Smith: Exact, dual description of equilibrium and compatibility for space frames, CTE 45, CMEST, 1982

5. Freitas, J.A.T. and D.L. Smith: Elastoplastic analysis of planar structures for large displacements, J. Struct. Mech., 12, 1984

6. Freitas, J.A.T. et al: Nonlinear analysis of elastic space trusses, Meccanica, 20, 1985

7. Denke, P.H.: Nonlinear and thermal effects on elastic vibrations, Report SM-30426, Douglas Aircr. Company, 1960

8. Argyris, J.H.: Recent advances in matrix methods of structural analysis, Progress of Aeronautical Services (Kuchemann, D. and L.H.G. Stern, Eds.) MacMillan, 1964

9. Haisler, W.E. et al: Development and evaluation of solution procedures for geometrically nonlinear structural analysis, J. AIAA, 10, 1972

10. Oliveira, E.R.A.: A method of ficticious forces for geometrically nonlinear analysis of structures, 14th Int. Congr. Technical and Applied Mechanics (Koiter,W.T., Ed.), 1974

11. Smith, D.L.: First-order large displacement elastoplastic analysis of frames using the generalized force concept, Engineering Plasticity by Mathematical Programming, (Cohn,M.Z. and G.Maier Eds.), Pergamon, 1979

12. Kohnke, P.C.: Large deflection analysis of frame structures by fictitious forces, Int. J. Num. Meth. Engrg., 12, 1979

13. Freitas, J.A.T. and D.L. Smith: Existence, uniqueness and stability of elastoplastic solutions in the presence of large displacements, SM Arch., 9, 1984

14. Freitas, J.A.T. and D.L. Smith: Energy theorems for elastoplastic structures in a regime of large displacements, J. Mécanique, 4, 1985

15. Freitas, J.A.T. and D.L. Smith: Finite element elastic beam-column, J. Engrg. Mechs. Proc. ASCE, 109, 1983

16. Manderla, H.: Die Berechnung der Sekundar Spannungen, welche in einfachen fachwerk infolge starrer Knotenverbindungen auftrefen, Allgemeine Bauzeitung, 1880

17. Berry, A.: The calculation of stresses in aeroplane spars, Trans. Roy. Soc., 1, 1916

18. James, B.W.: Principal effects of axial loads on moment distribution analysis of rigid structures, NACA TN 534, Washington DC, 1935

19. Jennings, A.: Frame analysis including change of geometry, J. Struct. Div., Proc. ASCE, 94, 1968

20. Mallet, R.H. and R.V. Marçal: Finite element analysis of nonlinear structures, J. Struct. Div., Proc. ASCE, 94, 1968

21. Powell, G.H.: Theory of nonlinear structures, J. Struct. Div., Proc. ASCE, 95, 1969

22. Maier, G.: "Linear" flow-laws of plasticity. A unified general approach, Rend. Accad. Naz. Lincei, 47, 1969

23. Zubov, L.M.: The stationary principle of complementarity work in non-linear theory of elasticity, J. Appl. Math. Mech., 34, 1972

24. De Veubeke, B.M.F.: A new variational principle for finite elastic displacements, Int. J. Eng. Sci, 10, 1972

25. Koiter, W.T.: On the principle of stationary complementarity energy in the non-linear theory of elasticity, SIAM J. Appl. Math., 25, 1973

26. Westergaard, H.M.: On the method of complementarity energy, Trans. ASCE, 107, 1941

27. Argyris, J.H. and S. Kelsey: Energy theorems and structural analysis, Butterwoths, 1960

28. Maier, G.: Incremental plastic analysis in the presence of large displacement and physical instability effects, Int. J. Solids Struct., 7, 1971

29. Contro, R. et al: Inelastic analysis of suspension structures by nonlinear programming, TR 16, ISTC, Politecnico di Milano, 1974

30. Corradi, L. and G. Maier: Extremum theorems for large displacement analysis of discrete elastoplastic structures with piecewise linear yield surfaces, JOTA, 15, 1975

31. Contro, R. et al: Large displacement analysis of elastoplastic structures - a nonlinear programming approach, SM Arch., 2, 1977

32. Corradi, L.: On a stability condition for elastoplastic structures, Meccanica, 12, 1971

33. Smith, D.L.: The Wolfe-Markowitz algorithm for nonholonomic elastoplastic analysis, Eng. Struct., 1, 1978

34. Corradi, L. et al: Developments in the imposed rotation method, TR 15, ISTC Politecnico di Milano, 1973

35. Freitas, J.A.T.: Structural analysis for nonlinear material behaviour, Mathematical Programming Methods in Structural Plasticity, CISM, Udine 1986

36. Freitas, J.A.T. and D.L. Smith: Plastic straining, unstressing and branching in large displacement perturbation analysis, Int. J. Num. Meth. Eng. 20, 1984

CHAPTER 20

ULTIMATE LOAD ANALYSIS BY STOCHASTIC PROGRAMMING

K. A. Sikorski and A. Borkowski
Polish Academy of Sciences, Warsaw, Poland

ABSTRACT

Usually a desired safety or reliability level of the structure is assessed in a semiprobabilistic way: the characteristic values of loads and material constants are taken from statistical analysis of available data, whereas the calculation of strength is performed in deterministic way. The present paper shows that at least in ultimate load analysis it is possible to evaluate the safety factor taking directly into account random scatter of loading and/or properties of material. Such an approach is based upon the theory of Stochastic Programming and upon the results obtained recently in the analysis of reliability of complex systems. It is demonstrated that stochastic problems representing the static and kinematic theorems of ultimate load analysis can be converted into equivalent deterministic Mathematical Programming problems. Usually the latter turn out to be non-linear even for piecewise-linear yield criteria. The second part of the paper deals with the bounds on reliability of hyperstatic structures. Numerical examples illustrate the conceptual power and numerical efficiency of the proposed models.

1. INTRODUCTION

Ultimate load analysis is widely used in engineering because it allows one to estimate the degree of safety of the designed structure. Based upon a simple rigid-perfectly-plastic model of material and upon an assumption of single-parameter loading, it is reducible to the solution of a dual pair of Mathematical Programming Problems (e.g. [1]). Depending upon the shape of the yield surface adopted, they may be Linear Programming Problems (LPP's) or Nonlinear Programmming Problems (NLP's). Usually piecewise-linear yield surfaces are assumed in order to keep the problem within the range of LPP's.

It is assumed in the conventional ultimate load analysis that such characteristics of the structure as loads, dimensions or properties of material have certain deterministic values. In fact, however, those characteristics have considerable scatter and should be treated rather as random quantities. Usually a semi-probabilistic approach is used: certain representative values of the loads, material constants, etc., obtained by probabilistic analysis of their random nature, are included into the design codes and the subsequent analysis is carried out deterministically. Such a procedure is simple and therefore apealing for engineering practice, but it does not allow a deeper insight into the reliability of the structure.

On the other hand, Mathematical Programming has been succesfully extended to cover problems with random constraints and variables [2] . Therefore, the structural reliability problems which are reducible to the LPP's or NLP's can be formulated and solved directly in the probabilistic formulation. The aim of the present lecture is to present an example of such an approach to the ultimate load analysis.

The probabilistic approach to plastic analysis of structures has been investigated by many authors (e.g. [3] to [7]). An exhaustive presentation of the general reliability analysis of structures can be found in the monographs [8] to [11] .

2. BASIC NOTIONS OF STOCHASTIC PROGRAMMING

The deterministic NLP is formulated in the following way: find $\underline{x}_*$ corresponding to the extremum of f attained over Ω . Here $\underline{x}_* \in R^n$ is the optimum solution, $f = f(\underline{x})$ is the cost function and Ω is the admissible domain. The latter is described by a finite number of inequalities and/or equations. Any $\underline{x}_* \in \Omega$ is called the admissible solution. In the sequel we confine our attention to a special case of the above problem, namely to the LPP:

$$\max\{\ \underline{\tilde{c}}\ \underline{x}\ |\ \underline{A}\ \underline{x} \le \underline{b},\ \ \underline{x} \ge \underline{0}\ \} \tag{2.1}$$

It has the linear cost function $f' = \underline{\tilde{c}}\ \underline{x}$ and its admissible domain is a convex polyhedron bounded by m hyperplanes ($\underline{b} \in R^m$). The maximization problem (2.1) can be associated with the minimization problem

$$\min\{\ \tilde{\underline{b}}\ \underline{y}\ |\ \underline{A}\ \underline{y} \geq \underline{c},\quad \underline{y} \geq \underline{0}\ \} \tag{2.2}$$

called its dual. The duality theory tells us that if the LPP (2.1) has a solution $\underline{x}_*$ corresponding to the finite value f'_* then there exists also a solution $\underline{y}_*$ of the LPP (2.2) such that $f''_* = \tilde{\underline{b}}\ \underline{y}_* = f'_*$.

Let us consider now what happens when the entries of matrices appearing in the LPP's (2.1), (2.2) become random. It is reasonable to assume a normal (Gaussian) distribution for those quantities, since other distributions (exponential, lognormal etc.) can be transformed into normal. In order to distinguish matrices of random quantities we place " ^ " over them. The mean value will be denoted by a bar over the symbol and the variance by a tilde under it. Thus $\hat{x}$ is a random variable, $\bar{x}$ - its mean value and $\underset{\sim}{x}$ - its variance. A single-column matrix $\underline{x} \in R^n$ has the mean value $\bar{\underline{x}} \in R^n$, whereas the variances of its entries enter the diagonal matrix

$$\underset{\approx}{x} = \lceil\ \underset{\sim}{x}_1\ ,\ \underset{\sim}{x}_2\ ,\ \dots\ ,\ \underset{\sim}{x}_n \rfloor \tag{2.3}$$

In general, the entries of $\hat{\underline{x}}$ can be statistically correlated. Therefore, we need also the complete (n x n)-matrix of covariance $\underset{\approx}{x}$ defined as:

$$\underset{\approx}{x} = \begin{cases} \mathrm{var}(\hat{x}_i) = \underset{\sim}{x}_i & \text{for } i = j; \\ \mathrm{cov}(\hat{x}_i\ ,\ \hat{x}_j) & \text{for } i \neq j\ . \end{cases} \tag{2.4}$$

Firstly, let us consider the problem (2.1) with random constraints: the matrices $\hat{\underline{A}}$ and $\hat{\underline{b}}$ are random, whereas $\underline{c}$ remains deterministic. The LPP's of this type, called chance-constrained, were investigated by A.Charnes and W. W. Cooper [12]. They proposed the method that allows one to reduce stochastic LPP's to equivalent deterministic problems (usually nonlinear). That method was further developed by S. S. Rao [13].

It is assumed in chance-constrained programming that the i-th constraint is satisfied with the probability not less than a given value ψ_i. Let ψ_i enter the matrix $\underline{\psi} \in R^m$, where m is the number of constraints in (2.1). Then the stochastic counterpart of (2.1) can be written as

$$\max\ \{\ \tilde{\underline{c}}\ \underline{x}\ \ |\ \mathrm{prob}(\hat{\underline{A}}\ \underline{x} \leq \hat{\underline{b}}) \geq \underline{\psi}\ ,\ \underline{x} \geq \underline{0}\ \} \tag{2.5}$$

Consider now a special case of (2.5): a stochastic LPP with random entries of the coefficient matrix but deterministic right-hand side:

$$\max\ \{\ \tilde{\underline{c}}\ \underline{x}\ \ |\ \mathrm{prob}(\hat{\underline{A}}\ \underline{x} \leq \underline{b}) \geq \underline{\psi}\ \ ,\ \underline{x} \geq \underline{0}\ \} \tag{2.6}$$

Let

$$\hat{\underline{d}} = \hat{\underline{A}}\ \underline{x} \tag{2.7}$$

Since the entries of $\hat{\underline{A}}$ are normally distributed and the unknowns x_j are deterministic, the entries of $\hat{\underline{d}} \in R^m$ must also obey the normal distribution defined by the mean values

$$\bar{d}_i = \sum_{j=1}^{n} \bar{a}_{ij} x_j \tag{2.8}$$

and the variances

$$\underset{\sim}{d}_i = \sum_{j=1}^{n} \sum_{k=1}^{n} x_j \, \underset{\approx}{a}^{(i)}_{jk} x_k \, . \tag{2.9}$$

Here $\underset{\approx}{a}^{(i)}$ is the covariance matrix of the i-th column of $\underline{\hat{A}}$.

The new matrix $\underline{\hat{d}}$ allows us to write the probabilistic constraint of (2.6) as

$$\text{prob}\ (\underline{\hat{d}} \leq \underline{b}) \geq \underline{\psi} \, . \tag{2.10}$$

This matrix inequality can be replaced by m constraints

$$\text{prob}\ (\hat{d}_i^0 \leq b_i') \geq \psi_i \, , \tag{2.11}$$

where

$$\hat{d}_i^0 = (\hat{d}_i - \bar{d}_i)/\sqrt{\underset{\sim}{d}_i} \tag{2.12}$$

is a standardized random variable ($\bar{d}_i^0 = 0$, $\underset{\sim}{d}_i^0 = 1$) and

$$b_i' = (b_i - \bar{d}_i)/\sqrt{\underset{\sim}{d}_i} \, . \tag{2.13}$$

Introducing the Laplace function Φ , we can replace (2.11) by the inequalities $\Phi(b_i') \geq \psi_i$ or, equivalently, by the constraints

$$\Phi(b_i') \geq \Phi(\beta_i) \, , \tag{2.14}$$

where

$$\beta_i = \Phi^{-1}(\psi_i) \, . \tag{2.15}$$

The inequalities (2.14) imply $b_i' \geq \beta_i$ which, after substitution (2.13) and a simple transformation, yields

$$\bar{d}_i + \beta_i \sqrt{\underset{\sim}{d}_i} \leq b_i \, . \tag{2.16}$$

Let us introduce the diagonal matrix

$$\sqrt{\underset{\approx}{d}} = \lceil \sqrt{d_1}, \sqrt{d_2}, \ldots, \sqrt{d_m} \rfloor \tag{2.17}$$

and the matrix $\underline{\beta} \in R^m$ that collects the parameters β_i known for each chance constraint (2.11) through given ψ_i and through the relation (2.15). Then the system of inequalities (2.16) can be written consisely as

$$\underline{\bar{A}}\, \underline{x} + \sqrt{\underset{\approx}{d}}\ \underline{\beta} \leq \underline{b} \, . \tag{2.18}$$

Thus the original chance constraint (2.10) has been replaced by the deterministic inequality (2.18). The latter is nonlinear since $\sqrt{\underset{\approx}{d}}$ depends in a nonlinear way upon $\underline{x}$. Hence the stochastic problem (2.6) has been reduced to the equivalent NLP:

$$\max \{ \tilde{\underline{c}}\, \underline{x} \mid \bar{\underline{A}}\, \underline{x} + \sqrt{\underset{\approx}{d}}\, \underline{\beta} \leq \underline{b},\ \underline{x} \geq \underline{0} \} \tag{2.19}$$

One might try to replace random coefficients $\hat{a}_{ij}$ by their mean values $\bar{a}_{ij}$. Then the stochastic problem (2.6) would be replaced by the deterministic LPP

$$\max \{ \tilde{\underline{c}}\, \underline{x} \mid \bar{\underline{A}}\, \underline{x} \leq \underline{b},\ \underline{x} \geq \underline{0} \} \tag{2.20}$$

However, such an approach is non-conservative: as shown in Fig. 2.1, the admissible domain Ω' described by the non-linear constraints (2.18) is smaller than Ω resulting from $\bar{\underline{A}}$.

The situation gets simpler if $\underline{A}$ is deterministic and only $\hat{\underline{b}}$ is random. Let the mean value $\bar{\underline{b}}$ and the variance $\underset{\approx}{b}$ be known. The initial stochastic LPP reads then:

$$\max \{ \tilde{\underline{c}}\, \underline{x} \mid \text{prob}\, (\underline{A}\, \underline{x} \leq \hat{\underline{b}}) \geq \underline{\psi},\ \underline{x} \geq \underline{0} \} \tag{2.21}$$

The random right-hand sides $\hat{b}_i$ can be standardized similarly as was done previously for coefficients $\hat{a}_{ij}$. Such a standardization allows us to replace the chance constraints in (2.21) by the set of inequalities

$$\Phi\Big(\sum_{j=1}^{m} a_{ij} x_j - \bar{b}_i \Big) / \sqrt{\underset{\sim}{b}_i} \leq \Phi(\beta_i) \tag{2.22}$$

with

$$\beta_i = \Phi^{-1}(1 - \psi_i) \tag{2.23}$$

Replacing the inequalities (2.22) by the equivalent constraints imposed upon the arguments of Φ and rearranging terms, we obtain finally

$$\underline{A}\, \underline{x} \leq \bar{\underline{b}} + \sqrt{\underset{\approx}{b}}\, \underline{\beta} \tag{2.24}$$

where

$$\sqrt{\underset{\approx}{b}} = \lceil \sqrt{\underset{\sim}{b}_1},\ \sqrt{\underset{\sim}{b}_2},\ \ldots,\ \sqrt{\underset{\sim}{b}_m} \rfloor \tag{2.25}$$

Thus, the stochastic constraint $\text{prob}\,(\underline{A}\, \underline{x} \leq \hat{\underline{b}}) \geq \underline{\psi}$ has been replaced by the deterministic inequality (2.24) and the initial problem (2.21) turns out to be equivalent to the LPP

$$\max \{ \tilde{\underline{c}}\, \underline{x} \mid \underline{A}\, \underline{x} \leq \bar{\underline{b}} + \sqrt{\underset{\approx}{b}}\, \underline{\beta},\ \underline{x} \geq \underline{0} \} \tag{2.26}$$

Geometrically it means that the original admissible domain is linearly reduced, as shown in Fig. 2.2. The bounding hyperplanes drawn for the mean values $\bar{b}_i$ must be shifted inwards in order to assure the required probabilities ψ_i of constraint satisfaction. The distance of such a shift is equal to $\beta_i \sqrt{\underset{\sim}{b}_i}$.

Finally, let us consider the case of deterministic constraints but random cost function. As an example, we take the stochastic counterpart of the dual LPP (2.2):

$$\min \{ \hat{\tilde{\underline{b}}}\, \underline{y} \mid \tilde{\underline{A}}\, \underline{y} \geq \underline{c},\ \underline{y} \geq \underline{0} \} \tag{2.27}$$

First of all, we must properly define the minimum of a random function. This can be done in such a way that one looks for a

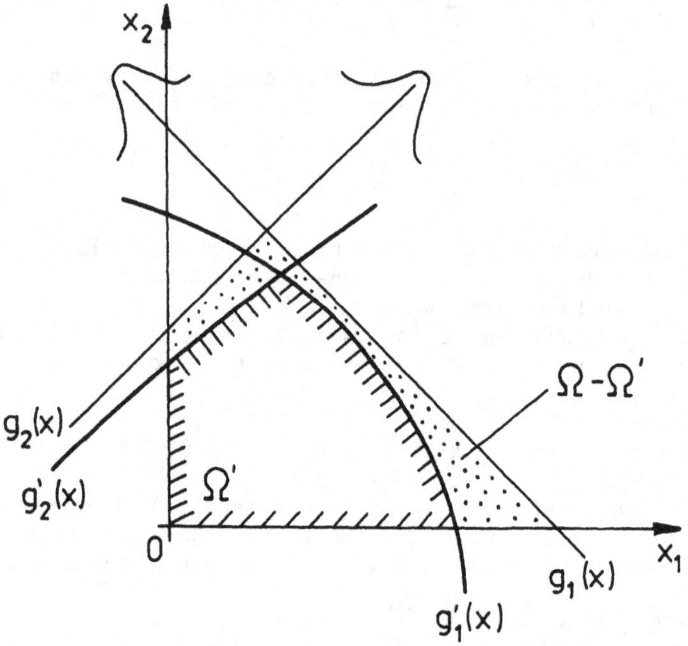

Fig. 2.1. Random coefficients replaced by mean values.

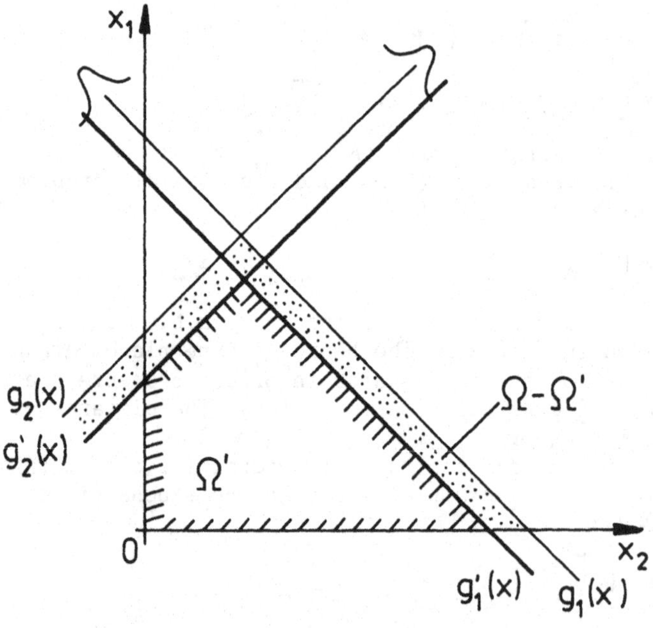

Fig. 2.2. Random righthand sides replaced by mean values.

minimum lower bound η on $\hat{f}'' = \tilde{\hat{\underline{b}}}\,\underline{y}$ under the constraint that the probability of violation of that bound is prescribed [14]. Then, the problem (2.27) is replaced by

$$\min \{ \eta \mid \mathrm{prob}\,(\tilde{\hat{\underline{b}}}\,\underline{y} > \eta\,) = \psi\ ,\ \tilde{\underline{A}}\,\underline{y} \geq \underline{c},\ \ \underline{y} \geq \underline{0}\ \}. \tag{2.28}$$

Since $\hat{b}_i$ are normally distributed and y_i are deterministic, the random value $\hat{f}''$ is also normally distributed. Its mean value and its variance follow from the standard formulae:

$$\bar{f}'' = \tilde{\bar{\underline{b}}}\,\underline{y}\ , \tag{2.29}$$

$$\underset{\sim}{f}'' = \tilde{\underline{y}}\ \underset{\approx}{b}\ \underline{y}\ , \tag{2.30}$$

where $\underset{\approx}{b}$ is the covariance matrix of $\hat{\underline{b}}$.

The probabilistic constraint that appears in (2.28) can be replaced by the equivalent deterministic one

$$\eta = \bar{f} - \beta\sqrt{\underset{\sim}{f}''} \tag{2.31}$$

in a similar way to that for the previously considered problems. Thus, the initial stochastic problem (2.28) can be replaced by the NLP

$$\min \{\ \tilde{\bar{\underline{b}}}\,\underline{y} - \beta\sqrt{\tilde{\underline{y}}\ \underset{\approx}{b}\ \underline{y}}\ \ \mid\ \tilde{\underline{A}}\,\underline{y} \geq \underline{c},\ \ \underline{y} \geq \underline{0}\ \}, \tag{2.32}$$

where

$$\beta = \Phi^{-1}(\ \psi\ )\,. \tag{2.33}$$

Comparing (2.32) with the LPP that would appear for $\hat{\underline{b}}$ replaced by $\bar{\underline{b}}$ in (2.27), we see that the additional nonlinear term emerges in the cost function. That term is the result of random scatter of the costs $\hat{b}_i$. It depends on the variances and correlations of those parameters, as well as of the assumed probability ψ which characterizes the lower bound on $\hat{f}''$.

3. DETERMINISTIC VERSUS STOCHASTIC ULTIMATE LOAD ANALYSIS

Deterministic evaluation of the ultimate load carrying capacity of a structure constitutes one of the most important topics in the theory of plasticity. This was also the area of the first applications of Mathematical Programming in Structural Mechanics. Since the problem is presented in detail in other lectures of the present course, we begin with a mere recapitulation of the deterministic approach.

Let the loading be proportional, i.e. let

$$\underline{F} = s\,\underline{F}_o \tag{3.1}$$

where $\underline{F}_o \in R^n$ is the reference load and s is the load factor. According to the known model (e.g. [15]), the ultimate value s_* of the load factor together with the stress state $\underline{Q}_*$ at the incipient plastic collapse and with the collapse mechanism $\underline{\dot{\lambda}}_*$, $\underline{\dot{u}}_*$ can be found by solving the following pair of dual LPP's:

$$\max\{\, s \mid \underline{\tilde{C}}\,\underline{Q} - s\,\underline{F}_0 = \underline{0},\ \ \underline{\tilde{N}}\,\underline{Q} \le \underline{R}\,\}, \tag{3.2}$$

$$\min\{\, \underline{\tilde{R}}\,\underline{\dot{\lambda}} \mid \underline{C}\,\underline{\dot{u}} - \underline{N}\,\underline{\dot{\lambda}} = \underline{0},\ \ \underline{\tilde{F}}_0\underline{\dot{u}} = 1,\ \underline{\dot{\lambda}} \ge \underline{0}\,\}. \tag{3.3}$$

A complete derivation of the model (3.2), (3.3), including the explanation of adopted notation, can be found in the chapter devoted to complementarity problems and unilateral constraints.

In order to eliminate equality constraints, we exploit the general solution of equilibrium equations

$$\underline{Q} = s\,\underline{\tilde{C}}^{-}\,\underline{F}_0 + \underline{\tilde{C}}^{0}\,\underline{Z}\,. \tag{3.4}$$

The substitution of that result into the LPP (3.2) gives the static theorem of ultimate load analysis expressed in terms of the redundant stresses:

$$\max\{\, s \mid \underline{\tilde{M}}\,\underline{Z} + s\,\underline{G} \le \underline{R}\,\} \tag{3.5}$$

Here

$$\underline{M} = \underline{C}^{0}\,\underline{N} \tag{3.6}$$

and

$$\underline{G} = \underline{\tilde{N}}\,\underline{\tilde{C}}^{-}\,\underline{F}_0. \tag{3.7}$$

The formal dual of (3.5) reads

$$\min\{\, \underline{\tilde{R}}\,\underline{\dot{\lambda}} \mid \underline{M}\,\underline{\dot{\lambda}} = \underline{0},\ \ \underline{\tilde{G}}\,\underline{\dot{\lambda}} = 1,\ \underline{\dot{\lambda}} \ge \underline{0}\,\} \tag{3.8}$$

It can be obtained also from the kinematic theorem (3.3) through elimination of the displacement rates $\dot{u}_j$.

Let us consider now the situation when the reference loading and/or the plastic moduli of the structure are random. Restricting ourselves to normally distributed parameters, we have the random matrices $\underline{\hat{F}}_0$ and $\underline{\hat{R}}$ characterized by the given statistical moments: the mean values $\underline{\bar{F}}_0$, $\underline{\bar{R}}$, the variances $\underset{\simeq}{F}_0$, $\underset{\simeq}{R}$ and the covariances $\underset{\approx}{F}_0$, $\underset{\approx}{R}$.

In the static approach it is reasonable to assume that the random stress distribution $\underline{\hat{Q}}$ must belong with a prescribed probability to the random admissibility domain $\hat{\Omega}$. Thus, instead of each deterministic stress admissibility constraint we introduce the stochastic inequality

$$\text{prob}\,\Big(\sum_{j=1}^{m} N_{ij}\,\hat{Q}_j \le \hat{R}_i \Big) \ge \psi_i \quad , \quad i = 1,2,\ldots,\ell \tag{3.9}$$

where $\hat{\psi}_i$ is given. Assuming that the proportionality rule (3.1) holds for $\underline{\hat{F}}$ and $\underline{F}_0$ as well, we are looking now for the maximum possible load factor that allows (3.9) to be satisfied. After transition to matrices and after elimination of the equilibrium equations, the stochastic counterpart of the LPP (3.5) reads:

$$\max\{\, s^s \mid \text{prob}\,(\tilde{\underline{M}}\,\underline{Z} + s\,\hat{\underline{G}} \le \hat{\underline{R}}\,) \ge \underline{\psi}\ \}. \tag{3.10}$$

Here $\hat{\underline{G}}$ has been calculated according to (3.7) with $\underline{F}_o$ replaced by $\hat{\underline{F}}_o$. Note that the load factor s^s as well as the residual stresses Z_i are deterministic in the above model.

The problem (3.10) belongs to the class (2.5) of stochastic LPP's. Hence, the Charnes-Cooper method can be applied in order to replace it by the equivalent deterministic problem. We are going to accomplish that for the three special cases of (3.10) taken separately.

Firstly, let the load be deterministic. Then the model (2.21) is relevant and the equivalent deterministic problem is analogous to LPP (2.26):

$$\max\{\, s^s \mid \tilde{\underline{M}}\,\underline{Z} + s^s\underline{G} \le \bar{\underline{R}} + \sqrt{\underset{\approx}{R}}\,\underline{\beta}\ \}. \tag{3.11}$$

Here

$$\sqrt{\underset{\approx}{R}} = \lceil \sqrt{\underset{\sim}{R}_1}, \sqrt{\underset{\sim}{R}_2}\ , \ \dots\ , \sqrt{\underset{\sim}{R}_\ell} \rfloor \tag{3.12}$$

and the entries of $\underline{\beta} \in R^\ell$ are to be computed from (2.23) for given probability levels $\underline{\psi_i}$.

Suppose that the load is random but the plastic resistances R_i are deterministic. Then the proper model is (2.6). However, only the last column of the matrix of coefficients is random now. Hence, the equivalent deterministic problem (2.19) reduces to the LPP:

$$\max\{\, s^s \mid \tilde{\underline{M}}\,\underline{Z} + s^s(\bar{\underline{G}} + \sqrt{\underset{\approx}{G}}\,\underline{\beta}\,) \le \underline{R}\ \}, \tag{3.13}$$

with

$$\sqrt{\underset{\approx}{G}} = \lceil \sqrt{\underset{\sim}{G}_1}\ , \sqrt{\underset{\sim}{G}_2}\ , \ \dots\ , \sqrt{\underset{\sim}{G}_\ell}\ \rfloor . \tag{3.14}$$

Finally, when both $\hat{F}_{oj}$ and $\hat{R}_i$ are random, the combination of (3.11) and (3.13) leads to the following NLP:

$$\max\{\, s^s \mid \tilde{\underline{M}}\,\underline{Z} + s^s\bar{\underline{G}} + \underline{P}\,\underline{\beta} \le \bar{\underline{R}}\ \}, \tag{3.15}$$

where

$$P_{ij} = \sqrt{s^{s2}\ \underset{\sim}{G}_{ij} + \underset{\sim}{R}_{ij}}\ . \tag{3.16}$$

The deterministic models (3.11), (3.13), (3.15) derived along the lines of the Charnes-Cooper method, differ substantially from the LPP that would be obtained by simple introduction of the mean values into (3.5). It is clearly seen from the above formulae that both the random $\hat{\underline{F}}_o$ and the random $\hat{\underline{R}}$ lead to the reduced admissible domain for stresses. The scale of such a reduction increases with the increasing ψ_i , i.e. with the increasing probability of constraints being satisfied.

In the kinematic approach we begin with randomization of the model (3.8). Since $\hat{\underline{R}}$ appears in the cost function, the problem must be reformulated according to the pattern (2.19), (2.28). That yields the following stochastic programming problem:

$$\min\{\, s^{\kappa} \mid \mathrm{prob}\,(\underline{\tilde{\hat{R}}}\,\dot{\underline{\lambda}} > s^{\kappa}) = \psi_d \,,$$

$$\mathrm{prob}\,(\underline{\tilde{\hat{G}}}\,\dot{\underline{\lambda}} < 1\,) = \psi_p \,,\; \underline{M}\,\dot{\underline{\lambda}} = \underline{0},\; \dot{\underline{\lambda}} \geq \underline{0}\,\}. \tag{3.17}$$

We are looking now for the plastic collapse mechanism $\dot{\underline{\lambda}}_*$ that fulfils the equation of kinematic compatibility $\underline{M}\,\dot{\underline{\lambda}} = \underline{0}$, minimizes the dissipated power $\underline{\tilde{\hat{R}}}\,\dot{\underline{\lambda}}$ and keeps the power of the reference load $\underline{\tilde{\hat{G}}}\,\dot{\underline{\lambda}}$ at the unit level. The last two conditions are to be satisfied with the given probabilities ψ_d and ψ_p, respectively.

The stochastic problem (3.17) can be replaced by an equivalent deterministic problem, as shown in the Section 2. The three following cases are to be distinguished:

a) when $\underline{R}$ is deterministic but $\hat{\underline{F}}_0$ is random, then the following NLP replaces (3.17) -

$$\min\{\, \underline{\tilde{R}}\,\dot{\underline{\lambda}} \mid \underline{\tilde{\bar{G}}}\,\dot{\underline{\lambda}} + \beta_p\sqrt{\dot{\underline{\tilde{\lambda}}}\,\underline{\underline{G}}\,\dot{\underline{\lambda}}} = 1,\; \underline{M}\,\dot{\underline{\lambda}} = \underline{0},\; \dot{\underline{\lambda}} \geq \underline{0}\,\}; \tag{3.18}$$

b) when $\underline{F}_0$ is deterministic but $\hat{\underline{R}}$ is random, then, according to (2.32), the equivalent problem reads -

$$\min\{\, \underline{\tilde{\bar{R}}}\,\dot{\underline{\lambda}} - \beta_d\sqrt{\dot{\underline{\tilde{\lambda}}}\,\underline{\underline{R}}\,\dot{\underline{\lambda}}} \mid \underline{\tilde{G}}\,\dot{\underline{\lambda}} = 1,\; \underline{M}\,\dot{\underline{\lambda}} = \underline{0},\; \dot{\underline{\lambda}} \geq \underline{0}\,\}; \tag{3.19}$$

c) when both $\hat{\underline{R}}$ and $\hat{\underline{F}}_0$ are random, then the equivalent deterministic problem is strongly nonlinear -

$$\min\{\, \underline{\tilde{\bar{R}}}\,\dot{\underline{\lambda}} - \beta_d\sqrt{\dot{\underline{\tilde{\lambda}}}\,\underline{\underline{R}}\,\dot{\underline{\lambda}}} \mid \underline{\tilde{\bar{G}}}\,\dot{\underline{\lambda}} + \beta_p\sqrt{\dot{\underline{\tilde{\lambda}}}\,\underline{\underline{G}}\,\dot{\underline{\lambda}}} = 1,$$

$$\underline{M}\,\dot{\underline{\lambda}} = \underline{0},\; \dot{\underline{\lambda}} \geq \underline{0}\,\}. \tag{3.20}$$

As usual, the probability levels β_d, β_p entering the above models are to be calculated from (2.33) for given ψ_d and ψ_p. Nonlinear terms that appear in the expressions of dissipated and external power account for the additional energy associated with the random scatter of plastic moduli and reference loads.

4. RELIABILITY OF STRUCTURE

We are in a position now to evaluate the overall reliability of the structure that has been analysed by means of the models introduced in the previous Section. Prior to doing that we recall some well known results of the probability theory.

The probability of an union of events can be estimated as

$$\max_i \;\mathrm{prob}\,(\,\Gamma_i) \leq \mathrm{prob}\,(\,\bigcup_{i=1}^{n} \Gamma_i\,) \leq 1 - \prod_{i=1}^{n}(1 - \mathrm{prob}\,(\,\Gamma_i\,)). \tag{4.1}$$

Here the lower bound is exact when all events are completely dependent and the upper bound is exact when they are completely independent. These bounds are more close to each other provided the correlation of events is known. For example, the bounds proposed by O. Ditlevsen [10] take into account the correlated pairs of events:

$$\text{prob}\,(\Gamma_1) + \sum_{i=2}^{n} \max_{i} \left[\text{prob}\,(\Gamma_i) - \sum_{j=1}^{i-1} \text{prob}\,(\Gamma_i \cap \Gamma_j),\ 0 \right]$$

$$\leq \text{prob}\,\Big(\bigcup_{i=1}^{n} \Gamma_i\Big) \leq$$

$$\sum_{i=1}^{n} \text{prob}\,(\Gamma_i) - \sum_{i=2}^{n} \max_{j<i} \left[\text{prob}\,(\Gamma_i \cap \Gamma_j) \right]. \tag{4.2}$$

The same formulae can be used for the evaluation of probability of intersection of events since

$$\text{prob}\,\Big(\bigcap_{i=1}^{n} \Gamma_i\Big) = 1 - \text{prob}\,\Big(\bigcup_{i=1}^{n} \Psi_i\Big) \tag{4.3}$$

where Ψ_i is the complementary event of Γ_i .

There are also numerical algorithms available (e.g. [21]) that allow one to find the probability of the union or intersection of events that depends upon several random variables.

In the frame of ultimate load analysis, the reliability of a structure has to be assessed through generalized theorems regarding the ultimate load. Such an approach leads to lower and upper bounds on the reliability.

According to the static theorem, the structure remains safe provided there exists a statically admissible stress distribution. One could choose, therefore, as a reliability measure the probability that all stress fields generated by the given load remain statically admissible. Unfortunately, no method exists to-day that would allow us to calculate effectively such a probability. Therefore, we must to be satisfied with a lower bound on the reliability obtained by considering a single stress distribution. Obviously, the choice of that stress field is crucial and there exists a certain optimum stress distribution corresponding to the best bound. For a fixed reliability level such an optimum stress field corresponds to the maximum load factor.

Hence, the statical assessment of reliability follows the next pattern. Firstly, we assign probability levels ψ_i for each statical admissibility constraint and then we solve the resulting problems (3.11), (3.13) or (3.15). That gives us s_*^s and $\underline{Z}_*$, which, in turn, allows to find the optimum stress distribution $\hat{\underline{Q}}_*$ from the relation (3.4) . As far as $\hat{\underline{Q}}_*$ is known, we can look for the lower bound on reliability of the entire structure defined as

$$\psi^s = \text{prob}\,\Big(\bigcap_{i=1}^{\ell} \hat{y}_i > 0\Big), \tag{4.4}$$

where

$$\hat{y}_i = \hat{R}_i - \sum_{j=1}^{m} N_{ij}\,\hat{Q}_j\,. \tag{4.5}$$

are the lower bound safety margins. Practically, it suffices to include active constraints only in the intersection evaluated in (4.4).

It must be pointed out that for hyperstatic structures the stress field $\hat{Q}_*$ depends upon the choice of the primary system. Thus, theoretically all possible primary systems should be considered in order to find the best lower bound on reliability. Practically it is possible for simple structures only. All examples given in the present paper were calculated with the usage of the Robinson-Haggenmacher procedure [22], which performs the choice of the primary system automatically. There is no ground to believe that such a choice is optimal but the obtained results confirm practical applicability of the method.

Let us turn now to the kinematic approach. According to the kinematic theorem, the structure collapses when the difference between the dissipated power and the external power becomes negative for a kinematically admissible distribution of displacement rates. The probability of such an event can be, therefore, taken as a measure of failure probability. For discrete, e.g. skeletal, structures there exists a finite, though usually large, number of all possible failure mechanisms. In practice, it is enough to consider dominant mechanisms, i.e. the mechanisms that have highest probabilities of occurence. Hence, the lower bound on the probability of failure γ^{κ} or the upper bound ψ^{κ} on the reliability of the entire structure can be found as

$$\gamma^{\kappa} = 1 - \psi^{\kappa} = \text{prob}\,(\bigcup_{i=1}^{r} \hat{t}_i < 0), \tag{4.6}$$

where

$$\hat{t}_i = \hat{\tilde{\underline{R}}}\, \dot{\underline{\lambda}}^{(i)} - \hat{\tilde{\underline{F}}}_o\, \dot{\underline{u}}^{(i)} \tag{4.7}$$

are the upper bound safety margins and i runs over the dominant mechanisms.

Several methods of the generation of dominant mechanisms were proposed [3], [6], [7]. All of them are considerably time consuming. The simplest solution is to determine the stochastically most relevant mechanism that gives

$$\gamma_*^{\kappa} = 1 - \psi_*^{\kappa} = \max_{\dot{u}^{(i)}} \left[\text{prob}\,(\hat{t}_i < 0) \right]. \tag{4.8}$$

For a given probability of failure, the most relevant mechanism corresponds to the minimum kinematic load factor. Such is the situation in the model (3.17). There hold the following relations between the assumed probabilities ψ_d, ψ_p and the bounds γ_*^{κ} and ψ_*^{κ}:

a) when $\hat{\underline{R}}$ is random and $\underline{F}_o$ is deterministic, then

$$\gamma_*^{\kappa} = 1 - \psi_*^{\kappa} = \text{prob}\,(\hat{\tilde{\underline{R}}}\, \dot{\underline{\lambda}} \le s_*^{\kappa}\, \hat{\tilde{\underline{G}}}\, \dot{\underline{\lambda}}\,) = \text{prob}\,(\hat{\tilde{\underline{R}}}\, \dot{\underline{\lambda}} \le s_*^{\kappa}) = 1 - \psi_d; \tag{4.9}$$

b) when $\hat{\underline{F}}_o$ is random and $\underline{R}$ is deterministic, then

$$\gamma_*^{\kappa} = 1 - \psi_*^{\kappa} = \text{prob}\,(\tilde{\underline{R}}\, \dot{\underline{\lambda}} \le s_*^{\kappa}\, \hat{\tilde{\underline{G}}}\, \dot{\underline{\lambda}}\,) =$$

$$\text{prob}\ (\tilde{\hat{\underline{G}}}\ \dot{\underline{\lambda}} \geq 1) = 1 - \psi_p\ ; \tag{4.10}$$

c) when both $\hat{\underline{R}}$ and $\hat{\underline{F}}_0$ are random, then

$$\gamma_*^K = 1 - \psi_*^K = \text{prob}\ (\tilde{\hat{\underline{G}}}\ \dot{\underline{\lambda}}_* \leq s_*^K \tilde{\hat{\underline{G}}}\ \dot{\underline{\lambda}}\ ) =$$

$$= \Phi\left(\beta \frac{\sqrt{\tilde{\dot{\underline{\lambda}}}_* \underset{\approx}{R}\ \dot{\underline{\lambda}}_*} + s_*^K \sqrt{\tilde{\dot{\underline{\lambda}}}_* \underset{\approx}{G}\ \dot{\underline{\lambda}}_*}}{\sqrt{\tilde{\dot{\underline{\lambda}}}_* \underset{\approx}{R}\ \dot{\underline{\lambda}}_* + (s_*^K)^2\ \tilde{\dot{\underline{\lambda}}}_* \underset{\approx}{G}\ \dot{\underline{\lambda}}_*}}\right), \tag{4.11}$$

where $\beta = \Phi^{-1}(\ \psi\ )$ and $\psi = \psi_p = \psi_d$.

Thus, for the first two cases the assumed probabilities ψ_p , ψ_d determine exactly the upper bound on reliability. In the third case, that bound can be computed after the stochastically relevant failure mechanism is already known.

It seems that the dominant mechanisms can be selected according to the proposal given by S. R. Parimi and M. Z. Cohn [5].One has to solve (3.18), (3.19) or (3.20) for each loading case separately, obtaining thus the stochastically relevant mechanisms. Those mechanisms can be then regarded as dominant for the entire loading and the upper bound on reliability can be computed according to (4.7).

5. NUMERICAL EXAMPLES

Numerical calculations were performed using the standard matrix model of the skeletal structure [16] . Linear problems were solved by means of the Simplex procedure available in the program "MATRIX" [17] . The "NLPQP" code developed by K. Schittkowski [18] was used for the solution of NLP's.

The first example is deliberately taken very simple in order to demonstrate various modifications of the proposed methodology. Let us find the ultimate load carrying capacity of the two span continuous beam shown in Fig. 5.1. Let each span have the random yield moment characterised by the identical coefficient of variation (i.e. the ratio of the standard deviation to the mean value) $\xi_1^M = \xi_2^M = 0.10$ and by the following mean values: $\bar{M}_1^0 = 2.0$ kNm, $\bar{M}_2^0 = 3.0$ kNm. The analogous statistical parameters of the loading are: $\xi_1^P = \xi_2^P = 0.05$, $\bar{P}_1 = 3.0$ kN, $\bar{P}_2 = 2.0$ kN. The common reliability level $\psi = \psi^S = \psi^K = 0.9999$ was assumed for all considered cases.

The results of calculations are given in Table 5.1. Each of the special cases considered in the Section 3 was taken into account. For the sake of comparison, the ultimate load factor computed deterministically for the mean values of loads and yield moments is also provided.

As mentioned in the Section 4, the assummed reliability level corresponds to the lower bound in the static approach, whereas in

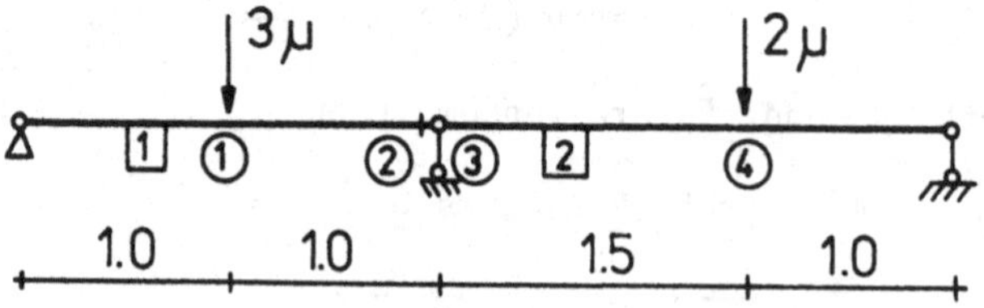

Fig. 5.1. Computational scheme of continuous beam.

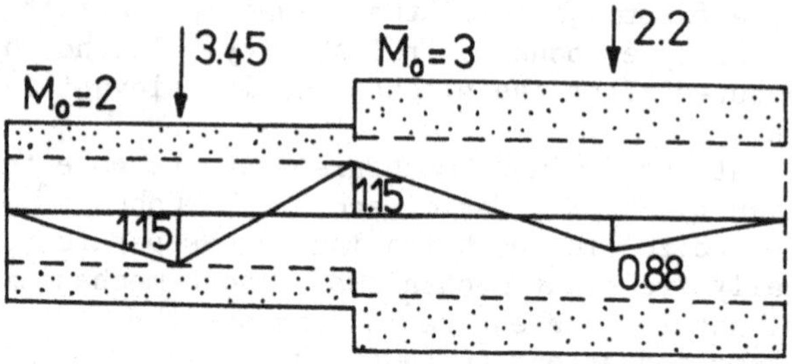

Fig. 5.2. Solution for random yield moments.

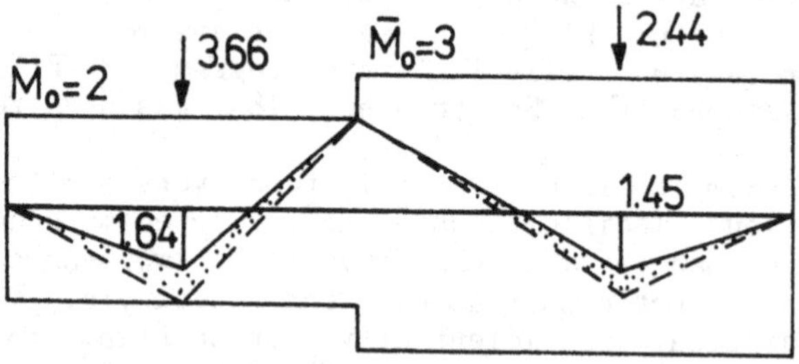

Fig.5.3. Solution for random load.

the kinematic approach it corresponds to the upper bound. Hence, stochastically computed static and kinematic load factors differ. That difference vanishes in the deterministic model, when both approaches are dual.

Table 5.1

No.	Formulation	Ultimate load factor	
		static	kinematic
1	deterministic	2.00	2.00
2	random $\hat{R}$	1.15	1.36
3	random $\hat{F}_o$	1.22	1.64
4	random $\hat{R}$ and $\hat{F}_o$	1.08	1.33

Fig. 5.2 shows the distribution of bending moments that correspond to the incipient collapse under the deterministic load $\underline{F}_* = 1.15 \; \underline{F}_o$ and random yield moments (the second row of Table 5.1). Shadowed area visualizes possible random decreases of the yield moments below their mean values. In the cross-sections 1 and 2 that area touches the bending moment diagram. Hence, plastic hinges are to be expected there. This was confirmed by the kinematic approach: the modal collapse mechanism occured in the left span. According to the recommendations of the Section 4, all other possible mechanisms were also checked. It turned out that they have much lower probabilities of occurrence.

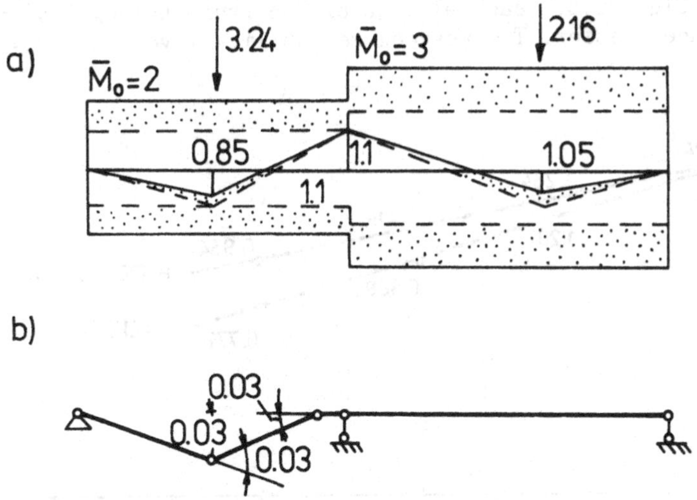

Fig. 5.4. Solution for random yield moments and random load.

Fig. 5.3 presents the results of the static approach for random loads but deterministic yield moments. According to (3.3), the resultant random bending moment is the sum of the random part $s_* \underline{\tilde{C}}^- F_o$ and the deterministic part $\underline{\tilde{C}}^0 Z_*$. The shadowed area shows unfavourable fluctuations of the bending moment. Similar results for random both loads and yield moments are shown in Fig. 5.4.a. The relevant plastic collapse mechanism is depicted in Fig. 5.4.b.

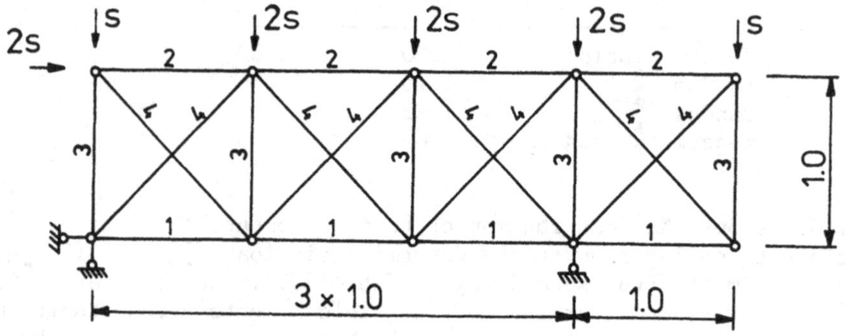

Fig. 5.5. Computational scheme of truss.

As the second example, consider the hyperstatic truss shown in Fig. 5.5. It will serve to demonstrate how the ultimate load factor depends upon the scatter of random plastic moduli. Let the mean values of yield axial forces be: $\bar{N}_1^o$ = 10.0 kN, $\bar{N}_2^o$ = 7.0 kN, $\bar{N}_3^o$ = 5.0 kN, $\bar{N}_4^o$ = 4.0 kN. As shown in Fig. 5.5, each element of the truss belongs to one of the four resistance classes. The assumed reliability level is ψ = 0.999 .

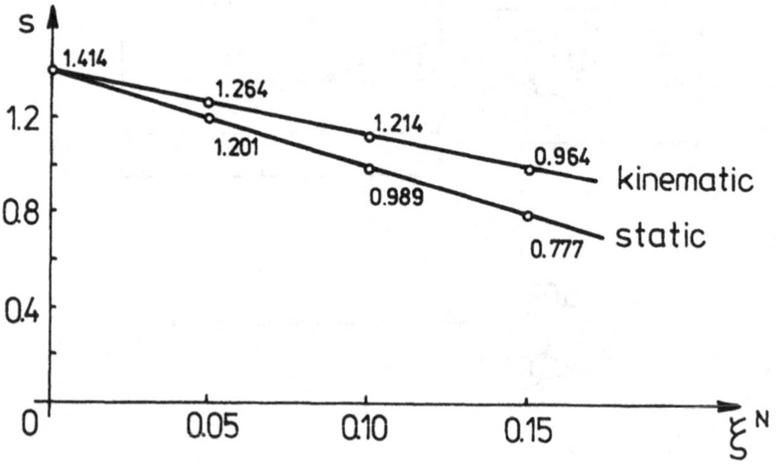

Fig. 5.6. Dependence of ultimate load factor upon coefficient of variation.

The dependence $s = s(\xi^N)$ mentioned before is plotted in Fig. 5.6: the load factors computed statically and kinematically decrease linearly with the growing coefficient of variation (ξ^N taken uniform for all bars). The point $\xi^N = 0$, where both lines meet, corresponds to the deterministic analysis. The distance between the static and kinematic estimates grows up with increasing ξ^N. The collapse mechanism for the case $\xi^N = 0.1$ is shown in Fig. 5.7 (solid lines visualize the bars at yield).

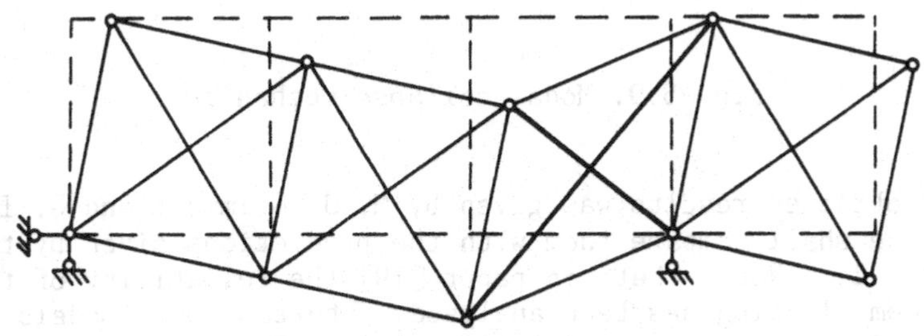

Fig. 5.7. Collapse mechanism for $\xi^N = 0.1$.

What happens if the degree of statical indeterminacy is reduced? The truss obtained from the previous one by cutting off one of the bracings is shown in Fig. 5.8. The probabilistic analysis shows clearly that the ultimate load carrying capacity of the new structure is less, even though

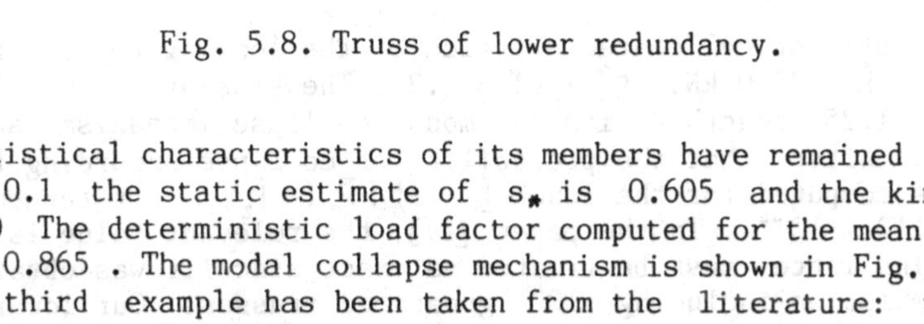

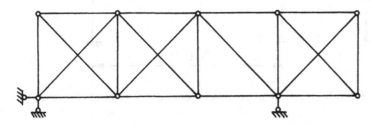

Fig. 5.8. Truss of lower redundancy.

the statistical characteristics of its members have remained unchanged. For $\xi^N = 0.1$ the static estimate of s_* is 0.605 and the kinematic one is 0.840 The deterministic load factor computed for the mean values is equal to 0.865 . The modal collapse mechanism is shown in Fig. 5.9 .

The third example has been taken from the literature: the portal frame depicted in Fig.5.10 was previously analysed by many authors. The

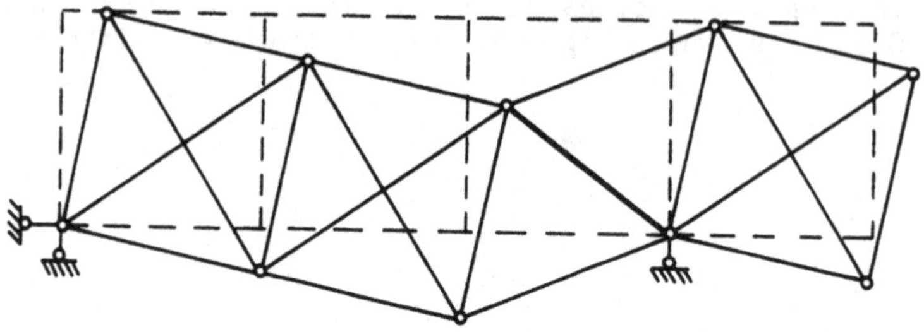

Fig. 5.9. Modal collapse mechanism.

summary of obtained results was given by M. J. Grimmelt and G. I. Schueller [19]. We shall compare them with the predictions given by the models (3.15), (3.22). Note that in paper [19] the reliability of the frame under random loading has been analysed, whereas our models evaluate ultimate load under a given reliability level. In accordance with [19] we assume that only the bending moments influence yielding of the frame and we take the uniform cross-section IPE 270 with $\overline{M}^{o} = 134.9$ kNm, $\xi^{M} = 0.05$.

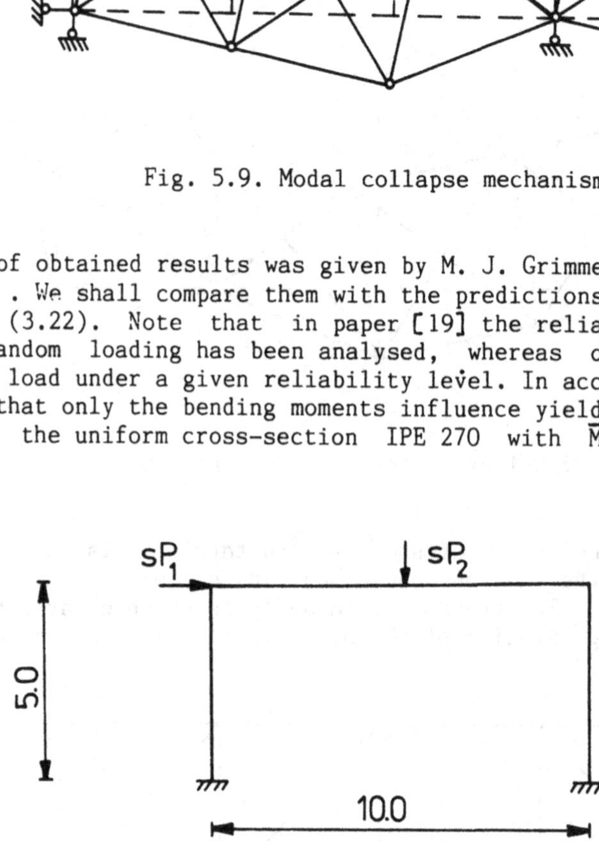

Fig. 5.10. Computational scheme of portal frame.

Let the statistical characteristics of the loading be as follows: $\overline{P}_1 = 40.0$ kN, $\overline{P}_2 = 32.0$ kN, $\xi_1^P = \xi_2^P = 0.3$. The kinematical model (3.22) yields $s_*^K = 1.25$ together with the modal collapse mechanism shown in Fig. 5.11.a. After that the probability of collapse according to that mechanism was computed for the load $\overline{F} = s_*^K \{ \overline{P}_1 , \overline{P}_2 \}$. It turned out to be $\gamma_*^K = 1.3112 \times 10^{-4}$. In the paper [19] the relevant value is 1.5625×10^{-4} . The latter must be treated as exact since it was obtained by numerical simulation including 10^4 generated trusses. Our lower estimate differs by 16 % from that exact value, which is quite acceptable for practical purposes.

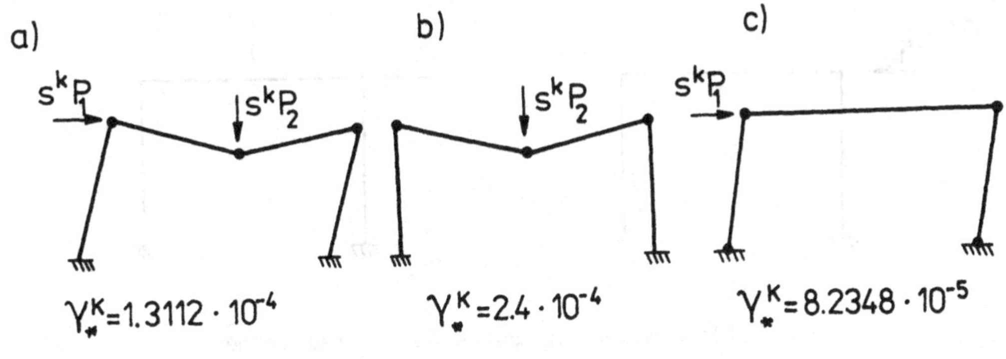

Fig. 5.11. Collapse mechanisms of portal frame.

As already mentioned in the Section 4, it is possible to obtain both upper and lower bound through the kinematic approach. In order to achieve that we have to solve separately two loading cases: the vertical load (Fig. 5.11.b) and the horizontal load (Fig. 5.11.c). After that the joint probability of collapse was found: $1.3112 \times 10^{-4} \leq \gamma_*^K \leq 2.1347 \times 10^{-4}$. However, despite the assumed statistical independence of yield moments and bending moments, upper bound safety margins $\hat{t}_i$ for each mechanism are mutually correlated, because they depend upon the same random variables. For the frame considered the correlation matrix (i.e. the normalized covariance matrix) of the collapse mechanisms shown in Fig. 5.11 reads:

$$\underset{\approx}{t} = \begin{bmatrix} 1.000 & 0.632 & 0.774 \\ 0.632 & 1.000 & 0.019 \\ 0.774 & 0.019 & 1.000 \end{bmatrix} \tag{5.1}$$

where

$$t_{ij} = \mathrm{cov}\,(\hat{t}_i, \hat{t}_j)/ \sqrt{\mathrm{var}(\hat{t}_i)}\ \sqrt{\mathrm{var}(\hat{t}_j)} \tag{5.2}$$

Applying the formula (4.2) that takes into account correlation of mechanisms, we obtain $\gamma_*^K = 1.3477 \times 10^{-4}$ (both bounds are identical).

The static approach gave $s_*^s = 1.18$ for assumed lower estimate of reliability $\psi^s = 0.9999$. As already pointed out in the Section 4, the lower estimate is exact only in the case of completely dependent lower bound safety margins. On the other hand, assuming them to be completely independent we obtain two sided estimate of the interval which contains the lower bound on reliability. For the considered example it turns out to be $0.9997 \leq \psi^s \leq 0.9999$. Thus, both approaches lead to correct results, though the kinematic one is more accurate.

In order to demonstrate the superiority of the stochastic approach over that of purely deterministic, we introduce the last example. Let the same portal frame as previously be subjected to the two loading cases

(Fig. 5.12). Their statistical characteristics are the following: $\bar{P}_1 = 65.0$ kN, $\bar{P}_2 = 32.0$ kN, $\xi_1^P = 0.05$, $\xi_2^P = 0.2$.

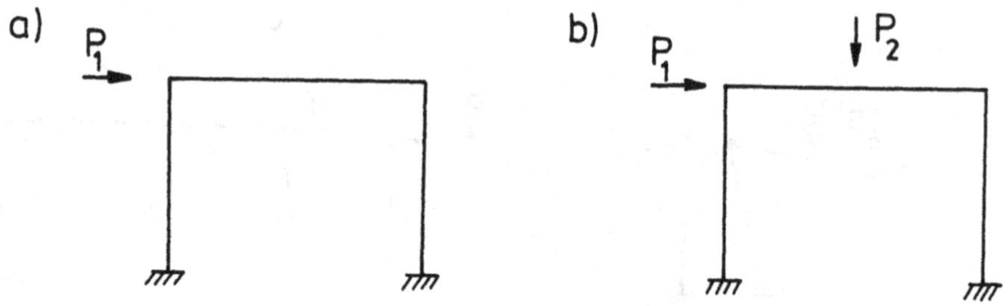

Fig. 5.12. Two loading cases for portal frame.

It is easy to check that the deterministic solutions for both loads are the same: $s_* = 1.66$ and the collapse mechanism shown in Fig. 5.11.c. However, it is clear that the case b) is less safe than the case a). The probabilistic analysis confirms that. Assuming the reliability level $\psi = 0.999981$ we obtain for the relevant loading cases: a) $s_*^S = 1.25$ and $s_*^K = 1.335$ and the side-sway modal mechanism (Fig. 5.11.c); b) $s_*^S = 1.13$ and $s_*^K = 1.248$, with the mixed modal mechanism (Fig. 5.11.a).

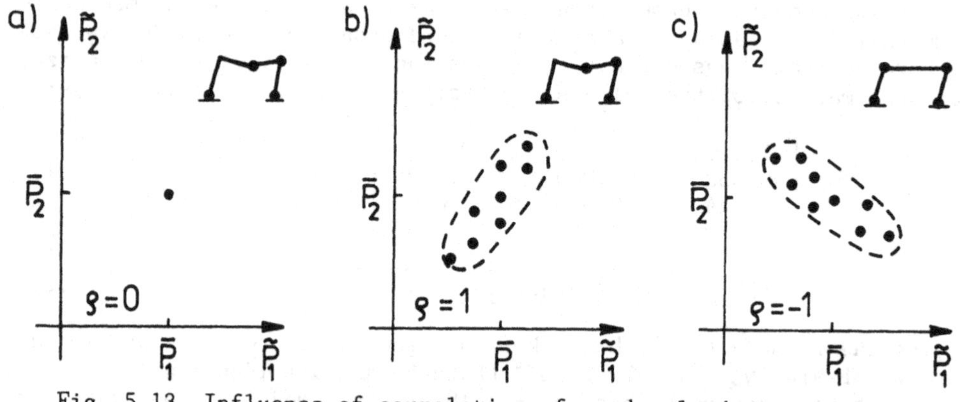

Fig. 5.13. Influence of correlation of random loads upon ultimate load factor: a) $s_* = 1.247$, b) $s_* = 1.16$, c) $s_* = 1.335$.

The previous results were obtained without the correlation of random variables. Fig. 5.13 shows the influence of such a correlation upon the the ultimate load factor for the frame subjected to the load depicted in Fig. 5.12.b. The reliability level is the same as previously. Different load factors were computed for the three values of coefficient of correlation between $\hat{P}_1$ and $\hat{P}_2$: $\rho = 0, 1, -1$. Points on the plane $0, \hat{P}_1, \hat{P}_2$ correspond to typical realizations of the random matrix $\hat{\underline{F}}$. The relevant modal collapse mechanism is presented in the right upper corner of each picture .

REFERENCES

1. Cohn, M. Z. and G. Maier, eds.: Engineering Plasticity by Mathematical Programming, Pergamon Press, New York 1979.
2. Kolbin, V. V.: Stochastic Programming, Reidel Publ. Company, Dordrecht 1977.
3. Ditlevsen, O. and P. Bjerager: Reliability of highly redundant plastic structures, Journal of Engineering Mechanics, 110 (1984), 671-693.
4. Bjerager, P.: Reliability analysis of structural systems, Reports of Department of Structural Engineering, Technical University of Denmark, Serie R, No. 183, 1984.
5. Parimi, S. R., and M. Z. Cohn: Optimal solutions in probabilistic structural design, J. de Mecanique Apliquee, 2 (1978), 47-82.
6. Ang, A. H.-S. and H.-P. Ma: On the Reliability of Structural Systems, in: Proc. 3rd Int. Conf. of Structural Safety and Reliability ICOSSAR81, Trondheim 1981, 295-314.
7. Murotsu, Y., Okada, H., Yonezawa, M., Kishi, M.: Identification of Stochastically Dominant Failure Modes in Frame Structures, in: Proc. 4th Int. Conf. of Statistics and Probability in Soil and Structural Engineering, Florence 1983, 1325-1338.
8. Augusti, G., Baratta, A., Casciati, F.: Probabilistic Methods in Structural Engineering, Chapman & Hall, London 1984.
9. Ang A. H.-S. and A. Tang: Probability Concepts in Engineering Planning and Design, Vol. 1, 2, Wiley, New York 1980, 1984.
10. Ditlevsen, O.: Uncertainty Modelling, McGraw-Hill, New York 1980.
11. Schueller, G. I.: Einführung in die Sicherheit und Zuverlässigkeit von Tragwerken, Ernst u. Sohn, Berlin 1981.
12. Charnes, A. and Cooper, W. W.: Chance-constrained programming, Management Science, 6 (1959), 73-79.
13. Rao, S. S.: Optimization: Theory and Applications, Wiley Eastern Ltd., New Delhi 1979.
14. Kataoka, S.: A stochastic programming model, Econometrica, 31 (1963), 181-196.
15. Čyras, A.: Optimization Theory in the Design of Elastoplastic Structures, in: Structural Optimization, ed. by Brousse, P., Cyras, A. and Save, M., (CISM course No. 237, Udine), Springer-Verlag, Berlin 1975, 81-150.
16. Borkowski, A.: Static Analysis of Elastic and Elastic-plastic Skeletal Structures, Elsevier-PWN, Warsaw 1987 (to appear).
17. Borkowski, A., Siemiątkowska, B., Weigl, M.: User's Guide for "MATRIX": a Matrix Interpretation System for Linear Algebra and Mathematical Programming, Adaptive Systems Laboratory, Institute of Fundamental Technological Research, Warsaw 1985.

18. Schittkowski, K.: User's Guide for the Nonlinear Programming Code "NLPQL", Report of the Institute of Informatics, University of Stuttgart, Stuttgart 1984.
19. Grimmelt, M. J., Schueller, G. I., Benchmark Study on Methods to Determine Collapse Failure Probabilities of Redundant Structures, Technische Universitat Munchen, Heft 51, 1981.
20. Hohenbichler, M., An Approximation to the Multivariate Normal Distribution, in: Reliability Theory of Structural Engineering Systems, DIALOG 6-82, Danish Engineering Academy, Lyngby 1982, 79-110.
21. Robinson J. and G. W. Haggenmacher: Optimization of redundancy selection in the finite element force method, AIAA Journal, 8 (1970), 1429-1433.

CHAPTER 21

FUZZY LINEAR PROGRAMMING IN PLASTIC LIMIT DESIGN

D. Lloyd Smith, P-H. Chuang and J. Munro
Imperial College, London, U.K.

ABSTRACT

Plastic limit design entails the designing of a structure so that it will just reach the limit state of plastic collapse under prescribed design loads, and, in addition, so that some design objective, often relating to weight or cost, is optimally achieved. When the geometry of the structure is fixed *a priori*, the design loading is known with complete precision and the objective function is linear, the problem can be expressed as one of linear programming.

In this chapter, the effect of uncertainty in the design loading is considered. Where the quality of the data is insufficient to represent the loading through a probability distribution, a simpler approach using fuzzy linear programming may be employed. This involves only a slight increase in the computational effort over that required when the loads are precisely defined. Two methods are described. In the first, the (possibly) several objective functions are recast as fuzzy goals and appended to the constraint set. The maximal value of the grade of membership of the design decision may be regarded as a measure of the degree of acceptablility of that decision in the face of uncertainty. In the second method, the original (single) objective function is retained, and an extra constraint is added to represent the designer's imposition of a lower bound on the acceptability measure of the optimal design.

§ Now at the Nanyang Technological Institute, Singapore 2263.

INTRODUCTION

Conventional methods of mathematical programming require that the data be precisely determined or deterministic. As an example, the ith main constraint of a linear programming problem is typically expressed through a linear inequality

$$\sum_j a_{ij}x_j \geq b_i, \qquad (j = 1,2,\ldots,n), \tag{21.1}$$

and the stipulation (b_i) is usually assumed to be known with complete certitude; in a real engineering problem, however, it may be known only approximately.

It may happen that the uncertainty in b_i is of a *random* nature; this implies that the operation of observing or otherwise determining a value for b_i is indefinitely repeatable. If a sufficiently large sample of values for b_i has been obtained for a reliable representation of its associated probability distribution to be established, then the problem may be cast legitimately as one of stochastic programming. Clearly, this representation of uncertainty demands a definite quality in the given data. Moreover, the increased computing effort in the solution process for stochastic programming over that for simple linear programming suggests that such an approach may only be worthwhile when the given data has the desired quality.

However, in an engineering context, the incertitude frequently stems from *imprecision* of numerical data (rather than randomness) or *vagueness* of linguistically represented information. For example, this may happen when b_i is fixed by a human expert through a subjective mental process that we may call "engineering judgement". Such data have for long provided an eminently respectable support for civil engineering decision-making, often the only immediate support when appropriate measurements either cannot be made or have not yet been made. In these circumstances, where the quality of the data is not adequate for a proper statistical representation, it may be considered more realistic to replace the crisp number b_i of the deterministic linear program by a fuzzy representation.

Similar imprecision may occur in the structural coefficients (a_{ij}) of the constraints and also in the cost coefficients (c_j) in the linear objective function to be optimised. Techniques for solving LP problems with such imprecise data have been summarised by Chuang and Munro[1,2].

PLASTIC LIMIT DESIGN BY LINEAR PROGRAMMING

The designing of a structure so that it will just reach the limit state of plastic collapse under the prescribed design loads, and, in addition, so that some (usually linear) design objective is optimally achieved, may be called *plastic limit design*. The simplest class of such problems is where:-

* the geometry and topology of the structure are fixed *a priori*,
* a single loading state is considered,
* statics and kinematics are founded on small displacements,
* the constitutive relations are those of perfect plasticity,

* plastic yielding is controlled by a single stress-resultant,
* a single linearised design objective function is utilised.

For the plastic limit design of frames and of reinforced concrete slabs by means of linear programming through use of the Simplex algorithm, the mesh description of statics and kinematics generally leads to less computational effort than the nodal description. Let **d** be the column vector of design variables $d_1, d_2, \ldots, d_n$ – possibly the relevant plastic moments of resistance – and $\boldsymbol{\ell}$ the column vector of lengths $\ell_1, \ell_2, \ldots, \ell_n$ of the members or segments of members associated with the corresponding design variables. Also, let us seek to minimise the linear objective function $z = \boldsymbol{\ell}^T \mathbf{d}$ where z is an approximation to the total cost or, alternatively, to the total weight of the structure. With the above restrictions, the static mesh linear program for plastic limit design takes the following form.

$$\text{Minimise } z = \begin{bmatrix} \boldsymbol{\ell}^T & \mathbf{0} \end{bmatrix} \begin{bmatrix} \mathbf{d} \\ \mathbf{p} \end{bmatrix}$$

$$\begin{bmatrix} \mathbf{J} & -\mathbf{N}^T\mathbf{B} \end{bmatrix} \begin{bmatrix} \mathbf{d} \\ \mathbf{p} \end{bmatrix} \geq \mathbf{N}^T\mathbf{B}_0\mathbf{F}$$

$$\mathbf{d} \geq \mathbf{0} \qquad (21.2)$$

STATIC MESH LP FOR PLASTIC LIMIT DESIGN

Matrix **J** is an incidence matrix which associates each of the design variables **d** with the appropriate critical sections. The remaining matrices are as defined in the previous lectures. It should be noted that the constraints of the above static LP represent static admissibility (equilibrium and the yield conditions). The elements of the matrix **J** are fixed *a priori* by the designer and the elements of the matrix **B** are determined from the frame geometry. The stipulations of the static LP are controlled by the vector **F** which constitutes the prescribed design loading for the collapse state.

In many design situations, it will be reasonable to consider that the part of the data which stems from the global geometry of the frame is known deterministically, whilst the magnitude of the applied loading can only be stated imprecisely. Thus the static LP becomes one in which the stipulations are imprecise, and it is this class of problems which can be most readily handled by the techniques of fuzzy mathematical programming. A brief summary of these techniques will be presented next before we modify them in order to apply them to the plastic design problem.

FUZZY LINEAR PROGRAMMING

In mathematics, a fundamental concept is that of a set or collection of objects. Let objects $x_1, x_2, \ldots, x_j, \ldots, x_n$ form a set U, and let A be a subset of U. It is usually supposed that membership of a set is crisply defined – the object x_j either is a member of the subset A or it is not. We can then define a *membership function* $\mu_A(x_j)$ which takes only two values: $\mu_A(x_j) = 1$ when x_j is a member of subset A, and $\mu_A(x_j) = 0$ when it is not.

Human intelligence, however, has a clear propensity for mental processes of reasoning in imprecise terms, terms which may include processes in which membership of a set is not so crisply defined. The theory of fuzzy sets has been developed by Zadeh[3,4] to handle the form of incertitude which stems from the imprecise definition of boundaries of classes of objects. To some degree, therefore, we may be unsure of whether object x_j is a member of the subset A. We may give quantitative expression to our own subjective opinion by allowing the membership function $\mu_A(x_j)$ to take its value from the continuous range [0,1]; a high value in this range represents a "strong" feeling that x_j belongs to the subset A, and a low value is correspondingly associated with a "weak" conviction that it does. The subset A is then said to be fuzzy.

Various methods whereby linear programming can be adapted in the face of imprecision have been discussed by Chuang and Munro[1,2], and the superiority of fuzzy programming over proximate programming and inexact programming has been indicated. An important advance in utilising this new theory for decision-making in a fuzzy environment was made by Bellman and Zadeh[5] who considered that a decision was the confluence of fuzzy goals and constraints. Consider the following deterministic LP,

$$\{ \text{Minimise } z = c^T x \;;\quad A\,x \geq b, \quad x \geq 0 \}. \tag{21.3}$$

The objective of minimising z may be converted to the goal that z should not exceed some aspiration level z'; thus

$$c^T x \leq z' \qquad \text{or} \qquad -c^T x \geq -z' \tag{21.4}$$

The problem is still deterministic and now consists of a system of inequalities

$$\{ E\,x \geq h, \quad x \geq 0 \}, \tag{21.5}$$

where

$$E = \begin{bmatrix} -c^T \\ A \end{bmatrix}, \qquad h = \begin{bmatrix} -z' \\ b \end{bmatrix}.$$

The designer might wish to introduce imprecision into some or all of the aspiration levels and stipulations h_i $(i = 1,2,\ldots,m)$ of the system (21.5). The ith inequality

$$\sum_j e_{ij}\, x_j \geq h_i, \tag{21.6}$$

can be softened or relaxed in the sense that it is no longer mandatory to satisfy the inequality (21.6) but it is necessary to satisfy the modified form

$$\sum_j e_{ij}\, x_j > h_i' = h_i - \sigma_i \tag{21.7}$$

where σ_i is a measure of the softness that has been introduced. Now the designer would prefer to satisfy the original inequality (21.6), but accepts – with diminishing enthusiasm – some contravention of that relation until the softer inequality (21.7) is fully taken up; at that point the designer will not accept any further relaxation. The designer's subjective and diminishing acceptability of the contravention of inequality (21.6) is indicated in Figure 21.1 in which μ_i is the membership function that he gives to the ith inequality.

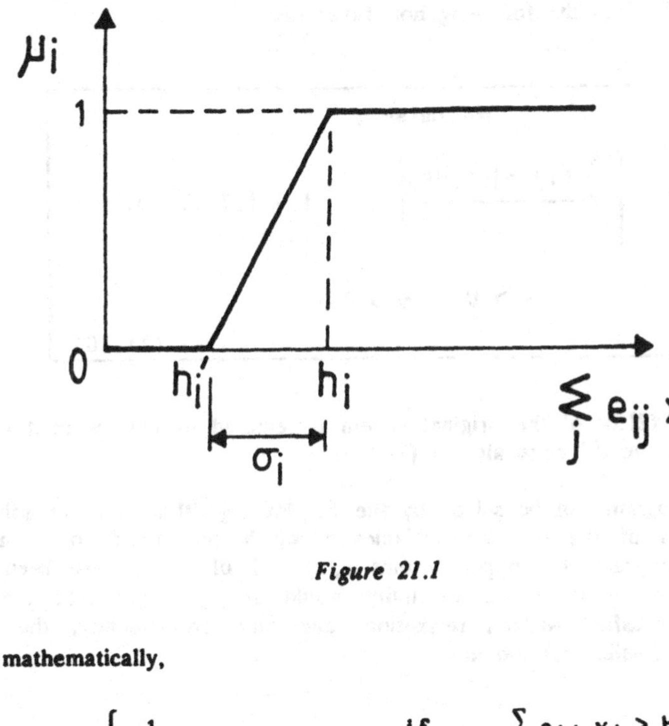

Figure 21.1

Expressed mathematically,

$$\mu_i(x) = \begin{cases} 1, & \text{if } \sum_j e_{ij}\, x_j > h_i \\ \left[\dfrac{\sum_j e_{ij}\, x_j - h_i'}{\sigma_i}\right], & \text{if } h_i' < \sum_j e_{ij}\, x_j < h_i \\ 0, & \text{if } \sum_j e_{ij}\, x_j < h_i' \end{cases} \tag{21.8}$$

Note that the goals are now included in the system of inequalities (21.5) and may be softened as well as the original constraints. Thus two different sources of incertitude – the designer's fuzziness regarding his design objectives and his fuzziness with respect to the constraints – have now been encoded in a single set of fuzzy inequalities.

For any set of fixed values of the decision variables $\mathbf{x}$, the grade of membership $\mu_D(\mathbf{x})$ associated with that decision is given by the minimum of the membership grades $\mu_i(\mathbf{x})$ as one spans across the ensemble, represented by the index i, of inequalities representing the goals and constraints. For all feasible values of $\mathbf{x}$, the membership function of the decision is

$$\mu_D(x) = \min_i \mu_i(x). \tag{21.9}$$

A new parameter (γ) is now defined to be such that $\gamma \leq \mu_D(\mathbf{x})$, and the "best" or most acceptable decision is then found by maximising γ with respect to all feasible decisions $\mathbf{x}$. In this manner, the original problem can be transformed [1,2], with the $\mu_i(\mathbf{x})$ given by (21.8), into the following non-fuzzy LP.

$$\text{Maximise } \gamma$$

$$\gamma \leq \left[\frac{\sum_j e_{ij} x_j - h_i'}{\sigma_i} \right] \qquad i = 1,2,\ldots,m.$$

$$x \geq 0, \qquad \gamma \geq 0 \tag{21.10}$$

If the ith inequality (21.6) of the original system is required to hold without relaxation, then it must replace the ith constraint in (21.10).

The above linear program can be solved by the simplex algorithm, and the solution will identify those values of the decision variables $\mathbf{x}$ which correspond to a maximising decision. In this program, the upper bounds $\mu_i(\mathbf{x}) \leq 1$ of (21.8) have been omitted; an optimal value for γ in excess of unity would simply imply that the original constraints can be satisfied without relaxation, and that, consequently, the goals or aspiration levels are possibly set too low.

PLASTIC DESIGN BY FUZZY LINEAR PROGRAMMING

A very simple example will be used to illustrate the procedure and the interpretation of the solution. The prismatic beam with fixed ends shown in Figure 21.2 is to be designed so that it resists the indicated single applied loading without collapsing plastically. The single design variable (d) is the plastic moment of resistance of the beam. The particular solution bending moment diagram ($\mathbf{m}_0 = \mathbf{B}_0\mathbf{F}$) and the complementary solution or self-equilibrating bending moment diagrams ($\mathbf{b}_1$, $\mathbf{b}_2$) are shown in Figure 21.3.

In order to minimise the amount of data for this illustrative example, the correct sense of the two complementary diagrams (b_1, b_2) has been anticipated, and this permits us to treat the mesh actions or hyperstatic forces p_1 and p_2 as non-negative.

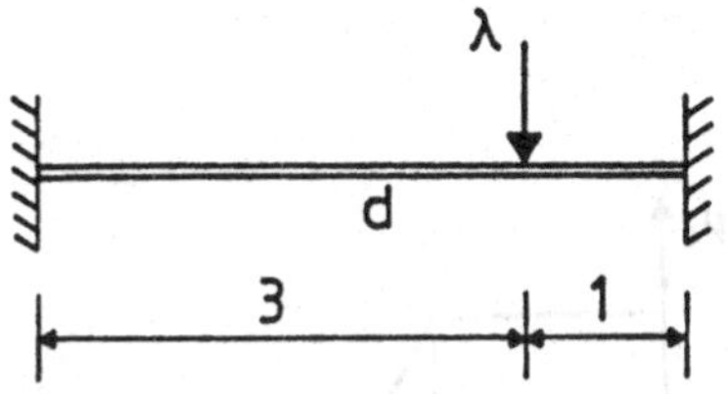

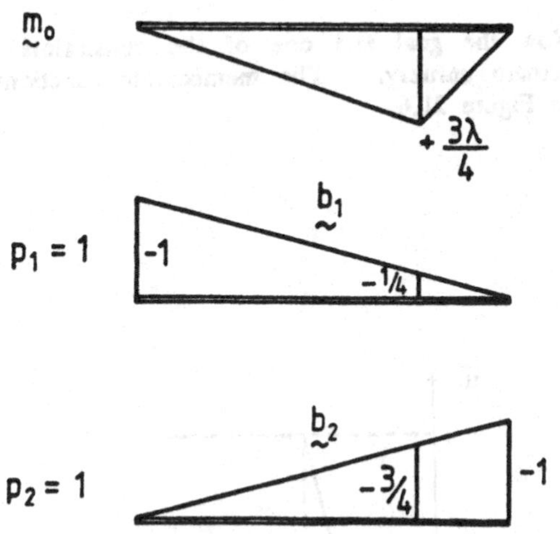

Figure 21.2 *Figure 21.3*

The deterministic LP(21.2) for plastic limit design then takes the following compact form:

$$\text{Min } z = \begin{bmatrix} 4 & 0 & 0 \end{bmatrix} \begin{bmatrix} d \\ p_1 \\ p_2 \end{bmatrix}$$

$$\begin{bmatrix} 1 & -1 & 0 \\ 1 & 0 & -1 \\ 1 & 0.25 & 0.75 \end{bmatrix} \begin{bmatrix} d \\ p_1 \\ p_2 \end{bmatrix} \geq \begin{bmatrix} 0 \\ 0 \\ 0.75\lambda \end{bmatrix}$$

$$d,\ p_1,\ p_2 \geq 0 \qquad (21.11)$$

Next the constraints are fuzzified. It is decided that, whilst the designer would prefer the structure to resist a load of 80 units, he would be prepared to lower the required collapse load as far as 60 with diminishing support, as indicated in Figure 21.4.

In a similar way, the designer would prefer the objective (weight) function (z) to be less than 100, but is prepared to let it be as much as 160 with diminishing support, as indicated in Figure 21.5. The goal therefore becomes

$$4d \lessdot z' \qquad \text{or} \qquad -4d \gtrdot -z' \tag{21.12}$$

Now the goal and one of the constraints are fuzzified, whilst two other constraints remain unfuzzy. The membership functions for the two fuzzy inequalities are shown in Figure 21.6.

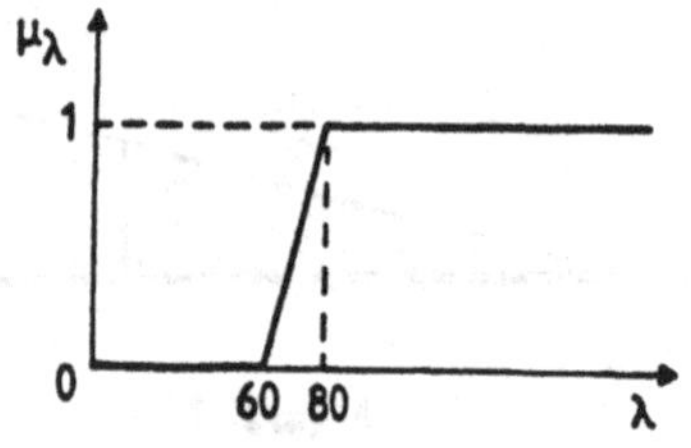

Figure 21.4

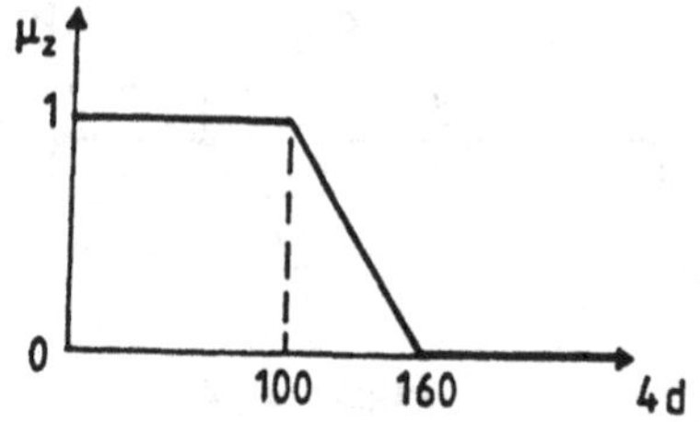

Figure 21.5

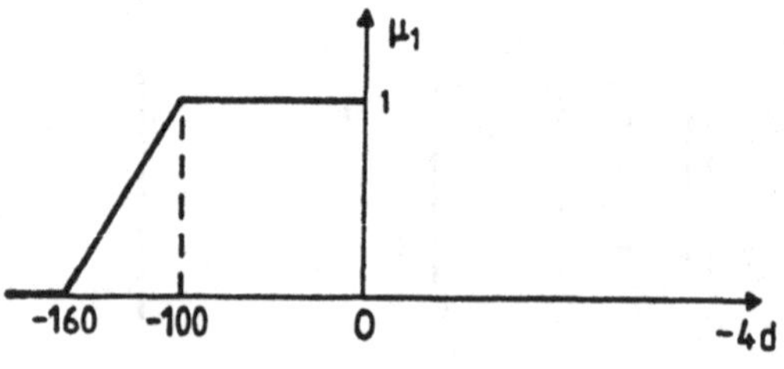

Figure 21.6(a)

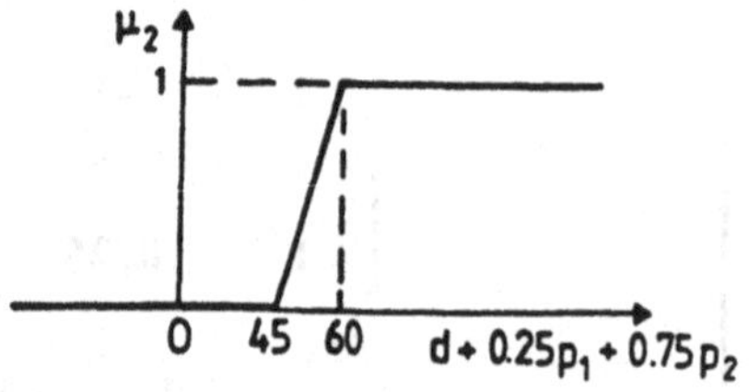

Figure 21.6(b)

The transformed (non-fuzzy) LP(21.10) corresponding to this problem takes the following form [2,6]:

$$
\begin{aligned}
&\text{Maximise } \gamma \\
&\gamma \le \frac{-4d + 160}{60} \\
&\gamma \le \frac{d + 0.25p_1 + 0.75p_2 - 45}{15} \\
&d - p_1 \ge 0 \\
&d - p_2 \ge 0 \\
&d,\ p_1,\ p_2,\ \gamma \ge 0
\end{aligned}
\tag{21.13}
$$

Defining the new variables **x** as

$$x_1 \equiv \gamma,\quad x_2 \equiv d,\quad x_3 \equiv p_1,\quad x_4 \equiv p_2,$$

and introducing the slack variables **s** and the artificial variable t, the LP can be written in minimising form as follows.

$$
\text{Minimise } Z = (-1)x_1
$$

$$
\begin{bmatrix} 15 & 1 & 0 & 0 \\ -60 & 4 & 1 & 3 \\ 0 & -1 & 1 & 0 \\ 0 & -1 & 0 & 1 \end{bmatrix}
\begin{bmatrix} x_1 \\ x_2 \\ x_3 \\ x_4 \end{bmatrix}
+ \begin{bmatrix} s_1 \\ -s_2 \\ s_3 \\ s_4 \end{bmatrix}
+ \begin{bmatrix} 0 \\ t \\ 0 \\ 0 \end{bmatrix}
= \begin{bmatrix} 40 \\ 180 \\ 0 \\ 0 \end{bmatrix}
$$

$$
x_1,\ x_2,\ x_3,\ x_4,\ s_1,\ s_2,\ s_3,\ s_4,\ t \ge 0 \tag{21.14}
$$

The optimal solution of this LP is

$$\bar{x}_1 = \bar{\gamma} = \frac{7}{9},\qquad \bar{x}_2 = \bar{d} = \bar{x}_3 = \bar{p}_1 = \bar{x}_4 = \bar{p}_2 = \frac{85}{3}$$

The original linearised objective function (z) is now given by

$$\bar{z} = 4\bar{d} = \frac{340}{3}$$

The optimal design in the face of the given incertitude has now been fixed deterministically. The "best" decision is to select a section whose plastic moment of resistance d is (85/3). The corresponding maximal value of γ is (7/9), and this may be considered as a measure of the degree of acceptability of this optimal decision. Multiple objectives $z_k(x)$ could be included without any conceptual difficulty, but their membership functions $\mu_k(x)$ would have to be linearised for the problem to remain one of linear programming.

PLASTIC LIMIT DESIGN BY MODIFIED FUZZY LINEAR PROGRAMMING

The fuzzy approach to design discussed in the previous section permits the simultaneous consideration of multiple objectives, since such objectives can be converted to fuzzy goals which in turn may be handled in the same way as fuzzy constraints. The problem is then transformed to the non-fuzzy LP (21.10) in which the new objective is to maximise the grade of membership of the decision.

It was suggested in the previous section that this maximal grade of membership is a measure of the acceptability of the design decision. If the maximal grade of membership is considered to be too small, then it might be necessary to modify the design goals to more modest aspirations. The new fuzzy problem could then be transformed to a non-fuzzy LP whose solution would then be sought through the simplex algorithm.

This process suggests a sequence of such modifications until an appropriate acceptability level is attained. However in many design problems, a single objective predominates; in such cases the deterministic objective can be retained whilst the fuzzy constraints are transformed as before. A single additional constraint can then be added to the transformed LP to specify the minimal level of acceptability. This modified form of the design problem can then be solved with a single run of the simplex algorithm.

Thus the optimal design of the beam with fixed ends takes the following modifed form

$$
\begin{aligned}
&\text{Minimise } z = 4d \\
&\gamma \le \frac{d + 0.25p_1 + 0.75p_2 - 45}{15} \\
&d - p_1 \ge 0 \\
&d - p_2 \ge 0 \\
&\gamma \ge \gamma_{min} \\
&\quad d_1,\ p_1,\ p_2 \ge 0
\end{aligned}
\tag{21.15}
$$

where γ_{min} is the minimal value of the grade of membership which is considered to ensure a sufficiently acceptable design. The LP(21.15) should be compared with LP (21.13). Introducing the variables x, as previously defined, the LP(21.15) can be transformed to the following form

Minimise z

$$\left[\begin{array}{cccc} 0 & 4 & 0 & 0 \\ \hline -60 & 4 & 1 & 3 \\ 0 & -1 & 1 & 0 \\ 0 & -1 & 0 & 1 \\ 1 & 0 & 0 & 0 \end{array}\right] \begin{array}{c} \\ \left[\begin{array}{c} x_1 \\ x_2 \\ x_3 \\ x_4 \end{array}\right] \end{array} \begin{array}{c} = \\ > \\ < \\ < \\ > \end{array} \left[\begin{array}{c} z \\ \hline 180 \\ 0 \\ 0 \\ \gamma_{min} \end{array}\right]$$

$$x_1,\ x_2,\ x_3,\ x_4 > 0 \qquad (21.16)$$

If γ_{min} is set at 0.5, the optimal solution of LP(21.16) is:

$$\bar{z} = 105 \qquad \bar{\gamma} = 0.5 \qquad \bar{d} = \bar{p}_1 = \bar{p}_2 = 26.25$$

It will be seen that, because the minimum level of acceptability was set as low as 0.5, the optimal value $\bar{z}$ of the structural objective function was reduced to 105. This should be compared with the previous computation which sought the maximum grade of membership within the specified fuzzy goal and constraint. A grade of membership of $\bar{\gamma} = (7/9)$ was obtained in that case with an associated structural objective function value of $\bar{z} = 113.33$. This extra weight would not be justified if that grade of membership were higher than was felt necessary for the operational circumstances which were visualised for this structure. Minimum weights associated with a range of acceptability levels can be computed and this gives an indication of the sensitivity of the decision.

REFERENCES

1. Chuang, P. H. and Munro, J., Linear programming with imprecise data, Civil Engineering Systems, 1 (1983) 37–41.

2. Chuang, P. H., Fuzzy Mathematical Programming in Civil Engineering Systems, Ph.D. Thesis, University of London 1986.

3. Zadeh, L. A., Fuzzy sets, Inf. Control, 8 (1965) 338–353.

4. Zadeh, L. A., Outline of a new approach to the analysis of complex systems and decision processes, IEEE Trans. Syst. Man. Cybern., SMC-3 (1973) 28–44.

5. Bellman, R. E. and Zadeh, L. A., Decision-making in a fuzzy environment, Management Science, 17 (1970) B141–B164.

6. Munro, J. and Chuang, P. H., Optimal Plastic Design with Imprecise Data, J. of Eng. Mech., Proc. ASCE, 112 (1986) 888–903.

Printed in the United States
By Bookmasters